CELLULAR AUTOMATA

PROSPECTS IN ASTROPHYSICAL APPLICATIONS

CELLULAR AUTOMATA

PROSPECTS IN ASTROPHYSICAL APPLICATIONS

Editors

J M Perdang
Institut d'Astrophysique, ULG, Belgium

A Lejeune
Institut de Physique, ULG, Belgium

Publié avec le concours de la Fondation Universitaire de Belgique

World Scientific
Singapore • New Jersey • London • Hong Kong

Published by

World Scientific Publishing Co Pte Ltd
5 Toh Tuck Link, Singapore 596224
USA office: 27 Warren Street, Suite 401-402, Hackensack, NJ 07601
UK office: 57 Shelton Street, Covent Garden, London WC2H 9HE

British Library Cataloguing-in-Publication Data
A catalogue record for this book is available from the British Library.

CELLULAR AUTOMATA: PROSPECTS IN ASTROPHYSICAL APPLICATIONS

ISBN-13 978-981-02-1346-6
ISBN-10 981-02-1346-8

CONTENTS

Part III: Astrophysical Applications

Dedicated to **Eugène Catalan**, teacher of mathematics at the University of Liège in the last century, whose abstract work in discrete mathematics laid the foundations to the Theory of Cellular Automata.

PREFACE

The idea of editing an introductory volume into the methodology of Cellular Automata (CA), specifically addressing the astronomical community, emerged during a Colloquium held in Han–sur–Lesse, Belgium, October 19-21, 1992, in celebration of the 175th anniversary of the University of Liège.

There are, at the moment, several excellent recent reviews available, addressing various isolated topics of CA techniques (such as, for instance, *Lattice Gas Methods*).

At the same time, there is also a notorious lack of works on an introductory level providing a fair coverage of the wide spectrum of currently existing applications on the one hand, and emphasising the prospects and capabilities of the CA environment on the other. Likewise, a synthetic review is wanting laying stress on the systematics of CA model construction for evolutionary phenomena whose general observational features are known — a subject of immediate relevance for astrophysical simulation. And further, we feel that an overview is currently lacking providing information both on the *theory* of CA behaviour (systematic procedures for deriving macroscopic equations associated with a given CA microscopic model, analytical prediction procedures for the space–time structure associated with a given CA dynamics, general analysis of time–behaviour resulting from CA rules, etc) and on *numerical experimentation* (demonstrating by explicit nontrivial illustrations the powerful CA capabilities).

Workers in the field of astrophysics in particular, and in applied physics in general, facing research subjects not easily handled with, or perhaps not even accessible to the conventional tools of the trade, are in the first place in need of broad information capable of triggering new ideas for handling their unsolved problems. They are not, at the outset, concerned with the specific technicalities of the method — which are accessible in the specialised literature in due time.

It is our hope that the present publication will help to close this gap, at least in part, and offer information currently not easily accessible, in a compact and yet readable form.

To make the collection of papers of this book essentially self–contained, and, perhaps more relevant, to make the subject accessible to a widest audience of astronomers not yet familiar with the cellular automaton concept, we felt that besides the more specialised review articles on a variety of topics of astrophysical interest a general tutorial–type introductory paper should be included. Focussing on astrophysical applications, and emphasising the distinguishing features of astrophysics, this review attempts to convey the particular philosophy of the CA method. It covers Part I of this book.

Part II is concerned with a variety of subjects which so far have been sucessfully investigated in the CA framework, and which, sometimes, have led to unexpected insights (in Reaction Theory, for instance). These topics have been selected for their obvious potential interest in astrophysics. They orbit around Fluid Dynamics (Lattice Gas Methods, Lattice Boltzmann Method), Diffusion and Reaction Simulation, Phase Transitions and Waves. This part also contains contributions to the simulation of multifractals (which are currently applied in the interpretation of the Large Scale Structure of the universe).

Part III centres around direct astrophysical applications. It deals with a variety of astronomical examples of self–organised criticality, with CA models of generation of space–time structure (Galactic Structure, Large Scale Structure), and with astrophysical hydrodynamics (Stellar Pulsations). It further includes a review on an alternative unconventional approach, namely the Smoothed–Particle Method, whose achievements are illustrated by numerical experiments on Supernova Explosions.

It finally deals with a model for a geophysical phenomenon (Earthquakes) which, we believe, could be directly translated into a corresponding proper astrophysical model (Starquakes).
The majority of the more specialised contributions of Parts II and III are extended versions of papers presented at the Han meeting.

Our purpose will be achieved if, through this contribution, the now standard fields of application of the CA framework (Lattice Gases, Diffusion, Reactions) will not just be taken over and adapted to the astrophysical environment, but if this publication helps specialists in some fields of astronomy devise really new and better adapted CA models for their specific area of research.

We thank the Belgian *Fonds National de la Recherche Scientifique* for sponsoring the Han-sur-Lesse meeting, and the Apple Education Centre, Liège, and IBM Belgium, for logistic support.

Part I :

Elements of Cellular Automaton Modelling

The Cellular Automaton Framework: A New Tool for Astrophysical Modelling

J Perdang

Institute of Astronomy, Madingley Road, Cambridge, CB3 OHA, UK
and
Institut d'Astrophysique, 5 Avenue de Cointe, Liège B-4000, Belgium *

Abstract: We review the cellular automaton methodology and analyse the major theoretical properties of the solutions of this class of discrete dynamical systems. Besides assessing the ability of the automaton environment for modelling geometrically complicated structures in space and time, we discuss two further potentialities of the automaton technique which are of relevance in the astrophysical context : The automaton may allow meaningful simulations of (a) phenomena resulting from large collections of only partially known physical processes, and (b) mechanisms strongly modulated by intrinsic fluctuations.

This paper further introduces various extensions of the conventional definition of a cellular automaton (non-standard automata, spatially nonuniform lattices, higher level evolution rules). The theoretical and practical interest of these generalisations is demonstrated by a few schematic astrophysical illustrations.

> Diese Regeln [des Glasperlenspiels], die Zeichensprache und Grammatik des Spieles, stellen eine Art von hochentwickelter Geheimsprache dar, ... welche die Inhalte und Ergebnisse nahezu aller Wissenschaften auszudrücken und zueinander in Beziehung zu setzen imstande sind. ...theoretisch ließe mit diesem Instrument der ganze Weltinhalt sich im Spiele reproduzieren.
>
> H. Hesse 1943

0. Introduction

The currently favoured dynamical or evolutionary classes of models for astronomical and astrophysical phenomena mainly belong into a continuous spectrum of categories which can be ordered by increasing physical complexity. The extreme representatives of this spectrum are perhaps best typified by a few examples.

On the *low complexity* side of the spectrum an illustrative example is Baker's one–zone model for the global vibrational stability of stars (Baker 1963) together with its more recent extensions to the study of global nonlinear stellar pulsations (Buchler and Regev 1982, Auvergne and Baglin 1985, Tanaka et al 1991). In this specific model the complexity of the genuine star is reduced to a bare minimum still capable of accounting for the chief physical factors responsible for the stability and for the mathematical nature of the oscillations (regular or chaotic) a genuine star can undergo. The interest of this type of modelling focuses on the qualitative (topological) behaviour, rather than on the detailed quantitative properties of the behaviour which in general are modulated by the full details of the structure. On the other hand, instead of reducing the star to a single zone, we may keep N zones; with increasing N we expect the model to reproduce the quantitative behaviour of a true star.

Although models of this category are typically derived on the basis of physical arguments and assumptions (geometrical symmetry assumptions, order of magnitude arguments, etc) simplifying and reducing the actual stellar equations — assumed to be exactly known — to a final 'model' form, it may happen that the mathematical structure of the latter remains correct under conditions much broader than actually anticipated from the assumptions (cf the conclusions drawn from Normal Form Theory, Perdang 1992). A further characteristic feature of this class of models is that they typically refer to small numbers of 'degrees of freedom': They are formulated in low

---

* Permanent address

dimensional phase spaces, i.e. described by low–order differential systems. Accordingly, by construction, the most complex time behaviour such systems are capable of exhibiting is ***deterministic*** (low dimensional) ***chaos***. The validity of models of this category completely breaks down as soon as within the real phenomenon to be modelled at least one 'degree of freedom' gets excited which was artificially frozen (as a result of a symmetry assumption for instance), or just not represented in the model environment : The one–zone model manifestly becomes inadequate as soon as 2 or more vibrational modes turn unstable.

More formally, the category of low complexity models consists in ***systems of equations whose algebraic structure is fully given***. The equations themselves are differential (or difference) equations of low order,

$$\frac{\mathrm{d}}{\mathrm{d}t}\, y(t) = r\ [\ y(t)\ ], \tag{0.1}$$

$y = (y_1, y_2, \ldots, y_n)$, n 'small' ($< 10^3$ say). The latter equations in turn are generated — adopting various simplifying physical and mathematical assumptions — from a well established physical theory (fluid dynamics in the above illustration), traditionally encoded in a system of partial differential equations,

$$\partial_t F(t, \mathbf{r}) = R\ [\ F(t, \mathbf{r}), \partial_{\mathbf{r}} F(t, \mathbf{r}), \partial^2_{\mathbf{r}} F(t, \mathbf{r}), \ldots\ ]; \tag{0.2}$$

$F(t, \mathbf{r}) = (\ F_1(t,\ \mathbf{r}),\ F_2(t, \mathbf{r}), \ldots,\ F_m(t, \mathbf{r})\ )$, collection of components of the 'field' variables (density field, velocity field, etc, in fluid dynamics). The assumptions, suggested by the observation of the phenomenon to be modelled, are selected to reduce the original general mathematical framework to a formal system of equations accessible to analytic or numerical discussion.

On a mathematically more fundamental level, models, in the sense of Eqs. (0.1), can be systematically obtained in their algebraically simplest form, as (multi–parameter) *Normal Form Families* associated with the 'correct' underlying partial differential set of equations (0.2) representative for the full physics of the system, provided that the latter exist i.e. ***provided that the number of 'degrees of freedom' which eventually remain excited is bounded.*** It should be kept in mind that the formalisation almost inevitably implies schematic assumptions which are only partially fulfilled in the real phenomenon the model is meant to duplicate.

High complexity can arise in two distinct ways.

On the one hand, even if a given natural phenomenon obeys standard laws leading to equations of evolution of the form (0.2) in which all parameters are known, it may happen that a reduction to a finite dimensional system (0.1) is not possible. This situation arises for instance in the treatment of turbulence in the interstellar medium, or in stellar convection. Under those circumstances we have no guarantee whatsoever that any conventional numerical method of solving the original Eq. (0.2) (finite difference approximation, finite element method, etc) can supply the solution we are looking for. Formal simplicity of the underlying mathematical equation does not secure the numerical availability of generic and hence physically meaningful solutions, although it typically does imply the existence of particular analytical solutions (to be traced by Lie–group methods for instance).

Besides this intrinsic complexity, there is another source of complexity illustrated by the problem of modelling the space and time behaviour of the metallicity in a galaxy. The galactic metal distribution results from the precise details of the chemical evolution of all stars in the galaxy; the chemical evolution of a single star in turn depends on the input chemicals — the chemical composition of the medium out of which the star is born, the thermonuclear reactions occurring in the stellar interior, the processes of mixing and transport of matter inside the star, the mechanisms of mass ejection, and eventually, the explosion of the star, if it is massive enough; the distribution of metals in the galactic medium is further affected by stellar dynamics, and by the dynamics of the galaxy itself, etc. Another problem that belongs into this category concerns the propagation of disturbances on the surface of the terrestrial planets. In the approximation in which the surface matter is just regarded as being a solid, this problem is again captured by a 'simple' model. In a more detailed treatment in which one attempts to take account for instance of the granular nature of the surface matter, the problem becomes intrinsically complex : The physics of the granular state is only poorly understood at the moment. There exists no thermodynamic formulation of the granular state of matter (Jaeger and Nagel 1992), so that there is no a priori reason that even the mathematical structure of the macroscopic continuum equations should survive in the fuller description of this medium.

The hallmark of complex modelling is that it is not, as a rule, captured by a small collection of well defined equations, whose solution would then supply the behaviour of the system under consideration (space and time pattern of the galactic metallicity). Instead, such a model appears as an aggregate of a large number of ad hoc and partially semi–empirical parametrised representations characterising the individual mechanisms entering the system. The input 'physics' — summarised in the parameters of the representation — if at all accessible theoretically, is itself the outcome of nontrivial calculations, based on 'simple' models and referring to isolated aspects of the problem to be analysed (for instance stellar evolution alone, etc). The treatment of the interactions between the different mechanisms is eventually the main objective of this type of modelling. As a rule, and by contrast with the systematic contruction of the models of the low complexity end, the derivation of a high complexity model system proceeds entirely heuristically.

On a formal level, this category of modelling thus addresses *systems whose underlying laws are incompletely understood.* The lack of knowledge is compensated for by substituting global phenomenological descriptions to an underlying unknown physics. A simple parametrised ansatz is used to capture the global contribution of a collection of individual physical effects operating on a level one is not directly concerned with. By analogy with the low complexity models, the trend has been to formulate these high complexity models too in the language of differential and partial differential equations, or sometimes as integro–differential equations.

The quantitatively most successful models are blends of simplicity and complexity, generally rooted on a basic simple framework on which complex effects are superposed. For instance in the current radial stellar pulsation calculations the underlying physics is regarded as given (radial fluid dynamics, etc), so that the pulsation model scheme (say an N–zone model, $N \approx 100$ for instance) basically belongs into the category of low complexity models. However, within this formally well defined framework a number of physical mechanisms are treated phenomenologically (convection, shocks, massloss, etc); the latter belong into the high complexity class. In fact, pure models of the simple class, to the extent that they provide accurate descriptions of real astrophysical phenomena, presumably exist under exceptional circumstances only. The model for the dynamic behaviour of globular clusters of stars (over timescales short in comparison with the lifetime of individual stars) identified with the classical gravitational N–body problem (cf Heggie 1989), may be one of these rare exceptions. Most 'simple' models, like the use of classic polytropic stellar models to interpret the location of the ZAMS, or the reference to even more rudimentary analytic stellar models to understand qualitatively the evolutionary tracks in the HR diagram (Stein 1966) are really mixtures of simplicity and complexity : While the chief underlying structure of the defining equations is well established, the auxiliary physical phenomena (heat transport, etc) are handled by parametrised heuristic representations concealing the physical complexity of the genuine problem to be modelled (polytropic index, etc).

Experience shows that standard modelling methods may generate misleading results due to several circumstances.

In the first place standard modelling entirely ignores *intrinsic fluctuations.* While it is true that fluctuations are both quantitatively and qualitatively without serious effect in a relaxed regime, provided that the dynamics be *regular and stable*, they may play a determining role in the relaxed phase when the dynamics is *chaotic.* They may further significantly modify the transient phase, even in the absence of instabilities of the subsequent stationary motion. In the problem of stellar pulsations the first point has been demonstrated by a 'model of ignorance', namely by algebraic maps disturbed by white noise meant to simulate the intrinsic fluctuations (i.e. a higher order Yule scheme); in the chaotic regime of the unperturbed map, the noise may generate qualitatively new evolutionary patterns (in particular a characteristic intermittent behaviour; cf Perdang 1990b,

1991). The second point is illustrated by an old and experimentally well known behaviour of reaction kinetics. Elementary theory predicts for the transient phase an asymptotic time behaviour in the form of a power law in time, t^{-b} for $t \to \infty$, for the overall concentration of the reactants; the exponent $b > 0$ is determined by the stoichiometric coefficients of the reaction. While experiment is indeed consistent with the power law behaviour, it turns out that the actually observed exponent can differ significantly from the theoretical b (Zumofen et al 1988).

For the reaction $\nu X + \mu A \to \ldots$, ($\nu > 1$), in a homogeneous medium in which the A concentration is held constant, the theoretical exponent in the asymptotic behaviour for the concentration of X is $b = \frac{1}{\nu - 1}$. For $\nu = 1$ the experimentally determined asymptotic behaviour obeys the Kohlrausch–Williams–Watt law, $\exp[-(\frac{t}{\tau})^a]$, where the exponent a may differ from the theoretical value 1.

The reason for the discrepancy is the lack of perfect spatial homogeneity in the distribution of the reactants, leading in turn to spatial and temporal fluctuations in the concentrations in the reaction products. In connection with chemical reactions, I mention yet another anomaly defying the conventional interpretation in terms of differential equations of reaction kinetics, and related again to a lack of spatial smoothness : The reaction rate of catalytic oxidation processes may exhibit a virtually self–similar time behaviour (Dress et al 1984).

Ideally then, an adequate model should have the ability of generating by itself the correct type of fluctuations.

In the second place — and this is relevant for models of higher complexity— as already mentioned, one typically attempts at squeezing the collection of involved interactions among many physical factors into the straitjacket of differential, partial differential or integro–differential equations. Such a procedure forces on the solutions smoothness properties which are not a priori warranted. Thus, in the problem of the distribution of metals, a smoothness requirement seems grossly inadequate: It has been observed for a long time (de Wijs 1951) that many minerals are roughly self–similarly distributed on Earth; more precisely iron ore has been shown to be concentrated over a fractal set of dimension 1.4-1.9 (cf Burrough 1981,1989).

The fact that a range of fractal dimensions has been determined for this distribution suggests incidentally that we are dealing here with a multifractal.

On scales up to several Megaparsecs, it is known for a long time that the universe is clumpy (cf Chincarini and Rood 1975) and the matter distribution — and therefore the metal distribution — is at least approximately fractal (cf Pietronero 1987, Coleman and Pietronero 1992); it would then be surprising if on the intermediate galactic scale a locally uniform, and globally smooth, metal distribution were to hold; this seems the more unlikely since simple models indicate that the precursors of supernovae, the sources of the metal content, are fractally distributed in the galactic environment (cf Lejeune and Perdang 1991).

In the third place, the use of difference schemes (0.1) approximating the original partial differential equations (0.2), may lead to *numerical instabilities* which are not easily detectable as such. Numerically unstable solutions could be mistaken for intrinsic chaotic behaviour.

There are a few model procedures, so far only occasionally referred to in the astrophysical literature, which do not directly belong into this linear classification. Among those I mention in the first place *algebraic maps*. In the astronomical context maps have been introduced not so much as independent models, but rather as 'models of models', to clarify and interpret the behaviour of conventional models.

For instance Hénon's iterative map for the interpretation of the hierarchical Saturn ring structure (Hénon 1988) is in fact analytically derived from the differential equations of motion for a ball (point mass) moving in the plane of an inclined billiard, whose lower boundary is specified by two intersecting circles; initial conditions : zero velocity; position $(x_o, y_o) = (a, H)$ in a coordinate system of origin through the intersection point, with H constant and a a parameter. The billiard model itself is introduced as an analytically tractable version of the topology of a restricted 3–body problem, designed to simulate the motion of an individual particle of the ring (Petit and Hénon 1986). The analytic study of the iterative map shows that the set of a values leading to periodic orbits of the ball (simulating trapped motions of the particle in the 3–body problem) is a Cantor set of structure reminiscent of the the structure of the real Saturn ring.

Likewise, the $2D$ iterative multi–parameter maps modelling stellar pulsations were originally derived from approximations to the differential pulsation equations of coupled modes (Perdang 1990); an independent and more powerful formal derivation relying on genericity arguments only was introduced later (Perdang 1990a). Again this iterative map plays the part of a model for the interpretation of the behaviour of a more conventional model, namely the results of detailed (N–zone) radial stellar pulsation calculations. The algebraic simplicity of this map lends itself to exhaustive numerical investigations of the nature of its solutions in the parameter space. It has led to the identification of a category of chaotic behaviour isolated in multi–mode stellar oscillations in the early 80s as Type II intermittency (Perdang 1990b). It has also allowed to predict the existence of fractal boundaries between zones of different pulsational behaviour in the HR diagram, a prediction which would have required a prohibitive amount of numerical experiments in the context of the conventional model. It has finally allowed to predict the effect of disturbing mechanisms on the pulsational behaviour near the fractal boundaries in the parameter space (a special intermittency generated by noise, already mentioned above).

On the other hand, stochastically perturbed maps, i.e. *autoregressions* were at some time explicitly proposed as models for irregular stellar variability (Ashbrook et al 1954; Deeming 1970). The more recently investigated differential oscillation systems perturbed by noise — thereby transforming the latter into higher order Langevin equations, as proposed by Vio et al (1992) as models for quasar variability also range into this category.

Pure *geometrical models*, which do not fit into the above linear classification, constitute a further class which have occasionally been proposed in the astronomical context. As examples I mention Voronoi tessalations of the plane or of the 3–dimensional space as simulations of a gravitational break–up of the cosmic or the interstellar medium, leading to galaxy or star formation respectively (Kiang 1966, Van Weygaert 1991). A variety of fractal models (galaxy distribution, Peebles 1989; distribution of crater sizes, etc) belong into this group as well. Besides their interest as descriptive phenomenological models, such representations may possess an interpretational interest, to the extent that the geometric construction (the iterative rule in the fractal case) reflects an actual physical mechanism.

Since the late 70s a novel modelling environment, namely the *Cellular Automaton* (CA) technique, has gained some popularity in fundamental physics (e.g. statistical mechanics: percolation), in applied physics (fluid dynamics, etc), and in fact in many scientific disciplines including biology and medicine (cf Liu 1991, Stauffer and Pandey 1992). Actually, the first astronomical CA simulation dealing with propagating star formation, by Seiden et al (1978; cf also Schulman 1993), was based on a reinterpretation of a medical — or geographical — model, namely the spread of epidemics, or the propagation of a forest fire (cf Stauffer 1985).

A cellular automaton is both a specialisation and a generalisation of a traditional iterative map : It is a particularisation in so far that it refers to *discrete variables* only, while maps are operating on continuous state variables; and it extends the framework of maps in so far that it is dealing with a very *large* — possibly infinite — *number of state functions*, while iterative maps are typically concerned with small numbers of state functions only (low dimensional maps). Moreover, in contrast with most iterative maps, in which the state variables describe the discrete time evolution of generally abstract magnitudes (an 'amplitude' of a signal in oscillation problems), the state variables have in general a concrete physical interpretation : they are typically 'occupation variables' (e.g. numbers of 'particles' specified by a given property) of a discrete spatial position at a discrete time. Exact rules of evolution for these state variables can then be set up if the nature of the local interactions of the occupation variables is fully understood. More important, even if only a partial knowledge of the interaction mechanism is available, it may be possible to devise meaningful evolution rules reflecting the chief effects of the interactions. It is precisely this flexibility of the CA framework — exploited in the applied sciences, and in biological modelling (cf Stauffer and Pandey 1992) — that should make it an attractive modelling tool for the category of complex astronomical processes.

Several further points of interest of the CA environment should be kept in mind.

In the first place, while the actual solution a PDE model becomes very often prohibitive once the configuration to be analysed takes on a complicated shape, the use of CA models requires virtually no extra labour in the treatment of ***arbitrary geometries***. A CA model of a molecular cloud, or of an SNR — if available — could perfectly well deal with the fractal shape of these objects; the precise shape could directly be read into the CA environment, from a digitised photograph of the structure. This remark is indicative that in astronomical problems, where complicated geometrical shapes abound, it is of interest to have a CA model at our disposal, even when a satisfying PDE model does exist.

In the second place, the typical CA environment is dealing with discrete elementary units, or 'particles' or 'states', so that this representation possesses a natural ability of generating ***intrinsic fluctuations***. There is a risk for the latter to be quantitatively exaggerated versions of the real physical fluctuations of the problem to be modelled, since the number of CA 'particles' will be, as a rule, orders of magnitude less than the actual number of true physical molecules; however, since qualitatively the origin of the CA fluctuations correctly simulates the origin of the fluctuations in the true problem, we are sure that the qualitative nature of the latter is properly duplicated in the CA simulation. The CA representation thus appears as a major improvement with respect to the procedure of adding white (or coloured) noise contributions to a differential model, in which the actual properties of the terms to be added are eventually arbitrarily chosen; they are not naturally generated in the model itself. I mention in this connection that Scalo (1987) precisely stressed the astrophysical interest of CA models of turbulence in the interstellar medium, in order to obtain information on the density fluctuations, velocity fluctuations, etc.

In the third place, the very algebraic formulation of a CA model implies that its direct solutions are discrete. Continuous properties, attached to a CA model, are obtained after averaging appropriate functions of discrete 'occupation variables' (state variables, particle numbers), the averaging being carried out over a statistical ensemble, or over time and space. The original discrete model solutions, as a result of the iterative process may exhibit self-similar, statistically self-similar or more general fractal space as well as space–time behaviour (Wolfram 1983, Willson 1984); the fractal-like structure in turn may then survive on the level of the averaged variables. It is this property that makes the CA environment a promising candidate for the generation of ***complex structures*** which especially in the astrophysical context are overabundant. Being hard to produce with conventional mathematical methods, these spatially complex structures, have just been ignored — unless their structure has been approximated by manifestly inadequate idealised shapes (spherical geometries for molecular clouds or supernova remnants, cylinders for galactic arms, etc).

We shall discuss in the following survey a few general properties of the CA technique which, we believe, highlight the specific advantages and, at the same time, the major drawbacks of this new approach to modelling of astrophysical relevance.

> In the world of human thought generally and in physical science particularly the most fruitful concepts are those to which it is impossible to attach a well-defined meaning.
>
> H.A. Kramers

1. Towards a Definition of Cellular Automata

The cellular automaton framework created by von Neumann and Ulam, and independently by Zuse in the late 40s (Ulam 1950; cf Burk 1970; Zuse 1972,1975) is a brainchild of electronic computers. Also early work and interest in CA centred around computation and computer technical applications. This is clearly spelled out in Codd's (1968) definition :

A CA is a system dealing with 'large collections (usually infinite in order to avoid boundary problems) of interconnected finite automata [in the computer theoretical sense], each finite automaton being thought of as a cell. The kinds of things investigated include conditions under which it is possible to exhibit : (1) The computation of all computable functions. (2) The construction of automata by automata, the offspring being at least as powerful (in some well-defined sense) as the parent.'

It is true, on the other hand, that both von Neumann and Ulam did have in mind scientific applications as well, specifically applications in biology. The former insisted on ***self-replication***, the latter on ***simplicity of a code capable of generating complexity*** (genes encode simple rules, and possibly meta-rules capable of changing the rules; Cooper 1988; cf a more recent version of this idea by Kaneko 1992). Ulam seemingly was the first to realise that the complex geometrical structures we encounter in the natural world may be reproducible by simple instructions executed in a CA framework. By a variety of numerical experiments on $2D$ and $3D$ lattices performed in collaboration with Schrandt (Schrandt and Ulam 1970), Ulam demonstrated the CA potentialities for creating ***patterns of growth figures***, emphasising that the geometries generated in the CA framework from the simple rules are 'intermediate in complexity between inorganic patterns like those of crystals and the more varied intricacies of organic molecules and structures'. Ulam's main purpose in pursuing these investigations remains however information theoretic : It is 'to show ... an enormous variety of objects which may be obtained by rather simple inductive definitions and to throw a sidelight on the question of how much "information" is necessary to describe the seemingly enormously elaborate structures of living systems' (Ulam 1962). And regarding the results of his experiments he states : 'The amount of "information" contained in these objects [the computed patterns] is ... quite small, despite their apparent complexity and unpredictability' (Schrandt and Ulam 1970).

The specific relevance of the CA framework to the ***modelling of genuine physics*** was apparently not spelled out by the earlier workers in this field. That idea seemingly emerged only progressively in the 70s (cf Toffoli and Margolus 1987). The shift in interest towards physical simulation was accompanied by a corresponding shift of emphasis on the ***dynamics*** in the concept of the CA itself, as is indicated by the following informal definition given in a recent review :

Cellular Automata are ***dynamical systems*** in which space and time are discrete. The states of cells in a regular lattice are updated synchronously according to a deterministic local interaction rule. Each cell obeys the same rule, and has a finite, usually small number of states.

(Gutowitz 1991; for a similar definition see Manneville 1990; a formalised definition is given in Culik et al 1990).

The lattice itself is implicitly understood to be a regular (n-dimensional) space covering lattice, generally a Z^n lattice, Z, set of integers; the cell states are encoded in a finite set of integers, for instance the set $Z_N = \{0, 1, \ldots, N-1\}$ (cf Goles 1990).

This definition actually appears to be too limited in scope to cover the full array of models which are currently listed as CA models. The main limitations are the following : (1) By insisting on ***determinism***, this definition ignores a broad class of popular models naturally referred to as CA models in the literature (diffusion, aggregation and percolation models, cf Stanley and Ostrowski 1986); in fact, it would miss out nearly all of what has been published so far under this title in astronomy. (2) By stressing strict ***locality*** of the evolution rules, the definition excludes a second interesting class of models which we would like to call CA models. (3) By imposing a ***crystal lattice***, a further category of important models is eliminated.

The earliest CA-type simulation proposed in astronomy is the model of propagating stochastic star formation regarded as a relevant mechanism for generating structure in galaxies, by Seiden et al (1979) and Gerola et al (1980), Seiden and Gerola (1982). This model actually violates the above definition on all 3 accounts. (1) It makes use of ***statistical evolution rules*** (random laws of star formation and decay), and (2) ***non-local evolution*** (simulation of differential rotation by global lattice operations). (3) It is also based on an uncommon lattice : In order closely to mimic the approximate geometry of actual disk galaxies, the $2D$ simulation adopts a ***polar grid***.

To meet with these objections we propose here a tentative more flexible definition of a CA which, we believe, matches more closely what most workers in discrete simulations have in mind when using this terminology. While the enlarged definition is chosen to cover most models discussed in the present Proceedings, it makes no claim of universality : In spite of its flexibility, it will need amendments whenever a basically new type of evolution problem is translated into a CA–like context (cf Kramers's quotation).

D : A cellular automaton is a dynamical system whose independent variables, time and spatial position, as well as its dependent variables, defining the state, are discrete; the evolution is fixed either by deterministic or stochastic rules.

More explicitly

(a) *Time* is represented by the sequence of integers $\tau \in \mathsf{Z}$.

(b) *Space* is identified with the set of cells of an arbitrary spatial network L carried by an n–dimensional manifold M^n; the network partitions the manifold into n–dimensional cells; the cells are labelled by a collection of integer coordinates $\mathbf{i} = (i_1, i_2, \ldots, i_n) \in \mathsf{Z}^{(1)} \times \mathsf{Z}^{(2)} \times \ldots \times \mathsf{Z}^{(n)}$ or Z_n (with $\mathsf{Z}^{(k)}$ $k = 1, 2, \ldots, n$, either the set of integers, the set of nonnegative integers, or a subset thereof).

(c) The state of cell $\mathbf{i}$ at time τ is defined by a field variable $F = F(\tau; \mathbf{i}) \in \mathbf{z}$ with the state space $\mathbf{z}$ being a subset of the integers; if $z = \mathrm{card}(\mathbf{z})$ is the number of different states per cell, then the automaton is referred to as a z–state CA; z is always assumed to be a *small* integer.

In some simulations it is necessary to attach an auxiliary counter, or integer label, $c(\tau; \mathbf{i}) = k \in \mathsf{Z}$, with the cells, such that groups of cells possessing same properties (defining an equivalence class) can be identified by the same label.

(d) The ***dynamics*** is specified by a synchronous application of a transition rule at each cell, the new state $F(\tau + 1; \mathbf{i}) \in \mathsf{N}(F)$ of cell $\mathbf{i}$ at time $\tau + 1$ depending on the states $F(\tau; \mathbf{i}')$ at time τ and the cells $\mathbf{i}' \in \mathsf{N}(\mathbf{i})$; here $\mathsf{N}(F)$ denotes a field neighbourhood in $\mathbf{z}$ of the F value at time τ and cell $\mathbf{i}$, and $\mathsf{N}(\mathbf{i})$ denotes a lattice neighbourhood of cell $\mathbf{i}$.

This dependence is specified mathematically by (i) a collection of functions, or ***evolution rules***, identified by the label c,

$$R_c\{F(\tau;\mathbf{i})\} \equiv R_c[\,\mathbf{i}; F(\tau;\mathbf{i}'), \mathbf{i}' \in N(\mathbf{i})\,], \quad c = 1, 2, \ldots, C, \qquad 1.1$$

taking values in $\mathbf{z}$, and explicitly depending on space $\mathbf{i}$, and on the states of the neighbourhood $\mathsf{N}(\mathbf{i})$; and (ii) a collection of probabilities

$$P_c\{F(\tau;\mathbf{i})\} \equiv P_c[\,\mathbf{i}; F(\tau;\mathbf{i}'), \mathbf{i}' \in N(\mathbf{i})\,], \quad c = 1, 2, \ldots, C, \qquad 1.2$$

obeying

$$0 \le P_c\{F(\tau;\mathbf{i})\} \le 1 \quad \text{and} \quad \sum_{c=1}^{C} P_c\{F(\tau;\mathbf{i})\} = 1, \qquad 1.3$$

and likewise depending on space $\mathbf{i}$, and on the states of the neighbourhood $\mathsf{N}(\mathbf{i})$; the evolution rule

$$F(\tau + 1;\mathbf{i}) = R_c\{F(\tau;\mathbf{i})\}, \quad c = 1, 2, \ldots, C \qquad 1.4$$

is then realised with probability $P_c\{F(\tau;\mathbf{i})\}$.

The CA is ***deterministic*** for $C = 1$ ($P_1 = 1$ if $C = 1$) , while for $C > 1$ the CA is ***stochastic***.

In the deterministic context, this definition admits of a further extension, by the following amendment of the dynamics (d):

(d') Let R (Eq.1.1) be one among the collection of possible evolution rules,..., $R_a^{(o)}$, $R_b^{(o)}$, $R_c^{(o)}$, ... Refer to the latter as ***level o rules***. Stipulate further that level o rules are to be changed according to a meta–rule

$R^{(1)} : \ldots, R_a^{(o)} \rightarrow R_{a'}^{(o)}, R_b^{(o)} \rightarrow R_{b'}^{(o)}, R_c^{(o)} \rightarrow R_{c'}^{(o)}, \ldots,$

in the sense that if a given rule $R_a^{(o)}$ is applied in the calculation of the evolution, then at the next call of a rule we have to select the rule $R_{a'}^{(o)}$, etc. Again, $R^{(1)}$ is one among the totality of possible 'rules for changing the rules', ..., $R_a^{(1)}$, $R_b^{(1)}$, $R_c^{(1)}$, ..., to be referred to as ***level 1 rules.*** This process is now iterated : If $R^{(l)}$ is a level l rule,

$R^{(l)}$: ..., $R_a^{(l-1)} \rightarrow R_{a'}^{(l-1)}$, $R_b^{(l-1)} \rightarrow R_{b'}^{(l-1)}$, $R_c^{(l-1)} \rightarrow R_{c'}^{(l-1)}$, ...,

in the sense that if a given rule $R_a^{(l-1)}$ applies to the determination of the rule of level $(l-1)$, then at the next determination of a level $(l-1)$ rule we have to choose $R_{a'}^{(l-1)}$, etc. The ***dynamics*** of the automaton is thus determined by a hierarchy of rules (actual automaton rules, or level o rules; meta-rules for changing the automaton rules, or level 1 rules; meta-meta-rules of a change of the meta-rules for changing the automaton rules, or level 2 laws, etc), up to a highest level Λ.

We conveniently refer to a CA subject to such a hierarchy of laws as a *level* Λ CA.

The deterministic CA in the sense of definition D is thus a level o automaton. The stochastic CA appears as a particular level 1 automaton; the rule $R^{(1)}$ consists in defining the probabilities of selection of individual automaton rules.

An equivalent and theoretically sometimes more convenient way of formulating the evolution rule (1.4) consists in defining a matrix of ***transition probabilities*** T, explicitly depending on the states of the neighbourhood $\mathsf{N}(\mathbf{i})$, and on space $\mathbf{i}$, of elements

$$T_{\phi\phi'}\{F(\tau;\mathbf{i})\} \equiv T[\mathbf{i}; F(\tau;\mathbf{i}'), \mathbf{i}' \in N(\mathbf{i})]_{\phi\phi'}, \quad \phi, \phi' \in z \qquad 1.5$$

such that if $F(\tau;\mathbf{i}) = \phi$then $T_{\phi\phi'}$ represents the probability for having $F(\tau+1;\mathbf{i}) = \phi'$. The counterpart of conditions (1.3) is here

$$0 \leq T_{\phi\phi'}\{F(\tau;\mathbf{i})\} \leq 1 \quad \textit{and} \quad \sum_{\phi' \in z} T_{\phi\phi'}\{F(\tau;\mathbf{i})\} = 1 \qquad 1.6$$

In the deterministic case the matrix element $T_{\phi\phi'}\{F(\tau;\mathbf{i})\}$ is either 0 (transition $\phi \rightarrow \phi'$ forbidden) or 1 (transition $\phi \rightarrow \phi'$ occurs surely).

A direct extension of the framework consists in regarding the dynamic functions (1.1) and the associated probabilities (1.2) as time dependent as well:

$$R_c\{F(\tau;\mathbf{i})\} \equiv R_c[\,\tau;\mathbf{i}; F(\tau;\mathbf{i}'), \mathbf{i}' \in N(\mathbf{i})\,], \qquad 1.1'$$

$$P_c\{F(\tau;\mathbf{i})\} \equiv P_c[\,\tau;\mathbf{i}; F(\tau;\mathbf{i}'), \mathbf{i}' \in N(\mathbf{i})\,], \quad c = 1, 2, \ldots, C, \qquad 1.2'$$

In the alternative formulation the transition matrix (1.5) becomes explicitly time dependent

$$T_{\phi\phi'} \equiv T[\,\tau;\mathbf{i}; F(\tau;\mathbf{i}'), \mathbf{i}' \in N(\mathbf{i})]_{\phi\phi'}, \quad \phi, \phi' \in z \qquad 1.5'$$

For reasons of notational simplicity it is often convenient to use a single positive integer i = 1,2,... as an identifier of the cells.

In its deterministic form definition D is essentially a translation of a familiar template for classical continuum physics — such as the PDEs of fluid mechanics — into a discrete setting. It involves therefore, in the first place, a specification of the spatial topology and geometry (a manifold M^n with an attached coordinate system) over which the physics is defined. The n-tuples $\mathbf{i}$ play the parts of cell identifiers; in an actual calculation it may be useful to introduce auxiliary coordinates expressed in the form of rational numbers (and represented for instance as p-adic numbers).

In the second place it requires a set of discrete dependent variables together with their space of definition and the topology of the latter; this choice is guided by the natural observables of the problem to be simulated. Next, the various specifications of neighbourhoods are the discrete-space counterparts of the 'infinitesimal' open sets associated with points in continuous spaces. The spatial neighbourhood $\mathsf{N}(\mathbf{i})$ fixes essentially the spatial order of a partial differential representation in a continuous model (the larger the neighbourhood, the higher the allowed order); the neighbourhood of a point in the space z of the dependent variables, $N(F)$, is a measure of the degree of smoothness of the dependent variables in time; if the number of states of the neighbourhood is small as compared to the total number of permissible states in z then these variables necessarily change smoothly in time; otherwise violent changes in time can be simulated. The precise nature of these changes is encoded in the algebraic form (1.1) of the evolution rule — the counterpart of the PDE or system of PDEs in continuum physics.

In the stochastic case, $C > 1$ alternatives for the change of state are allowed at each time step. The CA formulation then requires, besides the list of allowed changes (set of evolution rules), also a list of the associated probabilities. A simulation, starting with given initial conditions, then consists in the construction of *one realisation* of a solution.

The transition probability formalism (1.6) can be viewed as the counterpart of a Liouville type equation associated with an ODE (or PDE) in the deterministic case, or a Fokker–Planck type equation attached with stochastic differential equations (cf Schuss 1980).

A few further comments on the ingredients of the definition are in order.

The underlying spatial manifold in conventional automaton models is the Euclidean space E^n. But just as in continuous classical physics we may be led to adopt a different manifold (cf the circle S^1 which has the natural topology for the analysis of a $1D$ periodic motion), it may be technically more convenient to choose a specific manifold adapted to the physics simulated by the CA. A proper choice of M^n may automatically incorporate certain boundary conditions. Thus periodic boundary conditions on the inner walls of a slab of width L, $x_1 = 0$, $x_1 = L$, embedded in an n–dimensional Euclidean space, are automatically accounted for by choosing a cylinder $S^1 \times E^{n-1}$ (direct product of a circle by a lower dimensional Euclidean space) as the adequate manifold carrying the lattice; if the boundary conditions are reflections on the walls of the slab, then a cylinder becomes again the natural underlying manifold (containing now twice the number of cells of the original slab).

The motivation for leaving the spatial grid unspecified in our definition reflects the freedom we have in continuous representations to select a symmetry adapted coordinate frame. Thus CA models for the structure of isolated stars or planets, for isolated molecular clouds, and even for irregularly shaped nebulae (e.g. a model for the filamentary structure of the Crab nebula), seem to be most economically accommodated by a $3D$ spherical network (generalising the $2D$ polar grid of Seiden et al 1979). In a simulation of a star's structure on a cubic lattice for instance we generate errors due to the incompatibility of the lattice symmetry with the approximate global symmetry; in addition the cubic lattice would here be computationally inefficient since large parts of the lattice space would remain systematically unoccupied. Moreover, while conventional automata models make use of *regular lattices*, made up of identical cells (equidistant nodes on a straight line; square, triangular or hexagonal lattice in the plane; etc), it is of interest to allow for grids with spatially variable cell shapes generated by arbitrary continuous deformations of a regular lattice (nodes distributed according to a given law on the line; plane covered by n–gons , n=3 or 4, such that any vertex is common to n cells, etc). Wolfram (1986) has commented on the possible use of quasi–lattices, as well as random lattices in the context of gas simulations (cf also the discussion by Zuse 1975 on the geometry and topology of lattices).

The introduction of a counter, though not essential – at every timestep this parameter can be recovered from the knowledge of the field $F(\tau;\mathbf{i})$ — is time–saving when the evolution rules depend on class properties rather than on individual cellstates. Moreover, being a class label, the counter allows to deal easily with the statistics of the classes. For instance, if spatially contiguous cells (specified in terms of the selected spatial neighbourhood) are regarded as defining a same 'macroscopic entity (a spatial *cluster*), then the counter directly tracks this entity (cf Section 3).

The explicit space dependence in the evolution rules (1.1) and corresponding probabilities (1.2) (dependence on $\mathbf{i}$), enables us to simulate a *non–local effect* superimposed on the conventional local effect (dependence on the states in the neighbourhood $N(\mathbf{i})$). It is incorporated to account for the existence of space dependent 'forces' (e.g. the presence of a rotation field whose strength changes over the lattice space in the above mentioned galactic model). An explicit time dependence (Eqs

1.1',1.2') becomes relevant when problems involving *time delays* are envisaged (cf again the galactic model; Schulman and Seiden 1983).

By allowing stochastic evolution rules, we not only conform to a tradition which incorporates microscopic stochasticity in the physical processes to be modelled (cf the usual discrete models for diffusion-like processes mentioned above). We also have in mind a practical way of *handling deterministic mechanisms via probabilistic* CA *simulations* for the following reason. CA evolution by construction proceeds by jumps. To simulate then in a simple way a macroscopically continuous process, say the motion of a cloud of particles under the effect of an external gravitational force, it is expedient to adopt a description in which individual particles have their (discrete) velocities changed at a given timestep in accordance with a probability law; the latter must of course be chosen to comply with the laws of mechanics. Since the underlying physics is deterministic, we could dispense with a stochastic computation technique in the discrete simulation as well; in principle it suffices to change the velocity of a given simulated particle by one velocity step Δv every m^{th} timestep, the integer m being determined by the force and the velocity step (the implicit assumption is here that the true change of velocity per timestep resulting from the applied force is small as compared with the step in the velocity); in practice this leads to a problem of numerical precision, which is not easily dealt with. The probabilistic change of velocity has the advantage of avoiding this difficulty; the price to be paid in an actual experiment is the computation of a random number at every timestep and for every lattice cell occupied by a particle.

Explicit stochasticity is introduced for practical convenience. On a fundamental level it is well known that we can dispense with it : Random behaviour can be generated by a deterministic CA (Wolfram 1985,1986a; cf the explicit deterministic model of diffusion in Levermore and Boghosian 1990), requiring, however, a more involved programming scheme. Direct reference to stochastic rules alleviates the task of the programmer, though perhaps not necessarily the efficiency of the calculation process.

With the exception of stochastic CA, higher level CA as introduced by our amended definition have not been discussed in the literature. We conjecture, however, that deterministic level 1 CA could be usefully applied in the presence of mechanical forces.

The elementary idea is that an 'active' cellstate (cell occupied by a particle) at rest at time 0 is shifted, under the action of a uniform force, by 1 unit (say 1 spatial step) in the direction of the force (translation T^1); at time 1 a second application of the force on the active cellstate consists in a shift of 2 units in the same direction, i.e. the configuration now obeys a different rule (translation T^2); more generally, in this specific example we have a meta-rule expressible, with obvious notations, as $T^n \rightarrow T^{n+1}$. Observe that the meta-rule itself can be formulated analytically.

As will transpire from our survey of applications of the CA formalism, perhaps the majority of the astrophysically relevant automaton models supply a description of an auxiliary *unobservable microworld* — i.e. configurations we are not directly interested in. Within this context we are dealing with fictitious particle-like entities (states encoded in F) endowed with unusual dynamic properties described by the evolution rules; this microscopic picture is at best a crude schematisation of the true microphysics. The real physical phenomena we are interested in — the observable consequences of the model — occur on a *macroscopic level*, i.e. on spatial scales large in comparison with the typical lattice size, and times large in comparison with the timestep. It is therefore of some relevance to have at our disposal a mathematical apparatus that enables us to perform the direct translation from the microscopic level to the macroscopic in a systematic way.

The mathematical formalism that achieves this goal is Robinson's (1980) *Nonstandard Analysis*. We shall therefore rephrase definition D in nonstandard form. For the present purpose we provide a succinct summary of this language, adopting Nelson's axiomatic approach (cf Robert 1985; Lutz and Goze 1981).

With a given conventional set of elements S, the *standard set* (set of natural integers N; integers Z; reals R; n-dimensional manifold; etc), nonstandard analysis associates an enlarged set, $*$S, the *nonstandard set*. This extension is specified by Nelson's axioms of Internal Set Theory (IST), the axioms of *idealisation*, *standardisation* and *transfer*

Principle of Idealisation :

If $R(a, b)$ is an 'internal' relation among elements a, b, (i.e. involving no specification 'standard' or 'nonstandard') containing no quantifiers, then if the relation holds for *one* standard element a and *any* standard element b, then there is an ideal element $*a$ (*nonstandard element*) obeying the same relation for any standard b. In the case of sets of numbers the ideal element possesses properties reminiscent of Euler's 'infinite' or 'infinitesimal' numbers.

Principle of Transfer :

If $S(a, b)$ is an internal relation among elements a, b, containing no quantifiers, then if the relation holds for *any standard element* a, then it holds for *any* element $*a$ (standard and nonstandard element).

Principle of Standardisation:

If $P(*a)$ is a property of a nonstandard element $*a$, then there exists a standard set S whose elements possess this property (this set may be empty).

By idealisation, the enlargement of the set of integers Z, or nonstandard set $*$Z, thus contains, besides the usual integers, at least one ideal element (an *infinit*, in Robinson's terminology), and in fact, it must contain an arbitrary number of them, as follows from transfer. The enlarged set of reals contains besides the infinits also a second class of ideal numbers, namely the *infinitesimals* (by transfer, for instance from the relation $a = 1/b$ for $b \neq 0$; take for b an infinit). Denoting infinits by $*\omega, *\omega', \ldots$ and infinitesimals by $*\epsilon, *\epsilon', \ldots$, we have

$$*\omega^\alpha = *\omega', \text{ if } \alpha > 0, \quad \text{and} \quad *\epsilon, \text{ if } \alpha < 0;$$

$$*\epsilon^\beta = *\epsilon', \text{ if } \beta > 0, \quad \text{and} \quad *\omega, \text{ if } \beta < 0. \qquad 1.7$$

The product of an infinit and an infinitesimal can therefore be an infinitesimal, a finite, or an infinite nonstandard real.

It can be shown that any finite nonstandard real $*a \in *$R can be uniquely written in the form

$$*a = a + *\epsilon, \text{ with } a = \mathrm{st}(*a) \in R \qquad 1.8$$

the *standard part* of the nonstandard real. Obviously $\mathrm{st}(*\epsilon) = 0$.

A standard derivative of a standard differentiable function $f(x)$ is given by

$$\frac{\mathrm{d}f(x)}{\mathrm{d}x} = \mathrm{st}\left(\frac{*f(*x + *\epsilon) - *f(*x)}{*\epsilon}\right)$$

$$= \mathrm{st}\left(\frac{*f(*x) - *f(*x - *\epsilon)}{*\epsilon}\right), \qquad 1.9$$

where $*f$, $*x$ are the nonstandard generalisations of f and x. The extension to higher order derivatives is straightforward.

With these specifications we restate definition D in a nonstandard form.

*D : (a) *Time instants* are labelled by the sequence of nonstandard integers $\tau \in *$Z; *physical time* is given by the standard real

$$t = \mathrm{st}(*\epsilon T\ \tau), \qquad 1.10$$

where $*\epsilon T = \Delta t$ is an infinitesimal timestep (T, reference time).

(b) *Spatial location* is identified by a nonstandard network $*$L of infinitesimal cells carried by an n-dimensional nonstandard compact manifold $*\mathsf{M}^n$; the cells are labelled by a vector of nonstandard integers $\mathbf{i} = (i_1, i_2, \ldots, i_n) \in *\mathsf{Z}^{(1)} \times *\mathsf{Z}^{(2)} \times \ldots \times *\mathsf{Z}^{(n)}$, with $*\mathsf{Z}^{(k)}$, $k = 1, 2, \ldots, n$, a subset of the set of nonstandard integers; *physical position* of a point of the manifold in cell **I**, in a local reference system $\{O, x_1, x_2, \ldots, x_n,\}$ is then given by standard coordinates $\mathbf{r}\,[\,\mathbf{I}\,] = (\, x_1[\mathbf{I}], x_2[\mathbf{I}], \ldots, x_k[\mathbf{I}], \ldots, x_n[\mathbf{I}]\,)$ defined by

$$x_k[\,\mathbf{I}\,] = \mathrm{st}(\ \lambda_k(\mathbf{i}') + \lambda_k(\mathbf{i}'') + \ldots + \lambda_k(\mathbf{I})), \qquad 1.11$$

where $\lambda_k(\mathbf{i})$ is the infinitesimal size of cell $\mathbf{i}$ measured along the kth (curvilinear) coordinate line, ..., $\mathbf{i}'$, $\mathbf{i}''$, ..., being the row of cells joining the cell at the origin to the cell $\mathbf{I}$ along this coordinate line; the lengths $\lambda_k(\mathbf{i})$ obey

$$\lambda_k(\mathbf{i}) \;=\; *\epsilon^{\alpha} f_k(\mathbf{r}[\mathbf{i}])\; L\;, \qquad 1.12$$

where α is a positive real selected at the outset, $f_k(\mathbf{r})$ is (usually) a standard real-valued function, $*\epsilon$ is the (dimensionless) infinitesimal measure of the timestep (Eq. 1.10), and L a reference length.

(c) The ***state*** of a nonstandard cell i at the nonstandard timestep τ is defined by a standard-valued field variable $F = F(\tau;\mathbf{i}) \in \mathbf{z}$ a subset of the standard integers. (d) The ***dynamics*** is specified as in definition D, with the new state $F(\tau{+}1;\mathbf{i}) \in \mathsf{N}(F(\ \tau;\,\mathbf{i}\ ))$, a standard subset of $\mathbf{z}$; the neighbourhood of cell $\mathbf{i}$, $\mathsf{N}(\mathbf{i})$, is a nonstandard infinitesimal set containing (usually) a *finite* number of neighbour cells of cell $\mathbf{i}$. The evolution rules, or the different level rules, are as under D (appropriately translated in nonstandard form).

As a main difference with definition D we require here the lattice carrying nonstandard manifold to be *compact*. The relevance of this clause appears in the theoretical discussion of the allowed time behaviour (section 5).

The automaton operates on the observationally inaccessible nonstandard level, over which it manipulates counterfactual quantities (state variables F with a small number z of states defined over a nonstandard space lattice and a nonstandard time). True observable physics occurs on the associated standard level, and is therefore to be recovered from the fictitious state variables. Genuine classical physical variables take on real values over some continuous interval. To the extent that the CA representation is physically meaningful, it must be able to generate conventional real *physical variables*, say $N_q(t,\mathbf{r})$, $q = 1, 2, \ldots$, from the state variables $F(\tau,\mathbf{i})$. This can indeed be done if the variables $N_q(t,\mathbf{r})$ are defined in terms of an infinite number of state functions. We propose the following simple procedure.

Introduce auxiliary nonstandard functions, $*N_q(\tau,\mathbf{i})$, $q = 1, 2, \ldots$, as a space–time average of standard functions Q_q of the standard-valued state variables F; the averaging is carried out over a nonstandard spatial set $\mathsf{A}(\mathbf{i})$ containing an infinite number $*A(\mathbf{i})$ of cells, and over a time interval $\mathsf{T}(\tau)$ containing an infinite number $*T(\tau)$ of timesteps :

$$*N_q(\tau,\mathbf{i}) \;\equiv\; \frac{1}{*T(\tau)} \sum_{\tau' \in T(\tau)} \frac{1}{*A(\mathbf{i})} \sum_{\mathbf{i}' \in A(\mathbf{i})} Q_q(F(\tau',\mathbf{i}')). \qquad 1.13$$

The corresponding physical variables are then

$$N_q(t,\mathbf{r}) \;\equiv\; \mathrm{st}(*N_q(\tau,\mathbf{i})), \qquad 1.14$$

where the physical time t is related to the timestep number τ by (1.10) and $\mathbf{r}$ is the (standard) position vector of cell $\mathbf{i}$ (1.11,12). The choice of the functions Q_q is dictated by the nature of the observables N_q we want to construct.

If the total number f of cells of the automaton is $*\omega$(an infinite integer), and if the evolution is known over $*\omega'$ (another infinite integer) timesteps, then the averaging sets $*\mathsf{T}(\tau)$ and $*\mathsf{A}(\mathbf{i})$ are chosen to contain

$$*T(\tau) = (*\omega')^{d'} \quad \text{and} \quad *\,A(\mathbf{i}) = (*\omega)^{d} \qquad 1.15$$

intervals and cells respectively with d and $d' < 1$ (for instance 1/2). Such a choice secures that these numbers remain infinits, while their ratios to the total number of cells and timesteps respectively are infinitesimal.

Since CA models are devised to be run on a computer, and since real computers are capable of dealing with finite integers only, adopting a nonstandard version of a CA may appear as a practically irrelevant complication of an intrinsically simple framework. The main interest and practical advantage of this formulation lies less on the side of the CA model itself — although the nonstandard language enables us to keep exactly track of what is physically relevant at any time in the numerical results of a CA calculation (standard parts of actual results); infinitesimal deviations constitute 'numerical noise' related to the fictitious microscopic model. The advantage of a nonstandard formulation is that it simplifies the theoretical analysis, in particular the search for the connection between the CA behaviour and possible PDE formulations provided that the latter do exist.

We mention that in our nonstandard version the interrelation between the spatial and temporal steps is specified at the outset (Eq. 1.12). In a conventional CA the absence of such a relation leads to ambiguities in the search for a global PDE representation (cf Chopard et al 1993).

We complete this section with a remark on the parameters of a computational CA to be interpreted as the counterparts of the theoretically nonstandard quantities. We consider for the sake of definiteness a $1D$ finite square lattice CA with a finite large number f of cells, over which a finite large number N_p of 'particles' is defined; the calculations are followed over a finite large number N_t of timesteps. In accordance with definition *D, we identify these numbers with Robinson's infinits. Specifically, we choose a reference infinit $*\omega$ and identify the nonstandard number of cells with the latter: $*f \equiv *\omega$. Take for the parameter L the physical (finite) length of the automaton, assuming that we are concerned with a spatially confined problem.

If the problem requires a physically infinite space, a different choice of the identification of the number of cells $*f$ with the reference infinit $*\omega$ is required, mirroring the fact that $*f$ is then a higher order infinit (cf chapter 5); the physical length L is then taken as a characteristic finite length of the phenomenon to be modelled.

The infinitesimal cellsize of the lattice then becomes $\lambda = \frac{L}{\sqrt{f}} \rightarrow \frac{L}{\sqrt{*\omega}}$. To secure condition (1.11), the infinitesimal timestep must obey $\Delta t = \frac{T}{N_t} = TL^{-1/\alpha}\lambda^{1/\alpha} \propto *\omega^{-1/2\alpha}$, ($T$, total time); the nonstandard number of timesteps thus obeys $*N_t = *\omega^{1/2\alpha}$. We should also keep in mind that in the context of a given physical model we describe phenomena characterised by numerical parameters covering some preassigned range (say densities $< 10^7$ g cm^{-3}, temperatures $< 10^{10}$ K, etc) and specified with a preassigned precision. For the purposes of such a model values beyond the upper limit are 'infinite' — they cease to be meaningful because the physical theory is not valid outside the preassigned range; values less than the precision are 'infinitesimal' — they cease to be meaningful because other physical effects discarded within the formulation would introduce corrections exceeding the level of precision. The corresponding 'physical infinite' and 'infinitesimal' values are then mathematically formalised as nonstandard reals, namely as infinits and infinitesimals (cf Harthong 1981). For instance, a number density of 'particles' n_p is usually required to be physically finite; in the CA this implies that $\frac{N_p}{f} \rightarrow \frac{*N_p}{*\omega}$ is finite, i.e. the number of particles, N_p, is of the same order of infinity as the number of cells $*f$; if the spatial averaging is over $\sqrt{f}$ cells, and we average over $\sqrt{N_t}$ timesteps, then the relative density fluctuations are of order $\frac{1}{f^{1/4}N_t^{1/4}}$, or the infinitesimal attached with the density is $*\epsilon_n = \frac{1}{\sqrt{*\omega}}$.

Viajero, no hay caminos. Caminos se hacen caminando.

A. Machado 1940

2. CA Models for PDEs

In principle a trivial systematic procedure to convert any model given in the form of a system of PDEs, into an equivalent CA–like framework merely consists in interpreting the PDEs as the standard part of nonstandard difference schemes; the latter, rewritten as an iterative map, supply the CA evolution rules. Explicitly this is done along the following lines.

(a) Write the PDEs in normal form

$$\partial_t \underline{Y}(t, \mathbf{r}) = S_{op} \ \{\underline{Y}(t; \mathbf{r})\ \}, \tag{2.1}$$

where $\underline{Y}$ is the collection of dependent variables of the problem, and S_{op} a spatial differential operator (in general depending on position $\mathbf{r}$ and time t).

(b) Adopt a symmetry–adapted nonstandard spatial grid and regard (2.1) as the standard part of a nonstandard difference equation in space and time. With the notations of definition *D this is done by introducing a nonstandard $*\underline{Y}(t, \mathbf{r})$ such that $\underline{Y}(t, \mathbf{r}) = \text{st}\ (*\underline{Y}(\tau, \mathbf{i}))$; then $\partial_t \underline{Y}(t, \mathbf{r}) = \text{st}\ ([*\underline{Y}(\tau+1, \mathbf{i}) - *\underline{Y}(\tau, \mathbf{i})]/(\Delta t))$, with Δt the infinitesimal timestep defined under *D; the standard space derivatives are expressed in a similar fashion as nonstandard differences (obtained by a conventional difference scheme) cf Robinson (1980); the relation between the infinitesimal timestep and lattice size (exponent α in Eq. 1.12) is fixed by the structure of the original PDE.

(c) Rewrite the difference equation in the form of a (nonstandard) iterative map in time

$$*\underline{Y}(\tau+1, \mathbf{i}) = *\underline{R}\ \{*\underline{Y}(\tau, \mathbf{i})\ \}; \tag{2.2}$$

the curly brackets of the righthand side of (2.2) indicate that the variable $*\underline{Y}$ at cells $\mathbf{i}'$, $\mathbf{i}''$, ... in the neighbourhood of cell $\mathbf{i}$ is involved; the spatial neighbourhood N($\mathbf{i}$) is automatically specified by the nature of the difference equations.

(d) Discretise the nonstandard space of the dependent variables $*\underline{Y}$ by subdividing the latter into infinitesimal cells identified by nonstandard integers F (or by a vector of integer components, or by some other scheme making use of an integer representation); the map (2.2) then transforms into a nonstandard integer iteration

$$F(\tau+1, \mathbf{i}) = *R\{F(\tau, \mathbf{i})\}; \tag{2.3}$$

the function $*R$ is derived from $*\underline{R}$ in a natural way.

Eq. (2.3) thus represents the evolution rule (1.1) for a deterministic automaton. In this nonstandard version the standard part of the CA model solution yields the *exact* solution of the PDE model.

To illustrate this procedure we apply these steps to a 1–dimensional diffusion equation — the diffusion of a metal in a hydrogen–helium atmosphere of a star, disregarding gravitational effects and curvature. In its simplest form the starting PDE model is

$$\partial_t \rho \ = \ D\ \partial_x^2 \rho, \tag{2.4}$$

(D, diffusion coefficient; x height in a plane parallel atmosphere), which is already in normal form (2.1). The density of the diffusing matter, ϱ, plays the part of the dependent variable. The associated nonstandard iterative map is

$$*\rho(\tau+1, i) = *\rho(\tau, i) + *K\ [\ *\rho(\tau, i+1) - 2*\rho(\tau, i) + *\rho(\tau, i-1)\ ], \tag{2.5}$$

where $*K = D\ \frac{\Delta t}{\lambda^2}$, ($\Delta t$, infinitesimal timestep, and λ, infinitesimal constant cellsize). Since consistency requires the nonstandard parameter $*K$ to be finite and since the physical diffusion coefficient D is finite, we obtain for the exponent in relation (1.12) $\alpha = 1/2$. Next, by subdividing the allowed density range, say (0, $*\rho_{max}$), into a nonstandard integer number $*z$ of equal density intervals of infinitesimal width $\Delta\varrho = *\rho_{max}/*z$, labelled 0,1, ... , $*z$-1, we achieve a natural discretisation of the density space. With the nonstandard density value $*\rho$ associate the nonstandard integer F defined over the nonstandard space $*\mathbf{z} = *\mathsf{Z}_{*z}$.

$$*\rho \rightarrow F = \ \text{int}(\frac{*\rho}{\Delta\rho}) \tag{2.6}$$

(int x, integer part of the real x), and conversely, the nonstandard density corresponding to the CA variable F is

$$F \rightarrow *\rho \in [F\Delta\rho, (F+1)\Delta\rho). \qquad 2.6.a$$

The evolution rule becomes

$$F(\tau+1, i) = F(\tau, i) + \text{int}\{*K[F(\tau, i+1) - 2F(\tau, i) + F(\tau, i-1)]\}, \qquad 2.7$$

which completes the nonstandard CA–like model.

This discussion makes it clear that any numerical solution procedure of a PDE in terms of an explicit difference scheme is at the same time a computational approximation to a nonstandard CA–like representation (all numbers in a finite precision calculation being ultimately integers). There is, however one point in this construction that violates the specification of the CA definition D or *D: The definition stipulates that the dependent variables F have a *finite* and typically small number z of states. In the above nonstandard construction an *infinite* number $*z$ of states is required, implying that in any actual realisation of such a CA model an acceptable numerical precision in the computation of the physical variables requires the space of the dependent variables to be finely subdivided into a very large number of cells (2.6a). Numerical accuracy is thus inconsistent with the requirement of smallness of z of our definition.

For this reason it is not customary to view trivial numerical schemes of this nature as genuine CA models. Besides, a model equivalent in algebraic structure to the structure of the PDE cannot be expected to generate more information than the underlying PDE itself. In particular, features of interest, like fluctuations, and non–smooth behaviour, missed out in standard differential models, can manifestly not be captured either in such a representation.

Accordingly, genuine CA simulations of PDEs, complying with the condition of smallness of z, cannot directly deal with the actual dependent variables entering the PDE. CA models therefore typically introduce auxiliary dependent variables which refer to an underlying fictitious microscopic level characterised by a nonstandard discrete spatial lattice and a discrete time. This (formally nonstandard) CA then simulates the behaviour of this microscopic substratum. In turn the PDE formulation refers to a 'macroscopic', formally standard level, representing an ensemble average of the microscopic simulations.

In principle the procedure of setting up such a genuine CA model then consists in the following sequence of operations.

(a) Introduce a discrete–state 'particle' representation defined over an appropriately chosen spatial lattice; the representation comprises s species of particles, each particle having a small number σ of dynamic states; each lattice cell (or each node) can either be empty, or occupied by a finite total number (z-1) of dynamic particle states (obeying certain prescriptions); the states are labelled by integers, for instance: 0, empty cell; 1, single particle of species a in state α, ...; the labels then stand for the allowed values of the state function F (binary number representations, or other encodings may be technically more convenient); this procedure defines $\mathbf{z}$.

(b) Introduce evolution rules for the particle states : A given state in cell $\mathbf{i}$ may be spontaneously created, or result from a decay of another state at cell $\mathbf{i}$, or more generally it may result from a transformation of states in neighbour cells $\mathbf{i}' \in \mathsf{N}(\mathbf{i})$. In the general case these rules must be given explicitly in terms of a look–up table :

if $\mathbf{i}, \mathbf{i}', \mathbf{i}'', \ldots, \mathbf{i}^{(f'-1)}$ denote all the f' cells of the neighbourhood $\mathsf{N}(\mathbf{i})$, and the values $\phi, \phi', \phi', \ldots, \phi^{(f'-1)}$, range over the set $\mathbf{z}$ ($\equiv \mathsf{N}(F)$) then we have

$$F(\tau;\mathbf{i}) = \phi \wedge F(\tau;\mathbf{i}') = \phi' \wedge F(\tau;\mathbf{i}'') = \phi'' \wedge \ldots \wedge F(\tau;\mathbf{i}^{(f'-1)}) = \phi^{(f'-1)}$$

$$\begin{aligned} \rightarrow F(\tau+1; i) &= \Phi, \quad \text{with probability} \quad P, \\ &= \Phi', \quad \text{with probability} \quad P', \\ &= \Phi'', \quad \text{with probability} \quad P'', \end{aligned}$$

.... 2.8

Φ , Φ' , Φ'', ... being again values in z. Special rules may be summarised by analytical expressions.

(c) The final step consists in constructing the macroscopic level (PDEs) summarising the statistical behaviour of the microscopic CA model. This is done systematically as follows: (i) Introduce macroscopic ***physical variables***, $N_q(t, \mathbf{r})$, $q = 1,2, \ldots$, defined over the physical space and time, as discussed in the previous section; the physical nature of these variables will depend on the specific problem (for instance total number of particles of species a per unit volume; total number of particles in dynamic state α per unit volume; etc). In their most general form these variables will be ensemble averages of certain combinations of the microscopic CA variables F (often approximated by space–time averages, cf 1.13,14). The ensemble to be considered is a large number of replicas of the CA set-up, generating microscopic evolutions from different initial conditions consistent with the macroscopic initial conditions. (ii) Associate next formal nonstandard variables $*N_q(\tau, \mathbf{i})$ with the physical variables, as discussed above; taking account of all evolution rules (2.8), express $*N_q(\tau+1, \mathbf{i})$ as functions G_q (generally polynomials of low degree) in the collection of variables $*N_{q'}(\tau, \mathbf{i})$, $*N_{q'}(\tau, \mathbf{i}')$, $*N_{q'}(\tau, \mathbf{i}'')$, ... in cells $\mathbf{i}$, $\mathbf{i}'$, $\mathbf{i}''$, ... of the neighbourhood $\mathsf{N}(\mathbf{i})$,

$$*N_q(\tau+1, \mathbf{i}) \;=\; G_q\;[*N_{q'}(\tau, \mathbf{i}), *N_{q'}(\tau, \mathbf{i}'), *N_{q'}(\tau, \mathbf{i}''), \ldots\,], \quad q, q' = 1, 2, \ldots \qquad 2.9$$

(iii) Divide Eq. (2.9) by Δt and, possibly after rearranging the terms, take the standard part. If the resulting equations can be recast in the form of PDEs

$$\partial_t N_q(t, \mathbf{r}) \;=\; g_q[\,N_q(t, \mathbf{r}), N_{q'}(t, \mathbf{r}), \ldots, \partial_{x_1} N_q(t, \mathbf{r}), \partial_{x_2} N_q(t, \mathbf{r}), \ldots], \quad q = 1, 2, \ldots \qquad 2.10$$

(g_q, standard functions related to the functions G_q), and if a subset of these macroscopic equations coincide with the PDE we wish to simulate, then our goal is achieved.

The formulation immediately reveals a ***non-uniqueness*** property of the CA model for a given PDE equation : Since the latter results from a standard part of a nonstandard equation (Eq 2.9), only 'leading orders' in the nonstandard equation are preserved. Any two distinct microscopic CA models differing in higher order contributions to (2.9) belong to the same macroscopic equivalence class.

The procedure outlined actually supplies an answer to an ***inverse*** of the problem we are interested in here : The microscopic CA model is posited at the outset, and the procedure discusses the construction of the associated macroscopic PDE. It is worth noting that there is no guarantee that a proper PDE will be obtained in general; it may not be meaningful for the specific problem to introduce macroscopic variables in the first place; or it may simply not be possible to generate from Eq. (2.9) standard PDEs; the macroscopic variables may fluctuate so wildly that differentiation ceases to be a permissible operation. We shall mention below examples of evolution rules (2.8) for which an average description of the time behaviour is not relevant.

Regarding the ***direct problem*** — given a formal PDE, find a microscopic model such that the PDE represents some aspects of the statistical behaviour of the latter — it should be obvious that it can only be approached by trial and error. Fortunately, however, in astrophysical problems, when a PDE formulation is available it typically stands already for a representation of statistical properties of an underlying fine–grained discrete substratum. For instance, a fluid dynamic model for a globular cluster provides a global 'macroscopic' continuous density, a global velocity field, etc, of the cluster, i.e. it supplies already a statistical substitute for the actual microdynamic variables (linear momentum, etc), of the individual stars. A CA formulation for this problem directly mimics the dynamical behaviour of individual stars (the 'particles' of the CA model); it remains, however, always a challenge to find a microscopic model in which a ***minimal number*** of discrete dynamic particle states are needed to reproduce an adequate macroscopic formulation.

To illustrate the simplicity of the above general procedure, and to demonstrate also that often a surprisingly small number of states of the CA variable F is sufficient to simulate the physics, we briefly discuss two elementary examples : wave propagation and diffusion.

Wave Propagation

We discuss the linear scalar $1D$ wave equation in a homogeneous medium. To conform to (2.1) we write this second order PDE as two coupled first order equations :

$$\partial_t^2 \rho - c^2 \partial_x^2 \rho = 0 \quad \rightarrow \quad \partial_t \rho + \partial_x p = 0 \text{ and } \partial_t p = -c^2 \partial_x \rho. \qquad 2.11$$

The first order PDEs have physical interpretations as macroscopic continuity and momentum equations for a gas of matter density per unit length (or for the deviation of density from a uniform equilibrium density ϱ_o), $\varrho(t,x)$, and linear momentum density per unit length, $p(t,x)$, under the assumption that the pressure is a function of density alone, and the nonlinear terms are negligible; c is the sound speed evaluated in the equilibrium state, and therefore space and time independent.

To construct the simplest possible CA representation imagine a (nonstandard) gas of one–dimensional particles of unit mass, and of allowed velocities $+v$, or $-v$, v a finite positive constant. The (nonstandard) lattice is the straight line subdivided into segments of equal (infinitesimal) length λ, and the timestep $\Delta t = \frac{\lambda}{v}$ ($\alpha = 1$ in Eq 1.12). During one timestep a particle in cell i reaches either cell $i+1$, if its velocity is $+v$, or cell $i-1$, if its velocity is $-v$; this specification expresses the evolution rule of the gas. Moreover, we assume that the initial conditions are always chosen such that a cell contains at most one particle of each velocity state. The evolution rules then imply that the latter property is preserved for any later time. If we represent the cell states by the integers 0 (empty cell), 1 ($+v$ particle), 2 ($-v$ particle), and 3 ($+v$ and $-v$; pair of particles in cell), (cell space $\mathbf{z} = \{0,1,2,3\}$), then the look–up table for the evolution rules has 4^3 entries. In this particular case the look–up table may be condensed into the following set of instructions (R) :

Initialise $F(\tau+1,i)$ to empty cell value 0;

if $F(\tau,i-1)$ =1 or 3, and $F(\tau,i+1) \neq 2$ and 3, update $F(\tau+1,i)$ to 1;

if $F(\tau,i-1) \neq 1$ and 3, and $F(\tau,i+1) = 2$ or 3, update $F(\tau+1,i)$ to 2;

if $F(\tau,i-1) = 1$ or 3, and $F(\tau,i+1) = 2$ or 3 , update $F(\tau+1,i)$ to 3. (R)

For the purposes of computational efficiency we observe that it is the precise choice of the encoding of the cell states, i.e. the definition of space $\mathbf{z}$, which determines the complexity of the compressed form of the evolution rules — provided a condensed form does exist. For instance, in the present case, if we use the integers 1 (empty cell), 2 ($+v$), 3 ($-v$), 6 ($+v$ and $-v$), then the state of cell i at time τ+1 can actually be calculated by the following simpler instructions (R') :

Initialise $F(\tau+1,i)$ to empty cell value 1;

if 2 divides $F(\tau,i-1)$, update $F(\tau+1,i)$ by multiplying by 2 previous $F(\tau+1,i)$;

if 3 divides $F(\tau,i+1)$, update $F(\tau+1,i)$ by multiplying by 3 previous $F(\tau+1,i)$. (R')

Following now the steps listed under (c) we introduce two auxiliary nonstandard variables, $*\varrho_+(\tau,i)$ and $*\varrho_-(\tau,i)$, representing the *ensemble averaged* number of particles per unit length in cell i (now labelled by the continuous variable x) and moving in the positive (+), respectively negative (-), lattice direction. The ensemble is defined as a large collection of same CA models starting with initial conditions which differ for each representative on the scale of the individual cells of the automaton, but which are equivalent on a 'macroscopic' scale : The physical properties averaged over a large enough number of cells — the hydrodynamic neighbourhood — are the same for all realisations.

The microscopic evolution rules imply that these macroscopic variables obey

$$*\rho_+(\tau+1,i) = *\rho_+(\tau,i-1) \quad \text{and} \quad *\rho_-(\tau+1,i) = *\rho_-(\tau,i+1), \qquad 2.12$$

(cf Eqs. 2.9). Subtract $\rho_+(\tau,i)$ and $\rho_-(\tau,i)$ on both sides of Eqs. (2.12), divide by Δt and take the standard part, to obtain

$$\partial_t \; \rho_+(t,x) = -v \; \partial_x \; \rho_+(t,x),$$

$$\partial_t \; \rho_-(t,x) = +v \; \partial_x \; \rho_-(t,x), \qquad 2.13$$

where $*\rho_\pm(t,x) \equiv \text{st}\ (*\rho_\pm(\tau,i))$. The macroscopic variables $\rho_\pm(t,x)$ thus obey two linear partial differential equations (cf Eqs. 2.10). The linear densities $*\rho(\tau,i)$ and $*p(\tau,i)$, nonstandard 'enlargements' of the physical matter density and density of linear momentum, $\varrho(t,x)$ and $p(t,x)$, are related to $*\rho_+$ and $*\rho_-$ by

$$*\rho(\tau,i) = *\rho_+(\tau,i) + *\rho_-(\tau,i),$$

$$*p(\tau,i) = v\ (*\rho_+(\tau,i) - *\rho_-(\tau,i)\ ). \qquad 2.14$$

Introducing the standard parts of the latter variables (actual physical matter density and momentum density) into (2.13) we recover the two conservation equations (2.11), with $c \equiv v$.

We have thus shown that the one–dimensional 4–state (nonstandard) automaton specified by pure translational evolution rules of a $1D$ gas reproduces *exactly* the linear one–dimensional wave equation on the macroscopic level.

A slight geometrical modification converts this CA into a microscopic model of a less trivial PDE. Instead of adopting a *regular* (nonstandard) $1D$ lattice, we consider a lattice with *spatially variable* infinitesimal *cellsize*, $\lambda = \lambda(i)$. This lattice carries again a gas of $1D$ particles of unit mass, such that a particle in cell i has a finite velocity $v(i)$, $v(i) = \lambda(i)/\ \Delta t$, Δt a constant timestep (α is again 1). The evolution rules are similar to those of the previous formulation : A particle in cell i at time τ reaches, at time τ+1, either cell $i+1$, with finite (nonstandard) velocity $+v(i+1)$, if its velocity in cell i was $+v(i)$; or cell $i-1$, with velocity $-v(i-1)$, if its velocity in cell i was $-v(i)$. Introducing again the nonstandard densities $*\rho_+$ and $*\rho_-$ as above, these evolution rules now imply the following counterparts of Eqs. (2.12)

$$*\rho_+(\tau+1,i)\ \lambda(i) = *\rho_+(\tau,i-1)\ \lambda(i-1),$$

$$*\rho_-(\tau+1,i)\ \lambda(i) = *\rho_-(\tau,i+1)\ \lambda(i+1). \qquad 2.12'$$

The nonstandard matter and momentum densities are defined by Eqs. (2.14) in which the particle velocity is now space dependent, $v \equiv v(i)$. We have in this case

$$\frac{*\rho(\tau+1,i) - *\rho(\tau,i)}{\Delta t} =$$

$$-\frac{v(i)*\rho_+(\tau,i) - v(i-1)*\rho_+(\tau,i-1)}{\lambda(i)} + \frac{v(i+1)*\rho_-(\tau,i+1) - v(i)*\rho_-(\tau,i)}{\lambda(i)}, \qquad 2.15$$

$$\frac{*p(\tau+1,i) - *p(\tau,i)}{\Delta t} =$$

$$-v(i)\ \{\ \frac{v(i)*\rho_+(\tau,i) - v(i-1)*\rho_+(\tau,i-1)}{\lambda(i)} + \frac{v(i+1)*\rho_-(\tau,i+1) - v(i)*\rho_-(\tau,i)}{\lambda(i)}\ \}, \qquad 2.16$$

The standard part of Eq. (2.15) becomes, introducing the standard part of the velocity, $v(x)$ and the standard quantities $\rho_+(t,x)$, $\rho_-(t,x)$ (cf the two expressions of the derivatives 1.9),

$$\partial_t\ \rho(t,x) \ = \ -\partial_x\ v(x)\rho_+(t,x) \ + \ \partial_x\ v(x)\rho_-(t,x), \qquad 2.17$$

i.e. we recover the continuity equation (2.11). The standard part of Eq. (2.15) becomes, in a similar way

$$\partial_t\ p(t,x) \ = \ -v(x)\ \partial_x\ v(x)\rho(t,x). \qquad 2.18$$

The corresponding wave equation is

$$\partial_t^2\ \rho \ - \ \partial_x\ v(x)\ \partial_x\ v(x)\rho \ = \ 0, \qquad 2.19$$

where the particle speed $v(x)$ is an arbitrary function of x. Given the function $v(x)$, the shape function $f(x)$ (Eq. 1.12) of the lattice cells of the CA simulating the wave behaviour is determined by $f(x) \ \propto \ v(x)$.

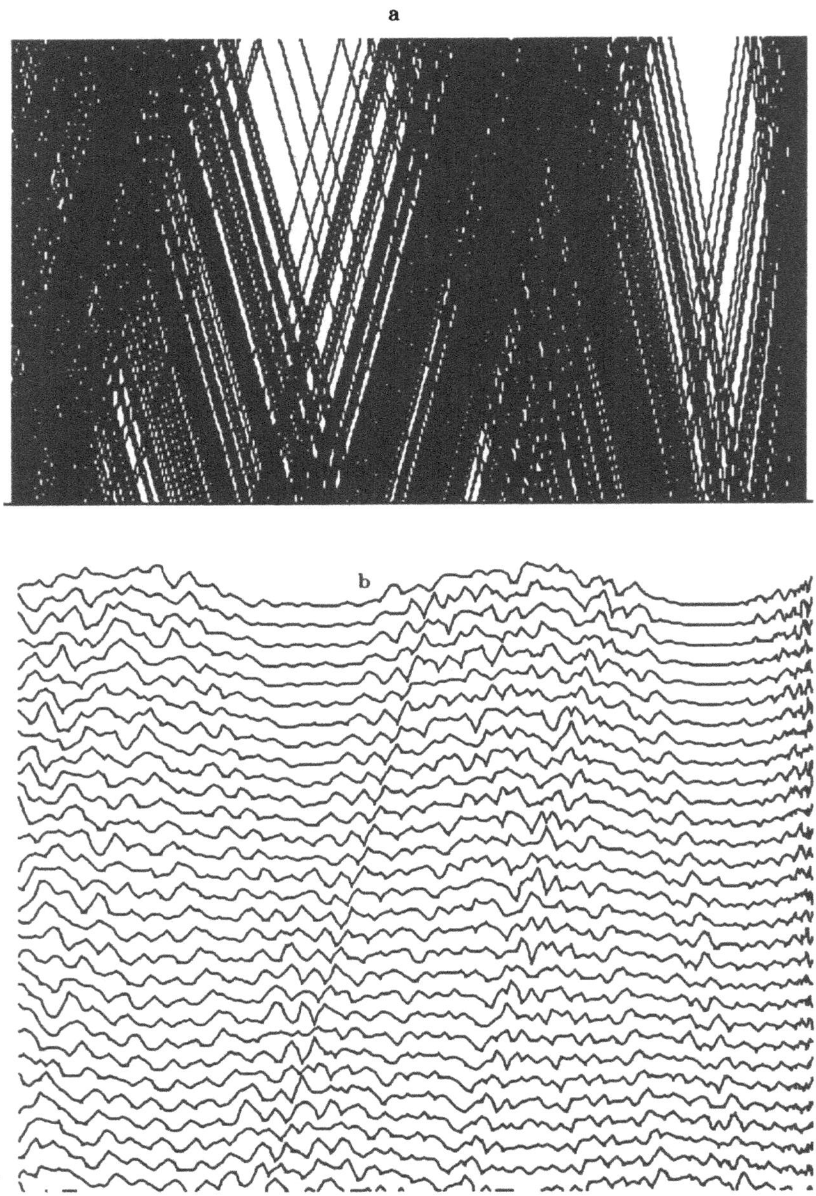

Fig. 1. – CA solution of a $1D$ wave equation with space dependent propagation speed. Initial condition : sine wave. (a) Plot of automaton states (Reproduced in Colour Section). (b) Run of macroscopic signal $\rho(t,x)$. Time increases downward in both representations.

Although Eq. (2.19) comes close to the general *linear radial oscillation equation* of a star (cf Rosseland 1949), it remains too special to be representative for realistic stellar models. An additional independent spatial 'degree of freedom' is required, say a function $\theta(x)$ multiplying the first term in the equation, in order to obtain the general form of the stellar wave equation.

Note that through separation of time, a change of the dependent variable, and a scaling of the independent space variable

$$\rho(t,x) = \rho_o(x)\exp(i\omega t) = \frac{R(x)}{v(x)}\exp(i\omega t), \tag{2.19.a}$$

Eq. (2.19) transforms into a standing wave equation whose closed form solution is

$$R(x) \;=\; C\sin\,[\omega\int\frac{\mathrm{d}x}{v(x)} \;+\; \phi], \tag{2.19.b}$$

(cf Kreith 1973), with C and ϕ integration constants.

Fig. 1 illustrates a result of the latter CA model for a grid of $f = 1800$ cells, with $v(x[i]) \propto 0.01 + (1-(i/f)^2)$, i=0,1, ... ,$f-1$, for reflecting boundary conditions; the initial condition is chosen close to a spatial sine wave pattern. The CA integration shows a propagation phenomenon of secondary bumps superimposed on the main signal (cf Fig. 1.b displaying $\rho(t,x)$, x, horizontal coordinate).

This example illustrates two points.

(1) In the first place, the microscopic CA model is entirely *fictitious*; it refers to one–dimensional non–interacting particles, whose dynamics is oversimplified (all particles have same speed in the original version of the model); in the generalised version we violate the laws of mechanics (non-conservation of momentum and energy). In spite of these idealisations, the collective behaviour of this crude microscopic physics does generate a physically correct model on a macroscopic level (in the present case, it represents exactly the one–dimensional adiabatic oscillations of a gas in a slab). The main lesson of this example is that the macroscopic behaviour just results from a few *significant features of the microscopic model*: Here it is the existence of microscopic translational motions.

On the other hand — and this is particularly relevant in astrophysical problems — the real interest of simulating a PDE by a CA model does not just consist in providing an alternative way for solving the PDE; what we ideally expect from the CA formulation is that it supplies meaningful information on the behaviour of the system on some level intermediate between the macroscopic level captured by the PDE and the genuine microscopic level we are unable to describe. In order to achieve this goal the CA simulation should mimic the true physical processes occurring on the microscopic level; this condition is fulfilled to some extent at least in our model. The validity of our CA wave simulation is therefore reasonably expected to extend over some ***submacroscopic level*** in the specific case of ***acoustic waves***; it is surmised in particular that at least some aspects of the fluctuations in the CA model reproduce the true fluctuations in a gas (ignoring collisional effects). In contrast, in the case of electromagnetic waves, for instance, the CA model presented here is meaningful to describe wave propagation on the macroscopic level only; in the absence of a specific ingredient of electromagnetic field theory, the above CA model is manifestly not capable of supplying any submacroscopic information, for instance, on electromagnetic fluctuations.

We mention at this point also that there is at the moment no theory available to assess the spatial or temporal scale of validity of a given CA model.

(2) Once a basic CA model covering a simple physical situation has been set up, an extension of the model to more complex physical situations is in general a trivial operation. One possible generalisation consists in changing the lattice (regular grid becoming space dependent, as illustrated above). Another generalisation consists in including additional significant microscopic mechanisms. In the above model we can introduce a ***particle creation*** and ***destruction*** mechanism, possibly depending on the presence or absence of particles in the neighbourhood of a given cell. A corresponding set

of CA rules is : A $(+v)$ or $(-v)$–state is created in cell i with probability Γ_+ or Γ_- per timestep provided that this cell does not contain already a $(+v)$ or $(-v)$– state respectively; and for a cell i containing a $(+v)$ or $(-v)$–state, the latter is destroyed with probability Δ_+ or Δ_- per timestep. Proceeding as previously, we notice that these additional steps generate a (positive or negative) damping term in the wave equation (2.11). We observe also that the resulting macroscopic representation is a ***deterministic*** PDE, in spite of the fact that the CA is stochastic.

In computer adapted form the algorithm R is completed by the following set of instructions:

r, r', random numbers uniformly distributed in [0,1] :

if $F(\tau+1,i)=0$, update $F'(\tau+1,i)=1$ if $r<\Gamma_+$ and $r'\geq\Gamma_-$;

if $F(\tau+1,i)=0$, update $F'(\tau+1,i)=2$ if $r\geq\Gamma_+$ and $r'<\Gamma_-$;

if $F(\tau+1,i)=0$, update $F'(\tau+1,i)=3$ if $r<\Gamma_+$ and $r'<\Gamma_-$;

if $F(\tau+1,i)=1$, update $F'(\tau+1,i)=0$ if $r<\Delta_+$;

if $F(\tau+1,i)=2$, update $F'(\tau+1,i)=0$ if $r'<\Delta_-$;

if $F(\tau+1,i)=3$, update $F'(\tau+1,i)=1$ if $r<\Delta_+$ and $r'\geq\Delta_-$;

if $F(\tau+1,i)=3$, update $F'(\tau+1,i)=2$ if $r\geq\Delta_+$ and $r'<\Delta_-$. R

A similar simple set of instructions holds under coding R'. We observe that while it is trivial to incorporate a variety of additional clauses in the coding of the rules (for instance in the above particle creation and destruction steps), very often these clauses are not easily, or not at all, translatable into PDE form.

Another generalisation of the microscopic model consists in swapping states with probabilities Σ_{+-} (state $+v\rightarrow-v$) and Σ_{-+} (state $-v\rightarrow+v$) per timestep. This simulates collisions of the particles with heavy obstacles. The coding R is obvious.

Diffusion

It is well known (cf Feller 1971) that for a large number of realisations of random motions of particles starting at point O, taken as the origin of a reference system in a Euclidean space of arbitrary dimension n, the probability density (or particle number density), $\varrho(t,\mathbf{r})$, obeys the standard diffusion equation in the long–time asymptotic limit. To model the one–dimensional diffusion PDE (Eq. 2.4) on a CA, therefore, we are naturally led to adopt a model simulating random motions of particles.

As already observed, there exists an equivalence class of microscopic models which all generate the same macroscopic PDE. In the specific problem of diffusion, the macroscopic equation (2.4) (a particular instance of the Kolmogorov equations) describes the probability density of ***any*** continuous Wiener–Lévy process, or Brownian motion (cf Bharucha–Reid 1960).

We consider therefore here the simplest (nonstandard) discrete process : A particle in an infinitesimal cell i of width λ at timestep τ has a probability $\frac{p}{2}$ to be found in cell $i-1$, a probability $\frac{p}{2}$ to be found in cell $i+1$, and a probability $q=1-p$ to remain in cell i at timestep τ+1. This probabilistic specification expresses the evolution rules of our CA. As a further specification of this problem we allow any cell to be occupied by an ***arbitrary number*** of particles. It is then convenient to choose for the dependent CA variable $F(\tau,i)$ the number of particles contained in cell i at step τ, $F=0,\,1,\,2,\,\ldots$.

In an actual computer model of diffusion the dynamically more natural procedure consists in encoding the velocity states (here $v=0$, +1 and -1) of the particles in the field variable, since from the point of view of mechanics the state of a particle is defined both by its position and its velocity or momentum; the methods adopted in lattice gases (cf next section) can be used. A lattice particle then obeys the following dynamics : (1) the velocity state $v(\tau,i)=0$, 1 or -1 in cell i at timestep τ transforms into the velocity state $v(\tau+1,i)=v(\tau,i)$ in cell i, $v(\tau+1,i+1)=v(\tau,i)$ in cell $i+1$, or $v(\tau+1,i-1)=v(\tau,i)$ in cell $i-1$ respectively at time $\tau+1$; since the corresponding particle suffers no force effect over the time interval $(\tau,\tau+1)$, it undergoes a translational motion; (2) at time $\tau+1$ the velocity $v(\tau+1,i)$ is 'instantaneously' changed into a new velocity state $v(\tau+1,i)$ =0 (with probability q), +1 (with probability $p/2$), or -1 (with probability $p/2$), as a result of the random collisions of the lattice particle with an underlying medium. In this model a lattice particle obeys classical mechanics exactly.

In this representation we need only one physical variable for our problem, namely the particle density $\varrho(t,x)$, whose nonstandard generalisation, $*\rho(\tau,i)$ has a simple physical interpretation : it is an ensemble average of $F(t,i)/\lambda$ over a large number of realisations (experiments carried out with *same* initial conditions); note that since the standard part of this average must be finite, the average of F is infinitesimal, i.e. the total number of particles per cell should remain 'small' (in agreement with our definition *D). The evolution rule thus yields

$$*\rho(\tau+1,i) \;=\; *\rho(\tau,i)\;q \;+\; \frac{p}{2}\;(*\rho(\tau,i+1)+*\rho(\tau,i-1)\,), \qquad 2.20$$

which can be rewritten as follows

$$\frac{*\rho(\tau+1,i)-*\rho(\tau,i)}{\Delta t} \;=\; \frac{p\lambda^2}{2\Delta t}\;\frac{*\rho(\tau,i+1)-2*\rho(\tau,i)+*\rho(\tau,i-1)}{\lambda^2}, \qquad 2.21$$

The standard part of this equation is seen to be the diffusion equation (2.4), provided that the diffusion coefficient $D = p\,\mathrm{st}\;(\frac{\lambda^2}{2\Delta t})$ is finite, i.e. $\alpha = 2$ in Eq. 1.12.
Observe again that although the CA representation is stochastic, the corresponding macroscopic behaviour is deterministic.

This model admits of two immediate generalisations which we carry out simultaneously. We introduce (1) a space dependent lattice ($\lambda = \lambda(i)$), and (2) space dependent probabilities ($p = p(i)$, $q = q(i)$, nonstandard probabilities). The counterpart of Eq. (2.20) then becomes

$$*\rho(\tau+1,i)\;\lambda(i) \;= \rho(\tau,i)\;\lambda(i)\;q(i)+$$

$$\frac{1}{2}(*\rho(\tau,i+1)\;\lambda(i+1)\;p(i+1) \;+\; *\rho(\tau,i-1)\;\lambda(i-1)\;p(i-1)). \qquad 2.20'$$

Rewriting this equation in the form

$$\frac{*\rho(\tau+1,i)-*\rho(\tau,i)}{\Delta t} = (\frac{\lambda(i)}{2\Delta t})^{\frac{1}{2}} \times$$

$$\frac{\{*\rho(\tau,i+1)p(i+1)\;(\frac{\lambda(i+1)}{2\Delta t})^{\frac{1}{2}} - 2*\rho(\tau,i)p(i)\;(\frac{\lambda(i)}{2\Delta t})^{\frac{1}{2}} + *\rho(\tau,i-1)p(i-1)\;(\frac{\lambda(i-1)}{2\Delta t})^{\frac{1}{2}}\;\}}{\lambda(i)^2}, \qquad 2.21'$$

we see that its standard part becomes

$$\partial_t\;\rho(t,x) = a(x)\;\partial_x^2\;b(x)\;\rho(t,x), \qquad 2.22$$

$$a(x) \;=\; \mathrm{st}(\frac{\lambda(i)}{2\Delta t})^{\frac{1}{2}},\;\; b(x) \;=\; \mathrm{st}\Big(p(i)(\frac{\lambda(i)}{2\Delta t})^{\frac{1}{2}}\Big). \qquad 2.22.a$$

We have to require the coefficients $a(x)$ and $b(x)$ to be continuous functions of x , which is secured if the nonstandard functions $\lambda(i)$ and $p(i)$ change 'smoothly' with i.
We mention in passing that by a separation of the time and space variables, $\rho(t,x) = \rho_o(x)\exp(st)$, and by a change of the independent variable ($x = x(z)$) and the dependent variable ($r_o(x) = A(z)$ $\Phi(z)$), Eq (2.22) reduces to a general *Sturm-Liouville equation* in $\Phi(z)$. In principle, therefore, such a formulation could be used to solve linear stellar pulsation equations. The warning formulated in the previous subsection should be kept in mind : Since the microscopic diffusion model adopted here simulates standing acoustic waves only formally, we are not entitled to expect this representation to provide relevant information on a submacroscopic level of the oscillations (for instance on the effect of fluctuations).

The extension of the above model to 2 or 3–dimensional (and more generally n–dimensional) diffusion on a square, cubic (hypercubic) lattice is straighforward. We note that in these cases again we can introduce space dependent lattice cells and space dependent probabilities. In the specific context of 2 (and 3)-dimensional regular lattices it has been shown that the elementary model does indeed correctly account for submacroscopic features : If diffusing particles are continuously injected along a straight line, then the diffusion front (external boundary of the cluster of diffusing particles) takes a fractal structure (observable on a submacroscopic scale) of dimension $D_f = \frac{7}{4}$; this value is established theoretically from percolation theory, and it is found by direct CA model simulations (cf Feder 1988); observation of the diffusion front in an experiment of silver chloride diffusing in a polymer has led to an observed value $D_f \approx 1.74$, in close agreement with the CA model result (reported in Gouyet 1992). Recall that the corresponding strictly macroscopic model, namely the 2-dimensional diffusion equation, does not account for a fractal structure at all.

The relevance of this observation transcends the special framework of simulation of diffusion by the above elementary model : It demonstrates that CA models in general, to the extent that they do faithfully mimic the actual physical micromechanisms, do indeed *provide information on a submacroscopic level* as well. By *faithful simulation* of the microphysics we mean that

(1) all basic physical laws (classical mechanics in the diffusion problem), and

(2) all physical symmetries (statistical isotropy of the velocities after collisions in our illustration) are maximally obeyed in the framework of the constraints imposed by the discrete lattice scheme (full isotropy of the discrete velocity states is not compatible with lattice geometries; the best approximation to this requirement is to impose invariance of the velocity distribution under the lattice symmetries); moreover,

(3) no conservation laws besides those flowing from the microphysics of the problem should be introduced into the model (a requirement not easily dealt with; there is no systematic method to trace hidden spurious invariances).

In order to recover an observationally accurate behaviour on a meso-scale, in between macroscopic physics (say the level of conventional fluid dynamics) and genuine molecular dynamics, the precise duplication of the 'real' microphysics is not required. We surmise in particular that *fractal sets*, when arising in the physical problem, are at least *qualitatively* recognised as such in even the most elementary CA simulations. This contrasts with the results of PDE models, since, in general, the latter tend to smooth out the fractal character. On the basis of the diffusion example we are even led to conjecture that fractal dimensionalities derived from CA simulations remain *quantitatively* meaningful parameters.

Lattice Gases

Over the past 10 years CA simulation of fluid dynamics, or *Lattice Gas* (LG) *Theory*, has developed into a field virtually independent of numerical hydrodynamics, and of CA theory (cf Hénon 1989). This is not the place to enter into a discussion of the theoretical developments of LGs, based on extensions of the classical continuous methods of statistical physics (Liouville and Boltzmann equations, moment method, Chapman–Enskog expansion, etc) to the discrete CA environment (cf Wolfram 1986, and Frisch et al 1987 for general reviews; recent results are summarised in van der Hoef 1993; for a more popular introduction see Boon et al 1993). We mainly wish to stress here that in *astrophysical fluids* several new features come in which, in typical laboratory scale fluid flows, remain of marginal interest only. As a rule, therefore, these aspects have received little attention in current LG theory. Astrophysical media very often deal with large *density gradients*; *self-gravitation* is important; and *compressibility* is a relevant property (cf wave propagation in stars and galaxies, or global pulsations of a star, where all of these three effects play a dominant part).

There remain, however, numerous astrophysical situations where these new features are not directly relevant. For instance in the context of the large-scale structure of the universe, density inhomogeneities in sticky cold matter are described by the usual Burgers equation (a nonlinear velocity diffusion equation), together with the (compressible) continuity equation (Shandarin and Zeldovich 1989). In the interstellar medium, thermal instabilities at the onset of condensations are captured by nonlinear diffusion equations (Regev 1990).

In order to indicate how one can incorporate high density gradients in the LG schemes we review the simplest lattice gas realisation introduced by Hardy, De Pazzis and Pomeau in 1976 (HPP gas).

This LG is defined on a $2D$ square lattice supporting, just like the simplest $2D$ diffusion simulation (without rest state), 4-velocity-state particles of unit mass in the absence of a force field. The traditional coding of the state vector $F(\tau; i, j)$ consists in using a 4-digit ***binary number***

$$F(\tau; i, j) = d_E(\tau; i, j)\, d_W(\tau; i, j)\, d_N(\tau; i, j)\, d_S(\tau; i, j); \qquad 2.23$$

the symbols d_a ($a = E$[east]; W[est]; N[orth]; S[outh]) stand for the binary digits, $d_a = 0$ or 1. The HPP gas is therefore said to be a 4-bit gas. Representation (2.23) means that cell **i** at timestep τ contains d_E (= 0 or 1) particles with discrete velocity $\mathbf{v} = (v_x, v_y) = (1{,}0) = \mathbf{v}_E$; d_W (= 0 or 1) particles with discrete velocity $(-1{,}0) = \mathbf{v}_W$; etc. The very encoding automatically enforces an *exclusion principle*, and confers the HPP gas the status of a *Fermi gas* : Any cell i, j at time τ cannot contain more than one particle in the same state.

In astrophysical modelling an exclusion principle has the drawback of setting an upper limit to the CA particle density (number of particles per cell ≤ 4 in HPP gas) — in addition to its unwanted thermodynamic side-effects. On the other hand, computer constraints set a limit to the actual size of the automaton, so that hydrodynamic averages of the CA properties are necessarily carried out over a relatively small spatial averaging set $A(i, j)$ (cf the standard version of Eq. 1.13). Statistical fluctuations then impose a limit of order $\frac{1}{\sqrt{A(i,j)}}$ to the smallest meaningful instantaneous hydrodynamic density (a $2D$ lattice even with 10^{12} sites is then not able to represent the actual density distribution of a star in the giant phase).

The exclusion property is shared by any σ-bit gas (σ different velocity states, with encoding of form 2.23). On the other hand, any one of the currently popular LG models belongs into this category (cf van der Hoef 1993). The advantage of representation (2.23), (and its generalisation to the σ-bit case) is that the use of Boolean variables d_a lends itself to a computationally efficient method of handling the CA evolution rules.

The microscopic description of the lattice dynamics of the HPP particles — obeying classical mechanics exactly — proceeds in two phases, characterising the ***translations*** (or ***streaming***) of particles, and the ***collisions*** respectively.

(1) ***Translation phase.***- Over the time interval (τ, τ+1) the HPP particle is not subject to a force field, so that its momentum is conserved. Hence, a particle at time τ+1 in cell (i, j) was in one of the 4 direct neighbour cells $(i', j') = (i-1, j)$, $(i+1, j)$, $(i, j-1)$, $(i, j+1)$ at time τ, with a velocity pointing towards cell (i, j); it suffices therefore to examine the neighbourhood cells (i', j') and extract from the field functions $F(\tau; i', j')$ the binary digit $d_a(\tau; i', j')$ referring to the state of motion a towards the central cell; we then have

$$d_a(\tau + 1; i, j) = d_a(\tau; i', j'). \qquad 2.24$$

Explicitly

$$F(\tau + 1; i, j) \equiv d_E(\tau + 1; i, j)\, d_W(\tau + 1; i, j)\, d_N(\tau + 1; i, j)\, d_S(\tau + 1; i, j)$$

$$= d_E(\tau; i-1, j)\, d_W(\tau; i+1, j)\, d_N(\tau; i, j-1)\, d_S(\tau; i, j+1). \qquad 2.24.a$$

(2) ***Collision phase.***- As in classical mechanics, LG particle collisions are regarded as occurring instantaneously; if, as a result of translation, cell (i, j) contains at time τ+1 a pair of particles with opposite velocity vectors directed along one lattice axis, then this pair is instantaneously transformed into a pair of particles in the same cell with velocity vectors pointing along the other lattice axis. This rule is fully consistent with the treatment of collisions in classical mechanics (due account being taken of the constraints imposed by the lattice). With representation (2.23) the effect of the collisions consists in the following swapping of adjacent pairs of identical digits :

$$F'(\tau + 1; i, j) = 1100 \rightarrow F''(\tau + 1; i, j) = 0011 \quad \text{and}$$

$$F'(\tau + 1; i, j) = 0011 \rightarrow F''(\tau + 1; i, j) = 1100 \qquad 2.24.b$$

(F', field function 'prior' to collision; F'', field function 'after' collision). The sequence of operations (1) and (2) — implying that the HPP gas is a deterministic CA — completes the calculation of the CA configuration at the new timestep $\tau+1$.

The full evolution rule for individual digits is

$$\begin{aligned} d_E(\tau+1;i,j) &= d_E(\tau;i-1,j) + \min_{d_E(\tau;i,j)=d_W(\tau;i,j)=0}\{d_N(\tau;i,j), d_S(\tau;i,j)\} \\ &\quad - \min_{d_N(\tau;i,j)=d_S(\tau;i,j)=0}\{d_E(\tau;i,j), d_W(\tau;i,j)\}, \end{aligned} \qquad 2.25$$

with

$$\begin{aligned} &\min_{d_E(\tau;i,j)=d_W(\tau;i,j)=0}\{d_N(\tau;i,j), d_S(\tau;i,j)\} - \\ &\min_{d_N(\tau;i,j)=d_S(\tau;i,j)=0}\{d_E(\tau;i,j), d_W(\tau;i,j)\} = \\ &d_N(\tau;i,j)d_S(\tau;i,j)[1-d_E(\tau;i,j)][1-d_W(\tau;i,j)] - \\ &d_E(\tau;i,j)d_W(\tau;i,j)[1-d_N(\tau;i,j)][1-d_S(\tau;i,j)] \ , \end{aligned} \qquad 2.25.a$$

and similar expressions for d_W, d_N, d_S. The first term on the RHS of Eq.(2.24.a) is the translation effect; the terms explicited in (2.25.a) correspond to the collisional contribution, the minimum being taken under the constraint indicated. The first collisional component expresses the *creation* of E–states, the number of E–states being increased by 1 unit if there is an NS pair ($d_N = d_S = 1$) but no EW pair ($d_E = d_W=0$) in cell (i,j); the second component corresponds to E–state *destruction*. Since the digits obey $0 \leq d_a \leq 1$, we can interpret $d_a(\tau;i,j)$ as the *probability of occupation* of cell (i,j) at time τ by an a–state. Eqs. (2.25) in conjunction with the analogous equations for d_W, d_N, d_S, play the part of a discrete version of a *Liouville equation*.

The encoding secures that the exclusion principle holds at any time. The rules in turn guarantee that we have exact *conservation of the number of particles* (since the binary digits are merely reshuffled at every timestep), and of *linear momentum* (since classical mechanics is obeyed) and *kinetic energy* (in 1–speed LGs kinetic energy is equivalent to the total number of particles).

The symmetry of the square lattice forces two unwanted features on the macroscopic behaviour. On the one hand, since particle momentum is changed in collisions only, and since in a collision a zero momentum pair of particles suffers a transformation from horizontal to vertical motion or vice versa, the vertical component of the total linear momentum of any vertical column i of cells, $P_y^v(i)$, and the horizontal component of the total linear momentum of any horizontal row j of cells, $P_x^h(j)$, are conserved; in an $n_x \times n_y$–cell CA realisation with periodic boundary conditions we thus have $n_x + n_y$ *spurious conservation laws*. On the other hand, the symmetry properties of the macrodynamics are affected by the symmetry of the square lattice, the latter being too low to ensure for instance isotropy of the pressure tensor (cf below). Both unsatisfying properties are eliminated by selecting a hexagonal lattice carrying 6 velocity states per cell (the 6–bit FHP LG), as was first shown in Frisch, Hasslacher and Pomeau (1986).

Before sketching an alternative formulation, we derive the macroscopic equations of the conventional HPP gas using the nonstandard language. Introduce as auxiliary functions $*N_a(\tau;i,j)$, $a = E, W, N, S$, the ensemble–averaged total (nonstandard) number of a–states in cell (i,j) at time τ

$$*N_a(\tau;i,j) = \langle d_a(\tau;i,j)\rangle; \qquad 2.26$$

alternatively, $*N_a(\tau;i,j)$ can be interpreted as a *probability of occupation* of cell (i,j) at time τ by an a–state, for a given initial probability distribution $*N_a(0;i',j')$, (i',j') ranging over the lattice. The ensemble is understood as a (nonstandard) infinite number of realisations starting with different initial conditions obeying the latter distribution, and corresponding to a fixed total number of particles and global momentum. The functions $*N_a(\tau;i,j)$ obey the ensemble–averaged discrete Liouville–type equation of evolution (2.25).

Introduce finally the fields

$$*N(\tau;i,j) = \sum_a *N_a(\tau;i,j), \quad *\mathbf{j}(\tau;i,j) = \sum_a *N_a(\tau;i,j)\mathbf{v}_a,$$

i.e

$$*j_x(\tau;i,j) \;=\; *N_E(\tau;i,j) \;-\; *N_W(\tau;i,j), \quad j_y(\tau;i,j) \;=\; *N_N(\tau;i,j) \;-N_S(\tau;i,j), \qquad 2.27$$

which define the macroscopic mass density ϱ and momentum density $\varrho\mathbf{u}$ respectively

$$\mathrm{st}(*N(\tau;i,j)) \;\equiv\; \rho(t,\mathbf{r}) \quad \text{and} \quad \mathrm{st}(\frac{\lambda}{\Delta t} * \mathbf{j}(\tau;i,j)) \;\equiv\; \rho(t,\mathbf{r})\mathbf{u}(t,\mathbf{r}) \qquad 2.28$$

(for st $(\frac{\lambda}{\Delta t})$ finite). From the discrete Liouville equation (2.25) in the auxiliary variables (2.26) we generate partial difference equations in $*N(\tau;i,j)$ and $*\mathbf{j}(\tau;i,j)$. The nonstandard partial difference equation in $*N(\tau;i,j)$ is conveniently written

$$[*N(\tau+1;i,j) - *N(\tau;i,j)] \;+\; [*j_x(\tau;i,j) - *j_x(\tau;i-1,j)] \;+\; [*j_y(\tau;i,j) - *j_y(\tau;i,j-1)]$$
$$+ \; \{ \; [*N_W(\tau;i+1,j) - *N_W(\tau;i,j)] \;-\; [*N_W(\tau;i,j) - *N_W(\tau;i-1,j)]$$
$$+[*N_S(\tau;i,j+1) - *N_S(\tau;i,j)] - [*N_S(\tau;i,j) \;-N_S(\tau;i,j-1)] \; \} \;=\; 0. \qquad 2.29$$

If we divide by the infinitesimal timestep, introduce the lattice size λ (cf Eqs. 2.28) and take the standard part, this equation collapses into the conventional continuity equation (the standard part of the contribution of the curly brackets vanishes).
The nonstandard partial difference equation in $*\mathbf{j}(\tau;i,j)$ becomes

$$*j_x(\tau+1;i,j) - *j_x(\tau;i,j) \;=\; -[*N_x(\tau;i,j) - *N_x(\tau;i-1,j)]$$
$$+ \; \{ \; [*N_W(\tau;i,j) - *N_W(\tau;i-1,j)] - [*N_W(\tau;i+1,j) - *N_W(\tau;i,j)] \; \} \,, \qquad 2.30$$

together with a similar equation in the y–component. We have set

$$*N_x(\tau;i,j) \;=\; *N_E(\tau;i,j) \;+\; *N_W(\tau;i,j) \qquad *\,N_y(\tau;i,j) \;=\; *N_N(\tau;i,j) \;+\; *N_S(\tau;i,j). \;\; 2.31$$

representing the average number of particles in motion along the x and y axes respectively. Taking again the standard part, after appropriate divisions (standard part of curly brackets vanishes again), we obtain a momentum conservation equation in general vector form

$$\partial_t \, \rho\mathbf{u} \;=\; -\partial_{\mathbf{r}}.\Pi, \qquad 2.32$$

with

$$\Pi \;=\; \mathrm{st}\Big((\frac{\lambda}{\Delta t})^2 \; \mathrm{diag} \; (*N_x(\tau;i,j), \; *N_y(\tau;i,j)) \; \Big), \qquad 2.32.a$$

the momentum flux tensor (diagonal in the present model).
The general form of this tensor in macroscopic fluid dynamics is

$$\Pi \;=\; \rho \, \mathbf{u} \, \mathbf{u} \;+\; (p\mathcal{E} \;-\; \sigma'),$$

$$\sigma' \;=\; \eta \; (\partial_{\mathbf{r}} \, \mathbf{u} \;+\; \partial_{\mathbf{r}} \, \mathbf{u}^T \;-\; \frac{2}{3}\partial_{\mathbf{r}} \, . \, \mathbf{u} \, \mathcal{E}) \;+\; \zeta \; \partial_{\mathbf{r}} \, . \, \mathbf{u} \, \mathcal{E}, \qquad 2.33$$

where p is the scalar pressure ($\mathcal{E}$, unit tensor), and σ' the viscous stress tensor (the superscript T indicates transposition); η ($= \frac{\nu}{\rho}$; ν, kinematic viscosity) and ζ are the viscosity coefficients; the coefficient ζ vanishes in the Navier–Stokes approximation (cf Landau–Lifshits 1966, sections 7,15; for a careful macroscopic analysis, see Truesdell and Toupin 1960, section 298). We recall that in astrophysical media (in particular in stellar pulsation problems) viscous effects are typically negligible. We discuss therefore the nonviscous contribution only.

A comparison of the HPP momentum flux tensor (2.32.a) and the hydrodynamic expression (2.33) is easily made under the following conditions.

(1) *Thermodynamic equilibrium of the LG*

For a fixed number of particles and a fixed total momentum it is seen that the averaged Liouville equation (2.25) admits of an exact time and space independent solution in the form of a Fermi–Dirac distribution

$$*N_a = \frac{1}{\exp(*\mu + *\nu.\mathbf{v}_a) + 1}; \qquad 2.34$$

the parameters $*\mu$ (scalar), $*\nu$ (vector) are fixed by the number of particles per cell $*N_o$ and the momentum per cell $*\mathbf{j}_o$

$$*N_o = \sum_a *N_a \text{ and } *\mathbf{j}_o = \sum_a *N_a \mathbf{v}_a. \qquad 2.34.a$$

In a thermal equilibrium state the macroscopic velocity field vanishes, or $*\mathbf{j}(\tau; i, j) = *\mathbf{j}_o = 0$. Hence from (2.34.a) $*\nu$ vanishes, so that from (2.34) $*N_a$ is independent of the state a, or

$$*N_a(\tau; i, j) = \frac{1}{4} N(\tau; i, j) \equiv \frac{1}{4} * N_o, \quad a = E, W, N, S \qquad 2.35$$

With $c = \text{st}\ (\frac{\lambda}{\Delta t})$ (constant finite particle speed), we have from (2.32.a), introducing the actual mass μ of an LG particle,

$$p(\tau; i, j) = N(\tau; i, j)\, \frac{1}{2}\, c^2 \mu \equiv N(\tau; i, j)\, kT \qquad 2.36$$

$\left(N(\tau; i, j) = \text{st}\ \left(*N(\tau; i, j)\right)\right)$. The temperature T is introduced by identifying the kinetic energy per degree of freedom with $\frac{1}{2}kT$ (k, Boltzmann constant). Hence in this equilibrium state the HPP gas is consistent with thermodynamics.

(2) *Local thermal equilibrium and small velocities*

If conditions (2.34) apply locally, with $|\mathbf{u}| \ll c$, then the equation of state (2.35) continues to hold. Combining the macroscopic equations (2.32,32.a,36) with the continuity equation we then obtain a density wave equation describing the $2D$ propagation of acoustic waves (sound speed $c_S = \frac{1}{\sqrt{2}} c$).

(3) *Nonlinear velocity regime* In the general case the structure of the momentum flux tensor of the HPP gas (Eq. 2.32.a) cannot match the hydrodynamic template (2.33). In fact, from (2.34, 34.a) it is immediately seen that in the low density approximation (Fermi–Dirac distribution reducing to Maxwell–Boltzmann distribution) the momentum flux tensor becomes, in the absence of viscous effects

$$\Pi = G(\rho)\, \rho\, (u_x^2 - u_y^2)\ \text{diag}(1, -1)\ + p\, \mathcal{E}, \qquad 2.33.a$$

with $G(\rho) = 1/2$. For arbitrary densities form (2.33.a) remains true with $G(\rho) = \frac{2-\rho}{4-\rho}$ (in CA units; cf Frisch et al 1987). Since a realistic fluid is locally *isotropic* (for a general discussion see Truesdell and Toupin 1960, and Truesdell and Noll 1965), the macroscopic momentum flux tensor must be isotropic as well. Therefore, knowing that Π depends on the macroscopic velocity, the isotropy requirement implies that the $O(|\mathbf{u}|^2)$ dependence must be of the form

$$a\, |\mathbf{u}|^2 \mathcal{E} + b\, \mathbf{u}\mathbf{u}; \qquad 2.37$$

a, b independent of $\mathbf{u}$. The fact that for the HPP gas this tensor violates (2.37) merely reflects the property that this LG possesses too low a symmetry — namely the symmetry of the square lattice — to secure the isotropy of this tensor.

The symmetry requirements are obtained by the following elementary method : The lattice is locally invariant under symmetry operations G_s (in the HPP case reflections σ and σ' with respect to a lattice axis and with respect to the bisector of the axes, and rotations $C(\theta)$ of angle $\theta = \frac{\pi}{2}$ in square lattice). Since the tensors Π and $\mathbf{u}\,\mathbf{u}$ are both symmetric, they can be written as 3–dimensional arrays for the $2D$ LGs (this method is adopted for instance in the determination of the stress–strain relation in elasticity; cf Sneddon and Berry 1958); the coupling tensor K between Π and $\mathbf{u}\,\mathbf{u}$ is then represented by a 3-dimensional matrix $K^{(3)}$; the lattice symmetry requires that the latter representation obeys

$$\left(R(G_s)^{(3)}\right)^{-1}.\,K^{(3)}.\,R(G_s)^{(3)} \;=\; K^{(3)},$$

where $R(G_s)^{(3)}$ is the 3–dimensional representation of the symmetry operation G_s. In the case of the square lattice 3 independent parameters survive in the coupling matrix $K^{(3)}$. In the case of the higher hexagonal lattice symmetry, on the other hand, only 2 free parameters survive, rendering the tensor isotropic (Eq. 2.37).

The explicit construction of the momentum flux tensor for general LG gases is discussed in Wolfram (1986) and Frisch et al (1987). The procedure follows the approach of conventional statistical mechanics. It consists in solving the Boltzmann–type equation, using a Chapman–Enskog expansion around the thermal equilibrium (Wolfram), or a more elegant multi–time expansion near the steady state of the Liouville equation (Frisch et al); for any 1–speed Fermi LG the local thermal equilibrium obeys the Fermi–Dirac distribution (2.34,34.a).

These symmetry conditions imply that for an LG defined over a hexagonal lattice, with 6 velocity states (the FHP gas of Frisch et al 1986), isotropy of the momentum flux tensor is indeed restored. The FHP gas is thus consistent with the behaviour of real gases.

We wish to stress at this stage that, although the HPP gas violates the basic requirement of isotropy, this does not imply that this LG is devoid of any practical relevance for simulating the behaviour of real fluids. In fact, for macroscopically $1D$ fluid flows — an assumption that holds *approximately* in various astrophysical problems — the HPP momentum equation (3.32) correctly pictures the flow provided that the flow direction is chosen to coincide with the x or the y lattice direction. This is in particular the case for a broad class of pulsations of variable stars. The practical relevance of a $2D$ (or $3D$) LG representation of approximately radial (i.e. $1D$) pulsations is that the simulation does include *nonradial* fluctuations. Even if the latter are not fully dealt with, such general fluctuations will excite any unstable nonradial degrees of freedom; provided that the latter contributions remain small as compared to the dominant radial motion, such a simulation will at least offer a first approximation to the qualitative effect of weak nonradial contributions.

We now briefly show how an alternative encoding enables us to deal with a *boson* LG gas of HPP type, thereby allowing us, in principle, to simulate higher density contrasts, as required in astrophysical applications. This can be achieved by replacing the binary representation of the state vector $F(\tau;i,j)$ (Eq. 2.23) by a *prime number encoding*

$$F(\tau;i,j) \;=\; \prod_a p_a{}^{\nu_a(\tau;i,j)}, \qquad 2.38$$

where the collection of symbols $p_a = p_E, p_W, p_N, p_S$ is the sequence of first primes 2, 3, 5, 7. The exponents $\nu_a(\tau;i,j)$ stand for the (now unrestricted) numbers of a–state particles in cell (i,j) at timestep τ . The translation rule remains the same as in the fermion representation (Eq. 2.34 with the change of notation $d_a \;\rightarrow\; \nu_a$). The collision rule expresses now that any pair of oppositely moving particles along one lattice axis in a cell (i,j) is replaced by a pair of oppositely moving particles along the other axis. Hence the corresponding discrete Liouville equation becomes (cf Eq. 2.25)

$$\nu_E(\tau+1;i,j) \;=\; \nu_E(\tau;i-1,j) + \min\,\{\nu_N(\tau;i,j),\nu_S(\tau;i,j)\} - \min\,\{\nu_E(\tau;i,j),\nu_W(\tau;i,j)\}, \qquad 2.39$$

with similar relations for ν_W, ν_N and ν_S. Note that there are no constraints in the evaluation of the minimum. The ensemble averaging of the numbers $\nu_a(\tau;i,j)$ now defines the non–standard distribution function of a–state particles,$*N_a(\tau;i,j)$ (2.33). The usual macroscopic continuity and momentum equations (Eqs. 2.32,32.a) are then reproduced as in the fermion case (the Fermi–Dirac distributions being changed into Bose–Einstein–type distributions by the substitution $+1 \rightarrow -1$ in Eqs 2.34).

With the exception of the advantage of allowing for higher densities, all weaknesses attached with the lattice geometry of the HPP gas survive in this model. We refer to van der Hoef (1993) for a catalogue of the (fermion) LGs which do not violate the isotropy constraint. We also mention that the inclusion of *external body forces* in the (fermion) Liouville equation is examined in Appendix D of Frisch et al (1987).

> ... je montrai comment la plus grande part de la matière de ce chaos devait, en suite de ces lois, se disposer et s'arranger d'une certaine façon qui la rendait semblable à nos cieux... Toutefois, je ne voulais pas inférer de toutes ces choses que ce monde ait été créé en la façon que je proposais ...[mais] leur nature est bien plus aisée à concevoir lorsqu'on les voit naître peu à peu en cette sorte ...
>
> R. Descartes 1637

3. Direct Construction of a CA

Given the individual mechanisms entering a physical configuration to be modelled, our definition D (or *D) essentially supplies a *method of construction* of the corresponding CA simulation. We shall substantiate this claim, and, by the same token, we shall illustrate and clarify several mathematical specifications of the definition, by explicitly setting up an admittedly oversimplified CA model for interstellar cloud formation. Since we are concerned with the construction of a *numerical algorithm* rather than with the construction of an associated macroscopic representation (in the form of PDEs), the analysis will be carried out in a standard language.

Modelling assumptions

Suppose we attempt to model the evolution of interstellar dust in a part of the Galaxy. We begin with listing a set of simplifying assumptions introduced to render the problem tractable :

(A1) The galactic region to be modelled is large enough to contain a total mass sufficient for the formation of several interstellar clouds.

Typical cloud masses, in the range $10 - 10^7$ M$_\odot$ approximately obey a power law distribution, $\propto m^{-\frac{3}{2}}$ (Elmegreen 1985); for instance the Orion clouds have a mass of 10^5 M$_\odot$ (Genzel and Stutzki 1989). These masses thus remain small as compared to the total galactic mass, $1.1\ 10^{11}$ M$_\odot$; a typical cloud dimension is 20 pc.

(A2) At the same time the region to be modelled is small as compared to the whole Galaxy (radius 12.5 kpc), so that self–gravitation can be neglected.

(A3) Rotation, magnetic fields and external gravitation are negligible.

(A4) As initial conditions we assume that the spatial zone to be modelled is filled with gas and dust, existing in the form of blobs of matter of same mass, which are uniformly distributed over the region.

(A5) The total amount of matter is conserved.

(A6) The blobs of matter are in random (thermal or turbulent) motion.

(A7) If two blobs collide, then they stick together with a probability P_s, thereby forming an *aggregate*; likewise, if a blob collides with an aggregate, or if two aggregates collide, then they stick together with the same probability P_s, thereby forming higher order aggregates; the component blobs of an aggregate are all in the same dynamic state. The aggregates, like the free individual blobs, are in random motion.

(A8) An aggregate may lose a blob by evaporation, at a rate r_e.

A *cloud* is then a large enough aggregate resulting from a sequence of collisional accretion steps.

Denoting by $a(n)$ an aggregate made up of n blobs,we can summarise assumptions (A7) and (A8) by stoichiometric schemes

$$a(n) + a(m) \rightarrow a(n+m), \quad n,m = 1,2,\ldots, \quad P_s, \tag{3.1}$$

$$a(n+1) \rightarrow a(n) + a(1), \quad n = 1,2,\ldots, \quad r_e, \tag{3.2}$$

where by extension the symbol $a(1)$ stands for a free blob.

In contrast with the usual chemical notation, $a(n)$ does not represent here a unique structure; it denotes an *equivalence class* of any connected sequence of n individual blobs, (activated cells of the CA), regardless of the resulting geometric shape. Moreover, from a macroscopic point of view — on the level of description on which we expect our model to be physically representative — for n 'large enough' this entity is not to be viewed as a point. It is a macroscopic structure whose typical shape is one of the factors our simulation is intended to reproduce.

The nonstandard language casts these ideas in a mathematically precise form. Regarding $^*n$ as a nonstandard integer, we interpret the aggregate $a(^*n)$ as a 'macroscopic point' if in the nonstandard CA lattice space the standard part of the geometrical pattern, st $[a(^*n)]$, reduces to a point; this happens necessarily for any finite nonstandard integer $^*n$. For $^*n$ an infinit, the geometrical pattern $a(^*n)$ refers to a (nonstandard) *cloud*. The nonstandard language thus supplies a precise formal definition of a cloud.

Although there are observational indications that gravity and magnetic fields are relevant factors in the cloud formation mechanism (cf the role of the Jeans instability, the magnetic Rayleigh–Taylor instability; Elmegreen 1985; Bally et al 1991), it remains of interest to analyse the physical implications of this simple accretion scenario (conditions for a stationary state, distribution of sizes of aggregates, etc). We now translate these physical assumptions, complemented by simplicity requirements, into a CA environment.

Choice of the CA geometry, topology and lattice

The approximate rotational symmetry of disk galaxies (A1) suggests that we choose for the portion of interest a segment of a 2–dimensional circular ring of radii $r_1 < r_2$ and angular opening $\Delta\theta$. From (A2) we have $\frac{r_2-r_1}{r_1} \ll 1$ and $\Delta\theta \ll 2\pi$; neglecting therefore the curvature of the boundaries, we are led to regard the area as a rectangle.

Boundary conditions compatible with matter conservation (A5) must express that matter leaving the rectangle at one boundary cell reappears in some other boundary cell. The simplest boundary conditions consistent with this property are *periodicity conditions* in both the θ and the r direction; the region thus acquires the topology $S^1 \times S^1$ ($= T^2$, torus).

The symmetry–adapted lattice for the rectangular region is the square lattice. We label the cells by pairs of integers $\mathbf{i} = (i, j)$, $i = 0, 1, \ldots, n_x - 1$ (tangential direction), $j = 0, 1, \ldots, n_y - 1$ (radial direction). The toroidal topology is directly taken care of if we define the coordinates as $(i, j) \in Z_{/n_x} \times Z_{/n_y}$, ($Z_{/m}$, set of integers modulo m).

Dependent variables

Regard as the 'particles' of our model the blobs of gas and dust. If we view these blobs as impenetrable, then a lattice cell is either empty or occupied by just one blob. A simple implementation of *motion* of the blobs (A6) is obtained by associating with each blob 5 dynamic states : a rest state (R), or an E, W, N, or S moving state, as in the HPP LG; a blob in an E state, occupying cell $\mathbf{i}$ at timestep τ, has a velocity $\mathbf{v} = \mathbf{v}_E = (1,0)$, so that at the next timestep τ+1 the same blob is shifted to cell $\mathbf{i}' = \mathbf{i} + \mathbf{v}_E = (i+1, j)$, etc. The state variable $F(\tau; \mathbf{i})$ for this problem, providing a complete dynamic description of cell $\mathbf{i}$ at time τ, is therefore a function specifying one among the 6 possible cell states (empty or quiescent state, or one of the 5 active states). A possible choice for the state space $\mathbf{z}$ is the set of integers $Z_6 = \{0, 1, 2, 3, 4, 5\}$, with $F = 0$ identifying the quiescent state, and $F = 1, 2, 3, 4$ and 5 the R, E, W, N and S state respectively. Computationally it may be more convenient to identify $\mathbf{z}$ with $Z_{[7]} = \{-3, -2, -1, 0, 1, 2, 3\}$, where $F = 0$ denotes the empty state; $F = \pm 1$ is the rest state; the values $F = +2$ and -2 correpond to E and W motions, and 3 and -3 correspond to N and S motions respectively (such a coding is useful if opposite motions are required in a same evolution step). Regardless of the precise coding adopted we shall write for an active state $F = V(\mathbf{v})$ if the state is characterised by a velocity $\mathbf{v}$.

Neighbourhoods

In a random motion a particle can switch from any one of its dynamic states to any other dynamic state; moreover an active cell at time τ may become inert (empty) at time τ+1. Hence *any* state of $\mathbf{z}$ can transform into *any* state of $\mathbf{z}$ in a single step. The neighbourhood of a cell state F , $\mathsf{N}(F)$, is therefore to be identified with the whole state space $\mathbf{z}$.

With our assignment of either a unit velocity in one of the 4 cardinal directions, or a zero velocity, a particle in cell (i, j) can reach anyone of the 5 cells of the *von Neumann neighbourhood* in one timestep; the von Neumann neighbourhood $\mathsf{N}_N(\mathbf{i})$ thus defines the spatial neighbourhood needed for handling the translational motions. For the purposes of defining aggregates a larger neighbourhood is needed; we shall use the 9–cell (*Moore*) neighbourhood $\mathsf{N}_M(\mathbf{i})$. And finally, in order to implement the constraint of matter conservation it may be necessary to use the 'diamond' neighbourhood $\mathsf{N}_D(\mathbf{i})$ formed by adding the 4 cells $(i \pm 2, j)$ and $(i, j \pm 2)$ to $\mathsf{N}_M(\mathbf{i})$. Observe that for specific steps in the simulation different geometric neighbourhoods $\mathsf{N}(\mathbf{i})$ are needed.

Evolution Rules and Probabilities

The full updating of the state functions at time τ+1 in a given problem is conveniently carried out sequentially in as many phases as there are different physical mechanisms affecting the states. Each phase consists in a loop over the whole cellspace, in which the automaton configuration (field F at every cell) is transformed into a new configuration (field F') by the specific mechanism the phase is dealing with.

This procedure compares with an analogous point in the numerical treatment of an ODE : The ODE simulation of the physics is of the form

$$\frac{\mathrm{d}}{\mathrm{d}t} x = \sum_k f_k(t, x),$$

with $f_k(x, t)$, $k = 1, 2, \ldots$ modelling the different 'forces' affecting the variable x; suppose $f_1(x, t)$ is a continuously acting force, $f_2(t, x)$ operates discontinuously, say at isolated instants $t_1, t_2, \ldots$, etc. Then the change in x per timestep, Δx, is conveniently evaluated as a sum of the changes due to the individual mechanisms, $\Delta x = \sum_k (\Delta x)_k$, with $(\Delta x)_k = f_k(x, t)\ \Delta t$. If an updating $x \longrightarrow x + (\Delta x)_k$ is made at every individual application of a component $f_k(t, x)$, then for any finite Δt the result depends on the order in which the components are applied. The same order dependence holds in CA simulations.

In our illustrative example we are led to include the following physical mechanisms :

(a) Any aggregate $a(n)$ carries out a uniform translational motion in the absence of collisions (rotational motion is not allowed by our specification of the state of motion of an aggregate under A7).

(b) The fusion process of aggregates $a(n)$ and $a(m)$ into $a(n+m)$, occurring if $a(m)$ intersects the spatial neighbourhood of $a(n)$, consists in assigning all cells of $a(n+m)$ the same state of motion $\mathbf{v}$ (A7).

(c) Any aggregate colliding with a particle of the underlying medium (not dealt with in the model) suffers a random reorientation of its velocity state (A6); we adopt different probabilities for the rest state, P_o, and for states of motion, P_m, (anisotropies in the medium can be incorporated in the model by a choice of different probabilities associated with different directions of motion). The probability parameter P_m simulates a uniform kinetic temperature of the original interstellar matter.

(d) A boundary cell $\mathbf{i}$ of an aggregate is allowed to acquire an independent velocity $\mathbf{v}$, so that it can detach from the aggregate in the translation phase (A8).

This sequence of operations, reminiscent of Kolb's (1986) diffusion–limited cluster–cluster aggregation, is applied at every timestep τ.

The initial matter density $n_o \ll 1$ (= number of active cells/total number of cells $f = n_x \times n_y$) being given, the program is initiated by randomly activating $N_o = n_o f$ cells (A4). The active cells are then all provisionally regarded as being in the rest state at the start (for all active cells $F(0, \mathbf{i}) = V(0)$). By going once through the cycle of phases, the actual initialisation of the states of motion is completed.

The rules describing these mechanisms are implemented in the following phases:

(a) *Translation.*– At time τ+1 cell $\mathbf{i}$ inherits the velocity state $\mathbf{v}$ of a neighbour cell $\mathbf{i} - \mathbf{v}$ ($\in \mathsf{N}_N(\mathbf{i})$) at time τ:

If $F(\tau, \mathbf{i} - \mathbf{v}) = V(\mathbf{v})$, update $F(\tau + 1, \mathbf{i}) = F(\tau, \mathbf{i} - \mathbf{v})$.

To obey the conservation requirement (A5), the assignment of the states at time τ must have been made such that in $\mathsf{N}(\mathbf{i})$ there is only one cellstate consistent with this premiss.

(b) *Fusion* (identification of aggregates).- To identify the aggregates we assign each cell $\mathbf{i}$ an auxiliary integer label α by the following sequence of operations. (b1) Scan the cellspace in a given direction (say first increment j, then i); scan the Moore neighbourhood of cell $\mathbf{i}$ in the same direction and assign cell $\mathbf{i}$, if active, the label $\alpha = \alpha'$ if α' is the label of an active cell in the part $\mathsf{L}(\mathbf{i})$ of $\mathsf{N}_M(\mathbf{i})$ in which the new labelling has already been carried out (an L–shaped subset of 4 connected boundary cells in the Moore neighbourhood); if $\mathbf{i}$ is the first cell to be labelled in the first run of the labelling loop, then set $\alpha = 1$; if $\mathsf{L}(\mathbf{i})$ contains no active cell, then α = previously assigned label + 1. All cells are (re)labelled when the scanning of the cellspace is completed, and all nonconnected cells then carry different labels. But not all contiguous cells are necessarily given the same label, so that a single scan is not enough to identify the aggregates. (b2) The initial scan of the cellspace is therefore continued by a second scan in which $\mathsf{N}_M(\mathbf{i})$ is examined; if for any $\mathbf{i}' \in \mathsf{N}_M(\mathbf{i})$ we have $\alpha' \neq \alpha$ then we relabel cell $\mathbf{i}$, choosing $\alpha_1 = \min(\alpha, \alpha')$ as the new label of $\mathbf{i}$; by the same token we change all labels $\alpha \rightarrow \alpha_1$ (requiring an extra scan of the cellspace). Repeat scan (b2) until all contiguous cells have the same label (cf the technique due to Hoshen and Kopelman 1976).

The procedure thus identifies an aggregate by the collection of cells carrying the same label α. As a by–product, at every timestep τ it provides all the information required for a statistical study of the distribution of aggregates (number of aggregates as a function of number of component cells, i.e. of area, or as a function of maximal x or y extension, etc).

(c) *Randomisation of velocities.*- If we merely assign the aggregates random velocity states (satisfying the probabilities P_o and P_m), then it may occur that two cells $\mathbf{i}+\mathbf{v}$ and $\mathbf{i}-\mathbf{v}$ in the von Neumann neighbourhood of $\mathbf{i}$ and belonging to different aggregates are given the velocities $-\mathbf{v}$ and $\mathbf{v}$ respectively. Such an assignment would lead to a loss of an active state in the translational phase (a). To save the conservation requirement (A5) we adopt the following constrained randomisation.

Scan the cellspace in a given direction. Three alternatives may occur. (i) Cell $\mathbf{i}$ is the first active cell encountered in the scanning. Then assign it a random velocity state (in agreement with P_o, P_m). Thereby all cells of the aggregate α to which $\mathbf{i}$ belongs are given the same velocity state (in a secondary scan). (ii) If $\mathbf{i}$ is a cell whose velocity state has not yet been determined in the principal scan (i.e. $\mathbf{i}$ belongs to a cluster not yet encountered in the scan) then examine the 2 cells $\mathbf{i}'$ and $\mathbf{i}''$ in $\mathsf{N}_D(\mathbf{i}) \setminus \mathsf{N}_M(\mathbf{i})$ already scanned (cells $(i-2, j)$ and $(i, j-2)$ if the scan proceeds by incrementing j and i); if $\mathbf{v}'$ and $\mathbf{v}''$ are the velocity directions from $\mathbf{i}'$ and $\mathbf{i}''$ to $\mathbf{i}$, then exclude from the random velocity choice of $\mathbf{i}$ the states $-\mathbf{v}'$ or $-\mathbf{v}''$ if $\mathbf{i}'$ or $\mathbf{i}''$ is in state $\mathbf{v}'$ or $\mathbf{v}''$ respectively. All cells belonging to the aggregate of cell $\mathbf{i}$ are then assigned the same velocity (in an auxiliary scan). (iii) If cell $\mathbf{i}$ has already been given a velocity state $\mathbf{v}$ in the current scan ($\mathbf{i}$ belongs to a cluster whose dynamic state has already been chosen) then examine again cells $\mathbf{i}'$ and $\mathbf{i}''$ in the part of $\mathsf{N}_D(\mathbf{i}) \setminus \mathsf{N}_M(\mathbf{i})$ already scanned; if an inconsistency is encountered ($\mathbf{v}$ equal to $-\mathbf{v}'$ or $-\mathbf{v}''$ for $\mathbf{i}'$ or $\mathbf{i}''$ in state $\mathbf{v}'$ or $\mathbf{v}''$ respectively as defined above), then we just assign a new velocity state $\mathbf{v} = 0$ to cell $\mathbf{i}$ (as well as to the corresponding aggregate).

This procedure eventually reduces the fraction of aggregates in motion in the actual experiment, P_m^{exp} with respect to the P_m (and $P_o^{exp} \geq P_o$) (in iii). Moreover, it creates an asymmetry in the motions as a result of the constraints (in ii). Both effects are inconsequential at low enough average densities.

(d) *Evaporation.*- We define a cell $\mathbf{i}$ of an aggregate as a ***boundary cell*** of the aggregate if at least one of its von Neumann neighbours is empty. Scan the cellspace. If cell $\mathbf{i}$ is a boundary cell of an aggregate in state $\mathbf{v}$, then we have the following alternatives. (i) $\mathbf{v} \neq 0$: If cell $\mathbf{i}+\mathbf{v}'$ together with its von Neumann neighbours (with the exception of cell $\mathbf{i}$) are empty, $(\mathsf{N}_N(\mathbf{i}+\mathbf{v}') \setminus \{\,\mathbf{i}\,\} = \emptyset)$, where $\mathbf{v}' = -\mathbf{v}$, then the state function of cell $\mathbf{i}$ is changed, $V(\mathbf{v}) \rightarrow V(\mathbf{v}')$, with probability r_e. (ii) $\mathbf{v} = 0$: Choose an empty neighbour cell $\mathbf{i}+\mathbf{v}'$, such that again $\mathsf{N}_N(\mathbf{i}+\mathbf{v}') \setminus \{\,\mathbf{i}\,\} = \emptyset$; then the state function of cell $\mathbf{i}$ is again changed, $V(\mathbf{v}) \rightarrow V(\mathbf{v}')$, with probability r_e.

The extra clause of an empty von Neumann neighbourhood $\mathsf{N}_N(\mathbf{i}+\mathbf{v}') \setminus \{\mathbf{i}\} = \emptyset$ avoids a violation of the conservation requirement in the translation phase. It could be replaced with a less restrictive set of conditions involving a test of the states of motion of the cells in $\mathsf{N}_N(\mathbf{i}+\mathbf{v}') \setminus \{\mathbf{i}\}$.

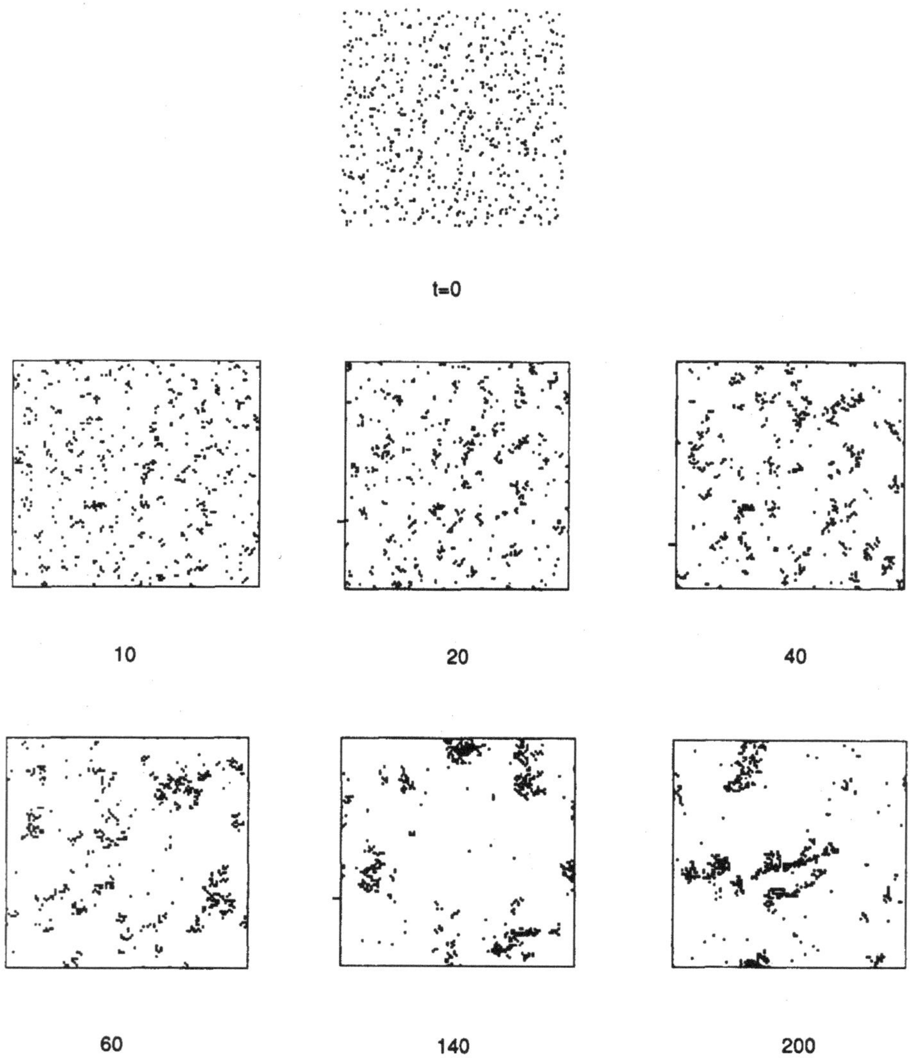

Fig. 2. – Aggregation of randomly moving particles.

Figs. 2 exhibit a few results of a run carried out on a 110 × 110 lattice, for an initial density n_o = 0.05 and in the absence of evaporation. In this experiment already at timestep 200 almost all of the matter is concentrated in 4 clusters, containing about 150 cells each. As for most multidimensional CA simulations, the actual formation of the complex spatial structure of this system is properly appreciated only in an *animation* of the time sequence (in the present case a series of 50-100 frames is sufficient to display this time behaviour). Density contours of these clusters obtained by averaging for instance over 7 × 7 cells indicate that the 'microscopic' highly complex geometry is virtually lost in the smoothing. In order to preserve some structure in the averaging process we need larger individual aggregates. The scale of our experiment is manifestly too small to be representative for a meaningful simulation of a structure like the Orion complex.

On the other hand, we observe that the simulation in Fig. 2 closely duplicates the geometrical pattern of experimental aggregates of gold colloids as obtained by Weitz and Lin (1986).

The model described here has the advantage of being open-ended. It can quite easily be extended to take account of the following more realistic physics. In the first place, instead of regarding the blobs as 'fermions', we can allow individual cells to accommodate arbitrary numbers of blobs ('boson'-type gas); in this way we can create high density gradients in the clouds. Next, instead of keeping a cluster artificially rigid, we can allow the individual cluster cells to move around under the constraint that the boundary can be deformed, but not broken up (in contrast with the evaporation process). A cluster can also be given a rotational motion to be perturbed at each step. The probabilities of random motion of an individual cluster may further be made cluster-size dependent. A treatment of self-gravity may also be incorporated in the model.

For a physically complex phenomenon such as the present aggregation process it seems unreasonable to look for a global description in terms of a PDE, as was done in the wave propagation and diffusion problems. The evolution rules rest here on an involved logic (class-dependence of the velocity state) which makes it unlikely that a system of reasonably simple alebraic equations for a complete collection of auxiliary variables can be set up (on the nonstandard level of the automaton). Even if such a formulation were available, it is manifest that a macroscopic averaging process (passage from the nonstandard equations to their standard parts) cannot generate in the present case a (low order) PDE.

On the other hand, it is clear that a mean field approximation can be formulated in terms of kinetic equations associated with reactions (3.1), (3.2),

$$\frac{d}{dt} N(n) = \sum_{k \leq n-k} A_{k,n-k}\, N(k)\, N(n-k)$$

$$- \sum_{r<n} A_{r,n}\, N(r)\, N(n) - 2\, A_{n,n}\, N(n)^2 - \sum_{r>n} A_{n,r}\, N(n)\, N(r)$$

$$- \lambda_n\, N(n) + \lambda_{n+1}\, N(n+1), \qquad n = 2, 3, \ldots, \tag{3.3}$$

$$\frac{d}{dt} N(1) = -\, 2\, A_{1,1}\, N(1)^2 + 2\, \lambda_2\, N(2)$$

$$- \sum_{n>2} A_{1,n-1}\, N(1)\, N(n-1) + \sum_{n>2} \lambda_n\, N(n), \tag{3.4}$$

$N(n)$ total number of aggregates of type $a(n)$; $A_{k,n-k}$, $k \leq n-k$ and λ_k, $k > 1$ 'reaction rates' and 'disintegration constants'. Equations similar to (3.4, 3.5) have been discussed in the literature of cluster formation processes under various simplifying assumptions (cf Vicsek 1991). Unfortunately, in the context of this mean field approximation we lose the most relevant part of the information provided by the CA simulation, namely the *information on spatial structure*.

Nel mezzo del cammin...

mi ritrovai per una selva oscura,

chè la diritta via era smarrita

Dante, Inferno I 1-3

4. Theoretical Analysis of the Microscopic CA Level Evolution: Generating Functions

In general there are no systematic direct ways for analysing and predicting the 'microscopic' behaviour of the CA if the rules of evolution, as formulated in definition D (or *D), are given. Rigorous techniques in achieving this microscopic description consist in mapping the automaton states and rules onto an auxiliary algebraic framework within which actual calculations can be carried out routinely. In this chapter we discuss a mapping technique based on *generating functions.* Since we do not attempt here to convert the microscopic description into a macroscopic (time and space averaged) description, we shall not insist on an explicit nonstandard formulation; however, all expressions derived here may be reinterpreted as nonstandard, in conformity with *D.

The practical purpose of this section is to emphasise on the one hand that certain global properties of the CA time behaviour (in particular global periods) are conditioned by number–theoretic properties of the size of the automaton (number of cells in x_1 ,x_2 ,... ,x_n direction). In actual simulations we should be aware that global periodic behaviour may arise as an artifact of the specific CA architecture adopted. Such spurious periodicities can be eliminated if care is taken in the design of the CA set up.

On the other hand, the theoretical analysis explicitly demonstrates the natural ability of the CA framework to generate complex patterns, in space as well as in space–time, which conceal an intrinsic regularity, namely fractal patterns whose properties may be studied analytically. These patterns survive on a macroscopic level, so that the formal analysis gives evidence that CA simulations quite naturally produce macroscopic fractal structures.

The technique allowing for a full analytic specification of the evolution of a nontrivial subclass of CA has been applied by Martin et al (1984). It consists in introducing a generating function A, depending on an n–dimensional array of formal variables $\mathbf{x} = (x_1, x_2, \ldots, x_n)$, whose expansion involves the field function $F(\tau;\mathbf{i})$ as coefficients, with $\mathbf{i} = (i_1, i_2, \ldots, i_n)$, ranging over its set of definition. More precisely we define the function A by the following formal Laurent–Taylor expansion

$$A(\tau;\mathbf{x}) \equiv \sum_{\mathbf{i}} F(\tau;\mathbf{i})\, \mathbf{x}^{\bullet \mathbf{i}}, \qquad 4.1$$

where the summation goes over the allowed values of $\mathbf{i}$. The *automaton function* $A(\tau;\mathbf{x})$ thus encodes the totality of cellstates, i.e. the configuration, or *automaton state* at timestep τ.

We shall make use in this chapter of the following compact notation : If $\mathbf{a}$, $\mathbf{b}$, ... are n–dimensional arrays, $\mathbf{a} = (a_1, a_2, \ldots, a_n)$, $\mathbf{b} = (b_1, b_2, \ldots, b_n)$, ..., then we denote by $\mathbf{a} \bullet \mathbf{b} \equiv \sum_{k=1}^{n} a_k b_k$ the scalar product. The array–product (or term–by–term product), $\mathbf{a} \circ \mathbf{b}$ is an n–dimensional array whose k^{th} element $(\mathbf{a} \circ \mathbf{b})_k$ is $a_k b_k$; the power α of an array, α a scalar, $\mathbf{a}^{\alpha}$, is the n–dimensional array whose k^{th} element obeys $(\mathbf{a}^{\alpha})_k = (a_k)^{\alpha}$; an m–fold iteration of the array–product can therefore be written $\mathbf{a} \circ \mathbf{a} \circ \ldots \circ \mathbf{a} \equiv \mathbf{a}^m$. The array $\mathbf{1} = (1, 1, \ldots, 1)$ is the unit element with respect to the term–by–term product : $\mathbf{1} \circ \mathbf{a} = \mathbf{a}$. The array–scalar product among three arrays, $\mathbf{a} \circ \mathbf{b} \bullet \mathbf{c} = \mathbf{a} \bullet \mathbf{b} \circ \mathbf{c} \equiv \sum_{k=1}^{n} a_k b_k c_k$, is a scalar. The array–power of an array, $\mathbf{a}^{\bullet \mathbf{b}}$, is the array whose k^{th} component is $a_k{}^{b_k}$; the scalar–power of an array, $\mathbf{a}^{\bullet \mathbf{b}} \equiv a_1{}^{b_1} a_2{}^{b_2} \ldots a_n{}^{b_n}$ is a scalar. It is convenient to associate with a scalar function, f, of scalar argument, x, $f(x)$, an array $f(\langle \mathbf{a} \rangle) \equiv (f(a_1), f(a_2), \ldots, f(a_n))$ (which will be abbreviated as $f(\mathbf{a})$ when the context makes it clear that this notation does not stand for $f(a_1, a_2, \ldots)$; for instance we have $\ln(\mathbf{a}^{\bullet \mathbf{b}}) = \mathbf{b} \bullet \ln \mathbf{a}$. When arrays of different dimensions are involved we distinguish the latter by different fontstyles $\mathbf{a}$, $\underline{a}$, ...; we likewise use different symbols $\bullet$, $\odot$, ..., and $\circ$, $\bigcirc$, ..., to denote the scalar and term–by–term products. Essentially a similar notation is used in the MATLAB expression language.

Consider the set $\mathcal{A}$ of all admissible automaton functions (for a given CA specification). Provided that this set is equipped with the algebraic structure of a *ring*, (essentially, addition and multiplication, with usual properties, are defined among the elements of $\mathcal{A}$, producing again elements of

$\mathcal{A}$), and ***provided that the evolution rules themselves can be translated into automaton functions***, the evolution is given by successive multiplications of elements of $\mathcal{A}$.

To secure the ring structure of the totality of automaton functions, a careful choice of the state space $\mathbf{z}$ is required (which may not be the most convenient for computational purposes); also, constraints are imposed on the allowed carrier geometry. Regarding the evolution rule, it depends on the specific nature of the latter whether it can be mapped onto an automaton function.

Let A, B, ... be elements of $\mathcal{A}$

$$A(\mathbf{x}) \equiv \sum_{\mathbf{i}} a(\mathbf{i})\, \mathbf{x}^{\bullet \mathbf{i}}, \quad B(\mathbf{x}) \equiv \sum_{\mathbf{i}} b(\mathbf{i})\, \mathbf{x}^{\bullet \mathbf{i}}, \ldots \tag{4.2}$$

$a(\mathbf{i}), b(\mathbf{i}), \ldots \in \mathbf{z}$. (We omit expliciting the dependence on the parameter τ whenever the time dependence is not essential). The ring structure requires that

(i) $A(\mathbf{x}) + B(\mathbf{x}) = S(\mathbf{x}) \in \mathcal{A}$

This condition is secured provided that the expansion coefficients $s(\mathbf{i})$ of $S(\mathbf{x})$ obey, for any $\mathbf{i}$,

$$a(\mathbf{i}) + b(\mathbf{i}) \equiv s(\mathbf{i}) \in z, \tag{4.3}$$

which holds true if $\mathbf{z} \equiv \mathsf{Z}$. However, our definition D requires the total number z of allowed cellstates to be finite and small. A convenient way to obey (4.3) then consists in choosing $\mathbf{z} = \mathsf{Z}_z = \{0, 1, \ldots, z-1\}$; the latter space in turn is formally identified with the cyclic space $\mathsf{Z}_{/z}$, within which addition (4.3) is understood as *modulo z addition*.

(ii) $A(\mathbf{x})\, B(\mathbf{x}) = P(\mathbf{x}) \in \mathcal{A}$

Formally this condition requires the expansion coefficients $p(\mathbf{i})$ of the product function $P(\mathbf{x})$ to obey, for any $\mathbf{i}$,

$$\sum_{\mathbf{j}} a(\mathbf{k}-\mathbf{j})\, b(\mathbf{j}) \equiv p(\mathbf{k}) \in z; \tag{4.4}$$

with the identification $\mathbf{z} \equiv \mathsf{Z}_{/z}$ we secure that $p(\mathbf{k})$ is an element of this space, provided that it is defined, i.e. provided that the algebraic operations among labels are meaningful. This is obviously the case if the label space is Z^n; it is also seen to hold if the label space is toroidal, $\mathsf{Z}_{/h} \times \mathsf{Z}_{/l} \ldots \mathsf{Z}_{/m}$ (implying that modulo h, l ,..., m addition is used for the first, second,..., n^{th} label component).

Since in numerical simulations the underlying topology of the automaton is very often toroidal, the ring structure of the automaton function space $\mathcal{A}$ covers in fact a practically important class of CA models.

Since any element $A \in \mathcal{A}$ uniquely characterises an automaton state (at a given time τ), and since multiplying a given element A by another element $R \in \mathcal{A}$, $RA \in \mathcal{A}$, characterises again an automaton state, the element R plays the part of a formal evolution rule mapped onto an automaton function. In the generating function scheme the rule itself thus represents an automaton state. This observation makes it clear that conversely not every conceivable evolution rule can be represented by an element of $\mathcal{A}$. In fact, only a small fraction of the possible rules can be mapped onto neighbourhood configurations, the fraction being the larger the smaller the number of cellstates z is, and the smaller the number of neighbourhood cells is.

For a finite state CA , with f cells, and z states per cell, the total number of configurations supported by the CA is $K = z^f$; the total number of conceivable rules, i.e. functions transforming automaton configurations into automaton configurations, is K^K, in the case of the largest spatial neighbourhood, $\mathsf{N}(\mathbf{i})$ identified with the whole lattice; for a neighbourhood of f' sites, the total number of rules is k^k, with $k = z^{f'}$, total number of neighbourhood configurations.

Instead of the above mere multiplication of states, we consider the following operation among automaton functions, whose meaning will be illuminated by a nontrivial example below. The class of CA we shall be dealing with is assumed to obey the following specifications:

(i) There is an s–dimensional array of integers $\underline{\kappa} = (\kappa_1, \kappa_2, \ldots, \kappa_s)$ not depending on time τ, such that at any time τ the automaton function $A(\tau; \mathbf{x})$ can be written as

$$A(\tau;\mathbf{x}) \;=\; \underline{\kappa} \odot \underline{A}(\tau;\mathbf{x}); \tag{4.5}$$

$\underline{A}(\tau; \mathbf{x})$ denotes an array of s functions

$$A_h(\tau;\mathbf{x}) \;\equiv\; \sum_{\mathbf{i}} a_h(\tau;\mathbf{i})\, \mathbf{x}^{\bullet\mathbf{i}}, \quad h \;=\; 1,2,\ldots,s, \tag{4.6}$$

with the coefficients $a_h(\tau; \mathbf{i}) \in Z_{/p_h}$, p_h a prime number, in general depending on h. With obvious notations we have

$$\underline{A}(\tau;\mathbf{x}) \;\equiv\; \sum_{\mathbf{i}} \underline{a}(\tau;\mathbf{i})\, \mathbf{x}^{\bullet\mathbf{i}}. \tag{4.6.a}$$

(ii) The evolution rule can be mapped onto an s–dimensional array $\underline{R}(\mathbf{x})$ of functions $R_h(\mathbf{x})$

$$R_h(\mathbf{x}) \;\equiv\; \sum_{\mathbf{j}} r_h(\mathbf{j})\, \mathbf{x}^{\bullet\mathbf{j}}, \quad h \;=\; 1,2,\ldots,s \tag{4.7}$$

with the coefficients $r_h(\mathbf{i})$ defined in the cyclic space $Z_{/p_h}$ of definition of the coefficients $a_h(\tau; \mathbf{i})$; ($r_h(\mathbf{j})$ is nonzero for a finite and small number of values $\mathbf{j}$ only). Again we write

$$\underline{R}(\mathbf{x}) \;\equiv\; \sum_{\mathbf{j}} \underline{r}(\mathbf{j})\, \mathbf{x}^{\bullet\mathbf{j}}. \tag{4.7.a}$$

(iii) The evolution is described by

$$\underline{A}(\tau+1;\mathbf{x}) \;=\; \underline{R}(\mathbf{x}) \bigcirc \underline{A}(\tau;\mathbf{x}). \tag{4.8}$$

The ***physical motivation*** for a decomposition of type (4.5) of the automaton state function is that the cellstates may be partitioned into independent classes, such that the evolution of each class is governed by an independent law. ***Mathematically*** the partitioning is given by the following number-theoretic considerations. For z identified with $Z_{/z}$, this cyclic space is a domain of integrity (zero possesses no divisors modulo z) provided that z is prime; it is only in a domain of integrity that every nonzero element has a unique inverse.

For instance in sets $Z_{/4}$, we have $2 \times 2 \equiv 0 \bmod (4)$, illustrating that this space is not a domain of integrity; 2 has no inverse, since $2 \times m \equiv 0$ or $2 \bmod (4)$, $m = 0,1,2$ or 3. More generally, for any $Z_{/h \times k}$, $h \times k \equiv 0 \bmod (hk)$.

If z is not prime, we use the prime factorisation

$$z \;=\; {p_1}^{\nu_1}\, {p_2}^{\nu_2}\, {p_3}^{\nu_3}\, \ldots, \tag{4.9}$$

(ν_k, multiplicityof the kth prime p_k); this decomposition enables us to regard $Z_{/z}$ as a direct sum of ν–fold integrity domains $Z_{/p}$, p the prime numbers, of multiplicity ν, in the factorisation (4.9). Let ν, ν', …, $\nu\sharp$ be the nonzero exponents of the actual prime factors p, p', …, $p\sharp$. Then we formally set

$$Z_{/z} \;=\; \underline{\kappa} \odot \underline{Z} \tag{4.10}$$

with

$$\underline{Z} \;=\; (Z_{/p}, Z_{/p}, \ldots, Z_{/p}, Z_{/p'}, Z_{/p'}, \ldots, Z_{/p'}, \ldots, Z_{/p\sharp}, Z_{/p\sharp}, \ldots, Z_{/p\sharp})$$

where $Z_{/p}$ occurs ν times, $Z_{/p'}$ occurs ν' times, etc; this notation symbolises that if we choose an arbitrary element (integer) (0, 1, …, or $p-1$), (0,1,…, or $p'-1$), etc, in each $Z_{/p}$, $Z_{/p'}$, etc, respectively, then by multiplying the element by the corresponding component κ_a, κ_b, etc, of the array $\underline{\kappa}$, we should generate an element of $Z_{/z}$, i.e. an integer among the collection (0, 1, 2, …,

$z-1$). We regard Eq. (4.10) as the defining relation for the array $\underline{\kappa}$. It is intuitively clear at the outset that a solution exists. A direct calculation yields

$$\underline{\kappa} = (1, p, p^2, \ldots, p^{n-1},$$
$$p^\nu, p^\nu p', p^\nu p'^2, \ldots, p^\nu p'^{\nu'-1},$$
$$\ldots,$$
$$p^\nu p'^{\nu'} \ldots, p^\nu p'^{\nu'} \ldots p\sharp, p^\nu p'^{\nu'} \ldots p\sharp^2, \ldots, p^\nu p'^{\nu'} \ldots p\sharp^{\nu\sharp-1}), \qquad 4.11$$

where the subset of μ array elements corresponding to the rth prime factor q is of the form M, Mq, Mq^2, ..., $Mq^{\mu-1}$, the multiplier M itself being a product of the preceding $r-1$ prime factors raised to their respective multiplicities, $M = p^\nu p'^{\nu'} \ldots$.

Each group of μ components κ_h corresponding to one prime factor q, and applied to the largest element of the corresponding $Z_{/q}$, (q-1), yields $M(1+q+q^2+\ldots+q^{\mu-1})(q-1) = M(q^\mu - 1)$; the multiplier M in front of the terms q^r, $r = 1, 2, \ldots, \nu - 1$, separates states belonging to different prime factors into independent 'layers'.

Define then the first collection of ν component functions $a_h(\tau;\mathbf{i})$ and $r_h(\mathbf{i})$, $h = 1, 2, \ldots, \nu$, in $Z_{/p}$, the second collection of ν' component functions $a_h(\tau;\mathbf{i})$ and $r_h(\mathbf{i})$, $h = \nu{+}1, \nu{+}2, \ldots, \nu{+}\nu'$, in $Z_{/p'}$, ..., and the last collection of $\nu\sharp$ component functions $a_h(\tau;\mathbf{i})$ and $r_h(\mathbf{i})$, $h = s-\nu\sharp+1, s-\nu\sharp+2, \ldots, s$, in $Z_{/p\sharp}$. Our choice of $\underline{\kappa}$ then secures that $\underline{\kappa} \odot \underline{a}$ or $\underline{\kappa} \odot \underline{r}$ is indeed defined in $Z_{/z}$.

For an initial automaton state represented by a generating function decomposed according to (4.5)

$$A_o(\mathbf{x}) = \underline{\kappa} \odot \underline{A}_o(\mathbf{x}) \qquad 4.12$$

we then have, for any time τ

$$A(\tau;\mathbf{x}) = \underline{\kappa} \odot [\underline{R}(\mathbf{x})]^\tau \bigcirc \underline{A}_o(\mathbf{x}). \qquad 4.13$$

More explicitly, from Eqs. (4.1) and (4.6.a) we have

$$\sum_{\mathbf{i}} F(\tau;\mathbf{i})\, \mathbf{x}^{\bullet\mathbf{i}} \equiv$$
$$\sum_{\mathbf{k}} \sum_{\mathbf{j}_1 \mathbf{j}_2 \ldots \mathbf{j}_\tau} \underline{\kappa} \odot \underline{r}(\mathbf{j}_1) \bigcirc \underline{r}(\mathbf{j}_2) \bigcirc \ldots \bigcirc \underline{r}(\mathbf{j}_\tau) \bigcirc \underline{a}_o(\mathbf{k})\, \mathbf{x}^{\bullet(\mathbf{k}+\mathbf{j}_1+\mathbf{j}_2+\ldots+\mathbf{j}_\tau)} \equiv$$
$$\sum_{i} \Big(\sum_{\mathbf{j}_1 \mathbf{j}_2 \ldots \mathbf{j}_\tau} \underline{\kappa} \odot \underline{r}(\mathbf{j}_1) \bigcirc \underline{r}(\mathbf{j}_2) \bigcirc \ldots \bigcirc \underline{r}(\mathbf{j}_\tau) \bigcirc \underline{a}_o(\mathbf{i} - \mathbf{j}_1 - \mathbf{j}_2 - \ldots - \mathbf{j}_\tau) \Big) \mathbf{x}^{\bullet\mathbf{i}} \qquad 4.13.a$$

Identifying the coefficients of same monomials (taking account that the identity becomes a congruence if the label space is toroidal) we obtain an expression for the field function $F(\tau;\mathbf{i})$.

We illustrate these points by two examples, each one defined by a $1D$ automaton with a finite number f of cells obeying periodic boundary conditions. The corresponding automaton function

$$A(\tau;x) = \sum_{i=0}^{f-1} F(\tau;i)\, x^i \qquad 4.14$$

is a polynomial of degree $f-1$. Identifying then the automaton function space $\mathcal{A}$ with the polynomials of degree $f-1$, whose coefficients are defined in appropriate $\mathbf{z} = Z_{/z}$ spaces, we must take care that powers x^{mf+v}, with m an arbitray nonnegative integer and v a nonnegative integer less than f, are identified with x^v. Formally this is achieved by the congruence

$$f(x) \equiv A(x) \mod(x^f - 1), \qquad 4.15$$

where $f(x)$ is an arbitrary formal Laurent–Taylor series in x, and $A(x)$ is a polynomial of degree $f-1$.

Translations

As a simple illustration consider a $1D$ gas whose particles are all in the same state of motion (motion in the positive x direction, with velocity $v = +1$). The state of a cell i is therefore represented by 0 (empty cell) or 1 (moving particle in cell i), so that $z = 2$ is prime. The initial configuration is given by the automaton polynomial

$$A_o(x) = \sum_{i=0}^{f-1} a_o(i)\, x^i, \quad a_o(i) = 0 \text{ or } 1, \quad i = 0, 1, \ldots, f-1. \qquad 4.16$$

The evolution rule describing the translational motion of a gas particle in the positive x–direction is represented by the polynomial

$$T_+(x) = x. \qquad 4.17$$

Hence from (4.13) we have

$$A(\tau; x) = [T_+(x)]^\tau A_o(x) = \sum_{i=0}^{f-1} a_o(i)\, x^{i+\tau} \bmod(x^f - 1). \qquad 4.18$$

Writing $\tau = mf + \theta$, m a nonnegative integer, with $\theta = \tau \bmod(f)$, we have

$$\begin{aligned} A(\tau;x) &= F(\tau;0) + F(\tau;1)\,x + F(\tau;2)\,x^2 + \ldots + F(\tau;f-2)\,x^{f-2} + F(\tau;f-1)\,x^{f-1} \\ &= a_o(f-\theta) + a_o(f-\theta+1)x + a_o(f-\theta+2)\,x^2 + \ldots \\ &\quad + a_o(f-2)\,x^{\theta-2} + a_o(f-1)\,x^{\theta-1} + a_o(0)\,x^\theta + a_o(1)\,x^{\theta+1} + \ldots \\ &\quad + a_o(f-\theta-2)\,x^{f-2} + a_o(f-\theta-1)\,x^{f-1}. \end{aligned} \qquad 4.18.a$$

From this expression we read off for instance the conservation of the number of particles

$$\sum_{i=0}^{f-1} a(\tau; i) = \sum_{i=0}^{f-1} a_o(i).$$

We also notice that for *any* initial condition $A(\tau + P; x) = A(\tau; x)$, if $P = P(f) = f$, i.e. the solution is periodic, of period f. For initial conditions $a_o(i) = 1$, if the cell labels obey $i = wf_o$, $w = 0, 1, \ldots$, (with $0 \le i \le f-1$), and $a_o(i) = 0$ for all other cells, the solution has a smallest period, $P(f) = f_o$, if f is a multiple of f_o; accordingly, if the size f of the automaton is taken as a parameter, if we let f increase from f_o to infinity, then the automaton switches ('bifurcates') from period-f_o to period-f and back to period-f_o behaviour, the latter occurring each time f has f_o as a divisor. Alternatively, for $f = p^\nu q^\mu \ldots$, ($p < q < \ldots$ sequence of primes, $\nu, \mu, \ldots \neq 0$), if we regard f_o as a parameter, and let it take on the sequence of integer values $f_o = p$, $p+1$, $p+2$, ...,$f-1$, then the automaton can undergo transitions between periodic behaviour of the type $a \to f \to b \to f \to c \to f \to \ldots$ where $a < b < c < \ldots$ is the sequence of successive divisors $p^\alpha q^\beta \ldots$ (α=1, 2, ..., ν; b = 0,1, ..., μ; ...) of f (with intermediate period–f behaviour omitted if $b = a+1$ or $c = b+1, \ldots$).

A detailed discussion of the periods (related to the multiplicative sub-order function of z modulo f) arising in the above $1D$ CA in the presence of an evolution rule encoded in the function

$$R(x) = x + x^{-1}, \qquad 4.19$$

(studied experimentally in Wolfram 1983, for $z = 2$) is given in Martin et al (1984).

These examples are meant to illustrate the general result that the time behaviour of a CA is regulated — at least in part — by number–theoretic properties. To minimise in actual simulations the effect of spurious periodicities related to an inadequate choice of the dimensions of the automaton, these examples indicate that the number of rows, (and columns) of cells of the automaton should be a *prime number*; this tends to render the CA geometry–related periods long enough to make them innocuous for any practical purposes; in multi–dimensional CA models the sizes in different directions (along the x_1, x_2, ..., axes) should be taken as different primes.

Wave propagation again

The generating function formalism allows us to solve the full microscopic behaviour of a CA simulating the $1D$ macroscopic wave propagation problem introduced in section 2.

We consider here a slightly more general problem. Take a CA containing particles moving in the positive and in the negative direction, as previously, as well as rest particles (zero velocity states). Adopt as cell space $z = Z_{/8}$, with 0, empty state; 1, positive velocity state; 2, negative velocity state; 3, positive and negative state; 4, rest state; 5, positive and rest state; 6 negative and rest state; 7 positive, negative and rest state. Since $z = 2^3$, we have by Eq. (4.11)

$$\underline{\kappa} = (1,2,4); \qquad 4.20$$

We encode the initial conditions in a triplet of polynomials, each polynomial being of the form (4.16)

$$\begin{aligned}\underline{A}_o(x) = (\, & a_o'(0) + a_o'(1)\,x + a_o'(2)\,x^2 + \ldots + a_o'(f-2)\,x^{f-2} + a_o'(f-1)\,x^{f-1},\\ & a_o''(0) + a_o''(1)\,x + a_o''(2)\,x^2 + \ldots + a_o''(f-2)\,x^{f-2} + a_o''(f-1)\,x^{f-1},\\ & a_o'''(0) + a_o'''(1)\,x + a_o'''(2)\,x^2 + \ldots + a_o'''(f-2)\,x^{f-2} + a_o'''(f-1)\,x^{f-1}), \qquad 4.21\end{aligned}$$

the coefficients being defined in $Z_{/2}$. If the coefficient values 0 or 1 in the first, second and third polynomial stand for the absence or presence of a positive velocity, a negative velocity, and a rest state respectively, then we see that the coefficients $a_o(i)$ of $A_o(x)$

$$A_o(x) = \underline{\kappa} \odot \underline{A}_o(x) = \sum_{i=0}^{f-1} \underline{\kappa} \odot \underline{a}_o(i)\, x^i = \sum_{i=0}^{f-1} a_o(i) x^i, \qquad 4.22$$

are indeed defined in $Z_{/8}$, with the interpretations of the cellstate values as specified. The evolution rule is represented by

$$\underline{R}(x) = (T_+(x), T_-(x), I) = (x, x^{-1}, 1), \qquad 4.23$$

($T_-(x)$, translation in negative direction; I rest operator, identity transformation). Hence we have from Eq. (4.13)

$$A(\tau;x) = \sum_{i=1}^{f-1} \underline{\kappa} \odot$$

$$\begin{aligned}& (a_o'(0)\,x^{\tau} + a_o'(1)\,x^{1+\tau} + a_o'(2)\,x^{2+\tau} + \ldots + a_o'(f-2)\,x^{f-2+\tau} + a_o'(f-1)\,x^{f-1+\tau},\\ & a_o''(0)\,x^{-\tau} + a_o''(1)\,x^{1-\tau} + a_o''(2)\,x^{2-\tau} + \ldots + a_o''(f-2)\,x^{f-2-\tau} + a_o''(f-1)\,x^{f-1-\tau},\\ & a_o'''(0) + a_o'''(1)\,x + a_o'''(2)\,x^2 + \ldots + a_o'''(f-2)\,x^{f-2} + a_o'''(f-1)\,x^{f-1}\,), \bmod (x^f - 1). \qquad 4.24\end{aligned}$$

Explicitly, with the notations of (4.18.a)

$$\begin{aligned}A(\tau;x) &= F(\tau;0) + F(\tau;1)\,x + F(\tau;2)\,x^2 + \ldots + F(\tau;f-2)x^{f-2} + F(\tau;f-1)\,x^{f-1}\\ &= [\, a_o'(f-\theta) + 2a_o''(\theta) + 4a_o'''(o)\,]\end{aligned}$$

$$+ [\ a_o'(f-\theta+1) \ + \ 2a_o''(\theta+1) \ + \ 4a_o'''(1)\ ]\ x$$
$$+ [\ a_o'(f-\theta+2) \ + \ 2a_o''(\theta+2) \ + \ 4a_o'''(2)\ ]\ x^2$$
$$+\dots$$
$$+ [\ a_o'(f-1) \ + \ 2a_o''(2\theta-1) \ + \ 4a_o'''(\theta-1)\ ]\ x^{\theta-1}$$
$$+ [\ a_o'(0) \ + \ 2a_o''(2\theta) \ + \ 4a_o'''(\theta)\ ]\ x^{\theta}$$
$$+ [\ a_o'(1) \ + \ 2a_o''(2\theta+1) \ + \ 4a_o'''(\theta+1)\ ]\ x^{\theta+1}$$
$$+\dots$$
$$+ [\ a_o'(f-2\theta-1) \ + \ 2a_o''(f-1) \ + \ 4a_o'''(f-1-\theta)\ ]\ x^{f-1-\theta}$$
$$+ [\ a_o'(f-2\theta) \ + \ 2a_o''(0) \ + \ 4a_o'''(f-\theta)\ ]\ x^{f-\theta}$$
$$+ [\ a_o'(f-2\theta+1) \ + \ 2a_o''(1) \ + \ 4a_o'''(f+1-\theta)\ ]\ x^{f+1-\theta}$$
$$+\dots$$
$$+ [\ a_o'(f-\theta-2) \ + \ 2a_o''(\theta-2) \ + \ 4a_o'''(f-2)\ ]\ x^{f-2}$$
$$+ [\ a_o'(f-\theta-1) \ + \ 2a_o''(\theta-1) \ + \ 4a_o'''(f-1)\ ]\ x^{f-1}, \qquad 4.24.a$$

(the middle expressions are explicited in the case $2\theta < f$; for $2\theta > f$ their forms are obvious). Conservation of the number of particles in each state is obeyed ($\sum_{i=0}^{f-1} \underline{a}(\tau;i) \ = \ \sum_{i=0}^{f-1} \underline{a}_o(i)$. Periodicity of the solutions is discussed as in the previous example. For *any* initial condition $A(\tau+P;x) = A(\tau;x)$, if $P = P(f) = f$. If the initial conditions are such that $a_o'(i+f_+) = a_o'(i)$, for cell labels $i = 0,1,\dots,f-f_+-1$, (spatially periodic distribution of the positive velocity states, with spatial period f_+), and $a_o''(i+f_-) \ = \ a_o''(i)$, for cell labels $i = 0,1,\dots,f-f_--1$, (spatially periodic distribution of the negative velocity states, with spatial period f_-), the coefficients $a_o'''(i)$ being arbitrary, then the solution has a smallest period, $P(f) = f_+f_-$, provided that f is a multiple of f_+f_-; the quantity f_+f_- plays the same part as f_o in the previous example. Alternatively, for f composite, having f_- as divisor, $f = (f_-)^{\phi}p^{\nu}q^{\mu}\dots$, ($\phi \neq 0$, $\nu,\mu,\dots$ integers), if we let f_+ take on the sequence of integer values $f_+ = 1, 2, 3, \dots$, then the automaton undergoes transitions between different types of periodic behaviour, as in the 1–velocity state automaton discussed before.

Pattern Formation

The generating function method also lends itself quite naturally to demonstrate the formation of complex space or space–time patterns in the CA framework.

For definiteness we consider a $2D$ square automaton defined over the infinite plane ($\mathbf{i}=(i,j) \in \mathsf{Z} \times \mathsf{Z}$), with corresponding automaton function

$$A(\tau;x,y) \ = \ \sum_{i,j} F(\tau;i,j)\ x^i\ y^j\,. \qquad 4.25$$

The field function is regarded as defined in $\mathsf{Z}_{/p}$, p a prime ($z = p$). The most general evolution rule compatible with the generating function formalism and involving the 9–site Moore neighbourhood is encoded in an automaton function of the form

$$R(x,y) \ = \ \sum_{\alpha,\beta=-1,0,1} r(\alpha,\beta)\ x^{\alpha}\ y^{\beta}$$

$$= r(0,0) \ + \ r(1,0)\ x \ + \ r(0,1)\ y \ + \ r(-1,0)\ x^{-1} \ + \ r(0,-1)\ y^{-1}$$
$$+ \ r(1,1)\ x\ y \ + \ r(-1,1)\ x^{-1}\ y \ + \ r(1,-1)\ x\ y^{-1} \ + \ r(-1,-1)\ x^{-1}\ y^{-1} \qquad 4.26$$

with coefficients $r(\alpha,\beta) \ \in \mathsf{Z}_{/p}$. It is clear that the presence of more than one nonzero coefficient $r(\alpha,\beta)$ implies a nonconservation of the number of excited cells, in other words, it corresponds to a rule capable of describing a *growth process*. For $A_o(x,y) = 1$, representing a single 'activated' state 1 at the initial timestep $\tau = 0$, the evolution of this CA is given by the following analytic expression, for any timestep τ

$$A(\tau;x,y) = [R(x,y)]^{\tau}$$
$$= \sum_{a,b,\ldots,i} \frac{\tau}{a!\ b!\ c!\ d!\ e!\ f!\ g!\ h!\ i!}$$
$$r(0,0)^a\ r(1,0)^b\ r(0,1)^c r(-1,0)^d\ r(0,-1)^e\ r(1,1)^f\ r(-1,1)^g\ r(1,-1)^h\ r(-1,-1)^i$$
$$x^{a+2b+c+e+2f+2h-\tau}\ y^{a+b+2c+d+2f+2g-\tau}, \qquad 4.27$$

where the summation extends over all a, b, ... ,i values subject to the constraints a, b, ... ,i nonnegative, with $a+b+\ldots+i=\tau$. We read off from this expression that the initial signal in cell (0,0) propagates with a speed 1, activating a fraction of the number of cells in the area bounded by the 'maximal square' of vertices (τ,τ), (-τ, τ), (τ,-τ), (-τ,-τ). It is obvious that the state of the vertex cell (τ,τ) is a function of the coefficient $r(1,1)$ alone; likewise the state of (-τ, τ) is determined by $r(-1,1)$ alone, and the states of (τ,-τ) and (-τ,-τ) by the coefficients $r(1,-1)$ and $r(-1,-1)$ respectively alone. In fact, setting for instance $f = \tau$ in the summation (4.27) shows that cell (τ,τ) is in state $[r(1,1)]^{\tau}$ mod z; $g = \tau$ yields the state $[r(-1,1)]^{\tau}$ mod z for cell (-τ, τ); $h = \tau$ yields the state $[r(1,-1)]^{\tau}$ mod z for cell (τ,-τ); and $i = \tau$ yields the state $[r(-1,-1)]^{\tau}$ mod z for cell (-τ,-τ).

For any coefficient $r(\alpha,\beta)$ of the evolution rule (4.26) which is equal to 0 or 1, we have $r(\alpha,\beta)^{\tau} = r(\alpha,\beta)$; therefore the vertex cell ($\alpha\tau$, $\beta\tau$) (α, $\beta = \pm 1$) then takes the value $r(\alpha,\beta)$. On the other hand, if $r(\alpha,\beta) = 2$, then $r(\alpha,\beta)^{\tau}$ mod 3 = 2 or 1 for $\tau = 2n + \theta$, θ =1 or 2 respectively; $r(\alpha,\beta)^{\tau}$ mod 5 = 2,4,3,1 for $\tau = 4n + \theta$, θ =1,2,3,4 respectively; the vertex cells thus go through all (activated) cellstates for z = 3 or 5; however $r(\alpha,\beta)^{\tau}$ mod 7 = 2,4,1 for $\tau = 3n + \theta$, θ =1,2,3 respectively, so that cellstates 3,5,6 are not attained. More generally, if $r(\alpha,\beta) \neq 0$ then $r(\alpha,\beta)^{\tau}$ mod $p \neq 0$, so that the vertex cell ($\alpha\tau$, $\beta\tau$) is always in an activated state. Any other cell in the maximal square may become inert at some timesteps; for instance, the state of site (τ, τ-1) is given by the coefficient of $x^{\tau}y^{\tau-1}$, which is equal to $\tau\ r(1,0)\ r(1,1)^{\tau-1}$ mod z ; this coefficient thus vanishes at timesteps $\tau = nz$, n = 1, 2, ..., regardless of the values of $r(1,0)$ and $r(1,1)$.

We mention that the largest area actually populated by active cells at time τ , or, more precisely, the ***convex hull*** of the active cells (smallest convex polygon enclosing all excited cells) coincides with the maximal square only if $r(1,1)$, $r(-1,1)$, $r(1,-1)$, $r(-1,-1)$ are all nonzero; for instance, if all of these 4 coefficients vanish but $r(1,0)$, $r(0,1)$, $r(-1,0)$ and $r(0,-1)$ are all nonzero, then the convex hull is the square of vertices (τ,0), (0, τ), (-τ,0) and (0,-τ). Fig. 3 (cf below) refers to a CA obeying (4.27) and exhibiting a *triangular* convex hull.

The general structure of expression (4.27) is made clear by the following geometric considerations: An *arc* [α, β] is the elementary spatial CA step, of 'direction' (α, β), a cellstate is allowed to carry out in one timestep (α, β= -1,0,1 in the above case); the arc [α, β] exists and connects two neighbour cells $(i,j) \rightarrow (i+\alpha, j+\beta)$ provided that $r(\alpha,\beta) \neq 0$; [0,0] defines a special arc, namely a *loop* connecting a cell with itself; a loop exists provided that $r(0,0) \neq 0$. A ***path*** of length τ connecting cell (i,j) to cell (i',j') is any sequence of τ arcs connecting successive cells $(i,j) \rightarrow (i+\alpha', j+\beta') \rightarrow (i+\alpha'+\alpha'', j+\beta'+\beta'') \rightarrow \ldots \rightarrow (i+\alpha'+\alpha''+\ldots+\alpha^{(\tau)}, j+\beta'+\beta''+\ldots+\beta^{(\tau)}) \equiv (i',j')$. With an arc [$\alpha$, β] we attach a ***weight*** $r(\alpha,\beta)$ ($\neq 0$ by definition). The ***weight of a path*** is defined as the product of the weights of its component arcs, modulo z; the weight of an individual path is therefore nonzero.

Each component

$$r(0,0)^a\ r(1,0)^b\ r(0,1)^c r(-1,0)^d\ r(0,-1)^e\ r(1,1)^f\ r(-1,1)^g\ r(1,-1)^h\ r(-1,-1)^i$$

in the sum (4.27) represents the weight of a path of length τ , made up of a loops [0,0], b arcs [1,0], c arcs [0,1], ... , and i arcs [-1,-1], connecting the origin (0,0) to cell $(i,j) \equiv (a+2b+c+$

$e + 2f + 2h - \tau, a + b + 2c + d + 2f + 2g - \tau)$; the coefficient $\tau!/[a!\ b!\ c!\ d!\ e!\ f!\ g!\ h!\ i!]$ is the number of distinct paths having the same number a of loops [0,0], b of arcs [1,0], ... , and i of arcs [-1,-1]; the summmation extends over the number of distinct paths made up of different numbers a of loops [0,0], b of arcs [1,0], ... , and i of arcs [-1,-1] ; this number is equal to the number $\sigma(\tau; i, j)$ of $a, b, \ldots, h$ values consistent with site (i, j) at timestep τ , i.e. the number of different solutions consistent with the constraints $0 \leq a, b, \ldots \leq \tau$,

$$a = A_1(\tau; i, j), \quad b = B_1(\tau; i, j), \ldots, \quad h = H_1(\tau; i, j),$$

$$a = A_2(\tau; i, j), \quad b = B_2(\tau; i, j), \ldots, \quad h = H_2(\tau; i, j),$$

$$\ldots$$

$$a = A_{\sigma(\tau;i,j)}(\tau; i, j), b = B_{\sigma(\tau;i,j)}(\tau; i, j), \ldots, h = H_{\sigma(\tau;i,j)}(\tau; i, j),$$

of the linear system

$$a + 2b + c + e + 2f + 2h - \tau = i,$$

$$a + b + 2c + d + 2f + 2g - \tau = j. \qquad 4.28$$

At timestep τ , the state of a cell (i, j) as read off from expression (4.27) is thus given by a (discrete form of a) 'Feynman path-integral', over the collection of paths connecting the origin (0,0) with cell (i, j) : The cellstate $F(\tau; i, j)$ is the sum of the weights modulo z of all possible (virtual) paths of lengths τ joining (0,0) to (i, j).

Although individual paths have nonzero weights, the superposition, or interference, of paths may add up to zero values of the resulting weight. These path interferences are responsible for pattern formation.

It is clear that this geometric formulation extends to evolution rules defined over arbitrary neighbourhoods in n-dimensional CA, provided only that a generating function representation is applicable.

Fractal space structures

Denote by $N(L; \tau; \zeta)$ the total number of cells in a given activated state ζ (=1, 2, ..., or $z - 1$) at time τ , in a subset of the convex hull of the actvated cells, of characteristic length L; as appropriate subsets we shall adopt reduced homothetic versions of the convex hull centered at the origin; these sets are thus completely fixed by the scale parameter L; we denote by L_H ($= L_H(\tau) \propto \tau$ for asymptotically large times) this length parameter corresponding to the convex hull at a given time τ. For the class of CA considered here the number obeys the inequality

$$N(L; \tau; \zeta) \leq G\, L^n, \quad \text{for any lengthscale } L, \text{ at fixed } \tau, \qquad 4.29$$

with n the spatial dimensionality of the automaton ($n = 2$ in the present examples), and G a finite constant depending on the geometry of the convex hull; the latter is conveniently chosen such that the right hand side of the inequality is the total number of cells enclosed by the convex hull, for $L = L_H$). Besides the extreme alternatives of *conservation* of the cellstate ζ ($N(L_H; \tau; \zeta)$ independent of τ), or regular asymptotic *growth proportional to the n-dimensional 'volume' of the convex hull* ($\propto \tau^n$), i.e. uniform distribution over the convex hull of the cells in state ζ ($N(L_H; \tau; \zeta)$ $= \rho(\zeta)\, G'\, \tau^n$, for τ large, $\rho(\zeta) \leq 1$ density of state ζ, G' again a numerical factor, and hence also $N(L; \tau; \zeta) = \rho(\zeta)\, G\, L^n$), we may have the alternative of a slower growth process, manifesting itself by the relation

$$N(L; \tau; \zeta) \rightarrow \rho_{F'}(\zeta)\, G\, L^{F'}, \quad \text{as } \tau \in T' = \{\tau'_1, \tau'_2, \ldots\} \subset Z; \qquad 4.30$$

here the coefficient $\rho_{F'}(\zeta)$ is finite and nonzero, independent of the admissible large enough τ values, $\tau'_1 < \tau'_2 < \ldots < \tau'_m < \tau'_{m+1} < \ldots$ of the set T', and independent of L; the exponent F' is subject to the inequality $F' \leq 2$, and likewise independent of L and the (large enough) values $\tau'_m \in T'$. We observe that we do not require the asymptotic behaviour (4.30) to hold for *any* large enough τ (i.e. $\tau \in Z$); we only require this behaviour to hold for the specific times of the list T'. Such an alternative arises if the cell states ζ distribute uniformly over a *fractal* subset of the convex hull at the (asymptotically large) times $\tau \in T'$; the *spatial pattern* exhibited by the collection of state ζ cells is then a *fractal set*, of mass dimension F' for any large enough $\tau \in T'$.

Relation (4.30) is formulated more precisely in a nonstandard form

$$st\left(\frac{*N(*L;\tau;\zeta)}{*L^{*F'}}\right) = \rho_{F'}(\zeta)\, G \qquad 4.30.a$$

for τ any infinit in the nonstandard set sets $*T' = \{\tau'_1, \tau'_2, \ldots\} \subset *Z$, $*F < n$, $\rho_{F'}(\zeta) \neq 0$ standard ; $F' =$ st $(*F')$, the standard parts F' and $\rho_{F'}(\zeta)$ being independent of the standard part L of a nonstandard scaling factor $*L$.
We note also that there may be different sets, T', T", T"', ... (T' ∩ T" = ∅, etc) such that the asymptotic behaviour (4.30) holds with distinct exponents F', F'', F''', etc, for τ taking on values in T', T", T"', ... respectively. Under those alternatives we observe transitions between different fractal space structures during the asymptotic evolution.

We include here a comment on fractal CA patterns. The existence of genuine fractals, in the sense of 'patterns preserving structure at any scale', is not consistent with the standard CA framework (definition D), since the latter possesses an intrinsic scale (the cellsize). The very concept of a fractal requires reference to the nonstandard CA definition (*D) (unless we adopt a definition based on a rescaling of the structures generated in the CA, as done in Perdang 1991a). The fractal set then becomes the standard part of the pattern actually generated by the CA. This observation implies that a computer realisation requires a large enough number of timesteps in order to achieve an acceptable approximation to the nonstandard pattern; therefore the operational definition of the fractal dimension provided by Eq. (4.30) supplies a meaningful result only for large enough times τ, and large enough scales L, L being measured in units of the cellsize λ.

As an illustrative example we discuss the propagating deterministic growth phenomenon generated by the 2–state automaton of evolution rule

$$R(x,y) = 1 + X, \quad \text{with} \quad X = (1+x)y^{-1}, \qquad 4.31$$

($r(0,0) = r(0,-1) = r(1,-1) = 1$, the remaining coefficients in Eq. 4.26 being 0). Taking account of the modulo 2 arithmetic of the coefficients entering the automaton functions we have at time τ

$$[R(x,y)]^\tau = \sum_{m=0}^{\tau} c_\tau^m\, X^m, \qquad 4.32$$

with

$$X^m = y^{-m} \sum_{k=0}^{m} c_m^k\, x^k, \qquad 4.32.a$$

the coefficients c_τ^m being the binomial coefficients modulo 2

$$c_\tau^m \equiv C_\tau^m \bmod 2\,, \quad \text{with} \quad C_\tau^m \equiv \frac{\tau!}{m!\,(\tau-m)!}; \qquad 4.32.b$$

(see Fine 1947 and Long 1981 for a discussion of the binomial coefficients modulo a prime). Explicitly, for timesteps $\tau = 2, 3, 4, 5, 6, 7$ we read off from Eq. (4.32)

$$R(x,y)^2 = 1 + y^{-2}(1+x^2),$$

$$R(x,y)^3 = 1 + y^{-1}(1+x) + y^{-2}(1+x^2) + y^{-3}(1+x+x^2+x^3),$$

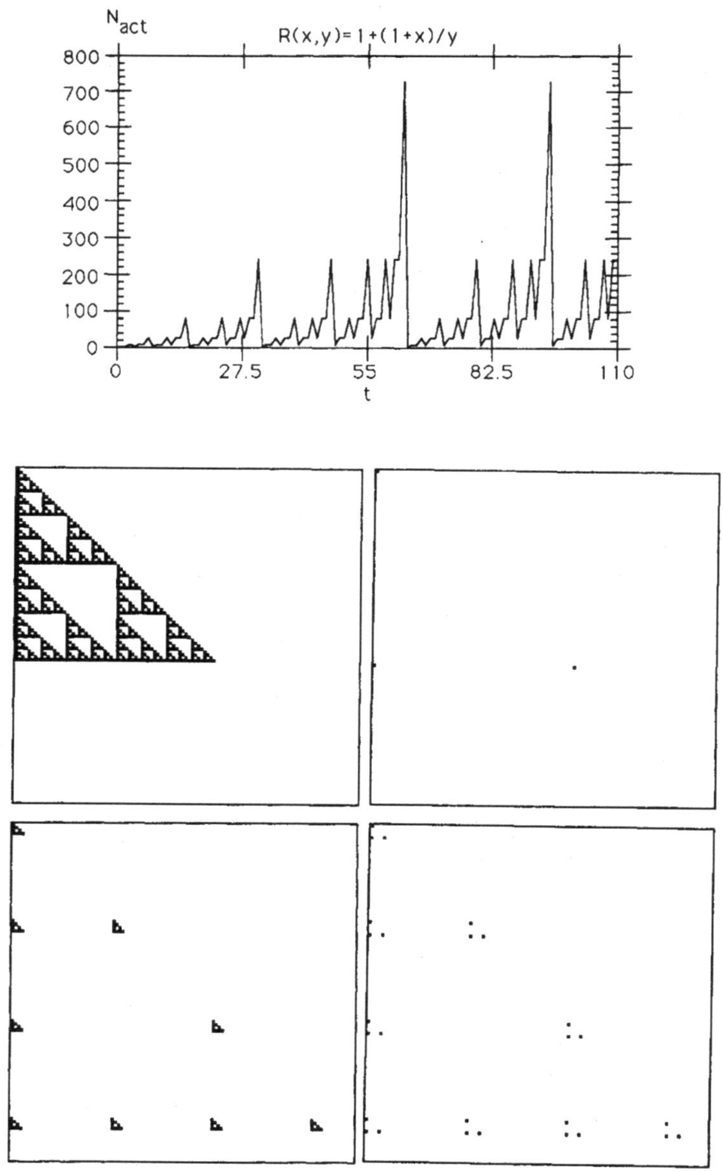

Fig. 3. – 2–state $2D$ CA of rule $R(x, y) = 1 + y^{-1}(1 + x)$. (a) Total number of active cells N against time. (b) Evolutionary snapshots and transitions.

$$R(x,y)^4 = 1+y^{-4}(1+x^4),$$
$$R(x,y)^5 = 1+y^{-1}(1+x)+y^{-4}(1+x^4)+y^{-5}(1+x+x^4+x^5),$$
$$R(x,y)^6 = 1+y^{-2}(1+x^2)+y^{-4}(1+x^4)+y^{-6}(1+x^2+x^4+x^6),$$
$$R(x,y)^7 = 1+y^{-1}(1+x)+y^{-2}(1+x^2)+y^{-3}(1+x+x^2+x^3)+y^{-4}(1+x^4)+$$
$$y^{-5}(1+x+x^4+x^5)+y^{-6}(1+x^2+x^4+x^6)+y^{-7}(1+x+x^2+x^3+x^4+x^5+x^6+x^7),$$
$$\ldots\ldots\ldots \qquad 4.33$$

The number of excited states in the convex hull, $N(L_H;\tau;1)$, is thus 3, 9, 3, 9, 9, 27, ... for $\tau =$ 2, 3, 4, 5, 6, 7, ...
It is found by induction that

$$R(x,y)^h = 1+y^{-h}(1+x^h), \text{ for } h=2^k, k=0,1,2,\ldots. \qquad 4.34$$

For τ ranging over the set T' = $\{2,4,8,\ldots,2^{m-1},2^m,2^{m+1},\ldots\}$ the number of activated cells is thus conserved ($N(L_H;\tau;1)=3$); hence the corresponding exponent F' in Eq. 4.30 vanishes (F' coincides with the topological dimension of the finite point set).

A more interesting situation arises if we consider the configuration of active cells at timesteps τ $\in$ T'' = $\{3,7,\ldots,2^{m-1}-1,2^m-1,2^{m+1}-1,\ldots\}$. The relevant automaton function is explicitly obtained by dividing (4.34) by $R(x,y)$

$$[R(x,y)]^{h-1} = \frac{R(x,y)^h}{R(x,y)} \text{ for } h=2^k,\ k=0,1,2,\ldots,$$
$$= 1+y^{-1}(1+x)+y^{-2}(1+x^2)+y^{-3}(1+x+x^2+x^3)+y^{-4}(1+x^4)+\ldots$$
$$+y^{-q}\sum_{r=0}^{q} c_q^r\, x^r+\ldots$$
$$+y^{-h+1}(1+x+x^2+\ldots+x^{h-1}). \qquad 4.35$$

This equation defines a triangular space structure containing on each row q (= 0,1, ..., $h-1$) a number $\chi(q)=\sum_{r=0}^{q} c_q^r$ of activated cells; the total number of activated cells is thus $N(h-1;1)$ $=\sum_{q=0}^{h-1}\chi(q)$. But this structure is Pascal's triangle modulo 2, which is a well known self-similar fractal; if we choose for L the height of the triangular structure, then for L_o = 4 rows (rows q = 0,1,2,3) we have a triangle containing N_o = 9 excited cells; doubling the size, $L_1 = 2L_o$ = 8 rows (q = 0,1, ...,7), we have a triangle containing $N_1 = 27 = 3N_o$ excited cells; hence we have from Eq. 5.30 $F'' = \frac{\ln 3}{\ln 2} \approx 1.58$. The structures appearing at timesteps $t \in$ T''' = $\{6,14,30,\ldots,2^{m-1}-2,2^m-2,2^{m+1}-2,\ldots\}$ can be studied by the same technique.

Fig. 3 illustrates a few aspects of the growth pattern associated with evolution rule (4.31) (pattern switches, cf timesteps 63 and 64, Fig. 3.a). The actual growth process is properly appreciated only in an *animation* (about 100 frames are needed). Notice also the intermittent behaviour of the total number N of activated cells (Fig. 3.b). From the latter we can estimate an *average mass dimension* $\langle F\rangle \approx 0.77$ (slope of $\log N - \log\tau$ linear regression).

By including more cellstates, we appreciate that the qualitative behaviour of rapid transitions between different spatial structures encountered in the above 2–state automaton is not only preserved, but it may become more differentiated. Different states may dominate the pattern at different timesteps (cf the sum of binomial coefficients mod p). We illustrate this behaviour just numerically, in the case of a z=5-state automaton, of evolution rule

$$R(x,y) = 1 + 2y^{-1} + 3y^{-1}x \qquad 4.36$$

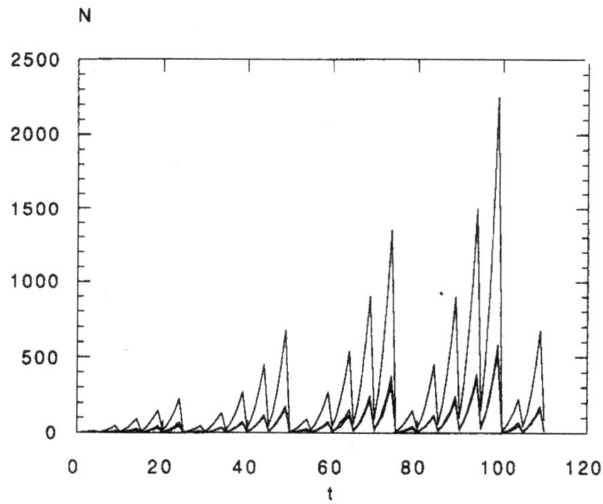

Fig. 4.a. – 5–state $2D$ CA obeying modulo 5 rule (Eq. 4.36). Initial state: single activated cell in state 1. Total number of active cells (upper curve) and numbers of cells in states 1 and 2 (lower curves) against time.

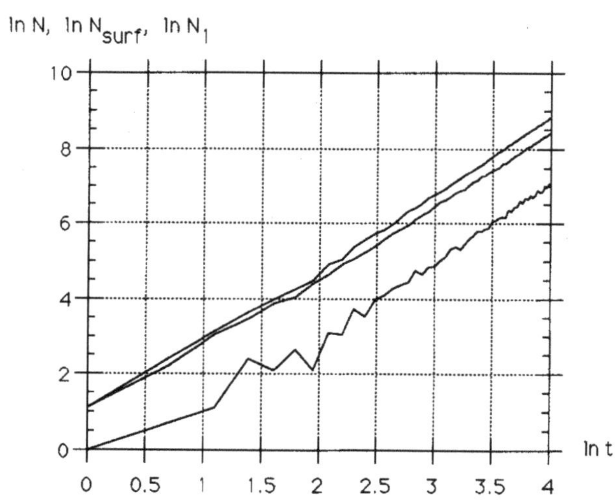

Fig. 5.a – 7–state level 1 2D automaton (complete randomness of level 1 rule). Initial configuration: single activated cell in state 1. Log-Log plot of total number of activated cells (upper curve), number of boundary cells (middle curve), and number of state 1 cells (lower curve) against time.

(Fig. 4), since a detailed analytic discussion (Eqs. 4.26,27) would be tedious. In this particular example the smoothed numbers of individual activated states $N_1(\tau)$, $N_2(\tau)$, $N_3(\tau)$, $N_4(\tau)$ (colour code adopted in Fig 4: state 1: yellow; 2: red; 3: green; 4: cyan) are practically equal, so that the total number of active sites is $\approx 4 \times N_\zeta$, $\zeta = 1, 2, 3$ or4. For the average dimension we find numerically $\langle F \rangle \approx 1$. We mention also that for a stochastic automaton in which the monomials in (4.31) are multiplied by random coefficients r the triangular hull is statistically evenly populated by all states. The intermittent behaviour disappears completely. The corresponding set of activated states then acquires a fractal dimension equal to the topological dimension 2.

These formal examples suffice to demonstrate the natural occurrence of *fractal space patterns* even in simplest CA models. They are indicative that fractal structures must be quite common in realistic CA modelling as well, since formally realistic modelling merely consists in combinations and superpositions of elementary evolution rules.

In the presence of growth processes specified by physically meaningful rules (including stochasticity), Eq. 4.30 can still be be applied to test numerically for the development of fractal structures. In the astrophysical context this has been done in a toy 3–state CA simulation of the distribution of luminous matter in the universe, by Bak and Chen (1989); Bak and Chen obtain fractal distributions of luminous matter of mass dimensions close to 1.6 and 1.7 in a $2D$ and a $3D$ CA respectively; we recall that while these values are significantly larger than the observationally determined dimensions of the distribution of galaxies (≈ 1.2), the simple model has at least the virtue of indicating that a fractal distribution can be accounted for by simple growth models (cf also Vicsek and Szalay 1987 for another CA model). A $2D$ CA model of the distribution of stars in the galaxy by Lejeune and Perdang (1991) suggests likewise that the space distribution of Red Giants is fractal, of dimension ≈ 1.8.

The latter experiments indicate that fractal growth patterns survive in the presence of stochastic evolutionary rules. However, if the stochastic element completely dominates the evolution, pattern formation disappears, and so does most of the fractal behaviour. This point is illustrated in Fig. 5 illustrating the evolution of a 7–state level 1 CA, in which the level 1 rule consists in randomly selecting the automaton rules (4.26) among ***all possible rules*** compatible with the generating function formalism, in the case of a 9–cell Moore neighbourhood. In the realisation of Fig.5 any coefficient $r(\alpha, \beta)$ has probability $\frac{2}{3}$ of being zero, and equal probabilities of taking the values 1,2,3,4). As transpires from the figure, the total number of activated cells, N, as well as the number of cells in a fixed activated state ζ, N_ζ, regularly increase like τ^2. The geometrical set of all active cells, or of cells in a given state of excitation ζ, at any time τ is approximately a disk whose radius (or size factor L_H) grows in proportion to τ and which is statistically uniformly populated. Moreover the density of all 6 states of activation is the same. Hence, by (4.30), the mass dimension is 2 for both the set of all activated cells, and the set of individual states ζ. The boundary of the zone occupied by excited cells keeps an irregular shape, of fractal dimension, estimated by a relation similar to (4.30), $F_b \approx 1.9$ (the number of boundary cells $N_b(\tau)$ at time τ increases like τ^{F_b}, and therefore like ${L_H}^{F_b}$; a boundary cell of the activated zone is understood here as an active cell having at least one empty cell in its neighbourhood). This high value is due to the existence of holes ('inner' boundary cells within the active zone). The 'hull' of the boundary cells has a lower dimension closer to 1.6. We observe in passing that the latter value is reminiscent of the dimension $\frac{7}{4}$ of the propagating hull of the diffusion front encountered in $2D$ diffusion in the presence of a source (cf Sapoval et al 1989). Essentially, complete randomness in the choice of the rules is equivalent with propagation in random directions, i.e. with standard diffusion, with superposed random excitation (statistically uniform creation of states within the occupied zone). In the propagating diffusion front problem creation of particles is concentrated over a fixed subset of the zone only (line or point).

Fractal space–time patterns

Besides fractal space patterns the CA framework supports also fractal space–time patterns (for $1D$ CA first systematically discussed in Wolfram 1983; the analytic discussion in terms of generating functions is introduced in Martin et al 1984).

Consider for definiteness an n-dimensional hypercubic space lattice (of unit meshsize) defined by Z^n. With CA time being again identified with Z (of unit timestep), the corresponding discrete space–time becomes an $(n+1)$–dimensional hypercubic lattice Z^{n+1} (of unit meshsize). If we denote the space–time coordinates by $\mathbf{s} = (s_o, s_1, s_2, \ldots, s_n) \equiv (\tau, x_1, x_2, \ldots, x_n) = (\tau, \mathbf{x})$, then the complete behaviour of the automaton is summarised by a single function defined in space–time, the ***space–time automaton function***, $B(\mathbf{s})$, given in terms of the array of automaton functions $(A(0;\mathbf{x}), A(1;\mathbf{x}), A(2;\mathbf{x}), \ldots) \equiv A(\langle \mathbf{t} \rangle; s_1, s_2, \ldots, s_n)$ at the ordered times $(0,1,2,\ldots) \equiv \mathbf{t}$, by

$$B(\mathbf{s}) = s_o^{\mathbf{t}} \bullet A(\langle \mathbf{t} \rangle; s_1, s_2, \ldots, s_n). \qquad 4.37$$

In the case of a $1D$ CA we can visualise this lattice by a 2–dimensional grid whose horizontal rows of cells (s_o constant) represent the $1D$ space at a fixed time, and whose vertical colums (s_1 constant) show the same space cell at successive times (representation adopted in Fig. 1, where the positive time axis is drawn downwards).

A convenient visualisation for a $2D$ CA is obtained by a stereoscopic display of the 3–dimensional space–time grid, for instance such that the space cells at a fixed time (s_o constant) form a layer parallel to the plane of the screen, while each column of space–time cells of base coinciding with a spatial cell (s_1, s_2 constants), and normal to the screen, displays the same space cell at successive times. Fig. 6 illustrates this type of plot for the growth pattern corresponding to rule (4.36).

The definition of the fractal mass dimension for a space–time pattern is a natural extension of definition (4.30) for a space pattern. Set

$$C(L;\tau;\zeta) = \sum_{t=0}^{\tau} N(L;t;\zeta); \qquad 4.38$$

this quantity represents the cumulated number of cells in state ζ (=1,2,..., or $z-1$), from the initial time 0 up to the current time τ, in an $(n+1)$–dimensional subset of the $(n+1)$–dimensional convex hull of the activated cells (chosen as a homothetic image of the convex active hull, and centred at the origin of space–time) ; L is again the characteristic 'length' (in space or in time, for instance, the largest width of the set) and L_H the corresponding 'length' referring to the convex hull at the current time τ. Suppose then that the total number of cells in state ζ scales with L as

$$C(L;\zeta) \rightarrow \Delta_F(\zeta)\, L^F, \qquad 4.39$$

for any large enough current times τ; the coefficient $\Delta_F(\zeta)$ is required to be finite and nonzero, and independent of the scale L, and likewise F is independent of τ and obeys the inequality $0 \leq F \leq n+1$. If the collection of cells in state ζ distribute uniformly over a fractal set in the space–time lattice, then F represents also the fractal dimension of the latter state. Note that in contrast with relation (4.30), we impose in (4.39) the stronger condition of independence of (asymptotically large) τ.

An illustrative analytically tractable example is provided by the 2–state $1D$ automaton of evolution rule

$$R(x) = 1 + x, \qquad 4.40$$

which describes ***propagation*** (component x), and ***replication*** and multiplication (presence of 2 components in R). For an initial condition at the origin of the $1D$ space, the space–time automaton function (4.37) becomes, with the previous notations

$$B(\mathbf{s}) = s_o^{\mathbf{t}} \bullet [R(s_1)]^{\mathbf{t}} = \sum_{t=o}^{\tau} s_o^t \sum_{m=0}^{t} c_t^m s_1{}^m$$

$$= 1 + s_o(1+s_1) + s_o{}^2(1+s_1{}^2) + s_o{}^3(1+s_1+s_1{}^2+s_1{}^3) + \ldots, \qquad 4.41$$

which is seen to be isomorphic with the $2D$ function $R(x,y)^{h-1}$ in Eq. (4.35). Therefore the space–time pattern obtained by rule (4.40) has the same fractal dimension as the space pattern of the $2D$ CA defined by rule (4.31), or $F = \frac{\ln 3}{\ln 2}$.

This example makes it clear that to any nD CA with fractal space pattern there corresponds an $(n-1)D$ CA with fractal space–time pattern. The ubiquity of fractal space patterns in CA models then implies in turn that fractal space–time patterns must be common features in the CA framework as well.

Fractal time behaviour

Finally, we indicate that CA models, in contrast with models defined by differential equations, can generate fractal time behaviour. This property logically flows from the existence of fractal space–time patterns : If F is the fractal dimension of a pattern in the $(n+1)$–dimensional space–time, then a generic intersection of this structure by a line will produce a set of dimension

$$f = F - n \leq 1 \qquad 4.42$$

(if $f \geq 0$ then the intersection is generically non–empty; see Perdang 1991a for an interpretation of negative dimensions). To the extent that the 1–dimensional time axis then behaves generically, we expect that a fractal space–time pattern implies a fractal time behaviour of dimension given by (4.42).

Explicitly, and regardless of genericity of the time axis, the fractal dimension of the time pattern is again obtained from a relation similar to (4.30). Let $N(L;\mathbf{i};\zeta)$ represent the number of times space cell $\mathbf{i}$ is in state ζ over the scale L (a time interval in the present case). If this measure obeys the scaling law

$$N(L;\mathbf{i};\zeta) \rightarrow \nu_f(\mathbf{i};\zeta)\, L^{f(\mathbf{i};\zeta)}, \quad \text{for } L \text{ large enough}, \qquad 4.43$$

with $\nu_f(L;\mathbf{i};\zeta)$ finite and nonzero, for some $f = f(\mathbf{i};\zeta) \leq 1$, then f defines the mass dimension of the time behaviour of cell $\mathbf{i}$; if this time structure is a fractal set, then $f < 1$. Note that this dimension depends, in principle, on the space cell selected. There is no guarantee that the time–axis (defined by the column of space–time cells) through a given space cell (the initial condition, i.e. a space cell at time 0) is generic, nor is there any guarantee that the axis so defined through *any* cell is generic, since a time–axis necessarily has a particular orientation in space–time. Space–dependence (cell–dependence) of the fractal dimensionality of a CA time behaviour is the counterpart of time–dependence of the fractal dimension of space patterns in CA evolution.

In the case of the 2–state $2D$ automaton of evolution rule (4.31) we read off from Eq. (4.34) that the origin remains in the activated state 1 at any time; hence $N(L;0,0;1) \propto L$ i.e. $f = 1$ (differing from the value suggested by Eq. 4.42). In the 2–state $1D$ CA defined by (4.40) the origin is activated at all times, i.e. it has a period–1 ($P1$) time behaviour; cell $i = 1$ is alternatively activated and inert, i.e. it shows a period–2 ($P2$) behaviour; cell $i = 3$ has the time pattern 00110011..., showing $P2^2$ behaviour, etc; in all of these cases $N(L;i;1) \propto L$, or f=1. The non–fractional value of f is indicative of the *regularity* of the time behaviour.

On the other hand, if we modify rule (4.40) setting

$$R'(x) = x + x^{-1}, \qquad 4.40.a$$

we generate a space–time pattern obtained from the space–time pattern of the previous example by a shear–like deformation, which does not modify the fractal dimension F of the space–time pattern; however now the origin is activated exclusively at times 0,3,7,..., $2^{m-1}-1$, 2^m-1, $2^{m+1}-1$, ..., m, integer. It follows that $f = 0$ for the cell at the origin, $i = 0$ (while 4.43 predicts $\frac{\ln 3}{\ln 2} - 1$).

These two examples suggest that a space–time pattern of fractal dimension F may actually generate a time pattern of *any* fractal dimension $0 \leq f \leq 1$, the reason being that lines of different orientations through a given cell, for instance the cell at the spatial origin at the initial time, intersecting the space–time patterns, may generate completely different structures. To recover

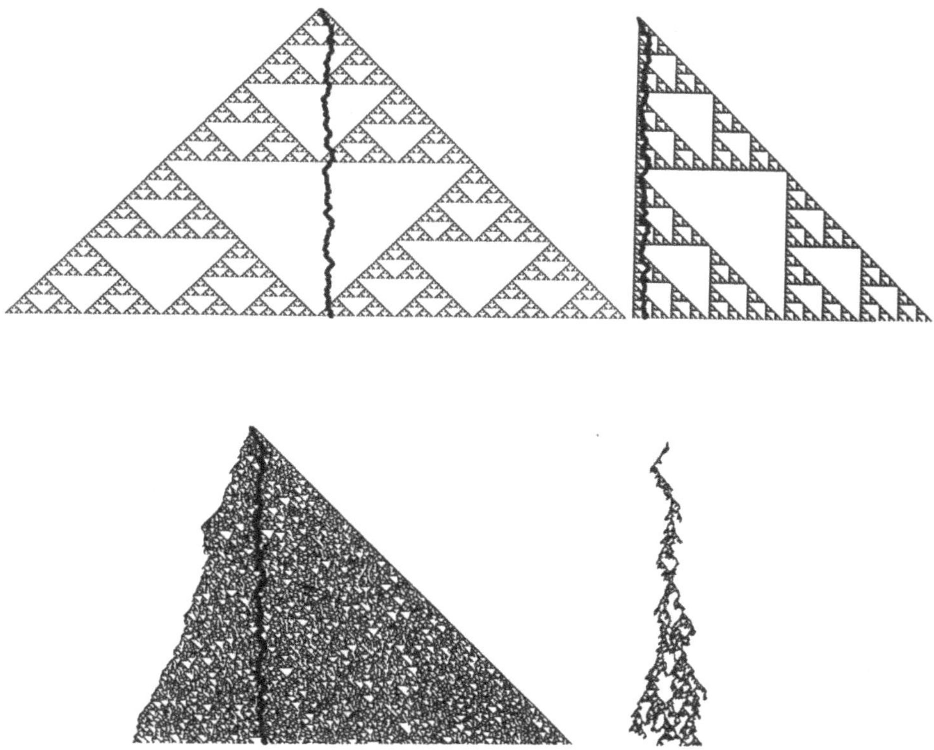

Fig. 7. – Space–time patterns of 2–state $1D$ deterministic automata corresponding to rules $R(x)$, and $R'(x)$ and to stochastic automata with mixed rules $R(x)$ and $R'(x)$, and $R(x)$ and $R''(x)$

genericity, we have chosen a formal time axis through a collection of cells connected by a random walk in space (heavy lines in Figs. 7; physically we consider the time–behaviour of different cells at different times). From several runs over 256 timesteps we have estimated fractal dimensions of the time–behaviour in the range 0.39 –0.82; these values are indeed in rough agreement with Eq. (4.42).

The lack of genericity of the true time axis in the above examples appears to be related to the high degree of geometric regularity of the space–time pattern. One may therefore expect that by introducing a stochastic element in the automaton rules genericity of the time axis may be restored. While this is essentially true, a stochastic level 1 rule operating on simple level 0 rules may lead to entirely new space–time patterns. This is demonstrated in Fig. 7 showing the space–time patterns of two 2–state $1D$ level 1 CA of automaton rules $R(x)$ (Eq.4.40), and $r(x) = R'(x)$ (Eq. 4.40.a) or $r(x) = R''(x) = 1 + x^{-1}$ respectively, the meta–rule in both cases being

$$R^{(1)} : \Big(R(x) \rightarrow (R(x) \vee r(x)) \Big) \wedge \Big(r(x) \rightarrow (R(x) \vee r(x)) \Big), \text{ with probability } \frac{1}{2}$$

Note that level 0 CA operating on the individual rules, $R(x), R'(x), R''(x)$ all generate the same space–time pattern. As is seen from the figure, in the $R(x), R'(x)$ stochastic automaton we obtain a compact approximately uniform distribution of activated states (the large triangular holes associated with the deterministic CA have disappeared). The mass dimension of the space–time pattern estimated from 256–step runs is always close to 2; the mass dimension of the time pattern measured along the true time axis through the space–time origin, or along a 'random walk path' (heavy lines) is always close to1 (in agreement with 4.42). In contrast, in the case of the $R(x), R''(x)$ automaton, the space–time pattern acquires a branching filamentary structure, which, in some of our experiments had a short lifetime (<200 timesteps). The formal space–time dimension as derived from Eq. (4.39), and the corresponding dimension of the time–pattern, of the cell at the spatial origin, or of a 'random walk path', fluctuate strongly from one realisation to the other, so that they do not seem to be meaningful characterisations of this type of structure.

> ... an understanding of the physical laws does not necessarily give you an understanding of things of significance in the world in any direct way. The details of real experience are often very far from the fundamental laws.
>
> R. Feynman 1967

5. Transition Matrix Formalism

A relevant technique allowing for a theoretical discussion of the general aspects of the behaviour of an automaton consists in mapping the *totality of possible automaton configurations*, onto an array, or *vector* F, with the evolution rules encoded in a *transition matrix* T acting in the space of configurations. This approach relies on algebraic methods (substitutional analysis, combinatorial methods, Markov matrix theory) to classify the allowed time patterns, and to estimate their relative frequencies. The matrix representation in turn can be interpreted in terms of graph theory (discussed in Robert 1986; Ito 1988), thereby providing a more intuitive support to the abstract algebraic formalism.

In contrast to the generating function technique which is efficiently applicable to a subset of the actually admissible CA evolutions only, namely under the conditions of evolution rules themselves expressible as automaton functions, the transition matrix description remains valid for a class of evolutions broader than permitted in definition D. To avoid notational complications we adopt the standard language, unless full precision of a statement requires a nonstandard formulation. Since we shall not refer to continuous representations in this chapter, we shall use the symbol t to denote discrete time.

We shall discuss the transition matrix technique in the context of level 0 automata. A shorter analysis of the extension to level 1 automata obeying a stochastic meta–rule will be presented in the final part of this chapter.

Phase space representation

The space cells, originally labelled by the vector index $\mathbf{i}$, will be regarded here as ordered according to some preselected scheme, such that we can relabel them by a scalar integer index i : -1,0,1,... For a finite number f of cells (case of toroidal topologies for instance) we conveniently number these sites by $0,1,...,f-1$. The *field coordinate* at site i is then denoted by $F(i) \in z \equiv Z_{/z}$. The counterpart of the automaton function in the previous generating function formalism is here the ordered array of field coordinates

$$\mathbf{F} = \Big(\ldots, F(-1), F(0), F(1), \ldots\Big), \qquad 5.1$$

which encodes a CA configuration.

Geometrically the vector $\mathbf{F}$ specifies a *phase point* (or a cell) of coordinates $\ldots, F(-1), F(0), F(1), \ldots$ in a (discrete) *phase space* Γ . The geometric representation bears out more clearly the duality between an ODE and a CA : The dependent variables of the ODE are the counterparts of the field coordinates; the number of dependent variables, or phase space dimension, of an ODE written in the form of a first order system, is the counterpart of the number of space–lattice cells.

Assume further that the *configurations are* also *ordered* according to some preselected scheme, so that we can label them by a scalar subscript k , $\mathbf{F}_k$: $\ldots, \mathbf{F}_{-1}, \mathbf{F}_0, \mathbf{F}_1, \ldots$. Then we denote by

$$F = (\ldots, \mathbf{F}_{-1}, \mathbf{F}_0, \mathbf{F}_1, \ldots) \equiv F^{(o)}, \quad \mathbf{F}_k \in \Gamma = \{\ldots, \mathbf{F}_{-1}, \mathbf{F}_0, \mathbf{F}_1, \ldots\} \qquad 5.2$$

the ordered array, regarded as a *column vector*, of *all* possible *configurations* of the CA (Γ being the set of all possible configurations); we refer to component $\mathbf{F}_k$ as the kth component of the vector of all configurations, or briefly, as *configuration k*, or *phase point k*. More generally, we represent by

$$\Phi = (\ldots, \mathbf{F}_q, \mathbf{F}_r, \mathbf{F}_s, \ldots), \quad \ldots, \mathbf{F}_q, \mathbf{F}_r, \mathbf{F}_s, \ldots \in \Gamma' \subset \Gamma, \qquad 5.2.a$$

any vector of same dimension as F (in the sense that the components of Φ can be paired off with those of F) whose components $\mathbf{F}_q$ represent again configurations.

The automaton evolution rule, R, is a mapping of the phase points into phase points, $R : \Gamma \to \Gamma$ (a *surjection*). Adopting the notation for substitutions, the rule is symbolised by

$$R \; : \; \begin{pmatrix} \cdots & \mathbf{F}_{-1} & \mathbf{F}_0 & \mathbf{F}_1 & \cdots \\ \cdots & \mathbf{F}_a & \mathbf{F}_b & \mathbf{F}_c & \cdots \end{pmatrix} \quad \text{or} \quad R \; : \; F^{(o)} \to F^{(1)} = (\ldots, \mathbf{F}_a, \mathbf{F}_b, \mathbf{F}_c, \ldots), \qquad 5.3$$

(phase point $\mathbf{F}_{-1}$ transforms into phase point $\mathbf{F}_a$, phase point $\mathbf{F}_o$ becomes $\mathbf{F}_b$, phase point $\mathbf{F}_1$ becomes $\mathbf{F}_c$, ...). In the second expression (5.3) $F^{(1)}$ denotes the vector of *first successor configurations* of $F^{(o)}$; it is a column vector of type (5.2.a) whose k^{th} component, ${F^{(1)}}_k$, (identical with some component $\mathbf{F}_l$ of F) is the successor configuration of configuration k (k^{th} component of the original vector F); in turn $\mathbf{F}_k$ is a *predecessor* of ${F^{(1)}}_k$; the *set* of all distinct components of the vector $F^{(1)}$ will be denoted $\Gamma^{(1)} \subset \Gamma$. In a similar way we represent by $F^{(2)}$ the vector of second successor configurations of $F^{(o)}$, and more generally, by $F^{(t)}$ the vector of tth successor configurations of $F^{(o)}$, and by $\Gamma^{(t)}$ the set of the distinct components of this vector. The relative loss of the number of phase cells after one iteration,

$$L = \frac{\text{card}\ (\ \Gamma\ )\ -\ \text{card}\ (\ \Gamma^{(1)}\ )}{\text{card}\ (\ \Gamma\ )},$$

if it exists, is an indicator of the degree of *irreversibility* of the CA; $L = 0$ corresponds to reversibility (bijection of the rule) (cf a similar characterisation of the contraction rate of the phase space in Hogg and Huberman 1985).

The components of the vectors $F^{(t)}$ are read off as usual from the substitutional notation (5.3). If α is an initial configuration (component $\mathbf{F}_\alpha$ of F), then the substitutional form of the evolution rule (5.3) supplies the sequence of successor configurations, $\alpha \to \beta \to \ldots \to \theta \to \iota \to \ldots$, symbolised by the string $(\alpha, \beta, \ldots, \theta, \iota, \ldots)$. In the phase space this ordered sequence represents an *orbit* of the CA, the ordering being achieved by the discrete time variable. Since a (unique) successor configuration exists for any configuration, the initial configuration α generates an orbit whose string symbol $(\alpha, \beta, \ldots, \theta, \iota, \ldots)$ involves an infinity of terms. If one configuration is repeated then the string symbol contains an indefinitely repeating substring. Let θ be the first repetitive configuration encountered in the orbit; the string then takes the form $(\alpha, \beta, \ldots, \theta, \iota, \ldots, \kappa, \theta, \iota, \ldots, \kappa, \theta, \iota, \ldots, \kappa, \theta, \iota, \ldots)$; denoting the repetitive fragment by $[\theta, \iota, \ldots, \kappa]$, we conveniently write the string symbol for the orbit $(\alpha, \beta, \ldots]\ [\theta, \iota, \ldots, \kappa]$. An orbit with a repetitive fragment is *closed.*

In general the rule R will be such that the set of configurations Γ can be partitioned into subsets $\Gamma_{(1)}, \Gamma_{(2)}, \ldots$

$$\Gamma = \Gamma_{(1)} \cup \Gamma_{(2)} \cup \ldots \cup \Gamma_{(g)} \cup \ldots \qquad 5.4$$

such that the configurations in any subset $\Gamma_{(g)}$ of the partition, $g = 1, 2, \ldots$, transform among themselves, (i.e. the subsets $\Gamma_{(g)}$ are ***invariant sets*** under the rule R), and such that (5.4) defines the 'finest' partition obeying this property. Similarly we then have, for any $t = 1, 2, \ldots$

$$\Gamma^{(t)} = \Gamma_{(1)}{}^{(t)} \cup \Gamma_{(2)}{}^{(t)} \cup \ldots \cup \Gamma_{(g)}{}^{(t)} \cup \ldots \qquad 5.4.a$$

The vector of all configurations $F^{(o)}$ is then a direct sum of vectors

$$F^{(o)} = F_{(1)}{}^{(o)} \oplus F_{(2)}{}^{(o)} \oplus \ldots \oplus F_{(g)}{}^{(o)} \oplus \ldots \qquad 5.5$$

such that the configurations of any vector $F_{(g)}{}^{(o)}$, with components in $\Gamma_{(g)}$, $g = 1,2, \ldots$, transform among themselves. Accordingly

$$F^{(t)} = F_{(1)}{}^{(t)} \oplus F_{(2)}{}^{(t)} \oplus \ldots \oplus F_{(g)}{}^{(t)} \oplus \ldots, \quad t = 1, 2, \ldots \qquad 5.5.a$$

with $F_{(g)}{}^{(t)}$ having again its components in $\Gamma_{(g)}{}^{(t)}$.

In matrix notation, the automaton rule is summarised by

$$T.F^{(o)} = F^{(1)}, \qquad 5.6$$

$$T.F^{(t-1)} = F^{(t)}, \ t = 1, 2, \ldots, \qquad 5.6.a$$

and hence

$$F^{(t)} = T^t.F^{(o)}. \qquad 5.6.b$$

The matrix T is the ***transition matrix*** whose row-vectors have one component equal to 1, the remaining components being zero. If row vector r has as nonzero element the component labelled c , ($T_{rc} = 1$, $T_{rk} = 0$ if $k \neq c$) then configuration $\mathbf{F}_r$ transforms into configuration $\mathbf{F}_c$. Under partition (5.4) the transition matrix splits up into diagonal blocks

$$T = T_{(1)} \oplus T_{(2)} \oplus \ldots \oplus T_{(g)} \oplus \ldots \qquad 5.7$$

such that $T_{(g)}$ acts on $F_{(g)}{}^{(o)}$ and its successors $F_{(g)}{}^{(t)}$, $t = 1,2,\ldots$

Minimal sets and eventual CA time behaviour

The sets of configurations $\Gamma_{(g)}{}^{(t)}$ obey the inclusions

$$\Gamma_{(g)} \supset \Gamma_{(g)}{}^{(1)} \supset \Gamma_{(g)}{}^{(2)} \supset \ldots \supset \Gamma_{(g)}{}^{(t)} \supset \ldots. \qquad 5.8$$

In terms of orbits in phase space relations (5.8) mean that if the initial point of the orbit $(\alpha, \beta, \ldots, \theta, \iota, \ldots)$, α , is a configuration in $\Gamma_{(g)}$, then the point β reached at timestep 1 is in $\Gamma_{(g)}{}^{(1)}$, etc. If Θ is the least integer (possibly infinite) such that

$$\forall t \geq \Theta \ : \ \Gamma_{(g)}{}^{(t)} = \Gamma_{(g)}{}^{(\Theta)} \equiv A_{(g)}, \qquad 5.8.a$$

then $A_{(g)}$ is a *minimal set* of the automaton.

This definition (essentially equivalent to the definition given in Nemytskii and Stepanov 1960) translated into the nonstandard form, shows that in the context of our CA definition *D a least nonstandard integer $*\Theta$ does always exist. The corresponding physical 'relaxation time', st ($*\Theta\Delta$ t, may be either finite, or it may be infinite (i.e. it may not exist); the latter alternative occurs in ODEs where the limit sets are asymptotically reached (cf Nemytskii and Stepanov 1960). Through the sleight of hand of excluding noncompact nonstandard manifolds as space lattice supports (cf definition *D), we always guarantee the existence of a total number of configurations, *K (as a nonstandard integer), so that all vectors or matrices defined in this section can be labelled by nonstandard integers. Moreover the nonstandard irreversibility index *L always exists and has a standard part in [0,1]. In the nonstandard version, since the nonstandard phase space is necessarily compact, the existence of a (nonstandard) compact minimal set follows from the theorem of nested compacts.

Under this condition, let θ be the first point of the orbit $(\alpha, \beta, \ldots, \delta, \theta, \iota, \ldots)$ lying on the minimal set $A_{(g)}$. Then the remainder of the orbit, itself an orbit generated by the initial configuration θ, $(\theta, \iota, \ldots)$, is carried by $A_{(g)}$. But by construction the latter set is not further decomposable into proper subsets transforming among themselves under the CA evolution rule; on the minimal set the evolution rule is thus bijective (it reduces to cyclic permutations of the configurations); we may say that this latter suborbit, conveniently denoted $[\theta, \iota, \ldots]$ 'covers' the minimal set $A_{(g)}$, in the sense that the set of points of the suborbit together with its limit points, is identical with $A_{(g)}$.

Again this specification becomes intuitively clearer in the nonstandard setting; there exists always a first configuration θ of a nonstandard sequence of configurations, at timestep $*\Theta$; the 'covering' then means that if ω is any configuration on the nonstandard minimal set $*A_{(g)}$, then there exists a nonstandard time t_ω such that the configuration ω is reached at step t_ω; the corresponding physical time, st $(t_\omega \Delta t)$, may be finite, or it may not exist; under the latter alternative the actual covering is never achieved.

The complementary set in $\Gamma_{(g)}$, $B_{(g)} = \Gamma_{(g)} \setminus A_{(g)}$, is the *basin of attraction* of the minimal set $A_{(g)}$. A configuration γ of the basin of attraction thus represents a *transient* state of the CA. The transient fragment of an orbit $(\alpha, \beta, \ldots, \delta][\theta, \iota, \ldots]$, is the ordered collection of transient states $(\alpha, \beta, \ldots, \delta]$ of the orbit, lying in $B_{(g)}$; the last configuration of the transient fragment, δ, is a predecessor of a configuration of the minimal set. The *relaxation time* Θ to the eventual minimal set evolution, for an initial state α, is equal to the length of the transient fragment of the orbit of initial point α.

The CA solution on the minimal set $A_{(g)}$ corresponds to a *periodic behaviour*, of period P_g, provided that the cardinality of the minimal set is finite

$$P_g = \text{card } (A_{(g)}). \tag{5.9}$$

To illustrate the nature of CA minimal sets consider the example of the 1D gas in translational motion (Eq. 4.17 in the generating function formalism) defined over the infinite line (infinite number of cells). For this CA the empty configuration (no 'activated' cell) defines a minimal set of cardinality 1; likewise, the configuration in which all cells are excited specifies a minimal set of cardinality 1; more generally, any configuration with a spatially periodic distribution of excited cells, of (finite) space period P (and therefore containing an infinite number of excited cells), is an element of a minimal set of cardinality P. On the other hand, for any initial configuration α involving a finite number of particles any iterate of the translation generates a configuration differing from any previous iterate; there is no finite integer P obeying (5.9); the minimal set itself is here made up of the collection of all successors as well as all predecessors of the configuration α. Alternatively, if the translational motion is defined over a circle, we have a finite total number f of cells, so that all configurations repeat; all minimal sets have finite cardinality.

A minimal set of cardinality 1 (periodic solution of period $P_g = 1$) is a fixed point of (5.6), or *equilibrium configuration*; the corresponding periodic suborbit is $[\theta]$. If $P_g = P$, a finite integer, the evolution on the minimal set is a period-P solution; the corresponding (closed) suborbit, of symbol $[\theta, \iota, \ldots, \kappa]$ has length P. If the cardinality of the minimal set is not finite, the eventual CA evolution is not periodic.

The nonstandard formulation supplies a substantially finer classification of the evolution on the (nonstandard) minimal set, $*A_{(g)}$. This classification relies on the behaviour of the *nonstandard* *period* of the evolution – whose existence is always guaranteed

$$*P_g \; = \; *\mathrm{card}(*A_{(g)}). \tag{5.9.a}$$

(a) For st $(*P_g)$ finite , and hence for a 'physical period' P = st $(*P_g\Delta t) = 0$, we have a *macroscopic equilibrium*.

(b) If st $(*P_g)$ does not exist, but P = st $(*P_g\Delta t)$ is finite, then the solution is *macroscopically periodic*.

(c) If st $(*P_g)$ does not exist, but there is a finite positive power m such that st $(*P_g\Delta t^m)$ exists, then the solution is referred to as *regular*.

(d) If st $(*P_g\Delta t^m)$ does not exist for any finite power m, but there is a pair of finite reals r, s such that st $(*P_g \exp -\frac{r}{\Delta t^s})$ exists, then the time behaviour corresponds to *deterministic chaos*.

(e) If st $(*P_g \exp -\frac{r}{\Delta t^s})$ does not exist for any pair of finite reals r, s, then we refer to the corresponding behaviour as *extended* or *generalised chaos*.

Continuing this scheme, we can break up the regime of extended chaos into a hierarchy of sub-regimes, differentiated by the the order k of the 'exponential tower',

$$\exp_k(-\frac{r}{\Delta t^s}) \equiv \exp\;(\;\exp\;(\;\ldots\exp\;(\;-\frac{r}{\Delta t^s})\ldots)\;),$$

needed to secure the existence of st $(*P_g \exp_k - \frac{r}{\Delta t^s})$.

We show that this classification is in agreement with the conventional terminology of the theory of dynamical systems.

A standard regular motion being a multiperiodic oscillation, consider for illustrative purposes a double periodic oscillation of incommensurate physical periods P_1, P_2 (finite reals). In nonstandard notation we can always set P_j = st $(*P_j)$, with $*P_j = \frac{*n_j}{*m_j}$, $j = 1,2$, where $*n_j, *m_j$ are nonstandard infinite integers (cf Robinson 1980); rewrite the latter in the form $\frac{*\nu_j}{\Delta t}$, $\frac{*\mu_j}{\Delta t}$, where $*\nu$, $*\mu$ are nonstandard reals of *finite* standard part, Δt is as usual the *infinitesimal* timestep. Hence, this motion is nonstandard periodic, with period (cardinality of the carrier of this solution $*\mathsf{A}_{(reg)}$)

$$*\mathcal{P}_{reg} \; = \; *n_2 * m_1 * P_1 \; = \; *n_1 * m_2 * P_2 \; = \; \frac{*a}{\Delta t^2}, \tag{5.10}$$

$*a$ a nonstandard real of *finite* standard part. Accordingly, criterion (c) consistently classifies this solution as regular.

On the other hand a conventional chaotic oscillation generated by the period doubling route has by definition a nonstandard period $P_o 2^{*n}$, with $*n$ an infinit; take the latter again of the form $\frac{*\nu}{\Delta t}$, $*\nu$ a nonstandard real of *finite* standard part; the corresponding nonstandard cardinality of the carrier $*\mathsf{A}_{(ch)}$ now becomes

$$*\mathcal{P}_{ch} \; = \; P_o 2^{\frac{*\nu}{\Delta t}} \; = \; *A \exp\frac{r}{\Delta t}, \tag{5.11}$$

$*A$ and r finite reals. Hence criterion (d) identifies this solution as deterministic chaos.

Our discussion thus shows that the CA formalism (in its nonstandard form) supports the types of time behaviour (a)–(d) encountered in systems of ODEs; class (e), referred to as extended chaos, not represented in differential systems of finite order, is allowed in PDEs (or systems of an infinite number of coupled ODEs).

This theoretical classification of general CA behaviour accommodates the more specific classification proposed by Wolfram (1983), and based on numerical experiments, under the following restrictions. On the one hand, the neighbourhood requirements of definition D (finite number f' of cells in spatial neighbourhood $\mathsf{N}(\mathbf{i})$) are respected, and on the other hand, further 'legality' requirements are imposed on the evolution rules in the following form : (a) von Neumann's *condition of quiescence* is assumed to hold, stating that if all neigbourhood cells of cell $\mathbf{i}$ are in a 'quiescent', or 'inert', state (field variable 0) at time t, then cell $\mathbf{i}$ remains quiescent at time t+1; and (b) ***spatial symmetry*** is imposed, requiring that the rule is invariant under the lattice symmetries. As a consequence, the total number k of evolution rules remains finite for a noncompact space manifold supporting the CA lattice.

Wolfram has carried out an exhaustive experimental and partially analytic survey of the 2–state $1D$ CA, ($F \in Z_{/2} = \{ 0,1\} \equiv N(F)$; 0, quiescent state; 1, activated state; spatial neighbourhood $N(i) = \{ i-1, i, i+1 \}$, $i \in Z$), subject to legality. He isolates 4 ***universality classes*** of increasing complexity (formalised in Culik and Yu 1988; cf also Culik et al 1990 where other classifications are mentioned). Characterised by their asymptotic behaviour, i.e. the minimal set behaviour as in the above general scheme, Wolfram's classes are : (1) asymptotic time independence; (2) asymptotic periodicity; (3) chaos; (4) other.

Class (1) and (3) behaviour correspond to the above categories (a) and (d) respectively; class (2) includes categories (b) and (c), regularity being viewed as a multi–periodicity with 'weak' rational dependence (in the sense that a double–periodic behaviour of the CA, of periods P_1, P_2 implies that there exist two 'very large' least integers m_1, m_2 such that $\frac{P_1}{P_2} = \frac{m_1}{m_2}$). Class (4) is described by Wolfram as asymptotic evolution to 'complex localised structures, sometimes long–lived'; in a subsequent paper (1984) class 4 behaviour is characterised as effectively unpredictible : this 'behaviour may be determined only by explicit simulation'. We mention here that the illustrations in Wolfram's papers referring to this behaviour are strongly reminiscent of the space–time patterns of the stochastic CA of component rules $R(x)$, $R''(x)$ (Fig. 7).

As observed in Chaté and Manneville (1990), a distinction between Wolfram's class (3) and (4) behaviour is 'difficult and probably impossible in certain cases'. Theoretically, the classification based on nonstandard CA provides a clear–cut partition between both types of behaviour.

Wolfram's experiments also show that in the $2D$ lattice space–time it is class (3) behaviour that generates ***fractal space–time patterns*** (cf our discussion in chapter 4; initial configuration : single excited cell) and ***spatio–temporal chaos*** (cf also Oppo and Kapral 1986 for an analysis of spatio-temporal chaos).

Extending his experimental survey to larger neighbourhoods and larger numbers of states per site, and to higher–dimensional CA, Wolfram (1984) observes a dominance of the number of 'legal' CA evolution rules, $\rho_{3,4}{}^{(legal)}(n, f', z)$, leading to 'complex' dynamics (class (3) and (4) behaviour), for a given lattice dimension n, a given number z of cellstates, and a given number of neighbourhood cells f'. Extrapolating these results we are led to conjecture that the fraction of 'legal' rules generating complex behaviour obeys

$$\frac{\rho_{3,4}{}^{(legal)}(n, f', z)}{\rho^{(legal)}(n, f', z)} \longrightarrow 1, \quad \text{if} \quad f' \longrightarrow \infty, \quad n \quad \text{fixed} \qquad 5.12$$

($\rho^{(legal)}(n, f', z)$, total number of legal rules). Accordingly, if we pick at random a 'legal' CA evolution rule, our odds of hitting a rule generating 'trivial' space–time behaviour of class (1) or (2) become progressively slimmer as we enlarge the spatial neighbourhood. In other words, in the CA environment 'enormously complex and intriguing behaviour' (Gutowitz 1990), becomes the rule rather than the exception, once we go beyond the elementary 2–state$1D$ CA with nearest neighbours. A combinatorial justification of this conjecture is given below.

Category (e) behaviour encompasses the concept of ***self–organised criticality*** as introduced more recently by Bak et al (1988; cf Bak 1993). Self–organised criticality refers to an instability leading to the activation of a 'large' number of degrees of freedom. This contrasts with conventional instabilities — including deterministic chaos — which involve small numbers of degrees of freedom.

Suppose first that we have a large enough number f of excited degrees of freedom, each one oscillating with a (finite) period P_j, j = 1,2, ... ,f, rationally independent of the remaining periods. Proceeding as above in the nonstandard language (cf Eqs. 5.10), and regarding f as an infinit, $*f$, the totality of degrees of freedom exhibits a global nonstandard period

$$*\mathcal{P}_{soc} = *m_1 * n_2 * n_3 \ldots * n_{*f} * P_1 = *n_1 * m_2 * n_3 \ldots * n_{*f} * P_2 = \ldots$$

$$= *n_1 * n_2 * n_3 \ldots * n_{*f-1} * m_{*f} * P_{*f} = \frac{*a}{\Delta t^{*f}} \qquad 5.13$$

($*a$ a nonstandard real with finite standard part); criterion (e) is then violated if we set $\frac{*b}{\Delta t} = *\omega$, and $*f = \exp *c * \omega$ ($*b$, $*c$ nonstandard reals with standard finite parts); this latter choice is adequate for a CA whose lattice is carried by a nonstandard compact manifold (cf also below). In the second place, we mention that a stronger type of irregularity is associated with self–organised criticality if we interpret this phenomenon as a superposition of an infinite number of *chaotic* pulsations (use Eq. 5.11 for the individual nonstandard periods and proceed as above).

If the evolution rule (5.4) is a genuine substitution or permutation (transition matrix whose row (or column) vectors are all different) then the CA is ***reversible***, every configuration having exactly one predecessor. For a reversible CA we have $\Gamma_{(g)} \equiv \mathsf{A}_{(g)}$. If the evolution rule is not a permutation (transition matrix with linearly dependent row vectors), the CA is ***irreversible*** or ***dissipative***.

Stability

Any model designed to simulate a natural phenomenon is a simplified idealisation of the real system in nature, from which it chops off an interacting 'environment', assuming the interaction is 'small'. Accordingly, in order to be observationally meaningful, the model must obey a condition of robustness : Solutions of the model are actually realisable in the natural phenomenon only to the extent that they are ***stable*** with respect to small perturbations (namely the disturbances generated by the discarded 'environment'). We discuss here two among the variety of stability concepts introduced in the theory of differential equations (cf Nemytskii and Stepanov 1960) which translate directly into the discrete CA context.

The notion of a ***small disturbance*** of a configuration $\mathbf{F}$ of a CA is made precise in terms of a specification of a neighbourhood of a cell in the phase space. The definitions of a neighbourhood of a spatial cell $\mathbf{i}$, $\mathsf{N}(\mathbf{i})$, and of a cellstate F, $\mathsf{N}(F)$, (cf D) induce a natural neighbourhood of a configuration $\mathbf{F}$ in the phase space. A configuration $\mathbf{F}'$ is a neighbour configuration of $\mathbf{F}$, or $\mathbf{F}' \in \mathsf{N}(\mathbf{F})$, if for any cell $\mathbf{i}$, in cellstate $\Phi = F(\mathbf{i})$ (component $\mathbf{i}$ of configuration $\mathbf{F}$; we omit here the time parameter) there is a cell $\mathbf{i}' \in \mathsf{N}(\mathbf{i})$, such that $\Phi' = F(\mathbf{i}')$ (component $\mathbf{i}'$ of configuration $\mathbf{F}'$) obeys $\Phi' \in \mathsf{N}(\Phi)$.

Any neighbour configuration $\mathbf{F}'$ of a configuration $\mathbf{F}$ will be referred to as a small disturbance of $\mathbf{F}$.

By ***neighbourhood of a set*** S ***of configurations*** in the phase space Γ, $\mathsf{N}(\mathsf{S})$, we then naturally understand the set made up of S itself and the union with the neighbourhoods of all of its elements,

$$N(S) = S \cup_{\mathbf{F} \in S} N(\mathbf{F}). \qquad 5.14$$

This construction of the neighbourhood of a set is similar to the operation of dilation of a set S by another set **s** (here the neighbourhood set of a configuration), as adopted in image analysis (Serra 1988).

We now have a natural definition of stability with respect to small disturbances of a minimal set, which duplicates the idea of infinitesimal stability in continuous dynamical systems :

The minimal set $\mathsf{A}_{(g)}$ corresponding to (5.8,8.a) is ***stable against small disturbances*** iff

$$N(A_{(g)}) \cap \Gamma_{(k)} = \emptyset, \text{ for any } k \neq g. \qquad 5.15$$

In fact, any small disturbance $\mathbf{F}'$ of a configuration $\mathbf{F} \in \mathsf{A}_{(g)}$ is in $\mathsf{N}(\mathsf{A}_{(g)})$; but the stability condition (5.15) implies that $\mathsf{N}(\mathsf{A}_{(g)}) \subset \Gamma_{(g)}$; hence it follows from relations (5.8,8.a) that $\mathbf{F}'$ reaches again the attractor after at most Θ timesteps (Θ possibly infinite).

If the stability condition (5.15) is not satisfied, then there are neighbour states of the minimal set $\mathsf{A}_{(g)}$ in a basin of attraction of at least one other minimal set $\mathsf{A}_{(k)}$ or on $\mathsf{A}_{(k)}$ itself; hence there exist small disturbances of the configurations of the minimal set which produce an evolution away from $\mathsf{A}_{(g)}$, and towards another minimal set $\mathsf{A}_{(k)}$.

A stable minimal set $\mathsf{A}_{(g)}$ is referred to as an ***attractor***.

Our definition of *reversible* CA implies that criterion (5.15) is never satisfied for the latter (cf the illustration of the translational gas, Eq. 4.17). It is appropriate, therefore, to introduce the weaker stability requirement of *orbital stability* (adapted from Hamiltonian mechanics):

A minimal set $A_{(g)}$ is *orbitally stable* if any small disturbance $\mathbf{F}'$ of any configuration $\mathbf{F} \in A_{(g)}$ eventually evolves on a minimal set $A_{(k)}$ contained in $N(A_{(g)})$, or

$$N(A_{(g)}) \cap A_{(k)} = A_{(k)}, \quad \text{for any} \quad \mathbf{F}' \in N(A_{(g)}). \qquad 5.16$$

The stronger definition of stability (5.15) requires that a phase point of the surroundings of the minimal set eventually returns to the same minimal set; the weaker definition (5.16) merely requires the phase point of the surroundings of the minimal set to return to the surroundings. A strongly stable minimal set is thus also weakly stable.

In order to rank as viable descriptions of natural phenomena, CA solutions must obey the weak *orbital stability* condition (5.16) which appears as the minimal stability requirement consistent with the process of idealisation of the CA physics.

There remains a major difference between linear stability analysis in continuous systems and stability as encoded in criteria (5.15) or (5.16). In ODEs or PDEs the existence of a single unstable 'mode' with respect to a minimal set $A_{(g)}$ (in the presence of a large, possibly infinite number of stable 'modes') implies that any arbitrary small enough disturbance, provided only that it is generic, evolves away from the attractor, and tends towards a new stable attractor $A_{(k)}$. Or if we select a disturbance 'at random', the latter evolves to some other attractor $A_{(k)}$ with probability 1. This property makes a complete stability analysis an easy task. In the CA case, we always have a non-zero probability that an arbitrarily selected neighbour state $\mathbf{F}'$ of the minimal set $A_{(g)}$ will eventually evolve on $A_{(g)}$.

Since for any physically meaningful CA model stability tests can only be carried out numerically, such tests always require an extended series of experiments using as initial conditions a broad sample of states of the neighbourhood $N(A_{(g)})$ ($f' \times z'$ experiments, $f' =$ card $N(\mathbf{i})$ and $z' =$ card $N(F)$ with initial condition close to one cell of the minimal set, if all cells of the minimal set can be regarded as equivalent).

The practical implication is that an isolated evolutionary sequence obtained from a (deterministic) CA simulation does not enable one to draw any strong conclusion on the actual behaviour of the natural phenomenon to be modelled. A large enough number of experiments ($f' \times z'$, under the above assumption) carried out with similar initial conditions, should always be performed before we can be sure that the features revealed by a solution are indeed realisable in nature.

We mention finally that a CA minimal set $A_{(g)}$ can in principle be characterised by a simple measure of nonlinear stability

$$s(A_{(g)}) = \frac{\text{card}(\Gamma_{(g)})}{\text{card}(\Gamma)}, \qquad 5.17$$

provided that this measure exists. This parameter gives the probability that any arbitrary perturbation $\mathbf{F}'$ of a state $\mathbf{F} \in A_{(g)}$, i.e. ***any*** randomly chosen allowed CA state, eventually evolves back towards $A_{(g)}$ (cf also Hogg and Huberman 1985 for estimates of the 'healing time' of a disturbance; cf Packard 1985 for a discussion of the CA counterpart of Lyapunov exponents).

Finite CA

The case of a finite number f of sites deserves a closer inspection on two accounts : From a practical point of view, in any 'realistic' CA model of a natural phenomenon (cf the simple aggregation problem of molecular cloud formation) the numerical handling of the evolution rules almost inevitably requires a finite lattice space. From a theoretical point of view, in the case of a finite number of sites we can establish a number of combinatorial results; the latter extend directly to our nonstandard CA definition, and thereby enable us to provide a few further general results on the CA dynamic behaviour.

With a finite number of cellstates z, the number of configurations is $K = z^f$ (finite). In the matrix scheme, the transition matrix T is turned into a $K \times K$ matrix acting on a K-dimensional vector F of all configurations. The total number of conceivable evolution rules (level 0 rules), and hence the total number of distinct deterministic CA (in the conventional sense), is therefore $\rho(o) = K^K$. Likewise the total number of conceivable level λ (=1, 2, ...) rules is finite, and given by $\rho(\lambda) = \rho(\lambda - 1)^{\rho(\lambda-1)}$. The total number of genuine substitutions among the automaton rules (generating a one–to–one correspondence between the states in $F^{(o)}$ and those in $F^{(1)}$, i.e. the total number of reversible CA of K configurations) is $\rho_{rev}(o) = K!$. Accordingly, for$K$ large enough (which is always the case), the fraction of rules which are genuine substitutions is e^{-K}. Therefore, if we select an automaton rule at random among the totality of possible rules, we have an exponentially small probability (= e^{-K}) of hitting a reversible rule.

The number of conceivable automaton rules producing cycles of length p (period–p solutions), $\rho(Pp)$, is equal to the number of transition matrices containing at least one cycle of length p; if $\rho(h; Pp)$ is the number of transition matrices containing exactly h cycles of length p, then $\rho(Pp) = \sum_{h \neq 0} \rho(h; Pp)$, with the summation extending over all numbers h of cycles. Thus $\rho(P1)$, the number of rules generating at least one equilibrium configuration, is equal to the number of transition matrices having at least one diagonal element equal to 1; since $\rho(h; P1)$ is the number of transition matrices with exactly h diagonal elements equal to 1, we have :

$$\rho(h; P1) = (K-1)^{K-h} C_K^h, \quad h = 1, 2, \dots, K,$$

$$\rho(P1) = \sum_{h=1}^{K} (K-1)^{K-h} C_K^h = K^K - (K-1)^K. \qquad 5.18$$

For K large enough, the asymptotic probability for a randomly prepared automaton to show at least one equilibrium state is independent of K and equal to (1-e^{-1}) = 0.632; (for K = 2,3,4 we have 0.75, 0.70, 0.68). The largest possible attractor, $\mathsf{A} \equiv \Gamma$, corresponds to a period–K solution, so that it occurs $\rho(PK) = K - 1$ times (transition matrices whose non–zero elements lie on parallels to the principal diagonal). Since a finite CA has at least one attractor, the transient times Θ obey the inequality $0 \leq \Theta \leq K - 1$; the lower limit 0 refers to reversible CA; the upper limit corresponds to the existence of a single attractor, the latter being a fixed point; this occurs $K(K - 1)$ times among all conceivable rules (transition matrix with one diagonal element 1, the remaining nonzero elements lying on parallels to the diagonal).

Other combinatorial parameters and their averages are estimated in Levy 1982 (cf also Harris 1960). For instance, the total number of conceivable rules leading to a cardinality α of the union of all attractors,

$$\alpha = \text{card}(\mathsf{A}), \qquad \mathsf{A} = \mathsf{A}_{(1)} \cup \mathsf{A}_{(2)} \cup \dots \cup \mathsf{A}_{(\gamma)}$$

(γ, total number of components in the partition (5.4), i.e total number of attractors), ρ_α, is given by

$$\rho_\alpha = \alpha \, K^{K-\alpha} \frac{(K-1)!}{(K-\alpha)!}. \qquad 5.19$$

The averages over all conceivable rules, for the total number of attractors, γ, the cardinality α of all attractors, the periods P, and the lengths of the transients Θ are, in the large K limit

$$\langle\gamma\rangle = \sqrt{\frac{\pi}{2}} \ln K, \quad \langle\alpha\rangle = \sqrt{K\frac{\pi}{2}}, \quad \langle P\rangle = \langle\Theta\rangle = \sqrt{K\frac{\pi}{8}}. \qquad 5.20$$

Note that both the average period $\langle P\rangle$ and the relaxation time $\langle\Theta + 1\rangle$ ($\rightarrow$ $\langle\Theta\rangle$ in the asymptotic case) are given by the geometric mean of the lower and the upper limit of P and $\Theta + 1$ respectively.

These average values of the statistical parameters also estimate the most probable values, since the frequency distributions of the statistical parameters are approximately Gaussian. Accordingly, a construction of a 'typical' rule (i.e. a rule neither too trivial, nor too contrived) should conform to these average laws, provided that 'typical' configurations only are considered, namely randomly generated initial states (all cellstates being equally probable).

Comments on the statistical estimates

The total number of 'conceivable' automaton evolution rules, $\rho(o)$, given above, does not take account of any constraints imposed by a specification of neighbourhoods (spatial neighbourhood, neighbourhood of cellstate). The estimates refer to the maximal neighbourhood of a spatial cell, namely the full CA lattice, and the maximal neighbourhood of a cellstate, namely the full range of z field values. With a selection of smaller neighbourhoods (f' space cells and z' field values), the number of neighbourhood configurations is reduced to $k = z'^{f'}$, so that the number of actually *independent* rules reduces to $\rho_{indep}(o) = k^k$. Therefore, given the neighbourhood specifications, in the above combinatorial averages (5.20) a large fraction of the a priori conceivable rules acquire zero weights (namely all those rules which map one configuration into a 'sufficiently different' configuration); the weights of the remaining fraction of rules are then fixed by their multiplicities of realisation when the independent rules are regarded as equally probable. To the extent that this latter restricted class of rules constrained by the CA specifications provide an approximate statistical sampling of the totality of conceivable rules, expressions (5.20) continue to supply at least order of magnitude estimates for the average CA properties.

A striking consequence of the statistical estimates is that the most likely period is an exponentially increasing function of the number of cells f. Even a small scale experiment, for instance on a 100×100 lattice, with a 5–state $2D$ automaton, has an expected period $\langle P \rangle \approx 10^{3500}$, i.e. for any practical purposes the most likely stationary evolution exhibits no periodicity. Moreover the relaxation time Θ towards this attractor is of the same order; the attractor is then never reached in any actual numerical experiment.

It is manifest, however, that in virtually any CA model designed to simulate a physical process, a privileged initial configuration is chosen in conformity with the specific goal of the simulation. This configuration does not qualify as 'generic'. An obvious illustration is the CA model of stellar pulsations, where care is taken to start the calculation with a configuration already close enough to the attractor over which we want to trace out the oscillatory motions of the star. The corresponding relaxation time Θ_{relax} is here substantially shortened with respect to the estimated $\langle\Theta\rangle$. On the other hand, the model rules themselves are designed here with the purpose in mind to generate a given type of attractor (although, as expressed by Feynman's quotation at the beginning of this chapter, we cannot a priori anticipate the macroscopic behaviour we would like to observe from our goal–oriented laws). The statistical estimate of the period will therefore not apply either. The observation of lack of genericity of the initial configurations also holds for CA simulations of 'order' emerging out of 'chaos', if the initial configuration is generated 'at random'. Thus in the kinematic aggregation process sketched in chapter 3, in galactic structure models, in CA models of the luminous matter distribution in the universe, etc, initial configurations are indeed randomly generated. But the random input in these models is not equivalent with genericity of the initial condition : The initial state typically refers to a featureless spatial distribution of privileged states of the field variable. True genericity of the initial configuration requires a statistically uniform distribution of all cell states (not just one type of active states, for instance). Therefore, again for these models an ensemble averaged relaxation time over a large number of initial configurations may be shorter than the statistical estimate $\langle\Theta\rangle$.

In the case of CA models designed to simulate realistic physical phenomena it is out of the question to run any actual numerical experiment over a number of timesteps comparable to $\langle\Theta\rangle$ as estimated by (5.20). This implies that we have no a priori statistical guarantee that when patterns are observed in a CA simulation, they do represent permanent features, in the sense that they correspond to an evolution proceeding on an attractor. The observed pattern may well be a transient phenomenon, dissolving again in the course of evolution, to be replaced by other transient patterns. The only way to decide whether an experimentally observed pattern is physically meaningful consists in repeating the experiment a large enough number of times with slightly different initial configurations. Accordingly, the numerical procedure of testing for stability discussed above supplies at the same time a test of significance for the formation of patterns.

While the above statistical estimates may not be of great practical relevance, they provide a theoretical complement to the classification derived from the nonstandard interpretation. In fact, in the context of the nonstandard CA definition *D, the above statistical estimates—with nonstandard symbols being substituted to the standard ones—continue to hold.

Thus the standard part of the probability of hitting a reversible rule in a nonstandard CA with $*f$ an infinit is zero, and the estimate of the probability of existence of at least one equilibrium configuration continues to hold. The average of the period $\langle *P \rangle = s\sqrt{*K}$ (5.20), s real, of finite standard part (close to 1), continues to provide an indication on the most probable type of behaviour in a generic CA. In fact, let $*\omega$ be a reference infinit (number of cells in a macroscopically observable part of the CA). Select the infinitesimal timestep of the form $\Delta t = \frac{1}{*\omega^b}$, b positive. We consider two alternatives for the 'order' of the total number of cells $*f$:

(i) infinit of the form $*f = *\omega^c$, c, positive real;

(ii) infinit of the form $*f = \exp\,(*\omega^d)$, d, positive real.

Under (i) the total number of configurations is an exponential of the reference infinit, while under (ii) it becomes an exponential of an exponential; the physical period $\langle *P\Delta t \rangle$ therefore remains an exponential of $*\omega$ under (i), or an exponential of an exponential under (ii), implying that the most probable behaviour of the CA is 'complex' behaviour satisfying criteria (d) or (e) respectively. These estimates provide a justification of conjecture (5.12). Note also that the behaviour corresponding to the maximum nonstandard period, $*P_{max} = *K$, belongs into the same complexity class as the behaviour associated with the most probable period $\langle *P \rangle$. Alternative (i), already sketched in chapter 1 (for the particular choice $c = 1$), is adequate for a *macroscopically compact support manifold* of the lattice (typically a torus in numerical experiments). Alternative (ii) refers to an *unbounded standard lattice* (standard part of a nonstandard compact lattice), since a 'higher order' of infinity is needed to embed standard unbounded manifolds.

These observations have motivated our choice of $*f$ in the discussion of self–organised criticality (Eq. 5.13).

Theoretical remarks on stochastic CA

The notation (5.6) describes the action of the transition matrix T on the totality of ordered configurations represented by the *column vector* $F^{(o)}$ (Eq. 5.2). Another convenient notation involving T consists in symbolising an individual configuration r by a *unit row vector* e_r whose components are all 0 except for component labelled r, which is 1. With the latter symbolism the evolution rule takes the form

$$e_r \,.\, T \;=\; e_k, \qquad 5.21$$

meaning that configuration r is is carried into configuration k after one timestep; this interpretation follows also naturally by right–multiplication of Eq. (5.12) by $F^{(o)}$. In a similar way

$$e_r \,.\, T^t \;=\; e_q, \qquad 5.21.a$$

describes the transformation of configuration r into q after t timesteps (cf 5.6.b). This formulation naturally extends to a probabilistic treatment of the CA evolution.

Suppose the initial state of the automaton is prepared by a random process, attaching a probability p_r to configuration r. The ordered collection of these probabilities defines the *row vector* $P(o)$ of initial probabilities

$$P(o) \;=\; \sum_r p_r\, e_r, \quad \text{with} \;\; P(o)\,.\,I \;=\; \sum_r p_r \;=\; 1, \qquad 5.22$$

(I, column vector whose elements are all 1). From Eq. (5.21) we then have

$$P(o)\,.\,T \;=\; P(1); \qquad 5.23$$

and hence, in general

$$P(t)\,.\,T \;=\; P(t+1), \tag{5.24}$$

$P(t)$ being the row vector whose component k, $p_k(t)$, represents the probability of realisation of the CA configuration k at timestep t, in a sample of runs prepared with the initial probability distribution (5.22). Eq. (5.24) is thus the general CA counterpart of the Liouville equation of mechanics.

This representation, allowing for the treatment of stochasticity in the initial conditions, directly extends to ***intrinsic stochasticity*** in the CA evolution rules. In fact, the evolution of a level 1 CA with a probability distribution defining the meta–rules is represented again by an equation of type (5.24), where T is a Markov matrix. For if ..., T_a, T_b, T_c, ... are the transition matrices encoding the level o rules, and the meta–rule assigns corresponding probabilities ..., q_a, q_b, q_c, ... to these rules, then the overall evolution is described by the transition matrix

$$T \;=\; \sum_h q_h\, T_h, \tag{5.25}$$

the sum extending over all level o rules. By construction the row vectors of this matrix have nonnegative elements whose sum is equal to 1 (Markov property). If we denote by t_{rk} the matrix elements of T, we have, for an initially 'pure' configuration r, $P(0) = e_r$,

$$e_r\,.\,T \;=\; (\ldots, t_{rk}, t_{rl}, \ldots) \;=\; P(1), \tag{5.26}$$

showing that t_{rk} represents the probability for configuration r to transform into configuration k in one timestep. The results on (denumerable) Markov chains apply (Kemeny et al 1976; Bharucha-Reid 1960; von Mises 1931; see also Kemeny and Snell 1960). Decomposition of the global transition matrix (5.25) into diagonal block matrices (5.7) corresponds, in the framework of the theory of Markov processes, to the concept of ***reducibility*** of the Markov chain into irreducible matrices $T_{(g)}$. The probability vector associated with a reducible chain (5.7) then becomes (cf 5.5.a)

$$P(t) \;=\; P_{(1)}(t) \oplus P_{(2)}(t) \oplus \ldots \oplus P_{(t)}(t) \oplus \ldots \tag{5.27}$$

i.e. , just as in the deterministic case, the states in any subspace $\Gamma_{(g)}$ transform among themselves and are not connected with the states of any other subspace $\Gamma_{(k)}$, $k \neq g$. In the stochastic case a further subdivision of the transition matrices is of relevance (von Mises 1931). If, after a reordering of the configurations, the irreducible matrix $T_{(g)}$ can be brought into the form of a lower triangular block matrix,

$$T_{(g)} \;=\; \begin{pmatrix} A_{(g)} & 0 \\ \Theta_{(g)} & Z_{(g)} \end{pmatrix}, \tag{5.28}$$

(with the submatrix $\Theta_{(g)}$ having by construction at least one nonzero element per row), then von Mises refers to it as ***one–sided reducible***; otherwise the matrix is ***totally irreducible***. In the case of a deterministic CA, one–sided reducibility implies that the diagonal submatrix $Z_{(g)}$ in (5.28) is zero. If $T_{(g)}$ is an $R \times R$ matrix, and $A_{(g)}$ a $P \times P$ matrix, then we relabel without loss the configurations such that the first, second, ..., R^{th} row of $T_{(g)}$ correspond to configurations 1,2,..., P, $P+1$, ..., R. In the case of infinite–dimensional matrices, the integers P and R are interpreted as nonstandard integers.

If α is any initial configuration, then provided that the transition probabilities from state α to β, from β to γ,..., from θ to ι,... are all nonzero, then any ordered sequence of configurations, α, β,..., δ,θ, ι, ... defines a ***stochastic orbit*** in the phase space Γ, symbolised as above by the string $(\alpha, \beta, \ldots, \delta, \theta, \iota, \ldots)$. Two states α and θ ***communicate*** if there is an orbit of origin α and traversing configuration θ, $(\alpha, \beta, \ldots, \delta, \theta, \iota, \ldots)$, and if simultaneously there is an orbit of origin θ and traversing configuration α, $(\theta, \psi, \ldots, \alpha, \pi, \ldots)$.

Periodicity in a deterministic CA has several counterparts in the stochastic version.

A configuration r is *recurrent* if the probability for r to repeat infinitely often is 1. The mean recurrence time of r, $\langle P_r \rangle$, is the expectation value of the recurrence time of r, P_r, which is the smallest number of timesteps for r to repeat in a given realisation of the stochastic process (i.e. r is the P_r^{th} successor of itself in the realisation, such that there is no 'earlier' successor coinciding with r). In the nonstandard CA this parameter always exists.

For a configuration r communicating with itself one further introduces a *stochastic period* Π_r. This parameter is the greatest common divisor among the exponents t of $(T^t)_{rr} > 0$. We have manifestly $\Pi_r \leq \langle P_r \rangle$, with the equality being realised in the deterministic case.

A configuration is *transient* if the probability for configuration r to repeat infinitely often is zero.

In the nonstandard setting a CA configuration is *ergodic* if it is recurrent.

In conventional Markov theory a configuration r is referred to as ergodic if it is neither transient, nor null (for a null configuration the mean recurrence time $\langle P_r \rangle$ does not exist; cf Bharucha–Reid 1960); in our nonstandard CA version null configurations are excluded.

Since in the framework of stochastic CA simulations we are especially concerned with the long–time behaviour of our models (cf the aggregation model), the following *limit theorems* are of interest. If all configurations in the irreducible phase space $\Gamma_{(g)}$ are ergodic, then

$$\lim_{t \to \infty} T_{(g)}{}^t = I \, P_{(g)}(\infty), \qquad 5.29$$

exists (I, the one–column vector introduced above; $P_{(g)}(\infty)$, a row vector of probabilities; both vectors are R–dimensional; the composition of these two vectors in (5.29) is an $R \times R$ matrix). Under the latter alternative we see from the evolution law (5.24) that the asymptotic probability vector $P_{(g)}(\infty)$ is the left–eigenvector of the matrix $T_{(g)}$ associated with the eigenvalue 1

$$P_{(g)}(\infty) = P_{(g)}(\infty) \, . \, T(g). \qquad 5.29.a$$

On the other hand, if configuration k is transient, then

$$\lim_{t \to \infty} (T_{(g)}{}^t)_{rk} = 0, \qquad 5.30$$

for every configuration r (Bharucha-Reid 1960; theorems 1.5, 1.6).

From the computation of successive powers $T_{(g)}{}^t$ of the transition matrix, using representation (5.28), it is seen that the upper diagonal matrix $A_{(g)}$ transforms into $A_{(g)}{}^t \to A \neq 0$, and the lower diagonal matrix $Z_{(g)}$ transforms into $Z_{(g)}{}^t \to 0$, for $t \to \infty$ (all eigenvalues of $Z_{(g)}$ being less than 1 in modulus); the off–diagonal matrix becomes

$$\Theta_{(g)} A(g)^{t-1} + (t-1) \, Z_{(g)} \, \Theta_{(g)} A_{(g)}{}^{t-2} + \ldots + Z_{(g)}{}^{t-1} \, \Theta_{(g)};$$

hence if configuration $k = P + s$, then $(T_{(g)}{}^t)_{r \; P+s} = (Z_{(g)}{}^t)_{r-P \; s}$ if $r > P$, or 0 (an element of the upper zero off–diagonal matrix) if $r \leq P$; hence the asymptotic behaviour (5.30). Such a configuration k is transient in the sense of the above definition.

Property (5.29.a), and the more general property

$$\lim_{t \to \infty} A_{(g)}{}^t = A \equiv I P'_{(g)}(\infty), \qquad 5.29.b$$

($P'_{(g)}(\infty)$ a probability row vector in the P–dimensional space in which the $P \times P$ matrix $A_{(g)}$ is operating), flow from the fact that the tth power of a matrix M with a single dominating eigenvalue λ_{max} and associated left row eigenvector v_L and right column eigenvector v_R (normalised such that $v_L \, . \, v_R = 1$) is asymptotically dominated by $\lambda_{max}{}^t \, v_R v_L$; in the present case the single dominating eigenvalue of $A_{(g)}$ is 1, all other eigenvalues being less than 1 in modulus; the corresponding right eigenvector is I (consequence of the Markov property); $P'_{(g)}(\infty)$ is the left eigenvector. In particular, if the matrix $A_{(g)}$ is symmetric, then the left and right eigenvectors coincide; hence, for the symmetric case

$$P'_{(g)}(\infty) = \frac{1}{P}(1,1,\ldots,1). \qquad 5.29.c$$

By analogy with the deterministic case we denote by $\mathsf{A}_{(g)}$ the set of recurrent (ergodic) configurations (cardinality P, according to decomposition 5.25) in the irreducible subspace $\Gamma_{(g)}$, and by $\mathsf{B}_{(g)}$ the set of transients (basin of attraction of the recurrent set). The stochastic orbit $(\alpha,\beta,\ldots,\delta,\theta,\iota,\ldots)$ is then made up of a transient fragment, $(\alpha,\beta,\ldots,\delta]$, in $\mathsf{B}_{(g)}$, and a recurrent fragment, $[\theta,\iota,\ldots)$, in $\mathsf{A}_{(g)}$; in contrast with the periodic fragment of a deterministic orbit, *the recurrent fragment is not repetitive.* It is then clear that if $\Gamma_{(g)}$ is an irreducible subspace for the stochastic CA resulting from a collection of deterministic evolution laws T_h (cf 5.25), then the recurrent set $\mathsf{A}_{(g)}$ must contain all minimal sets $\mathsf{A}_{(g')h}$ in $\Gamma_{(g)}$ of the component deterministic CA ($\Gamma_{(g)} \supset \Gamma_{(g')h} \supset \mathsf{A}_{(g')h}$, with $\Gamma_{(g')h}$ denoting the irreducible subspaces for the deterministic law T_h) :

$$A_{(g')h} \subset A_{(g)} \qquad 5.31$$

From the previous formal properties the time behaviour of the stochastic CA evolution may be summarised as follows :

For a given initial automaton configuration α, and a given realisation, the evolution proceeds through Θ_α transient configurations, before reaching the first recurrent state θ; the evolution then continues on the recurrent set $\mathsf{A}_{(g)}$ (an invariant set of the evolution), with a mean recurrence time $\langle P_\theta\rangle$. While this broad aspect of the behaviour duplicates deterministic CA evolution, the details of the stochastic evolution differ significantly from determinism :

(1) Although the mean relaxation time $\langle\Theta_\alpha\rangle$ understood as the average over all realisations has a well defined value, the individual relaxation time for a given realisation is a stochastic variable which can show large fluctuations. For instance, if there is a least integer $n = 1, 2, \ldots$ such that the diagonal element $(Z_{(g)}{}^n)_{\alpha-P\ \alpha-P}$, does not vanish, then the probability for having a relaxation time $\Theta_\alpha = nm$, m any integer, is nonzero.

(2) Any orbit of origin α in $\Gamma_{(g)}$ eventually covers the recurrent set $\mathsf{A}_{(g)}$ of the stochastic CA, with occupation probabilities $P'_{(g)}(\infty)$ (Eq. 5.29.b). In contrast with the deterministic case, where the occupation probabilities of all configurations of the minimal set are equal ($= \frac{1}{P_g}$, Eq. 5.9, or its nonstandard version), the occupation probabilities of different configurations are now in general different; an exception is the case of a symmetric submatrix $A_{(g)}$, as discussed above. The mean recurrence time for a given orbit of origin α and first configuration θ in the recurrent set, obeys the inequality $\langle P_\alpha\rangle \leq P$ (equality in the deterministic case). Again, this recurrence time itself fluctuates from one cycle to the next one; if there is a least integer $n = 1, 2, \ldots$ such that the diagonal element of $(A_{(g)}{}^n)_{\theta\ \theta}$, does not vanish, then there is a nonzero probability for having a recurrence time $P_\theta = nm$, m any integer.

Consider finally the ensemble–averaged behaviour of a CA (the ensemble corresponding to runs defined by different initial conditions constrained by global properties in the case of deterministic CA, or by identical initial conditions in stochastic CA). Then provided that a self–consistent global representation in terms of macroscopic variables exists, then the CA minimal set has an image in the phase space of these global variables, which acquires the structure of a *fuzzy set* in the sense of Zadeh (cf Zimmermann 1992).

Transition matrices for diffusion and aggregation

It is a consequence of the construction of the stochastic transition matrix as a weighted average of deterministic transition matrices (Eq. 5.25) that the resulting matrix T always acquires a 'special' (i.e. nongeneric) structure in actual physical models. This is due to the combination of two properties. On the one hand the deterministic transition matrices are sparse (number of nonzero elements = number of matrix dimension). On the other hand, even in the worst case of a 'completely random' law of motion for each cellstate, as happens in a stochastic diffusion CA, the number of deterministic laws among which the random choice operates remains small as compared to the dimension of the reduced matrices $T_{(g)}$.

This latter point is illustrated by the 1*D* ***diffusion*** problem already discussed in chapter 2. Take for definiteness a lattice of f cells, with periodic boundary conditions. Since there is no creation or destruction of particles, a phase space partition (5.4) always holds with $\Gamma_{(1)}$ corresponding to a 0–particle diffusion, $\Gamma_{(2)}$ corresponding to a 1–particle diffusion, ..., and $\Gamma_{(N+1)}$ corresponding to an N–particle diffusion problem. It suffices therefore to construct the transition matrices $T_{(g)}$, $g = 1,2, \ldots,$ relative to the latter spaces.

The alternative of 1–particle diffusion takes place in a subspace $\Gamma_{(2)}$ of dimension $R = f$; we have here 3 deterministic transition matrices, $T_{(2)o}$, $T_{(2)+}$, $T_{(2)-}$, correponding to the identity transformation (unit $R \times R$ matrix), a shift in the positive direction (all matrix elements vanish except elements $i, i+1$, $i = 1,2,\ldots, R-1$, and element $R, 1$), and a shift in the negative direction (all matrix elements vanish except elements $i, i-1$, $i = 2,\ldots,R$, and element $1, R$), respectively. The partial stochastic transition matrix relative to subspace $\Gamma_{(2)}$, $T_{(2)}$, then becomes (Eq. 5.25)

$$T_{(2)} = q\, T_{(2)o} + \frac{p}{2} T_{(2)+} + \frac{p}{2} T_{(2)-}, \qquad 5.32$$

with q, p the transition probabilities for a particle to remain in the same cell, or to move to a neighbour cell in one timestep (cf chapter 2). Hence, $T_{(2)}$, is totally irreducible ($= A_{(2)}$, with $P = R = f$, Eq. 5.28) and has as only nonzero elements the principal diagonal elements i, i ($= q$), the secondary diagonal elements $i, i \pm 1$, $i = 2,3,\ldots,R-1$ and the two elements $1, R$ and $R, 1$ ($= p$). This matrix is symmetric, and therefore, by (5.29.c), all f configurations are equally probable ($P_{(2)}(\infty) = \frac{1}{f}$).

The same conclusion follows from a more synthetic argument applicable under more general circumstances:

Compare the stochastic evolution with the component deterministic evolutions. For the deterministic law $T_{(2)o}$ the subspace $\Gamma_{(2)}$ is further reducible into R subspaces $\Gamma_{(2,1)}, \Gamma_{(2,2)}, \ldots, \Gamma_{(2,R)}$, each one being a minimal set of cardinality 1; on the other hand, the deterministic laws $T_{(2)\pm}$ represent permutations in the irreducible subspace $\Gamma_{(2)}$, so that the latter subspace coincides with the minimal sets $A_{(2)\pm}$ for both laws. Accordingly, since $\Gamma_{(2)} \supset A_{(2)} \supset A_{(2)\pm}$, (cf 5.31), we must have $A_{(2)} \equiv \Gamma_{(2)}$ as is indeed seen analytically. Equiprobability of all states is again inferred by symmetry.

The discussion of the alternative of 2–particle diffusion proceeds in a similar way. The corresponding subspace $\Gamma_{(3)}$ has dimension $R = \frac{f(f+1)}{2}$, (if we view the particles as indiscernible). Order now the configurations such that labels $i = 1, 2, \ldots, f$ refer to the states of both particles being in cell i; next, let labels $i = f+1, f+2, \ldots, 2f$ refer to two particles in nearest neighbour cells, one particle being in cell $i-f$, the other in cell $i+1-f$, (identifying 0 with f and $f+1$ with 1); let further labels $2f+1$ to $3f$ refer to two particles being separated by one empty cell, etc (note that if f is even, the last group of configurations contains $\frac{f}{2}$ elements only). We have here 9 distinct transition matrices for the underlying deterministic motions, $T_{(3)oo}$; $T_{(3)o+}$, $T_{(3)+o}$, $T_{(3)o-}$, $T_{(3)-o}$; $T_{(3)++}$, $T_{(3)--}$; $T_{(3)+-}$, $T_{(3)-+}$, corresponding to the identity transformation for both particles; the identity transformation for one particle and a positive (negative) shift of the other particle; a positive (negative) shift for both particles (all matrix elements vanish except elements $i, i+1$, $i = 1, 2, \ldots, f-1$, and element $f, 1$; or all matrix elements vanish except elements $i, i-1$, $i = 2, \ldots, f$, and element $1, f$, respectively); and a positive shift of one particle together with a negative shift of the other particle. $T_{(3)oo}$ is thus the unit matrix (fully reducible). $T_{(3)++}$ and $T_{(3)--}$ are made up of $\frac{f+1}{2}$ (odd f) or $\frac{f}{2}$ (even f) diagonal blockmatrices $T_{(2)+}$ and $T_{(2)-}$ respectively (for f even there is an additional diagonal blockmatrix of dimension $\frac{f}{2}$); in other words $\Gamma_{(3)}$ is reducible into $f' = \frac{f+1}{2}$ (f odd) or $1 + \frac{f}{2}$ (f even) subspaces $\Gamma_{(3,1)}, \Gamma_{(3,2)}, \ldots, \Gamma_{(3,f')}$, for these motions. The structure of the remaining matrices is easily worked out; for instance, the matrices $T_{(3)o\pm}$ transform the previous subspaces $\Gamma_{(3,s)}$ into $\Gamma_{(3,s\pm1)}$, so that the previous reducibility is destroyed in the superposition of these laws with the previous laws. Finally, the partial stochastic transition matrix in $\Gamma_{(3)}$ becomes (Eq. 5.25)

$$T_{(3)} = q^2\, T_{(3)oo} + q\frac{p}{2}\, T_{(3)o+} + q\frac{p}{2}\, T_{(3)+o} + q\frac{p}{2}\, T_{(3)o-} + q\frac{p}{2}\, T_{(3)-o}$$

$$+ (\frac{p}{2})^2\, T_{(3)++} + (\frac{p}{2})^2\, T_{(3)--} + (\frac{p}{2})^2\, T_{(3)+-} + (\frac{p}{2})^2\, T_{(3)+-} \qquad 5.33$$

which is irreducible, and manifestly symmetric. Therefore the recurrent set $A_{(3)} = \Gamma_{(3)}$ is asymptotically evenly populated (over a long enough number of timesteps all configurations occur with the same frequency, $\frac{2}{f(f+1)}$).

This algebraic discussion extends to arbitrary values of N, and leads to the seemingly paradoxical conclusion that *all automaton configurations of* $\Gamma_{(N+1)}$ *are equally probable*. This result requires a few comments. Since for an individual particle we have 3 independent allowed laws of motions, the collection of N particles is ruled by 3^N laws, or the stochastic transition matrix $T_{(N+1)}$ is a linear superposition of 3^N deterministic matrices whose weights are obvious (cf 5.33). The dimension of the space $\Gamma_{(N+1)}$, and hence the size of the matrix $T_{(N+1)}$ is $O(f^N)$ ($\gg 3^N$ in any physically meaningful simulation), so that the asymptotic occupation probability of one specific configuration is $O(f^{-N})$, the mean recurrence time being $O(f^N)$. The latter estimate indicates precisely that in any feasible numerical experiment exact recurrence can never materialise. Even for a small scale automaton of $f = 10^3$ cells, containing $N = 50$ particles initially concentrated in a neighbourhood of one cell, the recurrence time is of order 10^{150}. Any realisable experiment just reveals an irreversible behaviour towards a statistically spatially uniform distribution. Although theoretically the microscopic diffusion process over a circular cellspace involves no transient configurations ($\Gamma_{(N+1)} = A_{(N+1)}$, $N = 0,1,2,\ldots$), the coarse-grained macroscopic evolution — an average over a large enough number of neighbouring cells (cf the macroscopic equation of chapter 2) — mimics a transient behaviour towards a spatially and temporally featureless macro-configuration. The latter is an average over a large number of permanently evolving (fluctuating) microscopic automaton configurations. Since all (micro-) configurations are equiprobable, a Boltzmann-type statistical argument is applicable, which makes it clear that the macro-configuration is statistically uniform in space and time.

In this example of pure diffusion the transient spaces (basins of attraction) are empty sets. It suffices to incorporate for instance a mechanism of ***aggregation*** of particles, as was done in the molecular cloud formation CA (chapter 3), to modify the dynamics of the problem completely. In the circular $1D$ CA a rudimentary aggregation process is simulated by the following 'inner' and 'outer' dynamics :

(1) Any cluster of N_c particles, i.e a collection of N_c particles in a string of adjacent cells (each cell being allowed to accommodate any number of particles), cannot disintegrate.

(2) If two clusters of N_c and $N_{c'}$ particles collide, then they form a new cluster containing $N_c + N_{c'}$ particles.

(3) A cluster behaves dynamically as a single diffusing particle.

In the presence of a total number N of particles, it is then obvious that the recurrent set $A_{(N+1)}$ in subspace $\Gamma_{(N+1)}$ is now the collection of configurations containing a single cluster, whose spatial extension L = number of occupied cells) keeps undergoing 'statistical fluctuations', in the range $1 \le L \le N$. Since in a physically meaningful experiment we have $f \gg N$ the dimension of the recurrent set is Nf (f configurations with all particles in a single cell, f configurations with all particles in 2 cells, etc). Again, since under rule (3) governing the outer cluster dynamics, the asymptotic single cluster undergoes a 1-particle diffusion on the recurrent set, the transition matrix corresponding to the recurrent set is symmetric. Accordingly all configurations of the recurrent set are equiprobable. Hence statistics can be used to characterise the asymptotic stage (number of realisations of a cluster of N particles and size L : $\frac{N!}{n_1!\, n_2! \ldots n_L!}$, n_k , number of particles in k^{th} cluster cell, with $n_1 + n_2 + \ldots + n_L = N$); the average size obeys $\langle L \rangle \propto \sqrt{N}$ (cf remark about Eqs. 5.20).

Reaction-type processes

The rudimentary models of diffusion and aggregation are specified by a very small number of independent deterministic rules of evolution (3 for $1D$ diffusion, with the two rules of motion operating symmetrically, with same probability); for both models all configurations of the recurrent set are equiprobable. Genuine complex problems are governed by large numbers of alternative and competing physical processes (large numbers of deterministic rules in the simulation), which are selected in turn according to a hierarchy of probabilistic rules. For instance, the stochastic CA model of star formation and evolution in the galaxy, even in the schematic form discussed in Lejeune and Perdang (1991, cf also Lejeune 1993), takes account of some 30 individual mechanisms which are turned on and off by a scheme of conditional probabilistic choices (reflecting in part our ignorance of the exact physics). There is then no reason to expect equiprobability of all configurations of the recurrent sets to hold.

The following $1D$ toy analogue of the star formation model makes this point obvious. Consider a 3–state $1D$ circular CA, the cellstates being the inert state (empty cell), E, and two active states, dust, D, and the star state, S; ignore spatial motions and consider the following 7 'reactions' simulating the evolutionary processes

$$E \to E \ (a); \quad D \to D, \ q_D \ \text{or} \ D \to S, \ p_D \ (b);$$

$$D + S \to D + S, \ q_{DS} \ \text{or} \ D + S \to S + S, \ p_{DS} \ (c);$$

$$S \to S, q_S \ \text{or} \ S \to D, \ p_S \ (d); \qquad 5.34$$

(with $q_a + p_a = 1$, $a = D, S, DS$). Processes (b) and (c) mean that a dust cell with no star [at least one star] in its neighbourhood either remains a dust cell, with probability q_D [q_{DS}], or transforms into a star, with probability p_D [p_{DS}]. The remaining processes are self–explanatory. These laws immediately imply that not only the total number N of active states is conserved, but also *conservation of the distribution* ϕ of active cells over the lattice space holds (an active cell remains so forever), so that the collection of states $\Gamma_{(N+1,\phi)}$ of N active states of spatial distribution ϕ is an invariant subspace of the phase space. Moreover, the transformation laws among the active states imply that this subspace is recurrent, $\mathsf{A}_{(N+1,\phi)} \equiv \Gamma_{(N+1,\phi)}$. Due to the obvious *asymmetry* of the transition matrix $T_{(N+1,\phi)}$, we now have, in general, $P_{(N+1,\phi)}(\infty) \neq \frac{1}{R}(1,1,\ldots,1)$, the different components of the occupation probability vector depending on the independent model probabilities p_a, $a = D, S, DS$. This conclusion is readily confirmed analytically for the alternatives $N = 1$ and 2, the corresponding transition matrices, together with the asymptotic occupation probability vectors, being accessible explicitly. The result is physically obvious, since for p_D and $p_{DS} \ll 1$,and $q_S \ll 1$, virtually all active cells remain in the D state; the unique configuration in which all active cells are dust cells acquires a probability close to 1, while all the remaining configurations have vanishing probabilities.

Periodic time–behaviour and intermittency

The above illustrations should not leave an impression that stochasticity in a CA implies that the recurrent sets merely characterise macroscopically steady states just capable of suffering 'statistical fluctuations'. The fact that an asymptotic time independent frequency of occupation of the configurations on the recurrent sets can be established does not imply that on a macroscopic scale the behaviour is an equilibrium; recall in fact that for deterministic CA we even have the stronger property that all recurrent configurations have *same* frequency of occupation. The asymptotic occupation probability merely determines the average number of times $n(k)$ a given configuration k of the recurrent set is realised over an asymptotically large number of timesteps $t \to \infty$. As in the case of deterministic CA, the stochastic model itself may show *cyclic behaviour*. This flows directly from the rule of construction of the Markov transition matrix as an average over deterministic transition matrices; if one component matrix of the righthand side of Eq. (5.25) remains dominant, i.e. the corresponding probability coefficient q_o is close to 1, then the time behaviour of the automaton essentially inherits the deterministic behaviour resulting from the action of the latter matrix.

A trivial example is the superposition of a unity matrix and a permutation matrix of period P in a P–dimensional subspace; if the latter component dominates, then the CA time pattern is close to period-P behaviour. Long periodicity (interpretable by a mean field approach) has been observed in the star formation and evolution CA simulation, albeit under astrophysically contrived conditions (Lejeune and Perdang 1991).

A behaviour more interesting than approximate periodicity occurs if the dominant component matrix of $T(g)$ is a diagonal block matrix, made up of $\nu_{(g)}$ blocks $A_{(g,b)}$ of different sizes $P_{g,b}$, each block operating in a subspace $\Gamma_{(g,b)}$. If the CA evolution is started with a configuration in subspace $\Gamma_{(g,b')}$, then the ensuing evolution gets caught up in that subspace over an average time interval ΔT essentially of the order of $\frac{1}{1-q_o}$ (the factor $(1-q_o)$ playing the part of a coupling strength between the different subspaces); the evolution thus shows period–$P_{g,b'}$ behaviour, provided $\Delta T \gg P_{g,b'}$. The coupling between the subspaces $\Gamma_{(g,b)}$, due to the presence of the remaining transition matrices, eventually injects the configuration into a second subspace $\Gamma_{(g,b'')}$, so that the automaton then exhibits periodic behaviour with a different period $P_{g,b''}$ (provided again that $\Delta T \gg P_{g,b''}$; etc; note also that for periods $P_{g,b} \gg \Delta T$, the periodicity ceases to be observable and the behaviour appears 'chaotic'. The overall time evolution is a type of ***intermittency*** characterised by $\nu_{(g)}$ patterns of time behaviour (periodic or irregular) which alternate irregularly, the alternations being triggered by the stochasticity of the evolution laws. This intermittency has been observed in the star formation and evolution automaton, where bursts of well–defined short–period pulsations alternate with low amplitude 'fluctuations' (visible also in the space–averaged Red Giant or Main Sequence star population). This type of intermittency is of the same nature as the intermittency isolated in low–dimensional stochastically perturbed iterative maps (Perdang 1990b; 1991).

Finally, a further type of nontrivial time pattern can be predicted on the basis of the structure of the Markov transition matrix. Suppose now that we can rewrite the summation in Eq. (5.25) in the form

$$T = q_o T_o + (1-q_o)T_1 \qquad 5.35$$

with q_o again close to 1, but with T_o , the dominant contribution, itself a superposition of deterministic matrices. Consider again an invariant subspace $\Gamma_{(g)}$ of the phase space Γ. Suppose that the dominant contribution is a diagonal block matrix made up of $\nu_{(g)}$ blocks $A_{(g;b)}$ of different sizes $P_{g;b}$, each block operating in a subspace $\Gamma_{(g;b)}$. Suppose further that each block matrix can be represented as a dominant block matrix plus a correction

$$A_{(g;b)} = A'_{(g;b)} + \epsilon_1 A''_{(g;b)}, \qquad 5.36$$

such that $A'_{(g;b)}$ is made up of $\nu_{(g;b)}$ blocks $A_{(g;b,c)}$ (acting in $\Gamma_{(g;b,c)}$), ϵ_1 a small number. Likewise, suppose again each block $A_{(g;b,c)}$ decomposed in a similar fashion

$$A_{(g;b,c)} = A'_{(g;b,c)} + \epsilon_2 A''_{(g;b,c)}, \qquad 5.36.a$$

with now $A'_{(g;b,c)}$ made up of $\nu_{(g;b,c)}$ blocks $A_{(g;b,c,d)}$ (acting in $\Gamma_{(g;b,c,d)}$), ϵ_2 a small number; suppose this process continues until the final dominant blocks (acting in $\Gamma_{(g;b,c,d,\ldots,q)}$) are deterministic matrices. In this fashion, the dominant contribution is made up of a hierarchy of successively smaller block matrices within larger block matrices, and the phase space is subdivided into a ***hierarchy of nested subspaces.*** If the CA evolution is started with a configuration in subspace $\Gamma_{(g;b,c,d,\ldots,q,p)}$, the ensuing evolution remains frozen for some time within that space, before being sent into another $\Gamma_{(g;b,c,d,\ldots,q,p')}$ (same level subspace as $\Gamma_{(g;b,c,d,\ldots,q,p)}$), as in the intermittent behaviour discussed above; but now, on a longer timescale, there will be a shift into a 'lower level' phase space, $\Gamma_{(g;b,c,d,\ldots,q',p'')}$, etc. Therefore the corresponding time behaviour exhibits a ***hierarchy of intermittencies.***

That such a structure of the Markov transition matrix is indeed consistent with the CA framework is made clear from the fact that the individual deterministic components in Eq. (5.25) are built up from a small number of deterministic laws operating on the level of a cell, and therefore occur with weights following from the probabilities of these laws (cf Eqs 5.32,33). This implies that there will be, in general, a natural order of dominance of certain types of behaviour. Whether actual groupings of the form (5.36,36.a) are possible, depends on the nature and on the probabilities of the deterministic automaton laws.

We conclude this section with a warning. In the case of stochastic models there is always a risk that an individual realisation may not bear out the typical model features. Typical properties are those reproducible in a sufficient number of runs. An individual realisation does provide physically significant information on those stochastic parameters v only which obey

$$\frac{\sigma_v}{\langle v \rangle} \ll 1 \qquad 5.37$$

with $\langle v \rangle$ and σ_v the expectation and standard deviation of the stochastic variable v.

... modulus machinae exprimit machinam ipsam, scenographica rei in plano delineatio exprimit solidum ...

Leibniz 1678

6. Final Remarks

In this introduction to the CA methodology we have concentrated on those facets of the CA framework which we feel are of direct relevance for astrophysical applications.

(1) The faithful simulation of the details of a variety of microscopic processes involving particle dynamics, when sufficiently idealised so as to lend themselves to simple and efficient computer algorithms, is found, on a macroscopic level, to belong into different *universality classes* of behaviour captured by well–established systems of PDEs (diffusion equations, wave equations, Navier–Stokes fluid dynamics in various limit cases, etc). The simulation itself, therefore, serves the purpose of *solving the macroscopic* PDEs. The property of microscopic faithfulness of the CA model lends support to the idea that this model captures adequately the macrodynamics when conventional methods of solution fail (in particular discretisation techniques based on small numbers of degrees of freedom). Faithfulness secures that the CA plays the part of a schematic 'blueprint' of the macroscopic machinery (as expressed in the quotation by Leibniz). In particular, the subclass of the *high complexity models* formulated in terms of PDEs not reducible to low dimensional Normal Forms, can be approached by the CA method. The cascading β model of turbulence (Frisch et al 1978), which has been applied to the interpretation of the large scale structure in the universe (cf Provenzale 1991), is essentially an instance of such a procedure.

The process of generating the macro–level behaviour from the schematic microscopic picture (the CA model) proceeds via an intermediate level of representation in terms of *partial difference equations*. The latter in turn constitute a computer–adapted approximation scheme to the macroscopic PDEs. A major advantage of the natural CA difference scheme over conventional difference schemes directly set up from the macroscopic PDEs — and this is a point not explicited in the paper — is that the CA derivation is *numerically stable* (provided only that the parameters in the equations assume physically consistent values). In fact, the intermediate difference equations merely represent averages over the microscopic behaviour, which cannot generate spurious instabilities. Any instability traced within the averaged equations must correspond to a genuine physical instability.

An illustration of this point is provided by the diffusion process as represented by the difference equation (2.20) (interpreted as a standard equation). The classical stability analysis (cf Richtmyer 1957) indicates that the scheme is numerically stable provided that the parameter p obeys the inequality $p \leq 1$. But the latter condition is necessarily satisfied in the context of the microscopic model, since p is a probability (probability of motion of a lattice particle).

From this point of view, the CA framework thus offers in fact *two* computationally relevant schemes for solving macroscopic PDEs.

(2) Even though a schematic 'blueprint', and not a correct replication of the microphysics, a simulation of the elementary processes, to the extent that it remains faithful, necessarily conceals more

information than the corresponding macrodynamics encoded in the PDEs. The extra information supplied by the CA simulation manifests itself in the form of (space and time) ***fluctuations*** on a submacroscopic level. It is true that the simulation necessarily deals with a ***small*** number of microscopic 'particles' (in comparison with the real number of particles involved in the genuine astrophysical problem). The model therefore necessarily exaggerates the fluctuation amplitudes, so that for many workers in the field these fluctuations are regarded as a nuisance.

However, in the majority of astrophysical problems formulated in terms of PDEs we are forced to simplify the geometry, in order to render the treatment numerically manageable. A typical example consists in singling out a (mathematically exact or approximate) subclass of solutions, reducing for instance an originally $3D$ problem to a formally $1D$ problem. The calculation of an evolutionary sequence of a star or of a spherical cluster, or the treatment of stellar pulsations, is simplified in this way by restricting the admissible class of solutions to radially symmetric solutions (which are exact solutions of this problem). Or, by using for instance spatial expansions in a complete set of functions (such as Fourier expansions), and by truncating these expansions after a small number of terms, the original PDEs are approximated by a system of ODEs; the latter methods are favoured in studies of convection. Reductions of this nature carry with them the danger of artificially freezing and ignoring a broad class of collective degrees of freedom which may be excited in the real world as a result of instabilities in these modes.

A CA simulation avoids this physical inconsistency. Even though we may enforce simplifying statistical symmetries on the initial conditions (for instance approximate one–dimensionality in the input physics), if an unstable collective mode exists which is not excited by the mean field, the intrinsic fluctuations (of arbitrary symmetries) eventually lead to the excitation of this degree of freedom. The 'thermal' fluctuations correctly break an artificially introduced input symmetry.

It should be kept in mind, however, that the unrealistic amplitudes of the CA fluctuations can lead to qualitatively spurious side effects. For instance, a large CA fluctuation —resulting from the small numbers of particles involved in the numerical simulation — may be observed to transform one stable macroscopic solution into a second stable solution; if repetitive, such switches between different solutions manifest themselves as fluctuation–induced ***intermittency*** phenomena. In the corresponding real situation the true physical fluctuations may remain too small to produce the transition. Although intermittency generated by fluctuations is a perfectly conceivable phenomenon (especially in the neighbourhood of the transition towards chaos, cf Perdang 1990b, 1991), when traced in a particular CA realisation, a special analysis is needed before we can be sure that it is not just a small number artifact.

In this context a comment is in order on the more recent renewal of interest in stochastically perturbed macroscopic equations in astrophysics, the stochastic terms simulating intrinsic fluctuations (or possibly extrinsic unknown disturbances). In this method, both the amplitudes, and the spatial and temporal character of the fluctuations are unknown at the start; the fluctuations are specified by a priori selected statistical characterisations which therefore confer the stochastic approach a greater flexibility than the CA simulation. In the latter the nature of the fluctuations is rigidly determined by the model; only the fluctuation amplitudes can be modified by changing the size of the CA realisation or the size of the averaging box. On the other hand, to the extent that the microphysics is faithfully dealt with, it is precisely a virtue of the CA model that the real fluctuations are those generated in the simulation; physical fluctuations just do not obey ***any*** conceivable stochastic distribution. Moreover, and more important, in the stochastic models one keeps simplifying maximally the geometric setting of the PDEs (for instance one deals with a $1D$ instead of the real $3D$ problem, etc). Manifestly then, the stochastic representation cannot bring about transitions towards solutions not accounted for in the formulation of the problem (changes of symmetry, etc). On the other hand the stochastic formalism can induce intermittency phenomena (among the class of solutions actually captured by the deterministic part of the representation). But just as in the CA case, before we can assess the physical reliability of those phenomena, we must be sure that the required amplitude level is attainable in the real physical problem.

In the astrophysical context, the explicit inclusion of intrinsic fluctuations, as naturally provided by the CA framework, appears as a determining element for generating physically meaningful evolutionary results.

(3) Besides its relevance for simulating the solutions of an analytically given macrodynamics, the CA framework encounters an even more promising area of application in the category of ***high complexity problems*** which defy a concise global description in terms of the conventional formalism of macroscopic PDEs. Provided that the general nature of the basic (generally small scale) physical mechanisms is understood, even though the detailed mathematical description of these mechanisms is not accessible, a CA model seems always specifiable under the following conditions :

(a) The representative macroscopically observable quantities of interest, say N_q, $q = 1, 2, \ldots$, involved in each mechanism, are extensive variables distributed over space (and time), $N_q = N_q(t, \mathbf{r})$; on the level of the microscopic behaviour of the automaton, these physical variables are averages of corresponding microscopic variables F_q which are discrete (typically $F_q \in Z_k$, k small);

(b) the physical mechanisms to be modelled consist in processes of local production and destruction of the microscopic quantities, ruled by the physical conditions prevailing in a certain local neighbourhood; specifically, the rate of formation or decay of the qth quantity at position $\mathbf{r}$, and time t, $r_q(t, \mathbf{r})$, depends on the microscopic variables $F_{q'}$, $q' = 1,2, \ldots$ (or the corresponding macroscopic observables $N_{q'}$) at time t and at neighbour points $\mathbf{r}'$, $\mathbf{r}''$, ... of the local point $\mathbf{r}$,

$$r_q(t, \mathbf{r}) \;=\; r_q\;[\;F_1(t, \mathbf{r}'), F_2(t, \mathbf{r}'), \ldots; F_1(t, \mathbf{r}''), F_2(t, \mathbf{r}''), \ldots\;];$$

(c) these quantities are transported from neighbourhood points $\mathbf{r}^*$, $\mathbf{r}^{**}$, ... to point $\mathbf{r}$, the rate of transport, $c_q(t, \mathbf{r})$, depending again on the physics of a number of neighbourhood points $\mathbf{r}'$, $\mathbf{r}''$, ...

$$c_q(t, \mathbf{r}) \;=\; c_q\;[\;F_q(t, \mathbf{r}^*), F_q(t, \mathbf{r}^{**}), \ldots; F_1(t, \mathbf{r}'), F_2(t, \mathbf{r}''), \ldots; F_1(t, \mathbf{r}''), F_2(t, \mathbf{r}''), \ldots\;].$$

These conditions express that the model quantities obey Boltzmann–type equations (master equations). Barring exceptional situations, the latter are not necessarily expressible as PDEs, due to the complex character of the creation and transport mechanisms. As already realised by Ulam, the resulting model is effective in generating complex space and time patterns which the standard framework of PDEs is unable to account for (' ... the geometry of objects defined by recursions and iterative procedures ... produces a variety of sets different from those defined by explicit algebraic or analytical expressions or by the usual differential equations'; Schrandt and Ulam 1970).

Natural quantities which meet requirements (a), (b), (c) are 'particles of species q' i.e. physical objects or conditions possessing a certain degree of permanence, such as molecules, dust clouds, protostars, or superclusters of galaxies ($F_q(t, \mathbf{r}) = 0$ or 1 if space point $\mathbf{r}$ is occupied or not at time t by a type q object); other instances are more abstract concepts like momentum, entropy, or vorticity. In high complexity problems, as a consequence of the lack of a detailed knowledge (or the mere impracticality of a complete description) of the creation, decay and transport processes, our ignorance is parametrised by a probabilistic scheme. In many experiments it is found that the existence of typical space–time patterns (fractal structures, chaos, etc) is independent of the precise parametric description, although the exact occurrence of a specific pattern depends on the quantitative formulation. Such models of partial ignorance thus keep both predictive and interpretative value in spite of the fragmentary input physics on which they rely.

Besides the practical aspect of ***applicability*** of the CA framework to specific natural situations, we have analysed several mathematical formalisms capable of concisely capturing the time and space behaviour of the CA dynamics.

(4) In the first place, we have emphasised that the mere translation of the conventional CA discrete environment (spatial topology, spacetime geometry, evolution rules, etc) into the setting of ***nonstandard analysis*** supplies several nontrivial simplifications.

(a) It yields almost directly the averaged ***macroscopic evolution equations*** — provided that meaningful macroscopic evolution equations exist — from the set of the associated nonstandard difference equations.

(b) It further leads to a systematic characterisation of the different *classes of time behaviour* a CA is capable of supporting (in terms of different orders of infinity of the period of the solution).

(c) And finally, it has led to a natural extension of the conventional CA, defined over regular crystallographic lattices, to CA with underlying *continuously deformed space lattices.* The latter are relevant for simulating space dependent coefficients in the macroscopic equations (for instance, space dependent sound speed in a wave equation).

(5) In the second place we have stressed the theoretical interest of the *transition matrix formalism* (Markov chain representation of the CA evolution). While probably prohibitive from a computational point of view — in any realistic simulation the transition matrix is of such a high dimension that it defies the storage capabilities of any present day computer — this matrix does however lend itself to a theoretical analysis. A convenient systematic procedure consists in an inspection of alternatives of increasing complexity in the initial conditions; in fact, very often, having discussed explicitly the simplest alternatives (for instance 1, 2, 3 particle cases in problems in which a particle conservation law is obeyed), one can extrapolate to the general case. This type of analysis leads to a characterisation not only of the attractor, but also of the *occupation probabilities* of the attractor states. Consequently, it provides a main step towards a statistical mechanics of the asymptotic behaviour of the system. In particular the characterisation of the equilibrium state of general lattice gases of astrophysical relevance (not, or only partially subject to an exclusion principle, in the presence of forces) is expected to be accessible through this approach. We recall also that in the case of stochastic CA the Markov transition matrix representation clarifies the variety of intermittency processes.

Besides the various explicit models of fractal growth processes (cf the models reviewed in Stanley and Ostrowsky 1986, and in Gouyet 1992), the various types of percolation problems (cf Isichenko 1992), and besides applications of other theoretical methods to CA (renormalisation group theory applied to CA phase transitions; Bagnoli et al 1992; conformal invariance in CA phase transitions, Itzykson et al 1988) not touched upon in this paper, I briefly mention here the general purpose CA technique devised by Margolus (cf Toffoli and Margolus 1987). In the $2D$ case this approach is summarised as follows. An original square lattice is imbedded in two superlattices, of cells twice the linear size of the original cells, and such that the second superlattice is shifted with respect to the first superlattice by one cell in both the x and y direction; a cell of the original lattice is thus the intersection of a supercell of the first with a supercell of the second superlattice. Interpret now the occupation of a supercell as a 4–bit 'particle' model (each one of the 4 subcells of a supercell contains at most one 'particle'). The microscopic laws of evolution of the system are then formulated as rules of bit exchanges (particle exchanges among the subcells). The rules are applied at timestep 0 to the supercells of the first superlattice, and, at timestep 1, to the supercells of the second superlattice; then again at step 2 to the first, and at step 3 to the second superlattice, etc. Finally the first superlattice is regarded as the arena of the observable physical system, at the sequence of even times 0, 2, 4, ... (the secondary lattice, together with the sequence of odd times 1, 3, 5, ... merely plays the part of an auxiliary device). It is readily seen that translational motions along the diagonals are generated in this scheme by swapping diagonal subcells in a supercell; collisions, arising if a supercell contains two particles populating one diagonal, are simulated by swapping the diagonals, etc. This framework has been used to simulate LG gases (HPP gas), (Fermion–type) diffusion, and a variety of other physical phenomena (cf Toffoli and Margolus 1987).

One of the main currently missing elements in CA modelling is a satisfying treatment of long range *forces*. Although attempts at dealing with forces when the system is close to thermodynamic equilibrium have been discussed already for some time in the literature (Frisch et al 1987, Appendix D; cf the CA simulation of Rayleigh–Taylor instabilities in Clavin et al 1986), even such a trivial problem as the motion of a cloud of non–colliding particles under the effect of an external force is out of reach of present day CA simulation capabilities. But situations in which gravitational forces are acting on gases far from thermal equilibrium are the rule rather than the exception in the astrophysical context.

The lack of a general force treatment seems to constitute a major hurdle to a satisfactory application of the current CA framework to many astrophysical problems. It is our belief that the general setting of a higher level CA, as sketched in this paper, is required to provide a definitive solution to the force problem.

There are strong indications that a CA environment — whenever available — offers a logically more satisfying and, in general, also operationally more efficient computational framework for astrophysical modelling than the traditional continuous procedures. The latter approaches involve in fact at least two levels of approximation : In a first step, to set up the relevant macroscopic model equations, the mathematical properties of the physical variables of the phenomenon must be idealised (approximation by *smooth* field variables described in terms of characteristically nonlinear PDEs). In a second step, to solve these model equations, the latter in turn must again be approximated by some ad hoc discretisation scheme whose numerical stability is not guaranteed a priori.

As Hasslacher (1988) comments in the special case of modelling fluids by the Navier–Stokes dynamics : The relevant model equations 'are highly nonlinear; this is prototypical of field–theoretical descriptions of large physical systems. The nonlinearity ... forces one to solve the system by approximation techniques. Invariably these are some form of perturbation methods in whatever small parameters can be devised. Since there is no systematic way of applying perturbation theory to highly nonlinear partial differential systems, the analysis of [these equations must remain ...] a patchwork of ingenious techniques that are designed to cover special parameter ranges ...'. The approximation scheme eventually is solved by discretising its representative equations, which really amounts 'to introducing an *artificial microworld*'.

In the astrophysical context Hasslacher's remarks typify the current treatment of the variety of irregular space structures ranging from 2 and $3D$ astrophysical jet simulations to convective flows and supernova explosions (Müller 1991; Miyama and Nagasawa 1991). In all of these experiments the authors are dealing with models based on essentially well known, although sometimes involved physics (as in the case of supernova calculations) encoded in a well–established mathematical framework (conventional fluid dynamics). The complicated spatial patterns eventually emerge from these experiments, as the outcome of huge amounts of nontrivial numerical operations. For many theoretical astrophysicists the accuracy of these numerical operations is not taken for granted; for them there remains serious doubt whether the computed pattern formation is real or whether it is a mere artifact of the numerical approximation techniques.

If at the outset we are able to devise a proper CA 'microworld' adapted to the problem to be modelled, we automatically avoid any further approximations. All model variables being encoded in an integer number scheme, the evolutionary CA calculations are essentially *error–free*. Once the model itself is accepted, there remains no shred of a doubt about the validity of either the qualitative or the quantitative features of its solution. In the CA computational framework the burden of finding adequate solutions is placed on the construction of the microscopic model itself.

References

Ashbrook J., Duncombe R.L., Van Woerkom A.J.J. 1954 *Astronomical Journal* **59** 12

Auvergne M., Baglin A. 1985 *Astronomy and Astrophysics* **142** 388

Bagnoli F., Bulajich R., Livi R., Maritan A. 1992 *Journal of Physics* **A25** L1071

Bak P. 1993 (these Proceedings)

Bak P., Chen K. 1989 *Physica* **D38** 5

Bak P., Tang C., Wiesenfeld K. 1987 *Physical Reviews Letters* **59** 381

Baker N. 1963 in *Star Evolution*, International School of Physics E. Fermi, Varenna, Academic Press, New York

Bally J., Langer W.D., Wilson R.W., Stark A.A., Pound M.V. 1991 in *Mechanisms of Molecular Clouds and Star Formation*, IAU Symposium **147** 11; editors: Falgarone E., Boulanger F., Duvert G., Kluwer Academic

Bharucha-Reid A.T. 1960 *Elements of the Theory of Markov Processes and Their Applications*, McGraw-Hill, New York

Boon J.P., Frisch U., d'Humières D. 1993 *La Recherche* **24** 391

Buchler J.R., Regev O. 1982 *The Astrophysical Journal* **258** 426

Burrough P.A. 1981 *Nature* **294** 240

Burrough P.A. 1989 in *The Fractal Approach to Heterogeneous Chemistry*, p 383, editor: Avnir D., Wiley, New York

Burks A.W. (ed.) 1970 *Essays on Cellular Automata*, University of Illinois Press, Urbana, Ill

Chaté H., Manneville P. 1990 in *Cellular Automata and Modelling of Complex Physical Systems*, p. 298; editors: Manneville P., Boccara N., Vichniac G.Y., Bidaux R., Springer Proceedings in Physics 46, Springer, Berlin

Chincarini G., Rood H.J. 1975 *Nature* **257** 294

Chopard B., Drooz M., Cornell S., Frachtenbourg L. 1993 (these Proceedings)

Christy R.F. 1962 *The Astrophysical Journal* **136** 887

Clavin P., d'Humières D., Lallemand P., Pomeau Y. 1986 *Comptes Rendus de l'Académie, Paris* **303 II** 1169

Codd E.F. 1968 *Cellular Automata*, Academic Press

Coleman P.H., Pietronero L. 1992 *Physics Reports* **213** 311

Cooper N.G. (ed.)1988 in *From Cardinals to Chaos: Reflections on the Life and Legacy of Stanislaw Ulam*, Cambridge University Press, Cambridge

Culik K., Hurd L.P., Yu S. 1990 *Physica* **D45** 357

Culik K., Yu S. 1988 *Complex Systems* **2** 177

Deeming T.J. 1970 *Astronomical Journal* **75** 1027

de Wijs H.J. 1951 *Geologie en Mijnbouw* **13** 365

Dress A.W.M., Gerhardt M., Jaeger N.I., Plath P.J., Schuster H. 1984 in *Temporal Order*, p 67; editors: Rensing L., Jaeger N.I., Springer, Berlin

Elmegreen B.G. 1985 in *Birth and Infancy of the Stars*, p 215 Les Houches **XLI**, editors: Lucas R., Omont A., Stora R., Elsevier

Feder J. 1988 *Fractals*, Plenum Press, New York

Feller W. 1971 *An Introduction to Probability Theory and its Applications*, Vol. I, Wiley, New York

Feynman R. 1967 *The Character of Physical Law*, MIT Press, Cambridge, Massachusetts

Fine N.F. 1947 *American Mathematical Monthly* **54** 589

Frisch U., Hasslacher B., Pomeau Y. 1986 *Physical Review Letters* **56** 1505

Frisch U., d'Humières D., Hasslacher B., Lallemand P., Pomeau Y., Rivet J.P. 1987 *Complex Systems* **1** 649

Frisch U., Sulem P., Nelkin M. 1978 *Journal of Fluid Mechanics* **87** 719

Genzel R., Stutzki J. 1989 *Annual Review of Astronomy and Astrophysics* **27** 41

Gerola H., Seiden P.E., Schulman L. 1980 *The Astrophysical Journal* **242** 517

Goles E. 1990 in *Cellular Automata and Modelling of Complex Physical Systems*, p. 10; editors: Manneville P., Boccara N., Vichniac G.Y., Bidaux R., Springer Proceedings in Physics 46, Springer, Berlin

Gouyet J.F. 1992 *Physique et Structures Fractales*, Masson, Paris

Gutowitz H. (editor) 1991 *Cellular Automatata: Theory and Experiment*, MIT press, Cambridge, Mass.

Hardy J., De Pazzis O., Pomeau Y. 1976 *Physical Review* **A13** 1949

Harris B. 1960 *Annals of Mathematical Statistics* **31**

Harthong J. 1981 *Advances in Applied Mathematics* **2** 24

Hasslacher B. 1988 in *From Cardinals to Chaos: Reflections on the Life and Legacy of Stanislaw Ulam*, p 175, editor: Cooper N.G.

Heggie D.C. 1989 in *Dynamics of Dense Stellar Systems*, p 195, editor: Merritt D., Cambridge University Press, Cambridge

Hénon M. 1988 *Physica* **D33** 123 (for a more popular discussion see 1989 *La Recherche* **20** 490)

Hénon M. 1989 in *Kinetic Theory, Lattice Gases, and Foundation of Hydrodynamics*, editor: Monaco R, World Scientific

Hesse H. 1943 *Das Glasperlenspiel: Versuch einer Lebensbeschreibung des Magister Ludi Josef Knecht samt Knechts hinterlassenen Schriften*, Fretz und Wasmuth, Zürich, Band I und II

Hogg T., Huberman B.A. 1985 *Physical Review* **A32** 2338

Hoshen J., Kopelman R. 1976 *Physical Review* **B14** 3438

Ito H. 1988 *Physica* **D31** 318

Itzykson C., Saleur H., Zuber J.B. (editors)1988 *Conformal Invariance and Application to Statistical Mechanics*, World Scientific

Jaeger H.M., Nagel S.R. 1992 *Science* **255** 1523

Kaneko K. 1985 *Progress in Theoretical Physics* **74** 1033

Kaneko K., Ikegami T. 1992 *Physica* **D56** 406

Kemeny J.G., Snell J.L. 1960 *Finite Markov Chains*, Van Nostrand, Princeton, N.J.

Kemeny J.G., Snell J.L., Knapp A.W. 1976 *Denumerable Markov Chains*, Springer, New York

Kiang T. 1966 *Zeitschrift für Astrophysik* **64** 433

Kolb M. 1986 *Journal of Physics* **A19** L263

Kreith K. 1973 *Oscillation Theory*, Lecture Notes in Mathematics **324**, Springer, Berlin

Landau L.D., Lifschitz E.M. 1966 *Hydrodynamik*, Akademie–Verlag, Berlin

Lejeune A., Perdang J. 1991 in *Proceedings of the European Conference on Computational Physics*, World Scientific, Singapore

Levermore C.D., Boghosian B.M. 1990 in *Cellular Automata and Modelling of Complex Physical Systems*, p. 118; editors: Manneville P., Boccara N., Vichniac G.Y., Bidaux R., Springer Proceedings in Physics 46, Springer, Berlin

Levy Y.E. 1982 *Physics Letters* **88A** 1

Liu Y. 1991 *Physica* **D54** 147

Lutz R., Goze M. 1981 *Nonstandard Analysis*, Lecture Notes in Mathematics **881**, Springer, Berlin

Manneville P. 1990 in *Cellular Automata and Modelling of Complex Physical Systems* p. 1; editors: Manneville P., Boccara N., Vichniac G.Y., Bidaux R., Springer Proceedings in Physics 46, Springer, Berlin

Martin O., Odlyzko A.M., Wolfram S. 1984 *Communications in Mathematical Physics* **93** 219

Martinez V.J. 1991 in *Applying Fractals in Astronomy*, p 135, Lecture Notes in Physics **m3**, Springer, Berlin

Miyama S.H., Nagasawa M. (editors) 1991 *Symposium of Supercomputing*, Astronomy and Astrophysics in Japan , NAO, Mitaka

Müller E. 1991 in *Supernovae*, Les Houches Summer School, Session **LIV**, editors: Audouze J., Bludman S., Mochkovitch R., Zinn–Justin J.; Elsevier

Myhill J. 1964 in *Proceedings of the Second Systems Symposium at Case Institute of Technology*, p 106, editor: Mesarovic M.D., Wiley, New York

Nemytskii V.V., Stepanov V.V. 1960 *The Qualitative Theory of Differential Equations*, Princeton University Press, Princeton N.J.

Oppo G.L., Kapral R. 1986 *Physical Review* **A33** 4219

Packard N.H. 1985 in *Dynamical Behaviour of Automata : Theory and Applications*; editors: Demongeot J., Goles E., Tchuente M., Academic Press, New York

Peebles P.J.E. 1980 *The Large-Scale Structure of the Universe*, Princeton University Press, Princeton

Perdang J. 1979 *Stellar Oscillations: The Asymptotic Approach*, Lecture Notes, 3e Cycle Interuniversitaire en Astronomie et Astrophysique, FNRS, Brussels

Perdang J. 1990 *in The Numerical Modelling of Nonlinear Stellar Pulsations*, NATO ASI Series **C 302** 333

Perdang J. 1990a *Annals of the New York Academy of Sciences* **617** 87

Perdang J. 1990b *Radial Stellar Pulsations: Stability, Chaos, and Beyond*, p 311, in Cours de Structure interne 2e Session, Aussois, Octobre 1990, editors: Hubert A.M., Schatzman E, Observatoire de Paris, Meudon

Perdang J. 1990c *Vistas in Astronomy* **33** 371

Perdang J. 1991 in *Rapid Variability of OB Stars: Nature and Diagnostic Value*, ESO Conference and Workshop Proceedings **36** 349

Perdang J. 1991a in *Applying Fractals in Astronomy*, p 1, *Lecture Notes in Physics* **m3**, Springer, Berlin

Perdang J. 1992 in *Proceedings of the IAU Colloquium* **134** *Nonlinear Phenomena in Stellar Variability*, Mito, Japan

Pietronero L. 1987 *Physica* **A144** 257

Provenzale A. 1991 in *Applying Fractals in Astronomy*, p 97, Lecture Notes in Physics **m3**, Springer, Berlin

Rana N.C. 1991 *Annual Review of Astronomy and Astrophysics* **29** 129

Richtmyer R.D. 1957 *Difference-Methods for Initial Value Problems*, Interscience, New York

Regev O. 1990 in *The ubiquity of Chaos*, p 223, American Association for the Advancement of Science, Washington

Robert F. 1986 *Discrete Iterations*, Springer, Berlin

Robert A. 1985 *Analyse non standard*, Presses polytechniques romandes, Lausanne

Robinson A. 1980 *Non-Standard Analysis*, North-Holland, Amsterdam

Rosseland S. 1949 *The Pulsation Theory of Variable Stars*, Clarendon Press, Oxford

Sapoval B., Rosso M., Gouyet J.F. 1989 in *The Fractal Approach to Heterogeneous Chemistry*, p 227, editor: Avnir D.; John Wiley, New York

Saslaw W.C. 1985 *Gravitational Physics of Stellar and Galactic Systems*, Cambridge University Press, Cambridge

Scalo J.M. 1987 in *Interstellar Processes*, p 349; editor: Hollenbach D.J., Thronson H.A.; Reidel, Dordrecht

Shandarin S.F., Zeldovich Ya. B. 1989 *Reviews of Modern Physics* **61** 185

Schrandt R.G., Ulam S.M. 1970 in *Essays on Cellular Automata*, p 232 ; editor: Burks A.W., University of Illinois Press, Urbana, Ill

Schulman L. 1993 (these Proceedings)

Schuss Z. 1980 *Theory and Applications of Stochastic Differential Equations*, Wiley, New York

Seiden P.E., Gerola H. 1982 *Fundamentals of Cosmic Physics* **7** 241

Seiden P.E., Schulman L., Gerola H. 1979 *The Astrophysical Journal* **232** 702

Serra J. 1988 *Image Analysis and Mathematical Morphology*, Vol. 1, Academic Press, London

Sneddon I.N., Berry D.S. 1958 in *Handbuch der Physik* **VI** 1, Springer, Berlin

Stanley H.S., Ostrowski N. (editors) 1986 *On Growth and Form*, Nijhoff, Dordrecht

Stauffer D. 1985 *Introduction to Percolation Theory*, Taylor and Francis, London

Stauffer D., Pandey R.B. 1992 *Computers in Physics* **6** 404

Stein R. F. 1966 in *Stellar Evolution*, p 3 , editors: Stein R.F., Cameron A.G.W., Plenum Press, New York

Tanaka Y., Seya K., Takeuti S. 1991 (preprint)

Toffoli T. 1984 *Physica* **D10** 117

Toffoli T., Margolus N. 1987 *Cellular Automata Machines: A New Environment for Modeling*, MIT Press, Cambridge Mass.

Truesdell C., Noll W. 1965 in *Handbuch der Physik* **III/3** 1, Springer, Berlin

Truesdell C., Toupin R. 1960 in *Handbuch der Physik* **III/1** 226, Springer, Berlin

Ulam S.M. 1950 *Proceedings of the International Congress on Mathematics* **2** 264

Ulam S.M. 1962 *Proceedings of Symposia in Applied Mathematics* **14** 215

Valdarnini R., Borgani S., Provenzale A. 1992 in press

van der Hoef M. 1993 (these Proceedings)

Van Weygaert R. 1991 *Monthly Notices of the Royal Astronomical Society* **249** 159

von Mises 1931 *Vorlesungen aus dem Gebiete der angewandten Mathematik.* I. Band. Wahrscheinlichkeitsrechnung, Deuticke, Leipzig

Vichniac G.Y. 1984 *Physica* **D10** 96

Vicsek T. 1991 *Fractal Growth Phenomena*, World Scientific, Singapore

Vicsek T., Szalay A.S. 1987 *Physical Review Letters* **58** 2818

Vio R., Cristiani S., Lessi O., Salvadori L. 1991 *The Astrophysical Journal* **380** 351

Weitz D.A., Lin M.Y. 1986 *Physical Review Letters* **57** 2037

Willson S.J. 1984 *Discrete Applied Mathematics* **8** 91

Wolfram S. 1983 *Reviews of Modern Physics* **55** 601

Wolfram S. 1984 *Physica* **D10** 1

Wolfram S. 1985 *Physical Reviews Letters* **55** 449

Wolfram S. 1986 *Journal of Statistical Physic* **45** 471

Wolfram S. 1986a *Advances in Applied Mathematics* **7** 123

Zimmermann H.J. 1992 *Fuzzy Set Theory and its Applications*, Kluwer, Dordrecht

Zumofen G., Blumen A., Klafter J. 1988 in *Fractals, Quasicrystals, Knots and Algebraic Quantum Mechanics*, NATO ASI **C 235**

Zuse K. 1972 *Nova Acta Leopoldina* **37** (Nr 206) 129

Zuse K. 1975 *Nova Acta Leopoldina* **43** (Nr 220) 5

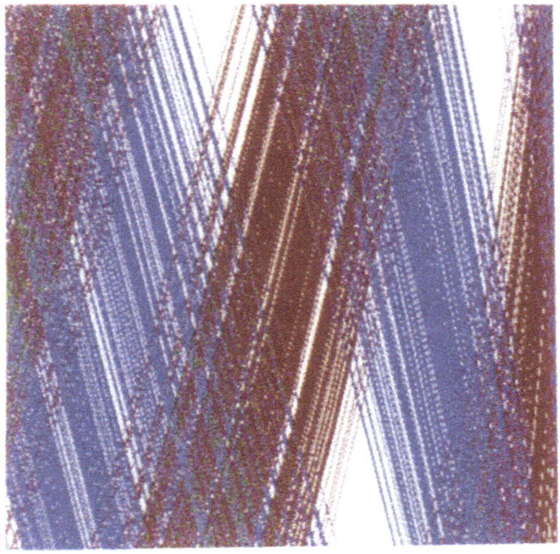

Fig. 1.a. – CA solution of a $1D$ wave equation with space dependent propagation speed. Initial condition : sine wave. (blue: $v < 0$; red : $v > 0$; green : pair of particles).

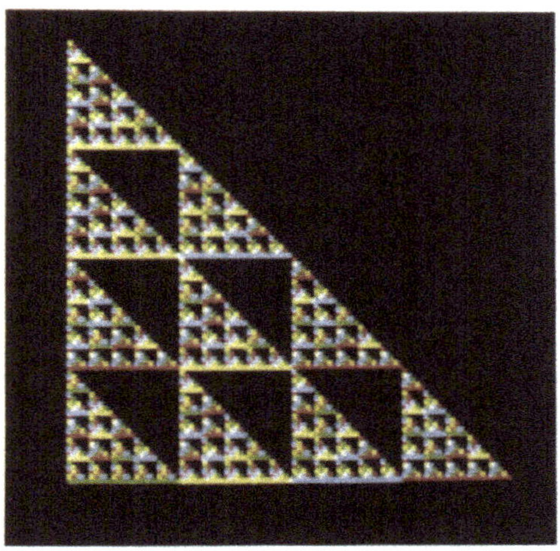

Fig. 4.b – 5–state $2D$ CA obeying modulo 5 rule (Eq. 4.36). Initial configuration: single activated cell in state 1. Configuration shown at timestep 99; at step 100 almost all cells are black (Colour code: state 0: black; 1: yellow; 2: red; 3: green; 4: cyan).

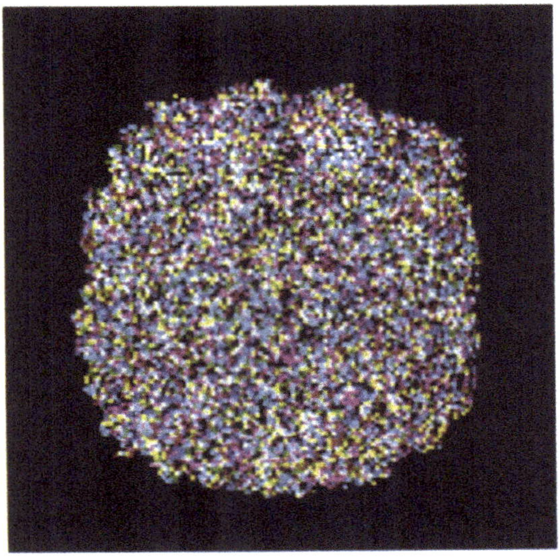

Fig. 5.b. – 7–state level 1 2D automaton (complete randomness of level 1 rule). Initial configuration: single activated cell in state 1. Configuration shown at timestep 55 (Colour code: as in Fig. 4.b; additional states 5: magenta; 6: blue)

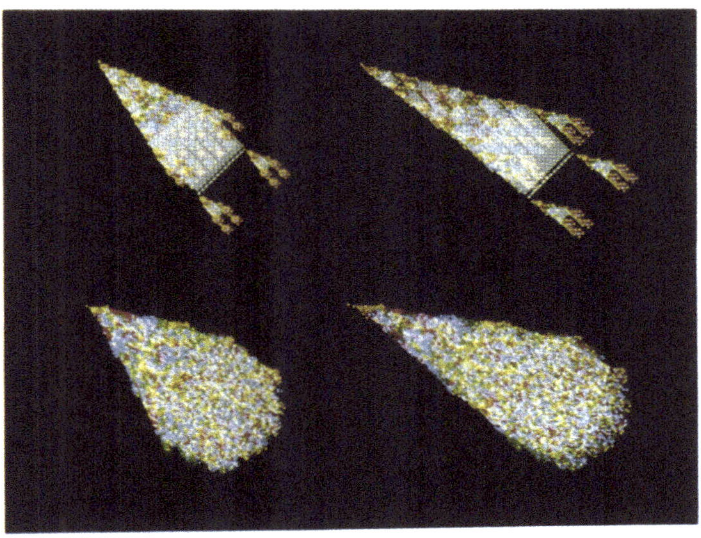

Fig. 6 – Stereoscopic display of growth pattern in $3D$ space–time generated by a 5–state level 1 $2D$ CA of level 0 rules $x + y$, $x^{-1} + y$, $x + y^{-1}$ and $x^{-1} + y^{-1}$. Rules alternating rules with time (upper frame); and randomly selected rules (lower frame).

Part II :

Models of Astrophysical Relevance

Introduction to Lattice Gas Cellular Automata as a Model for Simple Fluids

Martin A. van der Hoef[1]

F.O.M.-Institute for Atomic and Molecular Physics, Kruislaan 407, 1098 SJ Amsterdam, The Netherlands

Abstract

Lattice gas cellular automata, introduced by Frisch, Hasslacher and Pomeau in 1986 [1], can be considered as an extremely simplified form of molecular dynamics; for instance, both time and space are discretized. It can be shown that, in a well-defined limit, the behavior of these lattice gases is very close to that of a classical fluid, i.e. the macroscopic equations of motion of the model resemble the Navier-Stokes equations. The advantage of a lattice-gas model over other fluid models is that it is microscopic in character, so it contains all fluctuations and correlations in a natural way. In the present paper we give an introduction to lattice gas cellular automata in their most basic form, presenting the rules, some of the theoretical background, and one application for which these models have been successful.

[1]Present address: Physical Chemistry Laboratory, Oxford University, South Parks Road, Oxford OX1 3QZ, U.K.

1 Introduction

The behavior of a Newtonian fluid can be well described, on the macroscopic level, by the Navier-Stokes equations. These are equations which govern the time-evolution of the basic hydrodynamic quantities, which are the mass density $\rho(\mathbf{r}, t)$, the flow velocity $\mathbf{u}(\mathbf{r}, t)$ and energy density $e(\mathbf{r}, t)$. The Navier-Stokes equations follow from the continuity equations for mass, momentum and energy, combined with the phenomenological constitutive relations, which state that the gradients in the hydrodynamic variables are proportional to the corresponding fluxes. The proportionality constants are the transport coefficients of the fluid, which are the (collective) diffusion constant, the kinematic and bulk viscosities, and the thermal conductivity. In the above, the fluid is considered on the macroscopic level, and the hydrodynamic variables are continuous functions of the space variable $\mathbf{r}$ and time variable t. We also know that a fluid consists of many interacting molecules. Indeed, it is possible to derive the Navier-Stokes equation all the way from the microscopic dynamics of the appropriate many-particle system, by making use of the methods of statistical mechanics [2, 3]. Such a derivation make use the (approximate) Boltzmann equation as an intermediate level of description, and from the Boltzmann equation one arrives at the Navier-Stokes equation by the so-called Chapman-Enskog expansion. We stress however, that this derivation of the Navier-Stokes equations makes use of certain assumptions, and a rigorous derivation is still lacking. For a recent discussion, see [4].

If one wants to simulate fluid-flow, one option is to numerically solve the Navier-Stokes equation by standard techniques, like finite-element or finite-difference schemes, which may be rather time-consuming; moreover, such techniques can be impractical for problems with arbitrary shaped boundaries. Alternatively, we know that the mechanism underlying fluid motion is the movement of molecules. Thus, in a molecular dynamics simulation one should, on sufficiently large scales, observe hydrodynamic behavior. In the derivation of the equations of fluid mechanics from the microscopic dynamics it is implicit that there is large separation between microscopic scales (the mean-free time/path of a molecule) and macroscopic scales (time/length scales at which ρ, $\mathbf{u}$ and e vary). The technique of molecular dynamics, although straightforward, is relatively time consuming, and one is limited to small systems, where the separation of scales might not be sufficient. However, the computer simulations by Alder and Wainwright [5] demonstrated that a small system had already fluid-dynamic type behavior. Later it was demonstrated by others [6, 7, 8] that the flow patterns produced by molecular dynamics simulations tend to agree rather well with those predicted on the basis of the Navier-Stokes equations. The problem of a molecular dynamics model for simulating fluid flow is hence not so much that one cannot reach the fluid-dynamic regime, but rather the noise, due to insufficient 'coarse-graining' in time and space for small systems; this makes the molecular dynamics method not suitable for a study of true large-scale 'macroscopic' phenomena in the fluid, at least with the present-day computers. However, in the derivation of the equations of fluid dynamics from the microscopic level, it follows that the details of the collisions between the individual particles are not essential. In fact, the only requirement on the microscopic model to guarantee fluid-type behavior is colliding particles conserving energy, momentum and mass. One can therefore try to make the molecular dynamics method more efficient, by stripping it down to its barest essentials. The simplest model to think of would be one with a single species of particles moving on a lattice, with discrete velocities. The basic idea behind such a simulation model has been summarized by Richard Feynman in the following way [9]:

> "The behavior of a fluid depends very little on the nature of the individual particles in that fluid. For example, the flow of sand is very similar to the flow of water or the flow of a pile of ball bearings. We have therefore taken advantage of this fact to invent a type of imaginary particle that is especially simple for us to simulate. The flow of these particles on a large enough scale is very similar to the flow of natural fluids."

So in other words, these models contain the essential element of a microscopic fluid model, namely molecular motion of particles giving rise to correct macroscopic fluid-dynamic behavior, in a way that is most convenient computationally. Let us now be a bit more specific about these novel models, now generally called *lattice gas cellular automata* (LGCA).

The first type of such a lattice-gas model was put forward by Hardy, Pomeau and de Pazzis in the early seventies, and is known as the HPP model [10]. In this model, particles move in two dimensions on a square lattice, and can have only four possible velocities, along the four link directions in this lattice. At the lattice sites, particles can collide, i.e. change direction, provided that mass and momentum are conserved. However, the HPP model proved to have *not* the correct macroscopic behavior, as the 'hydrodynamic' equations were not isotropic, but still contained some structure of the underlying lattice. So apart from the constraint of conservations, one also has to consider symmetry. In 1985, an improved, two-dimensional model was introduced by Frisch, Hasslacher and Pomeau [1]. This model, generally denoted by the FHP model, is similar in spirit to the HPP model, but now the particles move on a *triangular* lattice, which has enough symmetry to guarantee isotropic macroscopic equations of motion. The rules of the FHP game are shown in Fig. 1. From the update rules it is clear that lattice gas cellular automata(LGCA) make an efficient simulation scheme, both on a parallel computer (local rules) and on a vector computer (simple 'unconditional' algorithm). Moreover, stability of the algorithm is guaranteed, as the update involves only bit manipulation, i.e. the update is exact with no round-off errors. The biggest advantage of LGCA models compared to molecular dynamics models is clearly the discretisation of the velocity space (there are typically some 24 different possibilities for single-particles velocities) which results in a extremely simplified micro dynamical behavior. It is for this reason that the model has been nicknamed "a caricature of a fluid" or "a poor man's molecular dynamics". Yet, in spite of their simplicity, the analogy of these models with the 'real' fluid models is very close, such that the theoretical framework of statistical mechanics of simple fluids can be applied, to a great extent, to these discrete fluids.

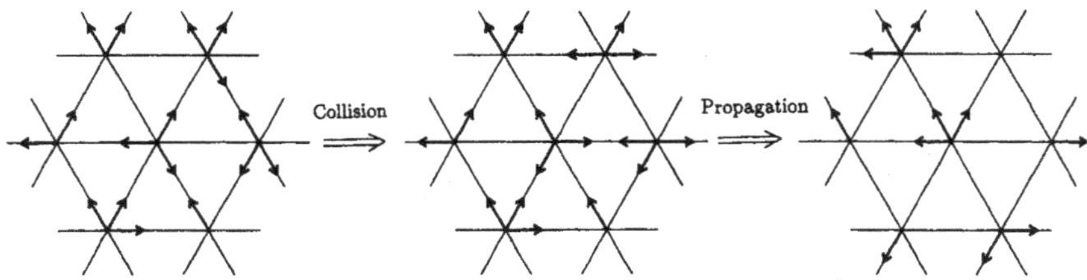

Figure 1: *Example of the time evolution in a small section of the FHP-model. In the figure, each arrow represent the velocity of a single particle. The particles are situated at the lattice sites. The update of the lattice consists of two steps. First there are local collisions at all sites, simultaneously, and such that on each site momentum and number of particles is conserved. Note that only some cases lead to a new configuration. The next step is a propagation step: all the particles move simultaneously according to their velocities to neighboring sites. Particles do not interact during this step. Note that the figure only represents a small section of the lattice; therefore we can only give the correct final configuration of the central site, as the state of the other sites after the propagation depends on neighboring sites which are not shown. Next there will again be a collision step, etc.*

This has been demonstrated initially by Frisch *et al* [11], who proved that the macroscopic equations of motions of these models are, in a well-defined limit, equivalent to the Navier-Stokes equations. Subsequently Ernst and coworkers [12]- [15] have rederived much of the basic equations of statistical mechanics for these LGCA. It is especially this solid theoretical basis which makes the LGCA method a serious numerical scheme, both in the study of hydrodynamic flow phenomena, *and* in the study and testing of theoretical models concerning the molecular motion in simple fluids. However, the first application is seriously hindered by one of the big drawbacks of LGCA: their inherent noisyness. Massive calculations are required to get accurate numbers; this makes the method less efficient compared to macroscopic schemes to model fluid flow, e.g. finite difference schemes. However, in the second application the noisyness is an asset, rather than a liability. This makes a LGCA an ideal model to test the validity of theories of molecular motion in simple liquids. The advantage is that the simplicity of the model not only facilitates the simulation, but also the theoretical evaluation. Verification of a theory in the LGCA fluid model then provides very strong support for the validity of the theory in more realistic fluid models.

A spin-off of the lattice-gas model is the lattice-Boltzmann model, first introduced by McNamara and Zanetti [16]. In this model one does a direct simulation of the lattice-Boltzmann equation, which describes the dynamics at some level in between the microscopic equations and the equations of fluid dynamics. The advantage is that at this level the data is essentially free of noise. It is now generally believed that for the study of fluid-flow, the LBE simulation method is superior to the boolean lattice-gas method, not only because of the absence of noise in the lattice-Boltzmann model, but also because it proves to be easier to achieve high Reynolds numbers, see for instance [17]. In the present paper we will only briefly touch upon the lattice-Boltzmann equation.

We will proceed with a more elaborate description of lattice gas cellular automata. The intention is not to present rigorous and formal definitions and derivations, but rather to give an introduction and show the essential ingredients. We will start in section 2 with the rules and definitions of the game. In section 3 we will present a derivation of the Navier-Stokes equation, from the microscopic equations of motion. Section 4 will be devoted to one of the essential features of LGCA compared to other methods, namely that it contains spontaneous fluctuations. In section 5 we present the specific models that have been used thus far, and mention some implementation problems. Finally in section 6 we discuss one application.

Most of the material below can be found in the basic paper by Frisch *et al*, which still remains the best and most complete reference. This paper, together with most of the other initial papers on lattice gases, has been reprinted in [18]. Other overviews of the lattice-gas method have been given by Ladd [19] and Ernst [20].

2 Rules and Definitions

We consider a system containing N particles. For simplicity we restrict ourselves to a single species of particles with mass m. More elaborate models are possible, but not discussed in this paper. In LGCA-models, time and space are discrete; this means that the model system is defined on a lattice and the state of the automaton is only defined at regular points in time, with separation Δt. The distance between near-neighbor sites in the lattice is denoted by Δl. So the model can be mapped onto a physical system by setting Δt, Δl and m equal to some value in respectively seconds, meters and kilograms. At the discrete times, particles are situated at the lattice sites, with b possible velocities $\mathbf{c}_i$, $i \in \{1, 2, ..., b\}$. The set $\mathbf{c}_i$ can be chosen in many different ways, although they are restricted by the constraint that

$$\mathbf{r}' = \mathbf{r} + \mathbf{c}_i \Delta t \,, \tag{1}$$

where $\mathbf{r}$ and $\mathbf{r}'$ must be lattice sites. This ensures that the particle will stay on the lattice. In the derivations below, it is assumed that all the velocities are non-zero. It is possible to have particles with zero velocity (normally referred to as 'rest' particles), but this would involve some nonessential complexity in the derivations; the final results can be modified to allow for rest particles [21, 22]. We can imagine that a particle at site $\mathbf{r}$ with velocity $\mathbf{c}_i$ occupies a link connecting the site $\mathbf{r}$ with the site $\mathbf{r}'$. In most cases, the velocities are such that only nearest-neighbor lattice sites can be reached in a time step. A next, quite crucial, restriction on the model is that no two particles can be at the same lattice site with the same velocity, hence the particles obey a kind of

Pauli exclusion principle. It will be convenient to define the mean square magnitude per dimension of the velocities

$$c_s^2 = \frac{1}{d}\frac{1}{b}\sum_{i=1}^{b} \mathbf{c}_i \cdot \mathbf{c}_i \; , \tag{2}$$

where d is the dimension of the system; later we will see that c_s can be identified with the speed of sound. Apart from (1) it will prove essential to have additional symmetry requirements on the velocity set, in order to get isotropic macroscopic behavior from these models. These requirements turn out to be that the even rank tensors we can construct from the velocity set are isotropic up to 4th rank, and the odd rank tensors are zero. If we denote the α-component of $\mathbf{c}_i$ by $c_{i\alpha}$, $\alpha = x, y, ..$, this implies that (see [24]):

$$\begin{aligned}
&\sum_{i=1}^{b} c_{i\alpha} = 0 \\
&\sum_{i=1}^{b} c_{i\alpha} c_{i\beta} = b c_s^2 \delta_{\alpha\beta} \\
&\sum_{i=1}^{b} c_{i\alpha} c_{i\beta} c_{i\gamma} = 0 \\
&\sum_{i=1}^{b} c_{i\alpha} c_{i\beta} c_{i\gamma} c_{i\delta} = \frac{bdc_s^4}{d+2}\left(\delta_{\alpha\beta}\delta_{\gamma\delta} + \delta_{\alpha\gamma}\delta_{\beta\delta} + \delta_{\alpha\delta}\delta_{\beta\gamma}\right) \; ,
\end{aligned} \tag{3}$$

where use has been made of (2). In the following we will make frequent use of the above properties of the velocities. One can verify that the above criteria are met by the triangular FHP model, but *not* by the square HPP model.

The time evolution of the LGCA consists of two steps:

1. Propagation: All particles move in one time step Δt from their initial lattice position $\mathbf{r}$ to a new lattice position $\mathbf{r}' = \mathbf{r} + \mathbf{c}_i \Delta t$.

2. Collision: The particles at all lattice sites undergo a collision that conserves the total number of particles and the total momentum at each site. The collision rules may or may not be deterministic.

The state of the automaton at time t can be completely determined by the 'boolean' variable $n_i(\mathbf{r}, t)$, which is equal to 1(0) if a particle is present(absent) on site $\mathbf{r}$ with velocity $\mathbf{c}_i$. From now on it will be more convenient to define time, space and mass (and hence related quantities like density, velocity etc.) *in units of the lattice parameters* Δt, Δl *and* m. The local microscopic mass density $\tilde{\rho}$ is then defined by

$$\tilde{\rho}(\mathbf{r}, t) = \sum_i n_i(\mathbf{r}, t) \; , \tag{4}$$

which is now in units of m per lattice site. The local microscopic velocity field $\tilde{\mathbf{u}}(\mathbf{r},t)$ is defined by

$$\tilde{\rho}(\mathbf{r},t)\,\tilde{\mathbf{u}}(\mathbf{r},t) = \sum_i n_i(\mathbf{r},t)\,\mathbf{c}_i\,. \tag{5}$$

where $\mathbf{u}$ is in units of $\Delta l/\Delta t$. The number of lattice sites is denoted by V. Normally a lattice gas simulation is set up in such a way that no particles or momentum can leak into or out of the system. From this, and from the above update rules follows that the total number of particles N, and the total momentum $\mathbf{P}$, are conserved. We define the average density ρ_0 as

$$\rho_0 = \frac{N}{V}\,, \tag{6}$$

and the average velocity $\mathbf{u}_0$ is

$$\mathbf{u}_0 = \frac{\mathbf{P}}{N}\,. \tag{7}$$

We stress that 'average' in this sense means the average over all lattice sites; this should not be confused with the ensemble average introduced in the next chapter. It is convenient to define the reduced average density f as:

$$f = \frac{\rho_0}{b}\,. \tag{8}$$

f is sometimes also referred to as the mean occupation number. Its value is between 0 and 1. We note that in these models, energy is either trivially conserved (energy is equal to number of particles), or not conserved at all. Hence the models discussed here should be considered as 'a-thermal', there is no proper definition of a temperature. However, there exist more sophisticated models [23] which do conserve energy in a non-trivial way, which is possible by having velocities of different magnitude.

3 From Microphysics to Macrophysics

3.1 Microscopic Equations of Motion

In order to describe the microscopic equations of motion it is convenient to define the state at two distinct moments, namely at time t, just before the collision, and at time t^+ just after the collision. 'Physically' these two times are the same as the collisions can be considered instantaneous. We next define the collision by means of a collision function. If we define the variables $s = \{s_1, s_2, ..., s_b\}$ and $s' = \{s'_1, s'_2, ..., s'_b\}$ as single-site states, then we define the collision by means of the function $\xi(s, s')$ which is equal to one only for states s, s' where s goes over into s' in a collision, and zero for all other possible pairs of states. In this, we have assumed a completely deterministic microscopic behavior, with a one-to-one map of input and output states. In the actual models used, there are in some cases several options for the output state (see for instance Fig. 3). In that case we use several collision functions ξ_k, and the symmetry of the underlying lattice determine which function should be used how often, see item 3 below. Sometimes it is possible to have still a deterministic model, by using prefixed rules deciding at which time step which function is going to be used; for example, if one has only two collision functions which both should be used with the same frequency, one can use one at the even time steps, and the other at the odd time steps. Also a probabilistic behavior is possible, by using a probability distribution to choose a collision function ξ_k. In case of several collision functions, one can describe the 'average' collision by the collision matrix $A(s, s')$

$$A(s,s') = \sum_k P_k \xi_k(s,s') ,$$

where in a deterministic model P_k is the fraction of time steps that ξ_k is used, and in a probabilistic model P_k is the probability distribution. There are several restrictions on the collision matrix A:

1. Conservation: mass and momentum should be conserved in a collision, and no other quantity. In general, conservation of a quantity in a collision implies that the collision function must be such that

$$\sum_i a_i(s_i - s_i')\, \xi(s,s') = 0 , \tag{9}$$

where a_i is a so-called collisional invariant. In a lattice gas, both density and momentum are conserved. This means that, for instance in three-dimensions, the collisional invariants are $1, c_{ix}, c_{iy}$ and c_{iz}; quite generally in a d-dimensional lattice gas there are $d+1$ collisional invariants.

2. Semi-detailed balance. By this we mean that if all states have equal probability before the collision, they should have equal probability after the collision, viz.

$$\sum_s A(s,s') = 1 ,$$

In other words, the collision should not 'favor' certain output states more than others.

3. Symmetry with respect to the lattice. The collision rules should be such that the collision matrix $A(s, s')$ is invariant under any isometry that leaves the lattice invariant. We will come back to this in the section on the viscosity and on the software implementation.

In the following, it will be more convenient to describe the microdynamics of the system with the use of a single collision function ξ. Once the system is described at the (ensemble-)average level, one can substitute $\xi(s, s')$ quite naturally by $A(s, s')$. With the use of $\xi(s, s')$, the microscopic equation of motion is as follows:

$$n_i(\mathbf{r} + \mathbf{c}_i, t+1) = n_i(\mathbf{r}, t^+) = \sum_{s,s'} s_i'\, \xi(s,s')\delta(n(\mathbf{r},t), s) , \tag{10}$$

where $\delta(n,s)$ represents the kronecker delta function, so $\delta(n,s)$ is only non-zero if the state n matches completely with state s, hence

$$\delta(n,s) = \prod_j \delta(n_j, s_j) = \prod_j n_j^{s_j}(1-n_j)^{1-s_j} .$$

By making use of the obvious relation $\sum_{s'} \xi(s,s') = 1$ we can rewrite (10) to

$$n_i(\mathbf{r}+\mathbf{c}_i, t+1) = n_i(\mathbf{r},t) + \Delta_i(n(\mathbf{r},t)) , \tag{11}$$

where

$$\Delta_i(n(\mathbf{r},t)) = \sum_{s,s'} (s'_i - s_i)\, \xi(s,s') \prod_j n_j^{s_j}(1-n_j)^{1-s_j} .$$

Using these equations of motion, the local conservation laws (9) can now be expressed as

$$\sum_i a_i n_i(\mathbf{r}+\mathbf{c}_i, t+1) = \sum_i a_i n_i(\mathbf{r},t) ,$$

or equivalently

$$\sum_i a_i \Delta_i(n(\mathbf{r},t)) = 0 .$$

All these equations are at the microscopic level. To deduce the macroscopic behavior of these systems we should go to a probabilistic description.

3.2 *Equilibrium Distributions*

In order to obtain macroscopic relations, one should go through the conventional procedure of statistical mechanics, i.e. perform ensemble averaging. This approach, in which the macroscopic behavior of the system is represented by the average over an ensemble of initial conditions, is basically the same for lattice-gas fluids and continuous fluids. We define the phase space Γ of the system as the set of all possible global configurations of the lattice compatible with the given $\{N, \mathbf{P}\}$. Thus a point in the phase space is given by the variable $S = \{s(\mathbf{r}_1), s(\mathbf{r}_2), ..., s(\mathbf{r}_V)\}$ which represents the states of all V lattice sites. We introduce the N-particle distribution function $P(S,t)$ which represents the probability that the system has configuration S at time t in the given ensemble. Clearly

$$\sum_{S\in\Gamma} P(S,t) = 1 .$$

If we have a dynamical variable $a(S)$, that is a variable that is completely determined by the state S of the system, then its ensemble-averaged value (or 'expectation' value) is given by

$$A(t) = \langle a(S) \rangle_t = \sum_{S\in\Gamma} P(S,t) a(S) . \tag{12}$$

An important expectation value is the average of the single-site occupation number, that is, of the variable $a(S) = s_i(\mathbf{r})$:

$$f_i(\mathbf{r},t) = \langle s_i(\mathbf{r}) \rangle_t , \tag{13}$$

which is the discrete equivalent of the single-particle distribution function. In the same way we can construct the average local density

$$\rho(\mathbf{r},t) = \langle \sum_i s_i(\mathbf{r}) \rangle_t = \sum_i f_i(\mathbf{r},t) \,, \tag{14}$$

and the average local velocity $\mathbf{u}(\mathbf{r},t)$:

$$\rho(\mathbf{r},t)\,\mathbf{u}(\mathbf{r},t) = \langle \sum_i \mathbf{c}_i s_i(\mathbf{r}) \rangle_t = \sum_i f_i(\mathbf{r},t)\mathbf{c}_i \,. \tag{15}$$

We get the equations of motion for $f_i(\mathbf{r},t)$ by ensemble averaging the microscopic equations of motion (11) which gives

$$f_i(\mathbf{r}+\mathbf{c}_i, t+1) = f_i(\mathbf{r},t) + C_i(f) \,, \tag{16}$$

where

$$C_i(f) = \sum_{s,s'} (s_i - s_i') A(s,s') \prod_j (f_j(\mathbf{r},t))^{s_j} (1 - f_j(\mathbf{r},t))^{1-s_j} \,. \tag{17}$$

This is the discrete equivalent of the Boltzmann equation. In this we have implicitly made use of the Boltzmann approximation, namely that there is no single-time correlation between particles arriving at the same site. Then we are allowed to write the average of the product of the occupation number as the product of the average occupation numbers. Note that we have used $A(s,s')$ instead of $\xi(s,s')$. If we ensemble average the conservation relations we get

$$\sum_i a_i f_i(\mathbf{r}+\mathbf{c}_i, t+1) = \sum_i a_i f_i(\mathbf{r},t) \tag{18}$$

and

$$\sum_i a_i C_i(f) = 0 \,.$$

We will next try to establish the form of the single-particle distribution function in global equilibrium, which we denote by f_i^0. In equilibrium, the distribution function should be independent of time and space (in absence of external forces etc.), and only dependent on the equilibrium density ρ_0 and velocity $\mathbf{u}_0$ given by (6) and (7). Clearly, if there is no drift velocity (i.e. $\mathbf{u}_0 = 0$), then f_i^0 is independent of the link direction i, and equal to f, defined by (8). For nonzero $\mathbf{u}_0$, it is a little less trivial. Lattice symmetry, and the absence of spatial correlations, imply that the total distribution function factorizes over all sites and links. It can then be shown that a minimum-entropy form for f_i^0 is of the Fermi-Dirac type:

$$f_i^0 = [1 + \exp(h + \mathbf{q}\cdot\mathbf{c}_i)]^{-1} \,. \tag{19}$$

In this equation, the exponent represents the most general form of a collision invariant, so $\mathbf{q}$ and h are arbitrary constants yet to be determined. This can be done in terms of ρ_0 and $\mathbf{u}_0$ by making use of the relations

$$\rho_0 = \sum_i f_i^0$$

$$\rho_0 \mathbf{u}_0 = \sum_i f_i^0 \mathbf{c}_i \, .$$

One finds, in a small-velocity expansion

$$f_i^0(\rho_0, \mathbf{u}_0) = \frac{\rho_0}{b}\left[1 + \frac{1}{c_s^2}(\mathbf{c}_i \cdot \mathbf{u}_0) + \frac{1}{c_s^4} G(\rho_0)\left((\mathbf{c}_i \cdot \mathbf{u}_0)^2 - c_s^2 u_0^2\right) + O(u_0^3)\right] , \tag{20}$$

with

$$G(\rho_0) = \frac{1}{2}\frac{b - 2\rho_0}{b - \rho_0} \, . \tag{21}$$

Note that this expansion is to order u^2, and thus only valid for small u.

3.3 *Equations of Fluid Mechanics*

In order to arrive at the equations of fluid mechanics one should consider the non-equilibrium distribution function. The first step is to consider the idea of local equilibrium: suppose that we have a situation where there is no complete homogeneity in space and time, but where ρ and $\mathbf{u}$ vary weakly. The idea is that this variation is so weak that, on a local scale, ρ and $\mathbf{u}$ can be considered as constant; one can then assume that the form of the distribution function in this 'local equilibrium' is the same as the form in global equilibrium, viz.

$$f_i^{le}(\mathbf{r}, t) = f_i^0(\rho(\mathbf{r}, t), \mathbf{u}(\mathbf{r}, t)) \, . \tag{22}$$

If this form is combined with the conservation relations, one will arrive at the Euler equations, which describe a fluid without dissipation. The above form for f is a first approximation, to lowest order in the variation of ρ and $\mathbf{u}$. A more complete approximation for the non-equilibrium function is an expansion around the local-equilibrium distribution function, where the expansion parameter is a measure of the variation of ρ and $\mathbf{u}$. Including one next order in the expansion will yield the Navier-Stokes equations, one order more the Burnett equations. In continuous systems this is know as the Chapman-Enskog expansion. In the same spirit a multi-scale expansion can be used for lattice gases [11], which goes as follows: first we write the finite differences into continuous differentials by means of a Taylor expansion:

$$f_i(\mathbf{r} + \mathbf{c}_i, t + 1) =$$

$$\left[1 + \partial_\mathbf{r} \mathbf{c}_i + \partial_t + \partial_t \partial_\mathbf{r} \mathbf{c}_i + \frac{1}{2}\partial_\mathbf{r}\partial_\mathbf{r}\mathbf{c}_i\mathbf{c}_i + \frac{1}{2}\partial_t^2 + ...\right] f_i(\mathbf{r}, t) \, . \tag{23}$$

Finally we want to arrive at the equations in terms of ρ and $\mathbf{u}$ in the continuous limit. We control this limit by means of a parameter ϵ that dictates new space and time scales

$$t_1 = \epsilon t$$

$$t_2 = \epsilon^2 t$$

$$\mathbf{r}_1 = \epsilon \mathbf{r} \ ,$$

where the new scale $\mathbf{r}_1$ is such that ρ and $\mathbf{u}$ vary on a scale 1 in these new units. Hence, in units of number of lattice sites, ρ and $\mathbf{u}$ vary on a scale ϵ^{-1}. Eventually ϵ should go to zero, so that the variation is over infinitely many sites, i.e. we have an infinitely fine meshed system. There are two time scales t_1 and t_2. On the 'short' time scale t_1 we expect propagation of sound waves, whereas on the 'long' time scale we expect diffusion of vorticity ('longer' and 'shorter' are in terms of lattice-gas time steps). The differentials in the new coordinates become

$$\partial_t = \epsilon \partial_{t_1} + \epsilon^2 \partial_{t_2} \tag{24}$$

$$\partial_{\mathbf{r}} = \epsilon \partial_{\mathbf{r}_1} \ . \tag{25}$$

It is natural to use ϵ as the expansion parameter in the expansion of the distribution function f around the local equilibrium:

$$f_i(\mathbf{r}, t) = f_i^0(\rho, \mathbf{u}) + \epsilon f_i^1(\rho, \mathbf{u}) + O(\epsilon^2) \ , \tag{26}$$

where f_1 is proportional to the linear gradient in ρ and $\mathbf{u}$ with respect to $\mathbf{r}_1$. The subsequent procedure is to substitute (24)-(26) into Eq. (23), and combine this with the conservation relation (18). Furthermore we use the small-velocity expansion (20) for the local equilibrium distribution function f^0, and make use of the fact that this lowest-order term in the expansion (26) correctly describes ρ and $\mathbf{u}$, so

$$\sum_i f_i^1 a_i = 0 \ . \tag{27}$$

Moreover, wherever necessary, we make use of the symmetry requirements (3) on the velocity set. The evaluation of the equations is rather technical, and can be found in the original literature. The final result is, to order ϵ:

$$\partial_{t_1} \rho + \nabla_1 \cdot (\rho \mathbf{u}) = 0 \ , \tag{28}$$

$$\partial_{t_1} (\rho \mathbf{u}) + \nabla_1 \cdot \mathcal{P}_0 = 0 \ , \tag{29}$$

where we have adapted the more convenient notation $\partial_{\mathbf{r}_1} \mathbf{c}_i = \nabla_1 \cdot \mathbf{c}_i$. $\mathcal{P}_0$ is the lowest-order, or 'dissipative free' momentum-flux tensor

$$\mathcal{P}_0 = c_s^2 \rho \mathcal{I} \left(1 - \frac{u^2}{c_s^2} \frac{2}{d+2} G(\rho) \right) + \frac{2d}{d+2} G(\rho)(\rho \mathbf{u}\mathbf{u}) \ ,$$

in which we have introduced the unity tensor $\mathcal{I}_{\alpha\beta} = \delta_{\alpha\beta}$. To order ϵ^2 we get

$$\partial_{t_2} \rho = 0 \ , \tag{30}$$

$$\partial_{t_2} (\rho \mathbf{u}) + \nabla_1 \cdot \left[\frac{1}{2} c_s^2 \mathcal{S} : \nabla_1 (\rho \mathbf{u}) + \sum_i f_i^1 \mathbf{c}_i \mathbf{c}_i \right] = 0 \ , \tag{31}$$

where $\mathcal{S}$ is the fourth rank tensor

$$\mathcal{S}_{\alpha\beta\gamma\delta} = \frac{d}{d+2}\left(\delta_{\alpha\gamma}\delta_{\beta\delta} + \delta_{\alpha\delta}\delta_{\beta\gamma} - \frac{2}{d}\delta_{\alpha\beta}\delta_{\gamma\delta}\right) , \tag{32}$$

Note that equation (30) implies that on the longer (t_2-) time scale there is no mass diffusion. To arrive at a closed set of equations in terms of ρ and $\mathbf{u}$, we should try to express f_i^1 in terms of these variables. As mentioned above, f_1 should be proportional to first order gradients of ρ and $\mathbf{u}$ with respect to $\mathbf{r}_1$. Symmetry arguments (see [11]) suggest that we should write f_i^1 in the following form:

$$f_i^1 = \psi(\rho)\left[\mathbf{c}_i\mathbf{c}_i - c_s^2\mathcal{I}\right]\nabla_1 \cdot (\rho\mathbf{u}) . \tag{33}$$

where $\psi(\rho)$ is a (still) unknown function. If we substitute this, and combine all the equations $O(\epsilon)$ and $O(\epsilon^2)$, by returning to the original space and time scale $\mathbf{r}$ and t, by use of (24) and (25), we get:

$$\partial_t\rho + \nabla \cdot (\rho\mathbf{u}) = 0 , \tag{34}$$

and

$$\partial_t(\rho\mathbf{u}) + \nabla \cdot (g(\rho)\rho\mathbf{u}\mathbf{u}) =$$
$$- c_s^2\nabla\rho\left(1 - \frac{g(\rho)}{d}\frac{u^2}{c_s^2}\right) + \nabla \cdot \nu(\rho)\nabla(\rho\mathbf{u}) + \nabla\left(\frac{d-2}{d}\nu(\rho)\right)\nabla \cdot (\rho\mathbf{u}) , \tag{35}$$

where

$$\nu(\rho) = \frac{c_s^2 d}{d+2}\left[-bc_s^2\psi(\rho) - \frac{1}{2}\right] , \tag{36}$$

and

$$g(\rho) = \frac{2d}{d+2}G(\rho) = \frac{d}{d+2}\left[\frac{b-2\rho}{b-\rho}\right] . \tag{37}$$

Note that this is very close to the Navier-Stokes equation with vanishing bulk viscosity. Inclusion of rest particles however, will give a nonzero bulk viscosity. One can arrive at the Navier-Stokes equation for incompressible fluids by ignoring all spatial variations in the density, except in the pressure term. This is justified at low Mach numbers ($u \ll c_s$) [11]. This gives

$$\nabla \cdot \mathbf{u} = 0 \tag{38}$$

$$\partial_t\mathbf{u} + g(\rho_0)\mathbf{u} \cdot \nabla\mathbf{u} = -\frac{c_s^2}{\rho_0}\nabla\rho\left(1 - \frac{g(\rho_0)}{d}\frac{u^2}{c_s^2}\right) + \nu(\rho_0)\nabla^2\mathbf{u} . \tag{39}$$

This is the final macroscopic equation in lattice-gas units of space and time, in the limit of low Mach numbers. To see the equivalence with the Navier-Stokes equation for an incompressible fluid, one has to do a rescaling of time. If one uses a new time variable $t' = g(\rho)t$, we get the Navier-Stokes equation with viscosity $\nu' = \nu/g(\rho_0)$ and pressure

$$P = \frac{c_s^2\rho}{g(\rho_0)\rho_0}\left(1 - \frac{g(\rho_0)}{d}\frac{u^2}{c_s^2}\right) .$$

Note that the viscosity depends on $\psi(\rho)$, which is yet to be determined. This function depends on the details of the collision rules, see next section. Although we have recovered the desired equation, there remain certain artifacts due to the discrete nature of the system, which should not be neglected. Firstly, the factor $g(\rho)$ in front of the convective term in the lattice version of the Navier-Stokes equation (39) reflects the fact that the system is not Galilean invariant. This is not surprising, as at the microscopic level particles can only have restricted velocities, hence one cannot relate areas in the system with non-vanishing flow to areas with vanishing flow by a simple Galilean transformation, which is the origin of the convective term. Although non-Galilean effects can be eliminated to second order in u by a rescaling of time, this does not imply that the system becomes truly Galilean invariant; if one incorporates other effects (multi-phase flow, surface tension etc.), one will in general *not* be able to eliminate the non-invariant effect by a simple rescaling. Secondly, in the derivations we have assumed that there are no other conserved quantities than mass and momentum. It turns out that this is not necessarily true lattice gas. In most LGCA there exist 'unphysical' spurious invariants that give rise to additional 'hydrodynamic' equations. As a consequence, the flow in LGCA may differ somewhat from the true Navier-Stokes type of flow. The extra conserved quantities will give rise to additional slowly decaying modes in the system, the so-called staggered modes, and hence to spurious transport coefficients. The spurious conserved quantities are relatively easy to find in the two-dimensions models [25], however there is no systematic way to find all of them. In particular in three-dimensional models it is very difficult to classify the spurious invariants [26]. However, numerical experience gathered thus far suggests that the effect of the spurious invariants on the hydrodynamic behavior of the lattice-gas fluid, is small. In particular, the observed flow patterns tend to agree rather well with those predicted on the basis of the Navier-Stokes equations.

3.4 *Lattice-Boltzmann Equation and Viscosity*

In the derivation of the equations of fluid mechanics, we found the Boltzmann equation as an intermediate level. As it turns out that a true boolean lattice gas simulation is rather noisy, it can be advantageous to do the actual simulation one level 'higher', on the level of the Boltzmann equation. This was first suggested by McNamara and Zanetti [16], and subsequently worked out by Higuera, Succi and coworkers [27, 28]. For a recent overview, see [29]. The idea is as follows. We start with the Boltzmann equation (16):

$$f_i(\mathbf{r} + \mathbf{c}_i, t + 1) = f_i(\mathbf{r}, t) + C_i(f) ,$$

and approximate f by (26). Expansion of the collision function $C(f)$ about f^0 gives

$$C_i(f^0 + \epsilon f^1) = C_i(f^0) + \sum_{j=1}^{b} \left(\frac{\partial C_i}{\partial f_j}\right)_{f_j = f_j^0} \epsilon f_j^1 + O(\epsilon^2) .$$

As $C_i(f^0) = 0$, we can write the Boltzmann equation, to order ϵ, as

$$f_i(\mathbf{r} + \mathbf{c}_i, t+1) = f_i(\mathbf{r}, t) + \sum_{j=1}^{b} \Omega_{ij}(f_j(\mathbf{r}, t) - f_j^0(\mathbf{r}, t)) , \tag{40}$$

where Ω is the linearized collision operator:

$$\Omega_{ij} = \left(\frac{\partial C_i}{\partial f_j}\right)_{f_j = f_j^0} .$$

From (17) we can explicitly evaluate this operator as

$$\Omega_{ij} = -\frac{1}{2} \sum_{s,s'} \left[(s_i' - s_i) A(s, s')(s_j' - s_j) f^{\tilde{\rho}(s)-1} (1-f)^{b-\tilde{\rho}(s)-1} \right] ,$$

where $\tilde{\rho}(s)$ is the number of particles of a microstate s. Equation (40) can be directly converted into an algorithm for simulation purpose. The advantage of a lattice Boltzmann simulation is that the system is essentially free of noise. Also the collision operator, which has only to be calculated once, is now a small $b \times b$ matrix, compared to $A(s, s')$ which is a $2^b \times 2^b$ matrix. We stress that the linearized collision operator need not necessarily be evaluated from an existing set of microscopic collision rules. One can define any operator which has the correct symmetry, and conserves momentum and number of particles; the eigenvalues of the operator define the transport properties of the system, for instance the viscosity. Hence another advantage of lattice Boltzmann is a much more flexibility in defining the collision rules. The disadvantage is that the update now involves floating number arithmetic, instead of 'exact' bit manipulation. The other disadvantage is, of course, that one has to give up one of the appealing features of LGCA, namely that it contains an 'atomistic' description of fluid flow, and hence all the fluctuations and correlations are incorporated in a natural way. Recently a new lattice-Boltzmann scheme was introduced by Ladd [30], which adds a fluctuating component to the fluid stress tensor. This might be a powerful method for systems where fluctuating hydrodynamic variables are important.

We will continue with establishing the relation between the viscosity and the eigenvalue of the collision operator. If we substitute for f^1 the form (33) that was used to arrive at the equations of fluid mechanics, and for f^0 the expansion (20) to order u, we get the following eigenvalue equation

$$\Omega \cdot \mathbf{Q}_{\alpha\beta} = -\mu \mathbf{Q}_{\alpha\beta} , \tag{41}$$

where the eigenvector $\mathbf{Q}_{\alpha\beta}$ has elements

$$(\mathbf{Q}_{\alpha\beta})_i = c_{i\alpha} c_{i\beta} - c_s^2 \delta_{\alpha\beta} ,$$

where α, β are equal to x, y,..etc. The eigenvalue μ is equal to

$$\mu = -\frac{1}{b c_s^2 \psi} ,$$

where ψ is defined by (33). This means that the viscosity, according to (36), is equal to

$$\nu_0 = \frac{c_s^2 d}{d+2}\left[\frac{1}{\mu} - \frac{1}{2}\right] . \tag{42}$$

where the subscript '0' indicates that this transport coefficient has been evaluated in the Boltzmann approximation. Note that not only the lattice must be of sufficient symmetry to guarantee isotropy, also the collision matrix $A(s, s')$ must be such that the eigenvalue μ is the same for all eigenvectors $\mathbf{Q}_{\alpha\beta}$. Otherwise the viscosity is not isotropic. If the linearized collision operator has the correct lattice symmetry, this implies that the operator contains only a few independent elements, and the eigenvalues can be directly expressed in terms of these elements, see for instance [27, 28].

4 Fluctuations

4.1 *Fluctuations and Transport Coefficients*

The transport coefficients basically describe the response of the system to a perturbation from the equilibrium state. For instance, the diffusion coefficient describes the relaxation of inhomogeneities in the density, and the viscosity the relaxation of an initial flow in the system. However, according to the fluctuation-dissipation theorem, the decay of fluctuations in the system is governed by the *same* transport coefficients. This makes it possible to study the transport coefficients of a system that is in equilibrium, by studying the spontanuous fluctuations. It has been demonstrated that the fluctuation-dissipation theorem is also valid for lattice-gas systems [11, 12, 31, 32]. Generally, the transport coefficients are equal to the time integral (summation for time-discrete systems) over the appropriate time correlation function; such a function expresses the deviation of a fluid property from its equilibrium value, in time. We start with the derivation for the self-diffusion coefficient of a tagged particle, which is related to fluctuations in the velocity.

First of all we have to introduce the concept of a tagged particle in lattice gases. In molecular dynamics simulations, labeled particles can be traced because each particle has a unique trajectory that does not intersect other trajectories. By contrast, in lattice gases, particles interact in point collisions. The collision rules described above only specify the occupation of the different velocity states after a collision, not the identity of the particles in those states. Hence, with such collision rules any initial labeling of particles is lost after one collision. Of course, it is possible to construct more elaborate rules that keep track of the identity of the colliding particles. For the study of self-diffusion one wants to follow one tagged particle, hence we should have rules deciding what happens to the tag after the collision. More formal, we define the tagged-particle occupation number $n_i'(\mathbf{r}, t)$ which is one if link $(\mathbf{r}, i)$ is occupied by the tagged particle at time t, and zero otherwise. The velocity of the tagged particle is given by

$$v_x(t) = \sum_{\mathbf{r}} \sum_i n_i'(\mathbf{r}, t) c_{ix} . \tag{43}$$

The collision rules should define both the outcome of n_i and n_i' in a collision. Of course, for a 'tag-blind' observer, these collisions should not be any different from the ordinary collision rules. An obvious choice is to distribute the tag by random selection among the outgoing particles. In the following, we will use these rules, and refer to them as *maximally random collision rules*, a term introduced by Ernst and Naitoh [33]. Once the rules are established, we can evaluate the self-diffusion coefficient. For this, we should consider the mean-square displacement $\Delta(t)$ of a tagged particle

$$\Delta(t) = \frac{1}{2d}\langle(\mathbf{r}(t) - \mathbf{r}(0))^2\rangle = \frac{1}{2d}\langle(\sum_{n=0}^{t-1} \mathbf{v}(n))^2\rangle =$$

$$\frac{1}{2}\langle\sum_{n=0}^{t-1} v_x(n) \sum_{m=0}^{t-1} v_x(m)\rangle = \frac{1}{2}\langle\sum_{n=0}^{t-1} v_x^2(n)\rangle + \langle\sum_{n=1}^{t-1}\sum_{m=0}^{n-1} v_x(n)v_x(m)\rangle =$$

$$\frac{1}{2}\sum_{n=0}^{t-1}\langle v_x^2(0)\rangle + \sum_{n=1}^{t-1}\sum_{m=1}^{n}\langle v_x(0)v_x(m)\rangle = \frac{t-1}{2}\langle v_x^2(0)\rangle + \sum_{n=1}^{t-1}(t-n)\langle v_x(0)v_x(n)\rangle \ .$$

Hence

$$\frac{\Delta(t)}{t} = \frac{t-1}{2t}\langle v_x^2(0)\rangle + \sum_{n=1}^{t-1}\frac{t-n}{t}\langle v_x(0)v_x(n)\rangle \ . \tag{44}$$

The function $\langle v_x(0)v_x(t)\rangle$ is generally called the *velocity autocorrelation function* (VACF). The diffusion coefficient is generally defined by the asymptotic value of $\Delta(t)/t$. If this limit exists, we get

$$D = \lim_{t\to\infty}\frac{\Delta(t)}{t} = \frac{1}{2}\langle v_x^2(0)\rangle + \lim_{t\to\infty}\sum_{n=1}^{t}\langle v_x(0)v_x(n)\rangle \ , \tag{45}$$

where the tagged particle velocity v_x is defined by (43). The zero-time value for the VACF is easily recognized as being

$$\langle v_x^2(0)\rangle = c_s^2 \ .$$

The kinematic viscosity is related to the fluctuations in the stress tensor σ in the same way; the derivation is a bit more involved, as one has to consider tensors instead of vectors. For a derivation, see for instance [12, 31]. We just quote the final result

$$\nu = \frac{1}{bc_s^2\, f(1-f)\, Vv_0}\left[\frac{1}{2}\langle\sigma_{xy}^2(0)\rangle + \lim_{t\to\infty}\sum_{n=1}^{t}\langle\sigma_{xy}(0)\sigma_{xy}(n)\rangle\right] \ , \tag{46}$$

where v_0 is the volume per lattice site, hence Vv_0 is the total volume of the system. The microscopic stress tensor σ is given by

$$\sigma_{xy}(t) = \sum_{\mathbf{r}}\sum_{i} n_i(\mathbf{r},t)c_{ix}c_{iy} \tag{47}$$

The zero-time value for the stress-stress correlation function is equal to

$$\langle \sigma_{xy}^2(0) \rangle = \frac{bdc_s^4}{d+2} f(1-f)V \,.$$

Equations (45) and (46) are very useful, as the correlation functions $\langle v_x(0)v_x(t)\rangle$ and $\langle \sigma_{xy}(0)\sigma_{xy}(t)\rangle$ can be directly measured in a simulation. In order to estimate the transport coefficients, one should in principle measure the correlation function for infinite times. However, in most cases the correlation function is a rapidly decreasing function, and the long-time contributions can be neglected, or estimated from the initial decay. This is true in particular in the Boltzmann approximation, that is, when correlations between colliding particles are neglected; it can be shown that in this approximation, correlation functions should decay *exponentially*, so knowledge of the decay from $t = 0$ to $t = 1$ is sufficient. In most cases, this initial decay can also be evaluated theoretically; this means that, at least in the Boltzmann approximation, one can have an exact theoretical prediction for the transport coefficients. We will proceed with this in the next section.

4.2 *Boltzmann Approximation*

Generally, we have the transport coefficient F given as the sum over a time-correlation function $\langle a(0)a(t)\rangle$

$$F = C\left[\frac{1}{2}\langle a^2(0)\rangle + \lim_{t\to\infty}\sum_{n=1}^{t}\langle a(0)a(n)\rangle\right] , \tag{48}$$

We introduce the normalized correlation function

$$Z_a(n) = \frac{\langle a(0)a(n)\rangle}{\langle a^2(0)\rangle}$$

Then, in the Boltzmann approximation (exponential decay), we can write:

$$Z_a(n) = Z_a(1)^n \,,$$

which means that (48) can be written as

$$F = C\,\langle a^2(0)\rangle \left[\frac{1}{\lambda_F} - \frac{1}{2}\right] , \tag{49}$$

where

$$\lambda_F = 1 - Z_a(1) \,. \tag{50}$$

One can associate a mean-free time t_0 with a correlation function, normally defined as the time at which the normalized function has decayed to e^{-1}. Hence

$$t_0 = \frac{-1}{\ln(1-\lambda_F)} \,. \tag{51}$$

The correlation function after one timestep $Z_a(1)$, and hence λ_F, can in general be derived from the microdynamics of the system. We will now give the explicit values for the diffusion coefficient and the viscosity.

The self-diffusion coefficient in the Boltzmann approximation is equal to

$$D_0 = c_s^2 \left[\frac{1}{\lambda_D} - \frac{1}{2} \right] \tag{52}$$

Ernst and Naitoh [33] have derived a very compact expression for λ_D, for the maximally random collision rules :

$$\lambda_D = \frac{(1-f)^b - 1}{(b-1)f} + \frac{b}{b-1} . \tag{53}$$

Note that λ_D is independent of the collision rules for untagged particles. Equation (53) has been checked by simulation [34], and was always found to be in perfect agreement.

The kinematic viscosity in the Boltzmann approximation is equal to

$$\nu_0 = \frac{c_s^2 d}{d+2} \left[\frac{1}{\lambda_\nu} - \frac{1}{2} \right] . \tag{54}$$

An explicit expression for λ_ν has been derived by Hénon [35]. He found

$$\lambda_\nu = \frac{d}{d-1} \sum_{s,s'} \left[(s_i - s_i') A(s,s') f^{\beta(s)-1} (1-f)^{b-\beta(s)-1} \sum_{j=1}^{b} s_j \cos^2 \theta_{ij} \right] , \tag{55}$$

where θ_{ij} is the angle between $\mathbf{c}_i$ and $\mathbf{c}_j$, and i can take an arbitrary value between 1 and b. The summation $\sum_{s,s'}$ is over all possible input and output states. For the other definitions, see section 2. We note that, if we compare (54) with (42), that λ_ν can be identified as the eigenvalue of the linearized collision operator. It can indeed be shown [36] that the expression by Hénon (55) satisfies the eigenvalue equation (42).

4.3 *Beyond Boltzmann: Long-Time Tails*

How true is the Boltzmann approximation? In Fig. 2 we show an extremely accurate measurement of the velocity autocorrelation function (points), compared with the theoretical prediction based on the Boltzmann approximation (solid line). We see that after some time there is a small but significant deviation from the exponential 'Boltzmann' decay.

Such behavior was first demonstrated more than twenty years ago by Alder and Wainwright [5], in a molecular dynamics simulation. They found, both for a system of hard discs and hard spheres, that the velocity autocorrelation function decays much *slower* than exponentially, like in Fig. 2. In fact, Alder and Wainwright could estimate the functional form of this so-called 'long-time tail'. In turned out that, for long enough times, the VACF decays as $t^{-d/2}$, where d is the dimensionality of the system. This behavior could be explained by a simple dimensional argument: suppose that we do not

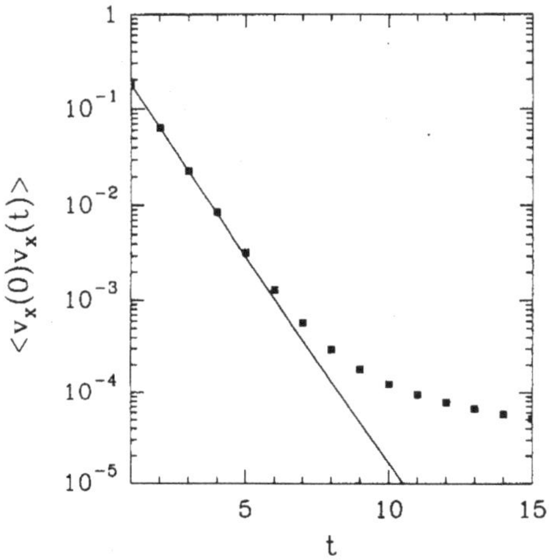

Figure 2: *Decay of the velocity autocorrelation function in a three-dimensional lattice gas. The points are the results of simulations, the solid line is the theoretical prediction in the Boltzmann approximation. We see that initially the agreement with 'Boltzmann' is very good, however, for longer times there are significant deviations.*

have a particle, but a small fluid element δV, which will be set into motion, while the surrounding fluid is at rest. According to hydrodynamics, the transverse momentum will diffuse away at a rate determined by the kinematic shear viscosity ν; after time t the total initial momentum will be contained by a volume $\sim (\sqrt{\nu t})^d$. The amount of momentum contained in the original fluid element is then $\sim \delta V/(\sqrt{\nu t})^d$, hence the velocity autocorrelation function of the fluid element decays as $(\nu t)^{-d/2}$. Thus far, we have assumed that the fluid element is not moving in time. If we take the diffusive motion of the fluid element into account (characterized by the diffusion coefficient D), we find that the long-time decay of the VACF should be

$$\langle v_x(0)v_x(t)\rangle \sim [(D+\nu)t]^{-d/2} , \tag{56}$$

which is in agreement with the Alder-Wainwright simulation results. Later more sophisticated theories, like mode-coupling theory [37], and kinetic theory [38] have rederived this result. It has also been demonstrated that the stress-stress correlation function has long-time tail of the same functional form. For an historic account of the discovery of the long-time tails, see [39]. Finally we note that, in three dimensions, the contribution of the long-time tail to the Boltzmann estimate of the transport coefficient is small; for instance, for the diffusion coefficient it turned out to be less then a percent [34]. Therefore, the Boltzmann estimates (52) and (54) are very good approximations for the transport coefficient in three dimensions. The situation is different in two-dimensions, because the long-time tail contribution diverges! However, for finite systems, the corrections will be small as well [40].

5 Some Models

Thus far, we have not been specific about the models themselves. Next we explicitly describe the models that are mostly used for simulations in two and three dimensions.

5.1 FHP-III Model (Two Dimensions)

The standard models in two-dimensions are the FHP models. They are all defined on a triangular lattice, as in Fig. 1; the only differences between the various versions concern the collision rules and the number of rest particles. The FHP-I model has six possible velocities, with no rest particles. The FHP-II and III models have an additional rest particle. The FHP-IV model has two rest particles. As an example we will explicitly describe the FHP-III model; for the other models, see for instance [21]. The velocities in the FHP-III model are defined as follows (in units of $\Delta l/\Delta t$):

$$\begin{aligned}
\mathbf{c}_1 &= (\ 1, & 0) \\
\mathbf{c}_2 &= (\ \tfrac{1}{2}, & \tfrac{1}{2}\sqrt{3}) \\
\mathbf{c}_3 &= (-\tfrac{1}{2}, & \tfrac{1}{2}\sqrt{3}) \\
\mathbf{c}_4 &= (-1, & 0) \\
\mathbf{c}_5 &= (-\tfrac{1}{2}, & -\tfrac{1}{2}\sqrt{3}) \\
\mathbf{c}_6 &= (\ \tfrac{1}{2}, & -\tfrac{1}{2}\sqrt{3}) \\
\mathbf{c}_7 &= (\ 0, & 0)
\end{aligned}$$

so

$$b = 7\,, \qquad c_s^2 = \frac{3}{7}\,.$$

The collision rules are shown, in a compact form, in Fig. 3. In terms of the reduced density f, the Boltzmann diffusion coefficient is given by

$$D_0 = \frac{3}{7\lambda_D} - \frac{3}{14}\,,$$

with

$$\lambda_D = \frac{(1-f)^7 - 1}{6f} + \frac{7}{6}\,.$$

The Boltzmann shear viscosity is given by

$$\nu_0 = \frac{1}{4\lambda_\nu} - \frac{1}{8}\,,$$

with

$$\lambda_\nu = f(7 - 15f + 16f^2 - 8f^3)\,.$$

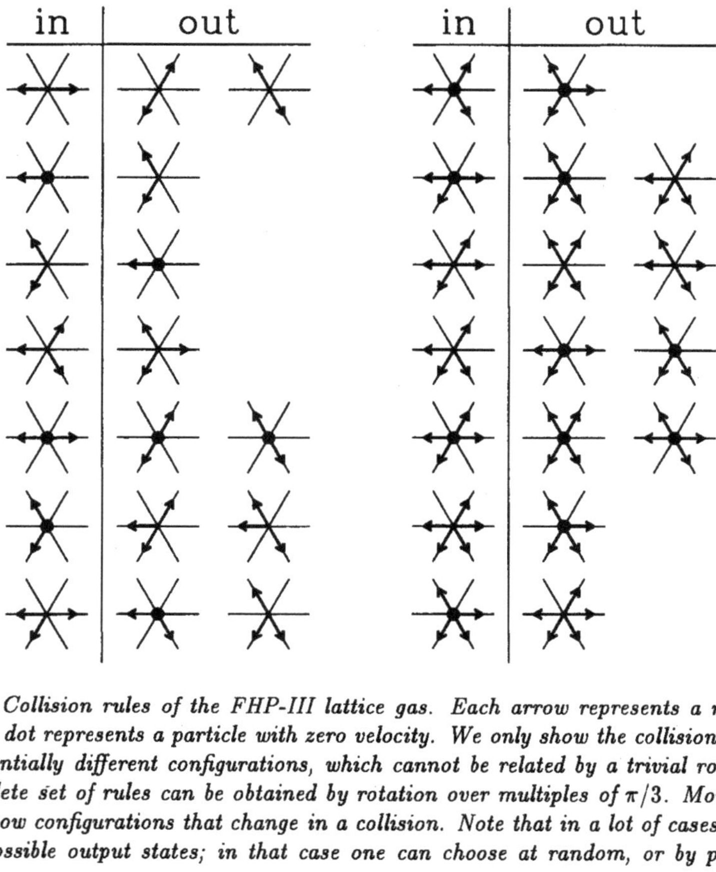

Figure 3: *Collision rules of the FHP-III lattice gas. Each arrow represents a moving particle, a dot represents a particle with zero velocity. We only show the collisions for a set of essentially different configurations, which cannot be related by a trivial rotation. The complete set of rules can be obtained by rotation over multiples of $\pi/3$. Moreover, we only show configurations that change in a collision. Note that in a lot of cases, there are two possible output states; in that case one can choose at random, or by prefixed rules.*

Note that these formula's are slightly different from those in section 2, which is due to the presence of a rest particle in the FHP-III model. The volume per site is

$$v_0 = \frac{1}{2}\sqrt{3} \, .$$

5.2 *FCHC Model (Three-Dimensions)*

The problem in three dimensions is that there is no lattice with sufficiently high symmetry to guarantee isotropy, i.e. there is no set of vectors that satisfies the conditions (3). However, there exists a satisfactory lattice in *four* dimensions [41]. This lattice is the four-dimensional version of the well-known three-dimensional FCC lattice, and is referred to as the Face-Centered Hyper-Cubic (FCHC) lattice. A three-dimensional model can then be obtained by projection. Hence there are 24 possible velocities, which are (in units of $\Delta l/\Delta t$)

$$\begin{array}{llllll}
\mathbf{c}_1 = (\ 1, & 0, & 0) & \quad \mathbf{c}_{13} = (\ 1, & 1, & 0) \\
\mathbf{c}_2 = (-1, & 0, & 0) & \quad \mathbf{c}_{14} = (-1, & -1, & 0) \\
\mathbf{c}_3 = (-1, & 0, & 0) & \quad \mathbf{c}_{15} = (\ 1, & -1, & 0) \\
\mathbf{c}_4 = (\ 1, & 0, & 0) & \quad \mathbf{c}_{16} = (-1, & 1, & 0) \\
\mathbf{c}_5 = (\ 0, & 1, & 0) & \quad \mathbf{c}_{17} = (\ 1, & 0, & 1) \\
\mathbf{c}_6 = (\ 0, & -1, & 0) & \quad \mathbf{c}_{18} = (-1, & 0, & -1) \\
\mathbf{c}_7 = (\ 0, & -1, & 0) & \quad \mathbf{c}_{19} = (\ 1, & 0, & -1) \\
\mathbf{c}_8 = (\ 0, & 1, & 0) & \quad \mathbf{c}_{20} = (-1, & 0, & 1) \\
\mathbf{c}_9 = (\ 0, & 0, & 1) & \quad \mathbf{c}_{21} = (\ 0, & 1, & 1) \\
\mathbf{c}_{10} = (\ 0, & 0, & -1) & \quad \mathbf{c}_{22} = (\ 0, & -1, & -1) \\
\mathbf{c}_{11} = (\ 0, & 0, & -1) & \quad \mathbf{c}_{23} = (\ 0, & 1, & -1) \\
\mathbf{c}_{12} = (\ 0, & 0, & 1) & \quad \mathbf{c}_{24} = (\ 0, & -1, & 1)
\end{array}$$

Note that this projection results in 12 single links, and 6 double links. We have the following values

$$b = 24 \, , \qquad c_s^2 = \frac{1}{2} \, .$$

The volume per site is, in units of $(\Delta l)^3$:

$$v_0 = 1 \, .$$

The collision rules cannot be given in any simple way. As there are 24 velocities, the full collision function is usually quite complex. However, for all maximally random collision rules, the value of the Boltzmann diffusion coefficient does *not* depend on the details of the collision rules. In terms of the reduced density f, the Boltzmann diffusion coefficient is given by

$$D_0 = \frac{1}{2\lambda_D} - \frac{1}{4} \, ,$$

with

$$\lambda_D = 1 - \frac{(1-f)}{23f}\left[1-(1-f)^{23}\right] .$$

In contrast to the diffusion coefficient, the Boltzmann shear viscosity *does* depend on the details of the collision rules. Basically there are two ways to do the collision. One option is to evaluate the output state from the input state 'on the fly', by some global rule, for example the isometric collision rules [42], in which the output state is obtained by applying one of the lattice isometries to the input state. Another example of a global rule is the random algorithm, in which one of the allowed output states is chosen at random ('allowed' means: with the same momentum and number of particles as the input state). The other option is to have detailed collisions, and tabulate the collision function. However, the construction of such a table is a non-trivial task: apart from conservation of momentum and number of particles, one should also consider the following:

1. The viscosity
Obviously the model is computationally more efficient if one can observe the desired phenomena with fewer lattice sites and in fewer time steps, i.e. one can use a larger grid. This efficiency can be expressed in terms of the Reynolds coefficient R_*, which is defined as
$$R_* = \frac{c_s g(\rho)}{\nu} .$$
The computational efficiency scales then as the $(d+1)$-th power of R_*, so it is desirable to have the lowest possible viscosity. To select the optimum rules in that respect, the formula by Hénon (55) is very useful: one starts by arranging all states in subgroups that have the same number of particles and momentum; then the tricky part is to select collision pairs within each group, such that λ in (55) is optimal. The rigorous option is to go over all possible combinations of pairs to select the optimal one, however this is impractical for groups that contain hundreds of elements; there are however some tricks to do this more efficiently. For more details on constructing full low-viscosity tables, see [43]. Still, a full table has a size of almost 17 million words, which is too large to fit into the memory of any but the largest computers. An option is to split the 24 bit word into smaller words, for instance into two 12 bits words, and do separate collisions on both words. Although this will decrease the efficiency (a substantial number of 'good' collisions are ruled out) the size of the table is dramatically smaller (2×2^{12} instead of 2^{24} entries). Another difficulty is that one is not completely free in choosing collision pairs because of the following consideration

2. Symmetry
The rules must be sufficient symmetric in order to have an isotropic viscosity (see section (3.4)). This implies that the collision rules must be invariant under any isometry which leaves the lattice invariant. In the FCHC lattice there are 1152 of such isometries which are independent. There are various ways to build the

required symmetry into the collision rules. A robust way was put forward by Somers and Rem [44]. In this method, the precollision state undergoes one of the basic 1152 isometries at random. The isometrized state is then the input of a collision table, which may be anisotropic. The output state undergoes the inverse isometry. The isometries guarantee that, on average, the complete update will be isotropic, without having any of such restrictions on the collision table itself. Another possibility is to built the full symmetry into the collision table, which is quite complicated for the 24-bit collisions. An option in between is to make use a partially symmetric table, and have the 'remaining' part of the symmetry taken care of by some of the isometries.

For a description of the various collision tables that have been used for the FCHC model we refer to [22]. We note that the optimum value of the viscosity is not always necessarily the smallest. In the simulations in which the lattice gas is being used as the solvent of a colloidal suspension [19, 45, 46], the optimum value is determined by the requirement of a proper time-scale separation of the various processes. In particular, the optimal value in 3D is found to be around $\nu = 0.10$ [19]. Also for the calculation of the velocity autocorrelation function of a tagged particle [34, 40], a table with a very high R_* is not required. This is because in these calculations the update of the lattice-gas takes only a small fraction of the total computation.

5.3 Software Implementation

In a lattice-gas simulation it is of course natural to have the state of one single link (occupied or empty) stored by a single bit in memory (1 or 0). There are basically two ways to organize this storage of bits.

The most straightforward way is to store the state of one site in the first b bits of a one word. The collision can then be done by simple lookup: the input state is the entry of an integer table, which contains the corresponding output state. Although the lookup is very fast this way, there are two drawbacks of this approach. Firstly, one does not use the full capacity of a word, only the first b bits. Secondly, the propagation step is relatively slow, as for the update of each word, one has to collect the b bits from b different words.

A second possibility is to store the bits related to all the *first* links of 32 sites in one word, say a_1, and the bits of the second links of the same 32 sites in a next word a_2, and so on. In this we assumed 32-bit integers. One advantage of this approach is that it makes full use of the capacity of the word. Another advantage is that the propagation is very fast now: the propagation of all particles on the first link of the 32 sites to their neighboring sites is achieved by a one single copy of a_1 to b_1, where b_1 stores the values of all the first links of the 32 neighboring sites. (this can only be done if the 32 sites which share a common word are *no* neighboring sites. Otherwise propagation would imply shifting bits within a word). However, for the same reason that the propagation was slow in the first representation, the collision is slow in this representation, if one makes use of a collision table. However, a fast collision procedure may be possible if the collision rules are expressed in terms of bit logic, i.e. the collisions are carried out by means of logical operations between the words a_1, a_2,.., a_b. As this happens on a bit by bit basis, 32 sites are updated simultaneously. It is possible to define the collision rules of the FHP-models in terms of bit logic, see the appendix of [21]. For the FCHC model the logical expressions themselves become unwieldy, and one has to make use of a collision table in any case. A fast solution would be if a piece of memory could be organized like a matrix of $b \times 32$ bits, in which the 32 bits of words $a_1, .., a_b$ are read in as columns, and then read out row-wise as new words of b bits, which can be the direct input of a collision table. The output of the table undergoes the inverse manipulation. Clearly, this regrouping of bits in different words requires special hardware. In fact, a chip that does precisely this has been designed by Bakker as part of a special purpose machine for LGCA simulations [47]. What still remains in such a representation is the construction a 24-bit collision table, a problem which has been addressed in the previous section.

6 Application: Diffusion in Two Dimensions

We will next discuss a prototypical example of a study in which a lattice gas has been extremely useful. This study concerns the diffusion of a single particle in two dimensions; in systems of such dimensionality, theories have predicted a renormalization of the self-diffusion coefficient. Although the predicitions have been made twenty years ago, the effect could only be observed recently with the use of lattice gas cellular automata [40, 48]. We will first briefly describe the basic idea behind the renormalization of the diffusion coefficient.

The diffusive properties of a particle are most conveniently described by means of the velocity autocorrelation function $\phi(t)$ introduced in section 4.1:

$$\phi(t) = \langle v_x(0)v_x(t)\rangle \ .$$

As was mentioned in section 4.3, mode-coupling theory predicts that $\phi(t)$ decays in two dimensions, for long times as

$$\phi(t) = \left(\frac{b}{D+\nu}\right) t^{-1} \ , \tag{57}$$

where b is some constant. However, this result is only true to lowest order; namely, the diffusion constant D, as used in this analysis, is defined by:

$$D = \lim_{t\to\infty} \int_0^t \phi(s)ds \ .$$

As a consequence, D diverges as $\ln t$ if $\phi(t)$ decays as t^{-1}. The same is true for the viscosity ν. This implies that it is inconsistent to assume that $\phi(t)$ decays as t^{-1}, as the divergence of D and ν will cause $\phi(t)$ to decay slower. Clearly this calls for a self-consistent theory. As was first suggested by Alder and collaborators, this can be done by using a 'time-dependent' diffusion coefficient, rather than one which is constant. In this self-consistent description, the long-time decay of $\phi(t)$ turns out to be slightly faster than t^{-1}, namely as $t^{-1}(\ln t)^{-1/2}$. The diffusion coefficient still diverges, but now as $\sqrt{\ln t}$ instead of $\ln t$. We refer to this as the renormalization of the diffusion coefficient. The logarithmic corrections to the 'lowest-order' t^{-1}-decay are extremely weak; they could not be observed in the original Alder-Wainwright simulations, and although the more recent simulations of the VACF of a 2-D fluid did corroborate the existence of the t^{-1} tail [49], faster decay could still not be seen.

As LGCA are a correct microscopic model of a fluid, the effects described above must be present in such systems as well. The advantage of LGCA over MD-models for measuring the VACF is not only that the update of the system is faster; in addition, for lattice gases a novel and very efficient algorithm, the moment-propagation method could be used to measure the VACF. The idea of this algorithm is as follows. The VACF is defined as the time-average of the product $v_x(0)v_x(t)$ along any possible trajectory of the tagged particle. Clearly in a simulation the statistical accuracy can be improved by averaging over different trajectories. Now, the indistinguishability of lattice-gas particles makes it possible to simultaneously compute *all* tagged particle trajectories rather than just one. In this way a much improved estimate of the VACF is obtained [50, 34]. Compared to 'single-trajectory' methods, as should be used in MD-simulations, the gain in computing speed amounts to more than 6 orders of magnitude. With the use of this method, extremely accurate measurements of the VACF could be obtained [34, 50]. Below we show one such a result for a tagged particle in the two-dimensional FHP-III lattice gas. The calculation was performed on a system of 2000×2000 lattice sites, and the correlation time interval was 2000 lattice-gas time steps. Periodic boundary conditions were used, and the reduced density was set to $f = 0.80$. The simulation result is shown in Fig. 4, where we plotted the velocity autocorrelation function multiplied by the time t, *vs* time, which is in units of the mean free time t_0. The points with errorbars represent the simulation result, where the solid line connects the successive points. Note that we have measured the VACF up to 6000 mean free times, without interference of the periodic images, and with an error of at most 5%. If the VACF decays faster than t^{-1} the measured curves should have a negative slope. This behavior is clearly observed in Fig. 4. However, there is not only a qualitative, but also a quantitative agreement with the prediction of self-consistent mode-coupling theory; the latter is indicated in Fig. 4 by the dashed line; this result was obtained from the mode-coupling expression (57) with the use of *time-dependent* transport coefficients, which could either be obtained from the simulation itself, or from theory. For comparison we have also plotted the predicition of lowest-order mode-coupling theory (dotted line), where we used the time-independent Boltzmann values, as introduced in section 4.3. From this figure we can draw two important conclusions. Firstly, the magnitude of the tail predicted by higher-order mode-coupling theory is in agreement with the simulations. Secondly, the faster-than-t^{-1} decay observed in the simulation is compatible with the decay predicted on basis of knowledge of the time-dependent transport coefficients in (57). We stress that the latter is the basic idea behind self-consistent mode-coupling theory. Finally we note that

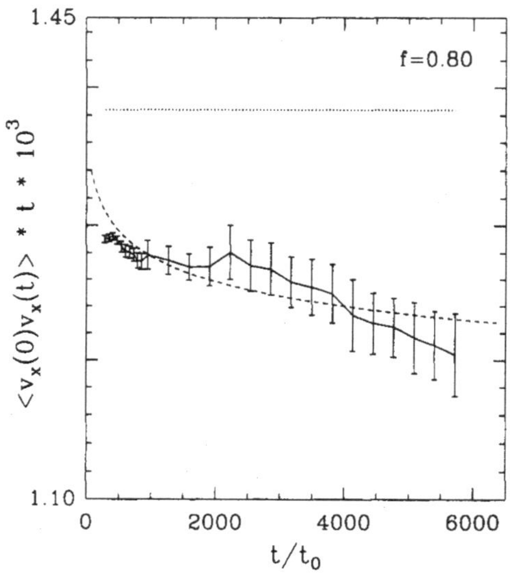

Figure 4: *Velocity autocorrelation function of a tagged particle multiplied by time, for density* $f = 0.80$. *Solid curve: simulation data; dotted line: prediction of lowest-order mode-coupling theory* (t^{-1} *decay). Dashed curve: prediction of self-consistent mode-coupling theory, by using the 'true' time-dependent transport coefficients. The time-dependent diffusion coefficient is obtained from the measured VACF, the time-dependent viscosity is estimated from lowest-order mode-coupling theory. We see that there is a very good agreement with the self-consistent theory.*

the same behavior has been observed for the diffusive properties of a *colloidal particle*, suspended in similar two-dimensional lattice gas [48].

7 Final Remarks

When LGCA were rediscovered as fluid models by Frisch, Hasslacher and Pomeau in 1986, it was hoped that, with the use of dedicated hardware, these models could be more powerful than existing techniques to study fluid-flow phenomena, and then in particular in the turbulent regime. This turns out to be not the case, as the rather high noise level of LGCA kills the efficiency; moreover, it seems to be impossible to get to really high Reynolds numbers. However, as we hope to have shown above, lattice gases can be very useful in the study of fundamental theories of molecular motion in fluids. This is not only true for existing theories, like the mode-coupling theory, but also for new theories: approximate schemes can be tried out first on the simple lattice-gas models, before trying them out on more realistic models of fluids and solids. LGCA have also been applied successfully to the study of colloidal suspensions [48, 45], in which the lattice gas represents the solvent. From both applications follows that the lattice-gas model is able to account for much of the underlying physics of fluids and suspensions, even more subtle aspects like renormalization of diffusion in two dimensions are implicit in the model. One of the real potentials of lattice gases is that often more complicated systems can be studied with only little extra effort. In the case of colloidal suspensions we think for instance of systems with arbitrary shaped particles or perhaps polymers, even in complex geometries. In these systems, precise theoretical predictions are extremely difficult to make, hence the lattice-gas colloidal-particle model can really help to gain a better understanding of these systems. It is however not at all clear that for such studies the boolean lattice-gas method is the best model. As has been discussed earlier, models based on the lattice Boltzmann equation may have most merits of lattice gases, but not the drawbacks; this may not only be true for the study of colloidal suspensions, but more generally for all applications where the presence of noise is not essential.

Acknowledgements

I would like to thank Matthieu Ernst, Tony Ladd and, in particular, Daan Frenkel for their cooperation over the past few years; and both Daan Frenkel and René van Roij for a careful reading of the manuscript. The work of the FOM institute is part of the scientific program of FOM and is supported by the Nederlandse Organisatie voor Wetenschappelijk Onderzoek (NWO).

References

[1] U. Frisch, B. Hasslacher and Y. Pomeau, Phys. Rev. Lett. **56**, 1505(1986).

[2] J.-P. Boon and S. Yip, Molecular Hydrodynamics (Dover, New York, 1991).

[3] P. Résibois and M. De Leener, Classical Kinetic Theory of Fluids (Wiley, New York, 1991).

[4] M. Mareschal and B.L. Holian, in Proceedings of the NATO ASI on *Microscopic Simulations of Complex Hydrodynamic Phenomena*, Alghero, Italy, July 15-27, 1991, edited by M. Mareschal and B. Holian (Plenum Press, New York, 1992).

[5] B.J. Alder and T.E. Wainwright, Phys. Rev. **A1**, 18(1970).

[6] D.C. Rapaport and E. Clementi, Phys. Rev. Lett. **57**, 695(1986).

[7] M. Mareschal and E. Kestemont, Nature **329**, 427(1987).

[8] M. Mareschal, M. Malek Mansour, A. Puhl and E. Kestemont, Phys. Rev. Lett. **61**, 2550(1988).

[9] R.P. Feynman, cited in Physics Today **42** no.2, 78(Feb. 1989).

[10] J. Hardy and Y. Pomeau, J. Math. Phys. **13**, 1024(1972); J. Hardy, Y. Pomeau and O. de Pazzis, J. Math. Phys. **14**, 1746(1973); J. Hardy, O. de Pazzis and Y. Pomeau, Phys. Rev. **A13**, 1949(1976).

[11] U. Frisch, D. d'Humières, B. Hasslacher, P. Lallemand, Y. Pomeau and J.-P. Rivet, Complex Systems **1**, 649(1987).

[12] M.H. Ernst and J.W. Dufty, J. Stat. Phys. **58**, 57(1990);

[13] M.H. Ernst, in Proceedings of the NATO ASI on *Liquids, Freezing and the Glass Transition*, Les Houches, France, July 3-28, 1989, edited by J.P. Hansen, D. Levesque and J. Zinn-Justin (North Holland, Amsterdam, 1991).

[14] M.H. Ernst, in Proceedings of the Summer School on *Fundamental Problems in Statistical Mechanics VII*, Altenberg, F.R. Germany, June 18-30, 1989, edited by H. van Beijeren (North Holland, Amsterdam, 1990).

[15] M.H. Ernst, Physica D **47**, 198(1991)

[16] G.G. McNamara and G. Zanetti, Phys. Rev. Lett. **61**, 2332(1988).

[17] G.G. McNamara and B.J. Alder, in Proceedings of the NATO ASI on *Microscopic Simulations of Complex Hydrodynamic Phenomena*, Alghero, Italy, July 15-27, 1991, edited by M. Mareschal and B. Holian (Plenum Press, New York, 1992).

[18] "Lattice Gas Methods for Partial Differential Equations : A Volume of Lattice Gas Reprints and Articles", Edited by G.D. Doolen (Addison Wesley, 1990).

[19] A.J.C. Ladd, in Proceedings of NATO ASI on *Computer Simulations in Material Science*, Aussois, France, March 25 - April 5, 1991, edited by M. Meyer and V. Pontikis (Kluwer, Dordrecht, 1992).

[20] M.H. Ernst, in Proceedings of the NATO ASI on *Microscopic Simulations of Complex Hydrodynamic Phenomena*, Alghero, Italy, July 15-27, 1991, edited by M. Mareschal and B. Holian (Plenum Press, New York, 1992).

[21] D. d'Humières and P. Lallemand, Complex Systems **1**, 599(1987).

[22] B. Dubrulle, U. Frisch, M. Hénon and J.-P. Rivet, J. Stat. Phys. **59**, 1187(1990).

[23] P. Grosfils, J.-P. Boon and P. Lallemand, Phys. Rev. Lett. **68**, 1077(1992).

[24] S. Wolfram, J. Stat. Phys. **45**, 471(1986).

[25] G. Zanetti, Phys. Rev. **A40**, 1539(1989).

[26] D. d'Humières, Y.H. Qian and P. Lallemand, in Proceedings of workshop on *Discrete Kinetic Theory, Lattice Gas Dynamics and Foundations of Hydrodynamics*, Torino, Italy, Sept. 20-24, 1988, edited by R. Monaco (World Scientific, Singapore, 1989).

[27] F.J. Higuera, S. Succi and R. Benzi, Europhys. Lett. **9**, 345(1989).

[28] S. Succi, R. Benzi, E. Foti, F.J. Higuera and F. Szelényi, in Proceedings of workshop on *Cellular Automata and Modeling of Complex Physical Systems*, Les Houches, France, Feb. 21-28, 1989, edited by P. Manneville (Springer, Berlin, 1989).

[29] S. Succi, R. Benzi, A. Cali and M. Vergassola, in Proceedings of the NATO ASI on *Microscopic Simulations of Complex Hydrodynamic Phenomena*, Alghero, Italy, July 15-27, 1991, edited by M. Mareschal and B. Holian (Plenum Press, New York, 1992).

[30] A.J.C. Ladd, to be published.

[31] J.-P. Rivet, Complex Systems **1**, 839(1987).

[32] L.P. Kadanoff, G. McNamara and G. Zanetti, Phys. Rev. **A40**, 4527(1989).

[33] M.H. Ernst and T. Naitoh, J. Phys. A: Math. Gen. **24**, 2555(1991).

[34] M.A. van der Hoef and D. Frenkel, Phys. Rev. **A41**, 4277(1990); Physica D **47**, 191(1991).

[35] M. Hénon, Complex Systems **1**, 763(1987).

[36] M.H. Ernst, private communication.

[37] M.H. Ernst, E.H. Hauge and J.M.J. van Leeuwen, Phys. Rev. **A4**, 2055(1971).

[38] J.R. Dorfman and E.G.D. Cohen, Phys. Rev. Lett. **25**, 1257(1970).

[39] B.J. Alder,*The Long-Time Tails Story*, in Proceedings of the NATO ASI on *Microscopic Simulations of Complex Hydrodynamic Phenomena*, Alghero, Italy, July 15-27, 1991, edited by M. Mareschal and B. Holian (Plenum Press, New York, 1992).

[40] M.A. van der Hoef and D. Frenkel, Phys. Rev. Lett. **66**, 1591(1991).

[41] D. d'Humières, P. Lallemand and U.Frisch, Europhys. Lett. **2**, 291(1986).

[42] M. Hénon, Complex Systems **1**, 475(1987).

[43] M. Hénon, in Proceedings of workshop on *Discrete Kinetic Theory, Lattice Gas Dynamics and Foundations of Hydrodynamics*, Torino, Italy, Sept. 20-24, 1988, edited by R. Monaco (World Scientific, Singapore, 1989).

[44] J.A. Somers and P.C. Rem, in Proceedings of the workshop on *Cellular Automata and Modeling of Complex Physical Systems*, Les Houches, France, Feb. 21-28, 1989, edited by P.Manneville (Springer, Berlin, 1989).

[45] A.J.C. Ladd and D. Frenkel, Phys. Fluids **A2**, 1921(1990).

[46] A.J.C. Ladd and D. Frenkel, in Proceedings of workshop on *Cellular Automata and Modeling of Complex Physical Systems*, Les Houches, France, Feb. 21-28, 1989, edited by P. Manneville (Springer, Berlin, 1989).

[47] A.F. Bakker, to be published.

[48] M.A. van der Hoef, D. Frenkel and A.J.C. Ladd, Phys. Rev. Lett. **67**, 3459(1991).

[49] J.J. Erpenbeck and W.W. Wood, Phys. Rev. **A26**, 1648(1982).

[50] D. Frenkel and M.H. Ernst Phys. Rev. Lett. **63**, 2165(1989).

Vortex Interactions in 2-D Lattice Gas Automata

P. Lavallée
Département de Physique
Université du Québec à Montréal
Montréal H3C 3P8 Canada

J.P. Boon
Faculté des Sciences , C.P. 231
Université Libre de Bruxelles
B-1050 Bruxelles, Belgique

A. Noullez
U.R.A. 1362, Observatoire de Nice, B.P. 229
F-06304 Nice Cedex 4, France

Abstract

We investigate the dynamics of dipolar vortex pairs with the lattice gas automaton method. For vortices with opposite polarity, three types of interactions are observed, depending on the value of the impact parameter of the colliding vortex pairs. For large values, the pairs interact in a scattering-like process ; for small values, the scattering involves an exchange of vortices between the two dipoles ; at intermediate values, tripole formation is observed. Our results are in agreement with experimental observations.

PACS numbers: 47.15.Ki, 02.70.+d

Keywords: Lattice gas automata, Vortex dynamics

1 Introduction

Many phenomena in fluid mechanics involve interactions of vortices in two dimensions : examples are found in geostrophic and oceanic flows, and, at a completely different scale, certain types of galaxies can be considered as two-dimensional vortices. The study of vortex dynamics is also closely related to two-dimensional turbulence ; the statistical behavior of vortices, in terms of the evolution of the distribution of their sizes and of the power spectrum of the enstrophy field, have been studied by several authors [1]–[9]. Observations and simulations have shown that, in some respects, vortex dynamics is similar to point particle mechanics.

Various aspects of the interaction of single vortices and of vortex pairs have been discussed by Christiansen [10], Christiansen and Zabusky [11], Overman and Zabusky [12], Mc Williams [5], Mc Williams and Zabusky [13][14], and by Basdevant et al. [6]. In the simplest situations, the theory of two-dimensional vortex interactions [15] asserts that the displacement of each vortex is governed by the velocity field of the others : in the case of vortex pairs, this implies that the two vortices rotate with respect to each other. When both vortices have the same absolute strength but opposite polarities (fig. 1), the trajectory of the pair is a straight line and the pair speed u is proportional to the reciprocal distance d^{-1} between vortex centres and to the vorticity Ω of each monopole. If two vortices having the same polarity come to interact, their velocity fields oppose and cancel each other in the contact zone ; after a transient period, the two vortices merge into a single one (fig. 2). If one neglects viscous dissipation during the interaction time, vorticity conservation implies that the vorticity of the resulting vortex is the sum of the vorticities of the initial vortices.

Numerical simulations of vortex pair dynamics have been performed using the quasi geostrophic vortex potential equation [13][16]. Laboratory experiments with two-dimensional vortices have been also reported by Heijst and Flor [17][18]. In this paper, we present a microscopic approach to vortex dynamics and we explore the different types of vortex pair collisions as a function of the impact parameter (i.e. the offset between pairs ; see fig. 3).

2 Lattice Gas Simulations

The Navier-Stokes equations for incompressible fluid flows can be simulated efficiently with lattice gas automata methods [19]. Here, we use the FHP-4 two-dimensional lattice gas model [20] to simulate vortex dynamics. We obtain a simulation speed of $1.5 \cdot 10^6$ site updates per second on a Sun IPC workstation using an efficient implementation of the velocity representation. Typical automaton universe sizes are of the order of 1000×1000 lattice units. The radii of the vortices range from $R = 50$ to $R = 100$ lattice units, giving Reynolds number between 75 and 150 ($R_\star = 2.65$ for the FHP-4 model [20]). Vortex pairs are generated by imposing a uniform velocity field over an area of $\approx 50 \times 50$ nodes for a few hundreds of iterations.

The main problem in these lattice gas simulations is the fast decay of the vorticity. Indeed, the vorticity ω has diffusive behavior, according to

$$\frac{D\omega}{Dt} = \nu \nabla^2 \omega \ .$$

The FHP-4 model has a quite low kinematic viscosity ν. Still, the limited sizes of the simulated vorticies restricts their observation times as the lifetime of isolated vorticies is proportional to the square of their radius. Very large scale simulations would be necessary to track vortical structures for long times or with high resolution. Another intrinsic problem of microscopic simulations is the presence of Monte-Carlo noise. Our simulations were set so that the outcome of a collision could be clearly identified in the velocity field representation before structures became blurred by the noise.

3 Simulation Results

Our results show three distinct types of interactions for the colliding vortices pairs as a function of the normalized offset Δ, the nondimensional parameter defined as the ratio of the impact parameter to the radius of the vortices (see fig. 3).

$0 \leq \Delta < 1.6$: In this range, the dipoles collide in a "scattering-like" repulsive process, and dipoles exchange partner (opposite polarity vortices from each dipole match each other before leaving the collison zone; see fig. 4). This result is in agreement with those found experimentally by Heijst and Flor [17][18]. The scattering angle β varies from 90° to nearly 180° as Δ varies from 0 to 1.6.

$1.6 \leq \Delta < 3.3$: As the offset becomes larger, the "contact" zone in the collision process involves only vortices with the same polarity. A fusion or coalescence process takes place between these two vortices which merge into a single, larger vortex, resulting in the formation of a tripole (see fig. 5). The larger vortex thus created is formed by the fluid from the two equal sign vortices from the original dipoles. The two other vortices remain attached to the large vortex by circling around, in agreement with experimental observation and other numerical simulations [21].

$3.3 \leq \Delta$: When the offset is further increased, the interaction induces an attractive-like process where the direction of propagation of the dipoles is modified (without exchange or merging of vortices) as illustrated in figure 6. As the offset further increases, the mutual influence of the dipoles decreases. When $\Delta > 4.5$, dipoles ignore each other ($\beta = \pi$).

Lattice gas simulation results are illustrated in figures 7 to 10 showing velocity and vorticity fields. The results are summarized in figure 11 which shows the three regions corresponding to the different interaction mechanisms. The small offset values correspond to scattering angles $\frac{\pi}{2} < \beta < \pi$, with vortex exchange between the dipoles. The intermediate offset range is the domain where rotating tripoles form as a result of the merging of two equal sign vortices into a single one. The third offset range corresponds to weak attractive interactions between dipoles. For $\Delta > 4.5$, the interaction between dipoles becomes negligible.

4 Concluding Comments

We have shown that lattice gas automata provides an operational method for the microscopic simulation of vortex interaction. Our results confirm the processes of scattering of dipoles and the merging mechanisms found experimentally for two-dimensional vortical flows. The whole range of interactions between equal strength dipoles has been explored. In particular, the limit value $\Delta = 3.3$ for the formation of tripoles agrees with the value obtained for the fusion of equal-sign monopoles into a single vortex [2]. This result suggests that the method should also be useful for the study of monopole interactions.

5 Acknowledgements

J.P.B. acknowledges support by the Fonds National de la Recherche Scientifique (F.N.R.S. Belgium). This work was supported in part by NATO (grant R.G. 910604) and by EC under contract SC1-0212.

References

[1] A Simple Point Vortex Model for Two-Dimensional Decaying Turbulence: R. Benzi, M. Briscolini, M. Colella and P. Santangelo, Physics of Fluids, in press (1992).

[2] Evolution of Vortex Statistics in Two-Dimensional Turbulence: G.F. Carnevale, J.C. Mc Williams, Y. Pomeau, J.B. Weiss and W.R. Young, Phys. Rev. Lett. **66**, 2735 (1991).

[3] On the Statistical properties of Two-Dimensional Decaying Turbulence: R. Benzi, S. Patarnello and P. Santangelo, Europhys. Lett. **3**, 811 (1987).

[4] Wave-Vortex Dynamics: R. Benzi and B. Legras, J. Phys. A: Math. Gen. **20**, 5125 (1987).

[5] The emergence of isolated coherent vortices in two dimensional vortices in turbulent flow: J.C. Mc Williams, J. Fluid Mech. **43**, 21 (1984).

[6] A Study of Barotropic Model Flows: Intermittency, Waves and Predictability: C. Basdevant, B. Legras, R. Sadourny and M. Beland, J. Atmos. Sci. **38**, 1230 (1981).

[7] Subgrid-Scale Modeling of Enstrophy Transfer in Two-Dimensional Turbulence: C. Basdevant, M. Lesieur and R. Sadourny, J. Atmos. Sci. **35**, 1028 (1978).

[8] Eddy Viscosity in Two and Three Dimensions: Kraichnan R.H., J. Atmos. Sci. **33**, 1521 (1976).

[9] On the Statistical Theory of Two-Dimensional Topographic Turbulence: Herring, J.R., J. Atmos. Sci. **34**, 1750 (1977).

[10] Numerical Simulation of Hydrodynamics by the Method of Point Vortices: Christiansen, J.P., J. Comp. Phys. **13**, 363 (1973).

[11] Instability, Coalescence and Fission of Finite-area Vortex Structures: Christiansen, J.P. and Zabusky,N.J., J. Fluid Mech. **61**, 219 (1973).

[12] Evolution and Merger of Isolated Vortex Structures: Overman,E.A. and Zabusky,N.J., Phys. Fluids **25**, 1297 (1982).

[13] Interaction of isolated vortices I. Modons colliding with modons: Mc Williams,J. and Zabusky,N.J., Geophys. Astrophys. Fluid Dyn. **19**, 207 (1982).

[14] Interaction of isolated vortices II. Modon generation by monopole collision.: Mc Williams,J., Geophys. Astrophys. Fluid Dyn. **24**, 1 (1983).

[15] G.K. Batchelor, An introduction to Fluid Dynamics, Cambridge Univ. Press, 1967.

[16] The dynamics of baroclinic and barotropic solitary eddies: Flierl,L.,Larichev, V., Mc Williams,J. and Reznik,G., Dyn. Atmos. and Oceans **5**, 1 (1980).

[17] Dipole Formation and Collisions in a Stratified Fluid: van Heijst G. J. F. and Flor J. B., Nature **340**, 212 (1989).

[18] La danse des tourbillons: Heijst, G.J. F. and Flor, J.B., La Recherche **218**, 224 (1990).

[19] Lattice Gas Hydrodynamics in Two and Three Dimensions: Frisch, U., d'Humières, D., Hasslacher, B., Lallemand, P.,Pomeau, P. and Rivet, J.P., Complex Systems **1**, 64 (1987).

[20] Fluid Dynamics with Lattice gases: d'Humières, D., Lallemand, P., Boon, J.P., Noullez, A., and Dab, D. in: Chaos and Complexity, S. Ruffo et al., eds. (World Scientific, Singapore, 1988), pp. 278-301.

[21] Tripolar Vortices in a Rotating Fluid: van Heijst, G.J.F. and Kloosterziel, R.C., Nature **338**, 569 (1989).

Figure Captions

Figure 1. Vortex pair propagation.

Figure 2. Merging of two vortices with opposite polarities.

Figure 3. Dipole collision configuration.

Figure 4. Vortex exchange in dipolar interaction.

Figure 5. Formation of rotating tripole.

Figure 6. Attractive interaction between dipoles.

Figure 7. Velocity field representation of a collision with vortex exchange (lattice: 1024×768); offset: 40 (lattice units); Re (Reynolds number)=75; T (number of time steps): 2000 (a), 4000 (b), 6000 (c), 10000 (d).

Figure 8. Velocity field representation of a collision with tripole formation (lattice: 1280×1024); offset: 190; Re=75; T: 1000 (a), 3000 (b), 7000 (c), 19000 (d).

Figure 9. Vorticity representation of a scattering process: Re=125; T: 7000 (a), 21000 (b).

Figure 10. Vorticity representation of a collision producing a tripole (lattice: 1280×1024); Re=125; T: 8000 (a), 16000 (b).

Figure 11. Dipole scattering angle β as a function of the offset Δ. Three regions of interaction are identified. The error bars are indicative of the noise in the simulations, particularly for offset values close to the border zone between dipoles and tripoles.

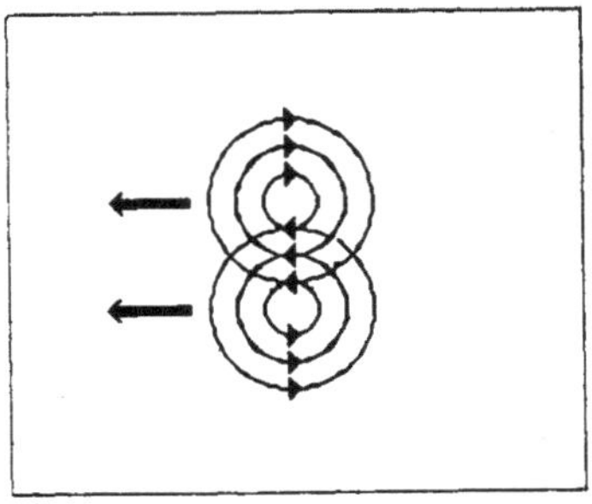

Figure 1

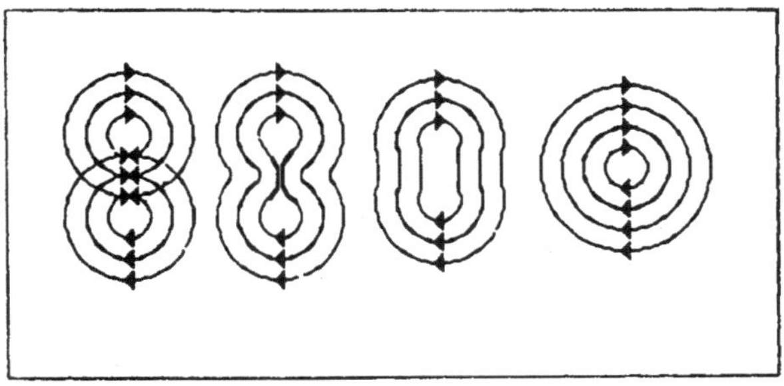

Figure 2

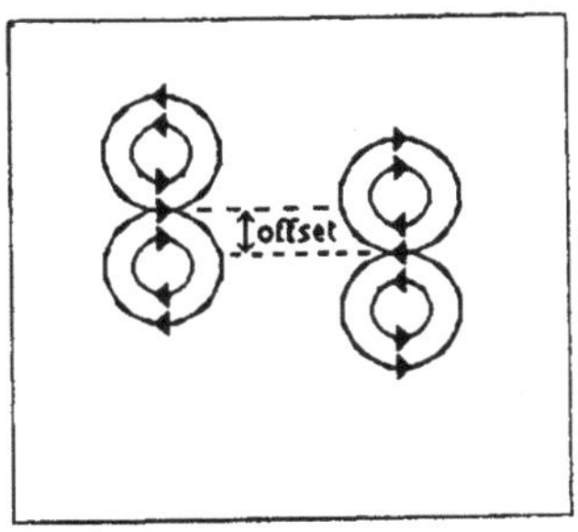

Figure 3

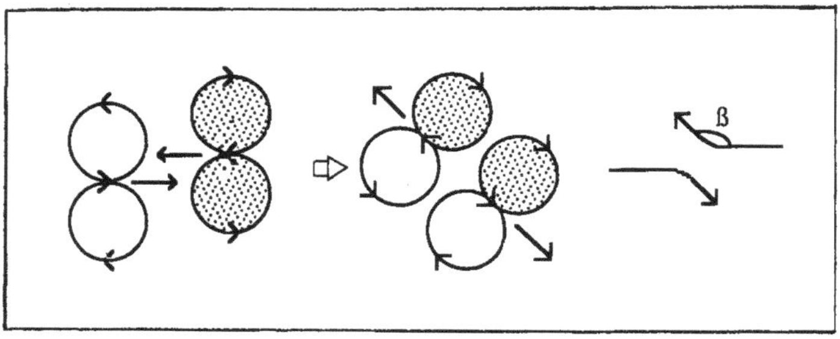

Figure 4

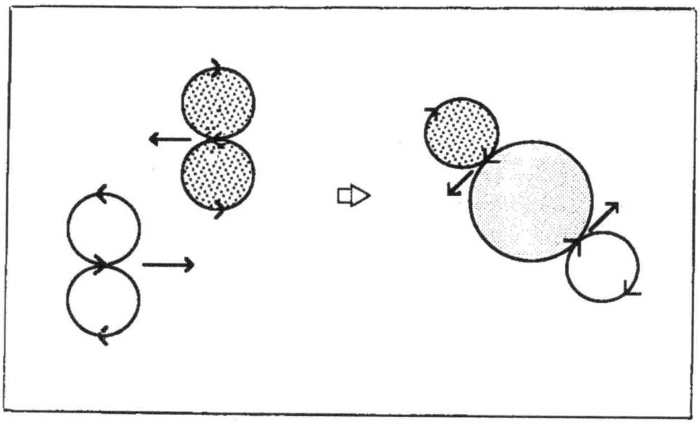

Figure 5

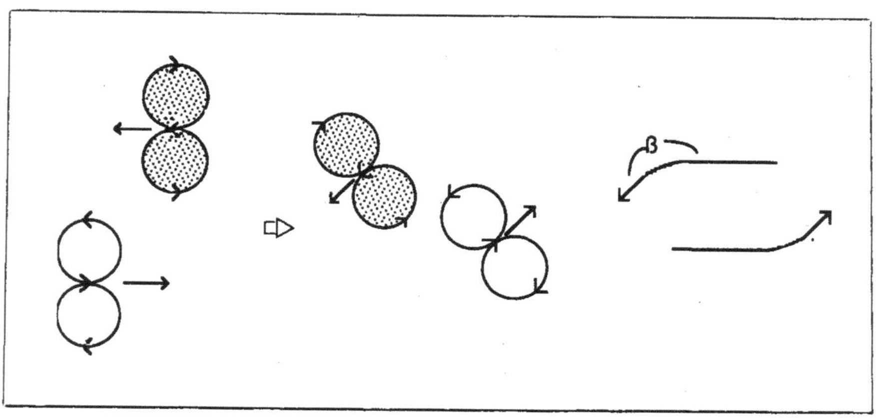

Figure 6

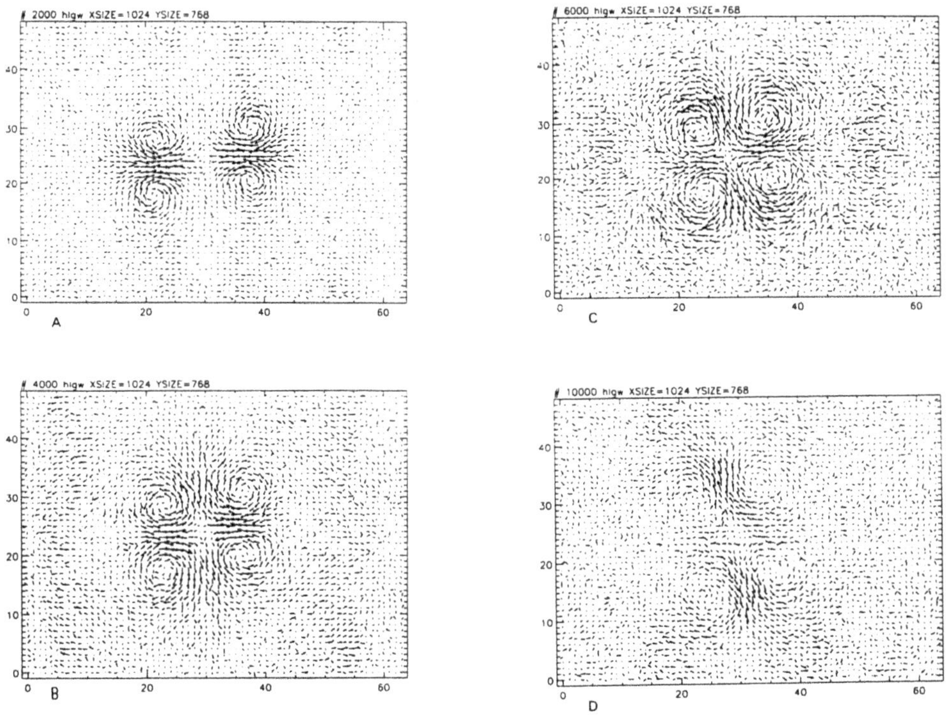

2000 higw XSIZE=1024 YSIZE=768
A
6000 higw XSIZE=1024 YSIZE=768
C
4000 higw XSIZE=1024 YSIZE=768
B
10000 higw XSIZE=1024 YSIZE=768
D

Figure 7

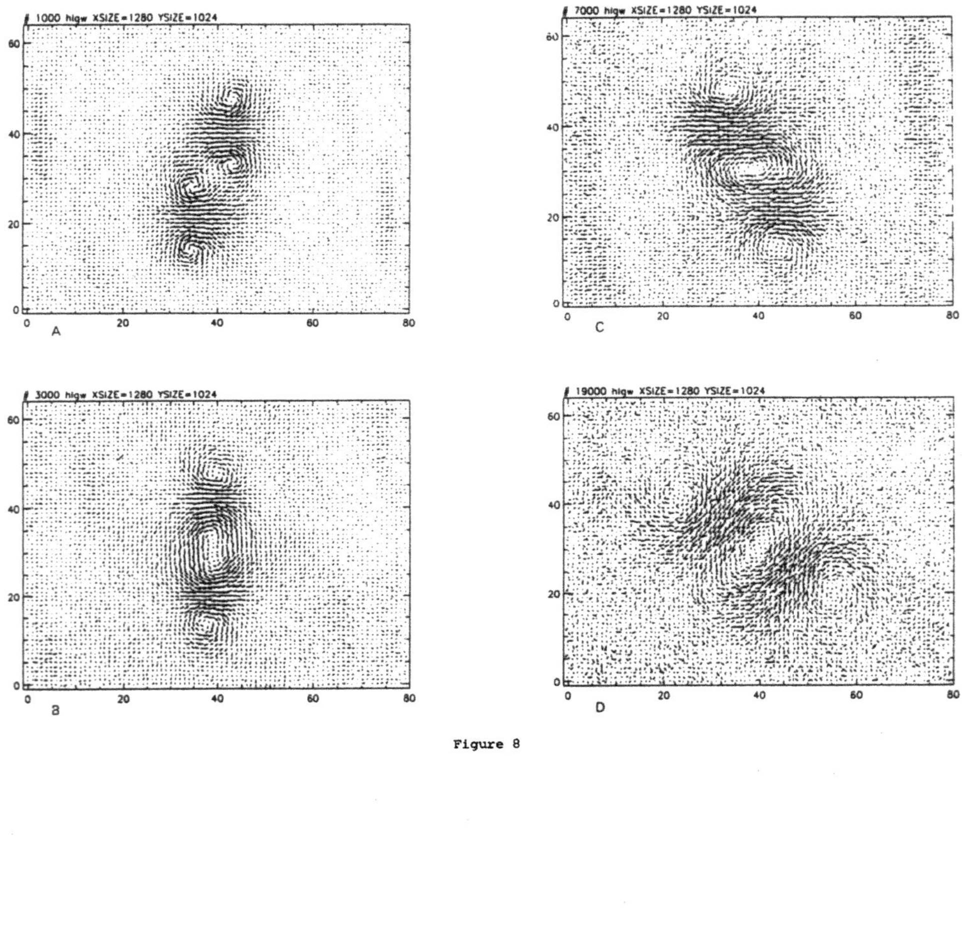

1000 higw XSIZE=1280 YSIZE=1024
7000 higw XSIZE=1280 YSIZE=1024
3000 higw XSIZE=1280 YSIZE=1024
19000 higw XSIZE=1280 YSIZE=1024
A
B
C
D

Figure 8

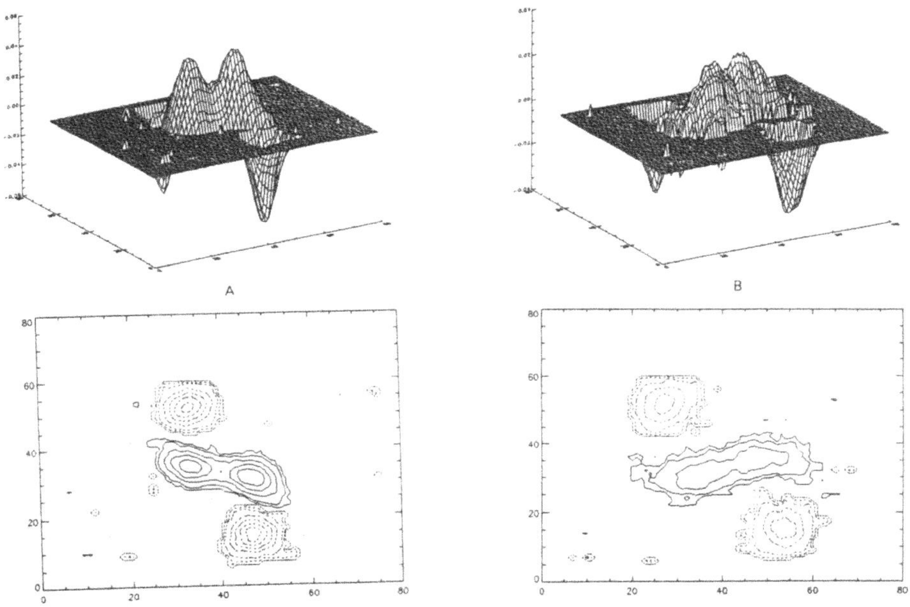

Figure 9

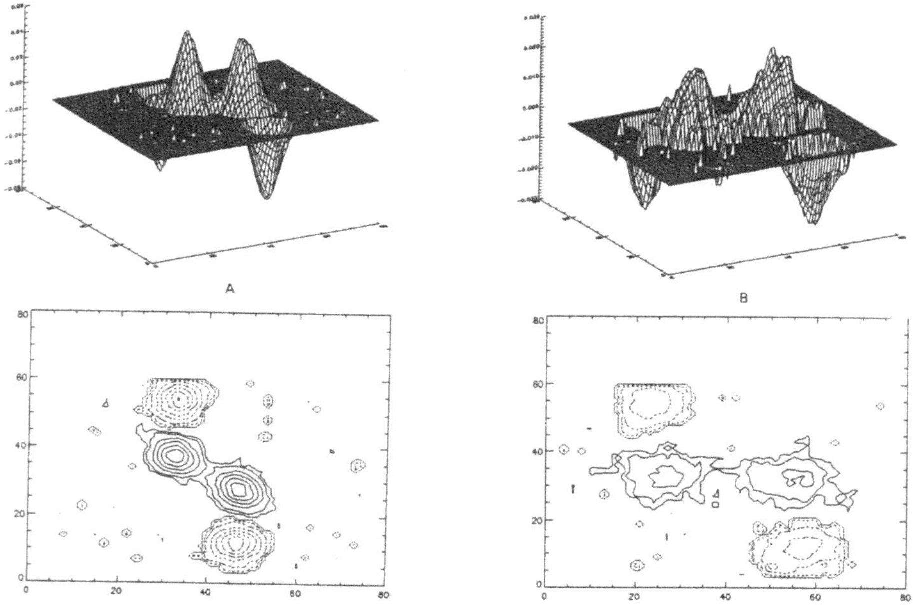

Figure 10

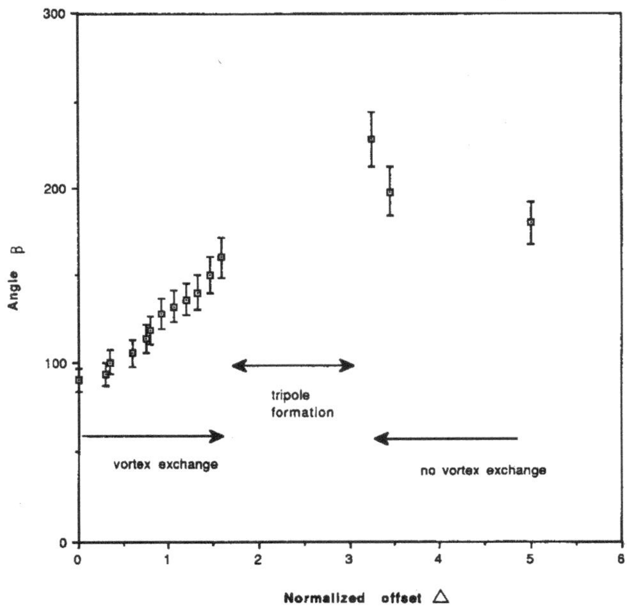
300
200
100
0
Angle β
0
1
2
3
4
5
6
Normalized offset Δ
tripole
formation
vortex exchange
no vortex exchange

Figure 11

LATTICE GAS MODEL OF A PARALLEL FLOW

LEESA BRIEGER AND ERNESTO BONOMI
Centro di Ricerca, Sviluppo e Studi Superiori in Sardegna
Via Nazario Sauro 10, I-09123 Cagliari, Italy

and

MARCO TOMASSINI
Centro Svizzero di Calcolo Scientifico
Via Cantonale, CH-6928 Manno, Switzerland

ABSTRACT

In the present contribution, we consider a variant of the stochastic lattice gas introduced in [1]. This model was shown to simulate the one-dimensional Burgers' equation including travelling shocks and rarefaction waves. However, the model predicts shock widening not characteristic of the actual solution. In the present variant shock widening is inhibited. In addition, in [1] we supposed decorrelated occupation probabilities in the lattice system in order to derive the approximate master equation. Here we incorporate the effect of the correlation into the derivation of the master equation. We find that the master equation is in fact a finite difference discretization of the one-dimensional Navier-Stokes equation for incompressible parallel flow. A highly parallel implementation of the model on the Connection Machine CM-2 is described.

1. Introduction

Lattice gases are dynamical systems in which particles move trough the lattice sites in discrete time steps. They have been found to be useful for describing several aspects of fluid motion. Lattice gases, and cellular automata in general, are very suitable systems for parallel computing. In fact, the evolution of those automata consist of a series of synchronous update phases in which every automaton does the same thing: look at some group of immediate neighbors and update its own state according to the state of those neighbors, using some predefined rules. Therefore, the dynamics is intrinsically parallel and can be mapped easily on an array of processors, as we will see in section 4.

The lattice gas considered in [1] is a one-dimensional model in which each particle takes a stochastic walk to the right: with probability α a particle chooses to move and with probability $1-\alpha$ chooses to remain stationary. In addition, an exclusion principle allows at most one particle per lattice site and is implemented in the following way: a particle which has chosen to move simply is not allowed to if the site to its right is occupied at the current timestep. The balance equation for this uni-dimensional system is

$$P_j^{n+1} - P_j^n = P_{j-1}^n W_{j-1,j}^n + P_{j+1}^n W_{j+1,j}^n - P_j^n (W_{j,j+1}^n + W_{j,j-1}^n) \qquad (1)$$

where P_j^n is the occupation probability of site j at moment n and W_{ij}^n is the conditional probability of transition of a particle from site i to site j at moment n, given that site i is occupied.

As in [1], if we suppose that occupation probabilities are independent, then the conditional probabilities W_{ij}^n can be treated as simple probabilities and the balance equation becomes

$$P_j^{n+1} - P_j^n = \frac{\alpha}{2}(P_{j+1}^n - 2P_j^n + P_{j-1}^n) + \frac{\alpha}{2}(2P_j^n - 1)(P_{j+1}^n - P_{j-1}^n). \tag{2}$$

Under appropriate conditions on α and with the change of variable $v_j^n = (-2P_j^n + 1)V$ where V is a characteristic velocity, this gives a finite difference discretization of Burgers' equation $u_t = \mu u_{xx} - uu_x$ describing the velocity of a fluid in uni-dimensional flow. Notice that particles represent negative velocity and holes positive velocity, and that our lattice gas behaves as an incompressible fluid.

2. Shock Widening

Fig.1 shows an example of the model described above in a case with periodic initial conditions; the dashed line is the finite difference solution, i.e. the results of Eq.(2) calculated numerically, and the symbols represent the lattice gas measurements. Shock position is captured accurately by the lattice system, but shock widening is noticeable. We mention here that the lattice gas measurements consist of ensemble averages: we have run several (512) simultaneous one-dimensional lattice experiments, measuring particle count on each column to produce the particle distribution. Configurations of particles are initialized stochastically, with resultant fluctuations around the desired initial velocity profile. A shock seems to be a point of accumulation of these initial particle fluctuations, apparently due to the uni-directional nature of the dynamics, Fig.3a.

Now if we introduce an additional degree of freedom to the particle displacement in the model, allowing particles to diffuse in the direction purpendicular to the flow, we eliminate this uni-directional nature. The resultant mixing of particles between the one-dimensional lattice systems prevents the accumulation of fluctuations around a shock and inhibits shock widening. The implementation is as follows. At each timestep in the lattice model a particle chooses with equal probability $1/2$ between the "convective" walk in the x-direction and a diffusive walk in the y-direction. The convective walk is as described above: probability α to move right, respecting the exclusion principle, probability $1-\alpha$ not to move. For the diffusive walk, a particle chooses with equal probability $\beta/2$ between moving up or down, again respecting the exclusion principle of at most one particle per site. Periodic boundary conditions are imposed in the direction of the diffusion. In order to avoid conflicts between diffusing particles, they do not choose their direction of diffusion independently but are synchronized to move up or down together. Macroscopically, the behavior remains diffusive even with this synchronization [2].

With this transition rule for the lattice gas and again supposing independent occupation probabilities, the two-dimensional master equation for the model is

$$\begin{aligned} P_{ij}^{n+1} - P_{ij}^{n} &= \tfrac{\alpha}{4}(P_{i,j+1}^{n} - 2P_{ij}^{n} + P_{i,j-1}^{n}) + \tfrac{\beta}{4}(P_{i+1,j}^{n} - 2P_{ij}^{n} + P_{i-1,j}^{n}) \\ &+ \tfrac{\alpha}{4}(2P_{ij}^{n} - 1)(P_{i,j+1}^{n} - P_{i,j-1}^{n}) \end{aligned} \tag{3}$$

Summing this equation over the columns, we average over y and return to a master equation which gives a description of the one-dimensional behavior of the model in x. Because the boundary conditions in y are periodic, the contribution of the Laplacian in y is annihilated and the one-dimensional balance equation is exactly as before, Eq.(2), the finite difference discretization of Burgers' equation. We have repeated the example of Fig.1 with the transition rule that allows for the two-dimensional mixing; the results are shown in Fig.2. The one-dimensional behavior is the same as that of the original model, but shock widening has been inhibited, Fig.3b.

3. Parallel Flow

The description furnished by Eq.(2) or (3) of the macroscopic behavior of the particle system is an approximate one following from the assumption that occupation probabilities are decorrelated. This approximation appears adequate when the solution exhibits a standing shock or rarefaction wave. However, for a travelling wave solution, there are values of α for which Eqs.(2) and (3) do not adequately describe the particle system, and the supposition of independent occupation probabilities is not justified. Thus in the following we reconsider the master equation of the particle system - *without* the hypothesis of decorrelated occupation probabilities.

The transition rule described above, with the convective walk in x and the diffusive walk in y, is programmed in our simulations as the following Boolean function

$$\begin{aligned} X_{ij}^{n+1} &= H^n[A_{i,j-1}^n X_{i,j-1}^n(1 - X_{i,j}^n + (1 - A_{i,j}^n)X_{i,j}^n + A_{i,j}^n X_{i,j}^n X_{i,j+1}^n] \\ &+ (1 - H^n)[K^n B_j^n X_{i+1,j}^n(1 - X_{i,j}^n) + B_j^n X_{i,j}^n X_{i-1,j}^n + (1 - B_j^n)X_{i,j}^n \\ &+ (1 - K^n)B_j^n X_{i-1,j}^n(1 - X_{i,j}^n) + B_j^n X_{i,j}^n X_{i+1,j}^n + (1 - B_j^n)X_{i,j}^n]. \end{aligned} \tag{4}$$

$X_{i,j}^n$ is the Boolean variable with the value 1 if a particle is present at site i, j at moment n and 0 otherwise. The other variables are also Boolean:

$$H^n = \begin{cases} 1 & \text{with probability } 1/2 \\ 0 & \text{with probability } 1/2 \end{cases}$$

$$K^n = \begin{cases} 1 & \text{with probability } 1/2 \\ 0 & \text{with probability } 1/2 \end{cases}$$

$$A^n_{ij} = \begin{cases} 1 & \text{with probability } \alpha \\ 0 & \text{with probability } 1-\alpha \end{cases}$$

$$B^n_j = \begin{cases} 1 & \text{with probability } \beta \\ 0 & \text{with probability } 1-\beta. \end{cases}$$

Averaging Eq.(4) over the ensemble of events, and setting $x^n_{ij} = < X^n_{ij} >$, we find the following balance equation in terms of the average occupation number

$$\begin{aligned} x^{n+1}_{ij} - x^n_{ij} &= \tfrac{\alpha}{2}\{x^n_{i,j-1} - x^n_{ij} + < X^n_{ij}(X^n_{i,j+1} - X^n_{i,j-1}) >\} \\ &+ \tfrac{\beta}{4}(x^n_{i+1,j} - 2x^n_{ij} + x^n_{i-1,j}). \end{aligned} \tag{5}$$

In order to simplify this expression, we use the identity

$$< AB > = < A >< B > + < (A - < A >)(B - < B >) > .$$

Then we let $\sigma^n_{ij} = X^n_{ij} - x^n_{ij}$ and use the following second-order approximation

$$\sigma^n_{ij} = \frac{1}{2}(\sigma^n_{i,j+1} + \sigma^n_{i,j-1})$$

to write the balance equation, Eq.(5), as

$$\begin{aligned} x^{n+1}_{ij} - x^n_{ij} &= \tfrac{\alpha}{2}(x^n_{i,j-1} - x^n_{ij}) + \tfrac{\alpha}{2}x^n_{ij}(x^n_{i,j+1} - x^n_{i,j-1}) \\ &+ \tfrac{\beta}{4}(x^n_{i+1,j} - 2x^n_{ij} + x^n_{i-1,j}) + \tfrac{\alpha}{4}(< (\sigma^n_{i,j+1})^2 > - < (\sigma^n_{i,j-1})^2 >). \end{aligned} \tag{6}$$

Now notice that $< (\sigma^n_{ij})^2 > = x^n_{ij}(1 - x^n_{ij})$. The balance equation can thus be written as

$$\begin{aligned} x^{n+1}_{ij} - x^n_{ij} &= \tfrac{\alpha}{4}(2x^n_{ij} - 1)(x^n_{i,j+1} - x^n_{i,j-1}) + \tfrac{\alpha}{4}(x^n_{i,j+1} - 2x^n_{ij} + x^n_{i,j-1}) \\ &+ \tfrac{\beta}{4}(x^n_{i+1,j} - 2x^n_{ij} + x^n_{i-1,j}) \\ &+ \tfrac{\alpha}{4}\{x^n_{i,j+1}(1 - x^n_{i,j+1}) - x^n_{i,j-1}(1 - x^n_{i,j-1})\}. \end{aligned}$$

Then with the appropriate change of variable, $v^n_{ij} = (-2x^n_{ij} + 1)V$, the balance equation takes the form

$$\begin{aligned} v^{n+1}_{ij} - v^n_{ij} &= \tfrac{\alpha}{4}(v^n_{i,j+1} - 2v^n_{ij} + v^n_{i,j-1}) + \tfrac{\beta}{4}(v^n_{i+1,j} - 2v^n_{ij} + v^n_{i-1,j}) \\ &- \tfrac{\alpha}{2V}v^n_{ij}\tfrac{v^n_{i,j+1} - v^n_{i,j-1}}{2} + \tfrac{\alpha}{4V}\{[V^2 - (v^n_{i,j+1})^2] - [V^2 - (v^n_{i,j-1})^2]\}. \end{aligned}$$

Since temperature, denoted by T, is given by the fluctuation of the velocity, the last term in this expression can be thought of as a gradient of $T = V^2 - v^2$. Multiplying and dividing by the right factors so as to introduce a discretization into the expression, we can then write the balance equation as

$$\begin{aligned} \frac{v_{ij}^{n+1}-v_{ij}^{n}}{\Delta t} &= \frac{\alpha\Delta x^2}{4\Delta t}\frac{(v_{i,j+1}^{n}-2v_{ij}^{n}+v_{i,j-1}^{n})}{\Delta x^2} + \frac{\beta\Delta y^2}{4\Delta t}\frac{(v_{i+1,j}^{n}-2v_{ij}^{n}+v_{i-1,j}^{n})}{\Delta y^2} \\ &- - \frac{\alpha\Delta x}{2V\Delta t}v_{ij}^{n}\frac{(v_{i,j+1}^{n}-v_{i,j-1}^{n})}{2\Delta x} + \frac{\alpha\Delta x}{2V\Delta t}\frac{(T_{i,j+1}^{n}-T_{i,j-1}^{n})}{2\Delta x} \end{aligned} \tag{7}$$

Now, consider the Navier-Stokes equation for parallel flow [3]

$$u_t = \frac{\mu}{\rho}(u_{xx} + u_{yy}) - uu_x - \frac{1}{\rho}p_x \tag{8}$$

for which the relationship between pressure and velocity fluctuation (or temperature) is the following: $p = p_0 - \rho$T.

Then we see that as long as the following relations between the parameters of the equation and the discretization hold,

$$\frac{\mu}{\rho} = \frac{\alpha\Delta x^2}{4\Delta t} \qquad \frac{\mu}{\rho} = \frac{\beta\Delta y^2}{4\Delta t} \qquad \frac{\alpha\Delta x}{2V\Delta t} = 1, \tag{9}$$

the master equation of the lattice system, Eq.(7), is just a finite difference discretization of Eq.(8). Thus it is a discrete form, not of Burgers' equation, but of the equation for parallel flow, complete with the pressure gradient. Fig.4 shows the laminar flow simulated by the model, with uniform velocity at the entrance on the left going towards fully developed flow to the right. Zero velocity is imposed at the upper and lower boundaries (not periodic boundary conditions), so that the contribution to the Laplacian in the y-direction does not disappear. In Fig.4, notice that the boundary layer grows with the distance from the entrance until the flow becomes fully developed, at which point the characteristic parabolic velocity profile can be seen, Fig.5.

4. An Outline of the CM-2 Architecture and Programming Model

We will limit ourselves to a brief description of the CM-2 system. A more complete account can be found in [4] and especially in [5].

The Connection Machine CM-2 is a single-instruction, multiple-data (SIMD) massively parallel computer. This means that there is a single control unit that broadcasts the same instruction to all processing elements. The instruction is then executed on independent data.

The parallel processing unit of the Connection Machine may contain 4K, 8K, 16K, 32K or 64K 1-bit processors. Each processor has direct access to a local memory that may range from 64K bits to 1M bits. Each group of 32 scalar 1-bit processors shares an interface to an optional floating-point chip. A fully configured CM-2 has 2048 floating-point chips. A front-end workstation is used for the programming development. The CM-2 acts as an attached processor; the workstation decodes the user program executing the sequential parts and issuing parallel instructions to the CM-2 control unit. The front-end can also read and write the 1-bit processors' local memories.

There is no globally accessible memory in the CM-2. However, any processor can communicate with any other processor via an interprocessor network. The most efficient form of communication occurs between neighbor processors in regular patterns, such as all processors in a line sending data to their, say, left neighbour (periodic boundary conditions apply). These patterns of communication are typical of Cellular Automata and other regular algorithms.

The CM-2 is designed to operate on large data sets. Many problems contain a high degree of *spatial* parallelism which can be exploited by performing the same operation simultaneoulsy on "arrays" of data spread throughout the processing elements memories called ***parallel variables***, one item per processor. If there are more array elements than physical processors then ***virtual processors*** (VP's), with each physical processor simulating a number of virtual processors, are used. Except for an increase in execution time, this arrangement is transparent to the user.

A collection of parallel variables is called a *VP set*. Each VP set has an associated *geometry* which must be specified by the user; this geometry is an organization of the virtual processors into an n-dimensional cartesian grid with dimensions that may range from 1 to 31. The grid is wrapped-around (periodic boundary conditions). Such a grid is also called *NEWS* grid — an extension of the concept of a square lattice in which each VP has four neighbors in the directions north (N), south (S), east (E) and west (W).

Besides the usual operations on parallel variables, there are ***reduction*** operations that reduce an array of values to a single value by repeated application of an associative binary operator. An example is the addition where all the values of a parallel variable are added together to give a single scalar result. The ***scan*** operations form another set of useful primitives. They combine computation and communication along an axis of a logical grid in a cumulative manner. An example might be the calculation of all the partial sums of n numbers.

Several programming languages with data-parallel extensions are available: C*, *Lisp, the low-level language C/Paris and especially CM Fortran, which looks like Fortran 90 for array operations. We wrote our programs mainly in C/Paris. A number of useful mathematical library routines is also provided.

5. Data Parallel Algorithms For The Parallel Flow Problem

In this section we give pseudo-code parallel algorithms for the lattice gas model described in section 3. The actual CM-2 C/Paris code is similar in spirit but much longer and difficult to understand. In fact, many inessential details such as initialisation, details of boundary conditions handling and I/O were purposely suppressed here for the sake of clarity. For all examples the geometry of the lattice is a two-dimensional grid with cyclic boundary conditions (a torus) and a particle (or a hole) is associated to each virtual processor.

```
for all processors do in parallel
   establish initial conditions
end parallel do
for k ← 1 to Max Number of Time Steps do
   for all processors do in parallel
      compute equation (4)
   end parallel do
end for
```

As another example of data-parallel programming in the context of lattice gases, the calculation of the velocity profile is sketched by the following pseudocode, in which ΔX is the sampling interval along the x axis and X is the virtual processor's coordinate along a row of the grid:

```
for all processors do in parallel
   Segment_Marker ← X mod ΔX
   scan_with_add particle velocity in each ΔX interval along rows
   The resulting array gives the velocity profile
end parallel do
```

6. Conclusions

The stochastic lattice model discussed, a two-dimensional system governed by a simple law, Eq.(4), captures the macroscopic behavior of a complex fluid. With certain parameter values and periodic boundary conditions in y, the model simulates the solution of Burgers' equation in one dimension. Strictly speaking, however, it is the solution of the two-dimensional equation of Navier-Stokes for parallel flow which is simulated by the model. With zero-velocity boundary conditions in y, we observe the formation of a boundary layer and development of the parabolic velocity profile characteristic of Poiseuille flow.

From the computational point of view, the CM-2 appears as a nearly ideal machine for these synchronous lattice gas simulations, to the point that today building specialized machines for cellular automata models might be questioned, if one has access to a CM. The model is rather easy to implement with the available programming primitives and there is a direct matching between the model topology and the virtual processor grid topology. Another good point is that global property calculations on the grid, like velocity distributions, are well supported by the scan and reduction operations of the CM-2. For all calculations we used the 8K CM-2 model of the University of Geneva.

7. References

1. L. Brieger and E. Bonomi, *J. Stat. Phys.* **69** (1992) 837.

2. L. Brieger and E. Bonomi, *Physica D* **47** (1991) 159.
3. H. Schlichting, *Boundary-Layer Theory* (McGraw-Hill, London, 1979).
4. E. Bonomi and M. Tomassini, *Int. J. Mod. Phys.* **C3** (1992) 709.
5. *Connection Machine CM-200 Series Technical Summary* (Thinking Machines Corporation, Cambridge, Massachussets, 1991).

CAPTIONS

FIG.1: Periodic initial conditions: finite-difference results (dashed curve) and automaton measurements. The evolution of the system illustrates shock formation and rarefaction wave dissipation. Shock widening is noticeable.

FIG.2: The same example as in Fig.1 with the transition rule that allows for the two-dimensional mixing. Shock widening has been inhibited.

FIG.3: Configuration snapshots a) and b) correspond to Figs.1 and 2 respectively:
a) shock widening,
b) no shock widening.

FIG.4: Laminar flow simulated by the model, with uniform velocity at the entrance on the left. Zero velocity is imposed at the upper and lower boundaries, $P = 0.5$. The boundary layer grows with the distance from the entrance until the flow becomes fully developed.

FIG.5: Velocity distribution:
a) close to the entrance on the left,
b) fully developed flow, parabolic fit, $V = -4.51 \times 10^{-3} + 3.47y - 3.51y^2$.

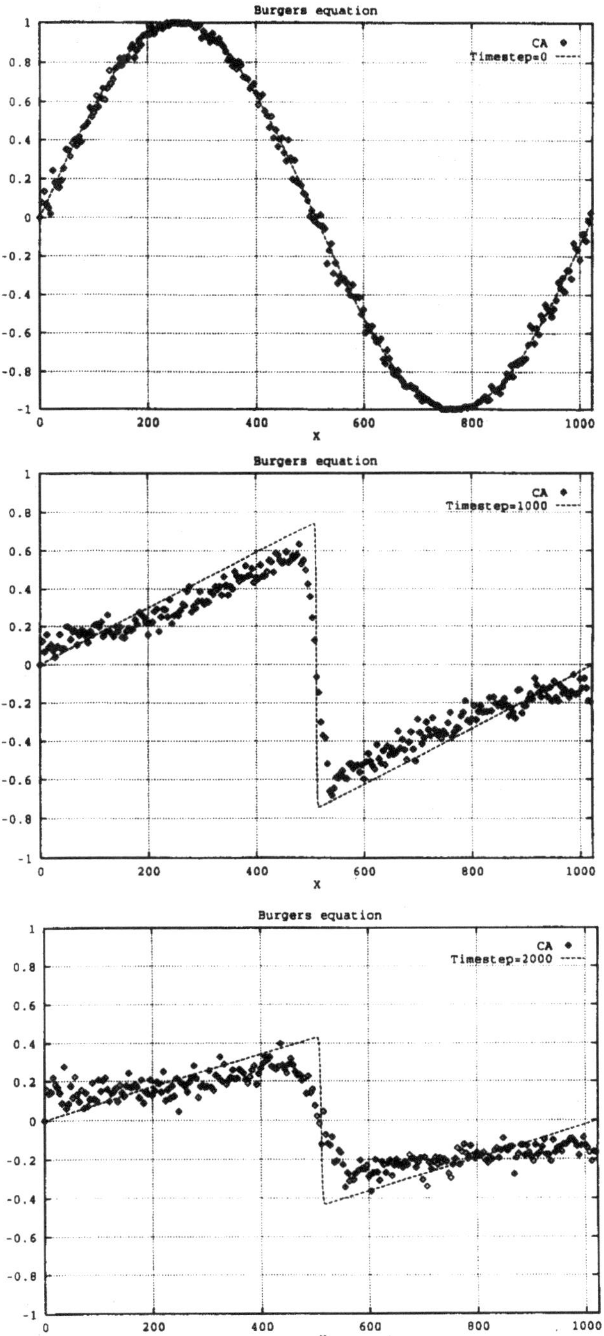

FIG.1: Periodic initial conditions: finite-difference results (dashed curve) and automaton measurements. The evolution of the system illustrates shock formation and rarefaction wave dissipation. Shock widening is noticeable.

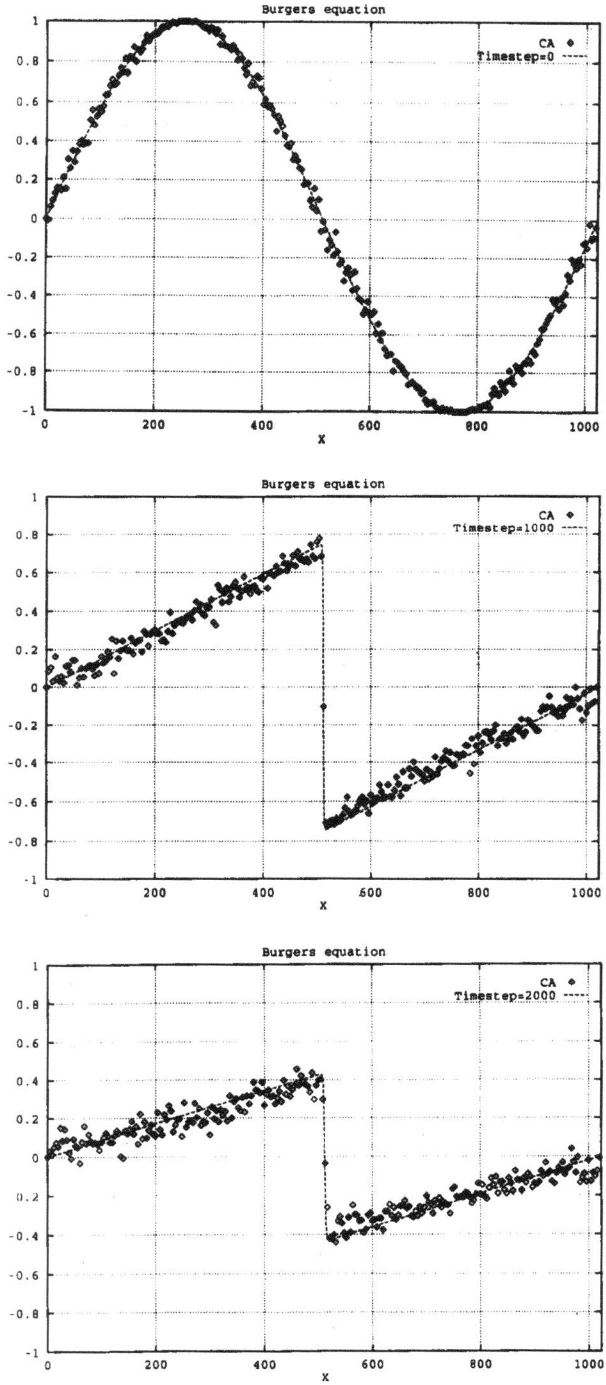

FIG.2: The same example as in Fig.1 with the transition rule that allows for the two-dimensional mixing. Shock widening has been inhibited.

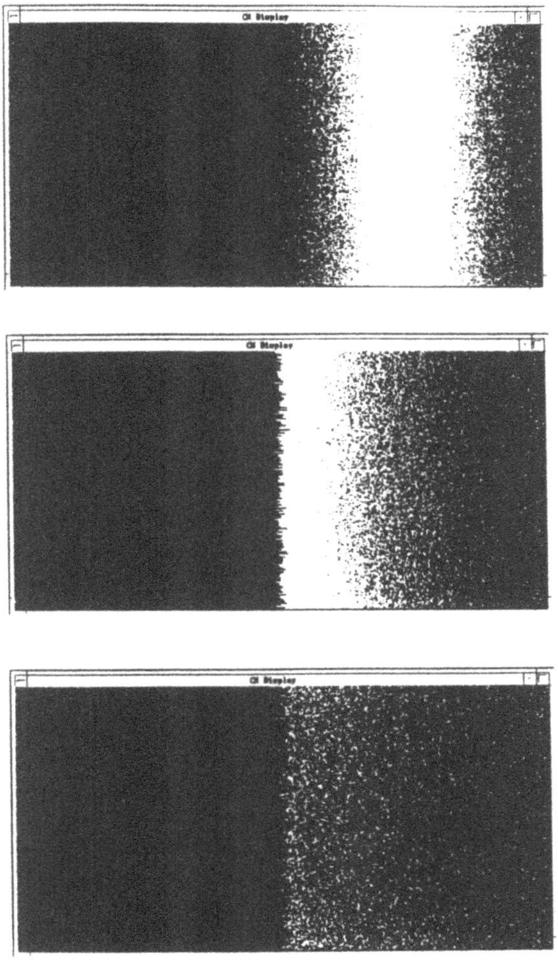

a)

FIG.3: Configuration snapshots a) and b) correspond to Figs.1 and 2 respectively:
a) shock widening,
b) no shock widening.

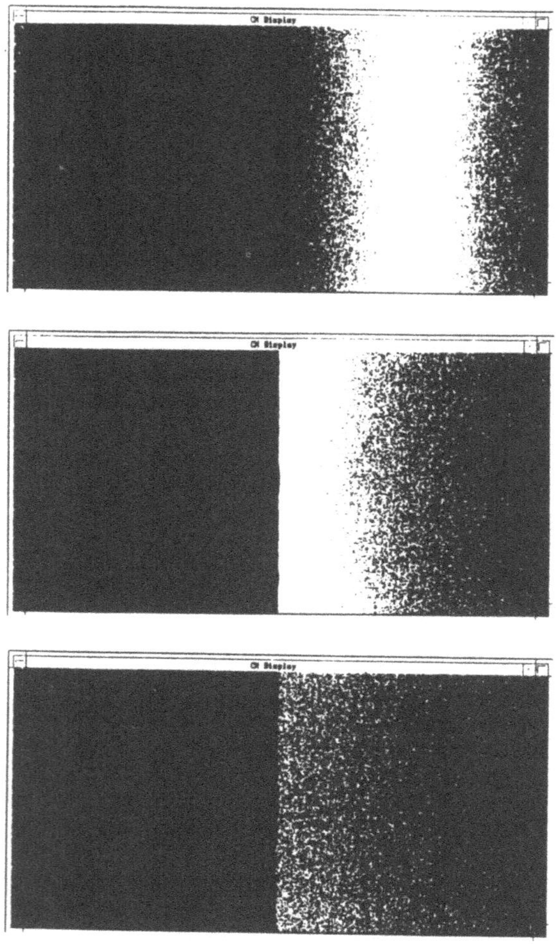

b)

FIG.3: Configuration snapshots a) and b) correspond to Figs.1 and 2 respectively:
a) shock widening,
b) no shock widening.

FIG.4: Laminar flow simulated by the model, with uniform velocity at the entrance on the left. Zero velocity is imposed at the upper and lower boundaries, P = 0.5. The boundary layer grows with the distance from the entrance until the flow becomes fully developed.

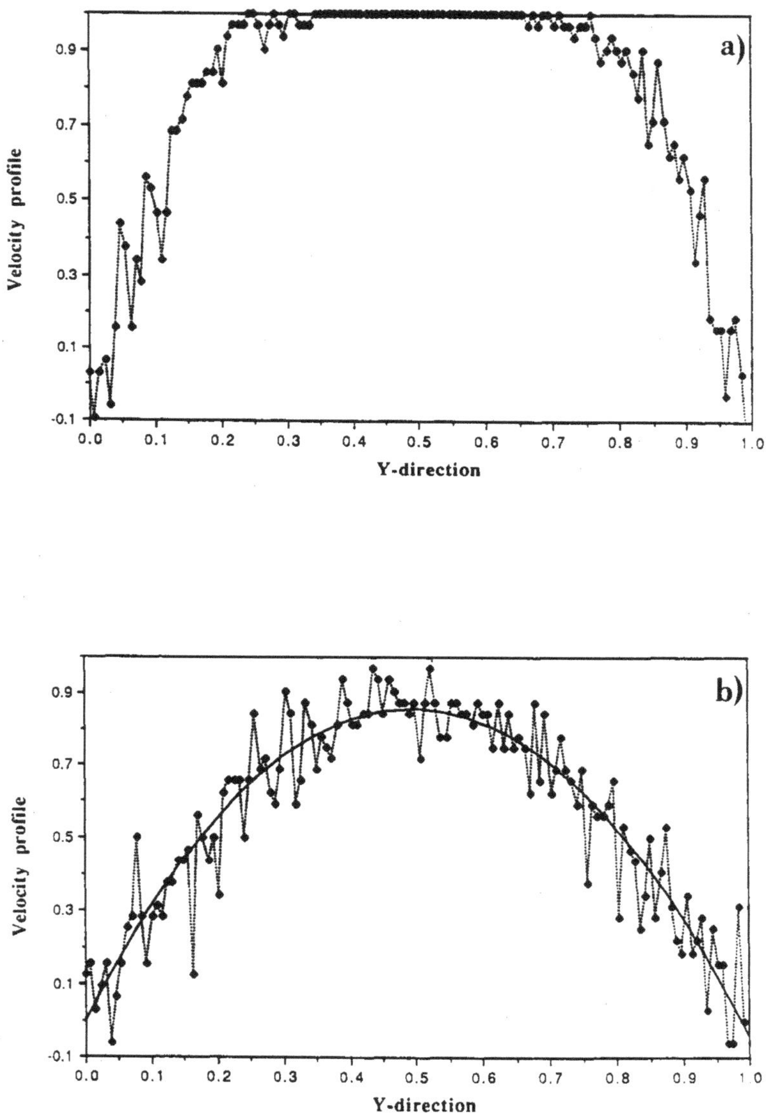

FIG.5: Velocity distribution:
a) close to the entrance on the left,
b) fully developed flow, parabolic fit, $V = -4.51 \times 10^{-3} + 3.47y - 3.51y^2$.

LATTICE BOLTZMANN METHOD: A REVIEW WITH A GLANCE AT ASTROPHYSICS

S. Succi
IBM European Center for Scientific and Engineering Computing, v. Oceano Pacifico 171 Rome, 00144, Italy

ABSTRACT

We present some recent developments of the Lattice Boltzmann method which may prove useful for the numerical study of astrophysical problems.

1. INTRODUCTION

Lattice-gas (LG) models obeying cellular automata rules and the related Lattice-Boltzmann (LB) technique have known a rapid expansion in the recent years mostly in connection with the simulation of complex hydrodynamic phenomena [1,2,3]. The driving force behind these methods is their outstanding amenability to parallel processing, a property which stems from the basic space-time locality of cellular automata rules. This aptness to parallel computing is nowadays witnessed by the wide span of different architectures on which the method has been succesfully implemented. So far, LG and LB methods have been directed to various hydrodynamic problems, but they definitely yield their best for the study of moderate-Reynolds incompressible flows in highly heterogeneus media. To the best of the author's knowledge, none of the aforementioned flows bear much astrophysical interest. Whether this is due to intrinsic limitations of the method or just to a lack of astrophysical focus, it's difficult to say. A very fact is, however, that in the recent years, extensions of the LB method have been developed which seem capable of dealing with many of the physical phenomena which characterize astrophysical flows, i.e. thermal effects, magnetic activity and compressibility. In this paper a synthetic account of some of these recent developments is presented.

2. THE LATTICE BOLTZMANN EQUATION

The Lattice Boltzmann Equation (LBE) is an outgrowth of Lattice Gas automata,[1] i.e. discrete dynamical systems which evolve according to the following rule

$$n_i(\mathbf{x}+\mathbf{c}_i,t+1)-n_i(\mathbf{x},t)=\Omega_i(n_j(\mathbf{x},t)) \quad i,j=1,\ldots,b \tag{1}$$

where n_i are Boolean variables representing the presence/absence of a particle with velocity $\mathbf{c}_i$ at site $\mathbf{x}$ at time t . By a suitable choice of the lattice crystallographic symmetries (the set of discrete speeds $\mathbf{c}_i$ has to ensure rotational invariance at a macroscopic level) and the constraint of local mass and momentum conservation, the large-scale limit of eq.(1) has been shown to reproduce the Navier-Stokes equations of continuum hydrodynamics. The first LG scheme possessing all of the aforementioned symmetries is a two-dimensional hexagonal lattice introduced as late as 1986 in a seminal paper by Frisch Hasslacher and Pomeau (FHP) [4]. The extension to the three-dimensional world is not straightforward. In three dimensions there is no space-filling lattice ensuring rotational invariance of the momentum-flux tensor. The way out is a 4-dimensional face-centered hypercube (FCHC) in which each node is surrounded by $b = 24$ neighbours. Projections of FCHC to three, two and one dimensions results in 18,9 and 3 independent populations respectively (see figure 1). The equation (1) is very appealing from the computational viewpoint because it lends itself to massive vector and parallel computing. In addition, all the operations can be performed in boolean logic so that the resulting simulations are automatically free of the round-off and truncation errors. This simplicity is counterbalanced by two main drawbacks: statistical noise (large ensembles of automata need to be used to extract smooth hydrodynamic signals) and exponential complexity of the collision rule as a function of the number of bits needed to encode the automaton state at each lattice site.

The LBE was first generated in response to the first of these problems [5]. Within LBE the occupation numbers n_i are replaced by corresponding mean populations $N_i = \langle n_i \rangle$ where the brackets denote ensemble averaging. By supplementing this averaging procedure with further statistical assumptions (molecular chaos), a Lattice Boltzmann Equation is derived which has exactly the same form as the equation (1) with the occupation numbers n_i replaced by the mean populations N_i. This yields

$$N_i(\mathbf{x} + \mathbf{c}_i, t + 1) - N_i(\mathbf{x}, t) = \Omega_i(N_j(\mathbf{x}, t)) \quad i, j = 1, \ldots, b \tag{2}$$

We refer to the mean populations N_i as to **mesoscopic** fields, to stress the fact that they represent an average over microscopic degrees of freedom.

The advantage of the equation (2) over equation (1) is that the unknowns N_i (no longer Boolean variables) are now smooth functions of space and time because the single-particle fluctuations have been averaged out. The other side of the medal is the loss of physical information due to the fact that all 'non-Boltzmann' effects, i.e. particle correlations, have been swept under the carpet.

It is to be noted that, at variance with the standard Boltzmann equation of continuum theory, which usually accounts for only two-body collisions, the factorized collision operator Ω_i still contains the whole tree of interactions up to b-body collisions. This means that this operator entails of the order of 2^b (floating-point) operations, which makes its implementation in three-dimensions by no means any easier than its boolean counterpart. This difficulty was subsequently bypassed by Higuera and Jimenez [6], with the introduction of a Chapman-Enskog expansion of

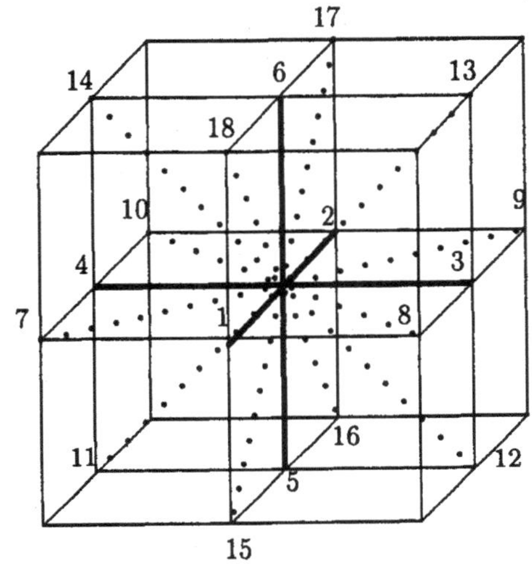

Figure 1: The FCHC lattice

the collision operator.

2.1 LBE-CE: LBE in the Chapman-Enskog Limit

The key observation is as follows: since the validity of the LG method is limited to low speed (i.e. low Mach numbers) regimes, a quadratic expansion of the non-linear collision operator is still compatible with the aim of a large-scale description of the flow. Technically, the procedure consists of expanding the collision operator Ω_i around the uniform steady-state $N_i = d$, d being the density per link of the discrete fluid, and retaining terms up to second and first order in the Mach and Knudsen numbers respectively.

By using the fact that, by definition, $\Omega_i(N_j^{eq}) \equiv 0$ and further symmetry arguments, one is led to the following LBE with a quadratic collision operator

$$N_i(\mathbf{x} + \mathbf{c}_i, t+1) - N_i(\mathbf{x}, t) = \sum_{j=1}^{b} \Omega_{ij}(N_j - N_j^{eq}) \tag{3}$$

where Ω_{ij} is the scattering matrix between state i and j and N_j^{eq} are the local equilibrium populations, expanded to second order in the flow speed u to retain convective effects. The equation (3) can be regarded as a finite-difference approximation to a model Boltzmann equation of the BGK type.

This equation is a real breakthrough from the computational point of view because it turns the former exponential complexity into a quadratic one.

2.2 LBE as a self-standing hydrodynamic model

In the work by Higuera&Jimenez, the LBE was still in a one-to-one correspondence with the underlying Lattice Gas dynamics, in the sense that the scattering matrix Ω_{ij} and the e.d.f. (equilibrium distribution function) N_i^{eq} were derived by the microdynamic equations of motion.

A further step forward was marked by the observation that neither the scattering matrix nor the e.d.f. need be related to any underlying microdynamics; in other words eq.(2) can be regarded as a 'self-standing' model of hydrodynamics, where Ω_{ij} and N_i^{eq} can be chosen with a wide degree of latitude, the only constraint being that mass and momentum be locally conserved [7].

$$\sum_{i,j=1}^{b} \Omega_{ij}(N_j - N_j^{eq}) = 0 \ , \ \sum_{i,j=1}^{b} \Omega_{ij}\mathbf{c}_i(N_j - N_j^{eq}) = 0 \tag{4}$$

Breaking this 'correspondence principle' between LG and LB provides a lot of additional freedom, which permits on the one hand to perform a complete analytical study of the transport coefficients, and on the other hand, to insert new physics in a well controlled fashion.

On the analytical side one can prove the existence of an H-theorem which guarantees numerical stability in the **linear** regime (flow speed $u \ll 1$) provided the spectrum of Ω_{ij} be confined to the strip $-2 < \lambda < 0$.

The theoretical insight into LBE-CE can be further deepened by inspecting the spectral properties of the matrix Ω_{ij} . In the case of FCHC, the spectrum consists of four distinct eigenvalues $(0, \lambda, \sigma, \tau)$, with multiplicities $(5, 9, 8, 2)$ respectively. The eigenvectors $\mathbf{V}_0$ and $\mathbf{V}_\alpha$ associated with the null eigenvalue correspond to the collision invariants and generate $1 + D$ hydrodynamical fields

$$\rho = \sum_{i=1}^{b} N_i V_{0i} \ , \ J_\alpha = \sum_{i=1}^{b} N_i V_{\alpha i} \tag{5}$$

D being the spatial dimension.

The eigenvalue λ is associated with non-equilibrium transport of momentum, and the corresponding eigenvectors generate the stress tensor $S_{\alpha\beta}$, corresponding to $D(D+1)/2 - 1$ hydrodynamic fields (-1 comes from the fact that this tensor is traceless). As a result, the eigenvalue λ can be tuned to minimize the viscosity according to relation

$$\nu = -\frac{1}{3}\left(\frac{1}{2} + \frac{1}{\lambda}\right) \tag{6}$$

The eigenvalues σ and τ are related instead to the relaxation of a different type of fields which cannot be given any direct interpretation at the macroscopic scale. They take part to the LBE dynamics but are unable to emerge from the mesoscopic to the hydrodynamical scale, being only active at the shortest scales of the order of the lattice spacing.

Summarizing,

- Lattice gas automata are intrinsically parallel and free of round-off errors. However, they suffer of two blemishes: statistical noise and exponential complexity of the collision rule.
- The non-linear LBE, eq.(2), was introduced to cure the problem of statistical noise. Due to the exponential complexity of the collision operator, this equation is however impractical for three-dimensional hydrodynamics.
- The Chapman-Enskog LBE, eq. (3), reduces the complexity of the collision operator to quadratic terms, thus opening up the way to three-dimensional simulations.
- LBE-CE needs **not** be related to any underlying microscopic dynamics. Once this independence is realized, one can play with the collision matrix and the e.d.f. to tune the viscosity of the LBE fluid and also to insert additional physics into the scheme.

It is worth reminding that some basic assets of the LG technique are retained by LBE, namely:

- ideal amenability to vector/parallel computing
- flexibility in the handling of grossly irregular geometries

3. Applications of LBE

To date, LBE has been applied to a large number of fluid-dynamics problems. These comprise low Reynolds mono and multiphase flows in porous media, including hydrodynamic dispersion, as well as highly turbulent flows, possibly in the presence of magnetic fields. Many of these applications are described in a recent review paper [8], to which we direct the the reader interested to further details. Here we shall just attempt a cursory summary of those which can bear a certain interest to the astrophysics community.

3.1 Homogeneus incompressible turbulence

Although the LBE merits as a fluid solver become especially apparent when complex boundary conditions have to be dealt with, the scheme has nonetheless been extensively used to investigate two-dimensional homogeneus incompressible turbulence.

One of the early questions was on whether LBE can support a better near-subgrid behaviour than 'conventional' numerical schemes. The question was motivated by eq. (6), from which one could be tempted to guess the possibility of reaching virtually infinite values of the Reynolds number by choosing the eigenvalue $\lambda = -2$. This "mirage" was investigated in a series of papers aimed to ascertain the ability of LBE to reproduce the statistical properties of fully developed turbulence, as well as its computational efficiency with respect to other well established techniques, mainly the pseudo-spectral method. These experiments substantially confirmed the ability of LBE to handle the physics of highly turbulent flows, although a number of pitfalls were also clearly identified. The basic conclusions can be summarized as follows

- LBE is able to reproduce the dynamics of coherent structures in decaying and forced turbulence
- good convergence to Navier-Stokes is achieved within a few (about 2) lattice spacings. This means that a factor two is lost in terms of the maximal resolution offered by the numerical grid
- zero viscosity is unattainable in a finite mesh because when modes at the lattice spacing are excited, LBE convergence to Navier-Stokes breaks down
- in the aforementioned situation LBE undergoes Gibbs phenomena (violation of positivity constraint) which may trigger catastrophic instabilities.

The conclusion is that, with the present computational bounds (about 256^3 in three dimensions) LBE does not offer any significant adkvantage with respect to spectral methods. This situation is going to change, however, with the advent of massively parallel computers which are expected to dominate the scene of large-scale scientific computing in the near future [9].

3.2 Turbulent convection

The LBE algorithm lends itself to straightforward extensions that allow its application to a number of fluid-related problems, such as tracer diffusion, magnetohydrodynamics and thermal convection. These extensions rest on a simple but powerful property of the LBE method: if the propagation phase of a D-dimensional LBE scheme is confined to the $(D-1)$-dimensional projection of the original lattice, the current J_D along the D-th dimension behaves like a passive scalar advected by a (D-1) dimensional fluid. As a result, D-dimensional LBE schemes can straightforwardly be used as (D-1)-dimensional passive scalar trackers.

As is well known, the fluid temperature is an active scalar, in the sense that in addition to being advected by the flow, it reacts back on it via buoyancy forces

$$F_B = \alpha g \Delta T \tag{7}$$

where α is the thermal dilatation coefficient, g the gravity acceleration and ΔT the local departure from the average temperature profile.

To model thermal fluids we need three ingredients: 1) track the temperature field, 2) define a thermal conductivity χ independently of the viscosity ν, 3) model the buoyancy term.

All of these requirements can be met with minor changes to the collision matrix of the 4D FCHC model.

To define a thermal conductivity independently of momentum diffusivity we need to remove the degeneracy of the eigenvalue λ discussed in section 2.2. Calling μ the eigenvalue associated with S_{xz} and S_{yz}, the modified collision matrix can be generated as

$$\Omega\prime = \Lambda\Omega\Lambda^{-1} \tag{8}$$

where Λ is the diagonal matrix made out of the eigenvalues $\{\lambda, \mu, \sigma, \tau\}$.

The introduction of body forces is simply accomplished by adding a momentum source to the r.h.s. of equation (2). For instance, in the case of a two-dimensional flow with gravity along the y axis, one has:

$$F_B = \frac{\alpha g}{12}\frac{1}{24}\sum_{i=1,24} c_{iz}N_i \tag{9}$$

where the sum over the 24 discrete speeds corresponds to the current $J_3 \equiv \rho T$. This term is added to all populations with $c_{iy} > 0$ and subtracted to those with $c_{iy} < 0$.

This technique has been recently applied to thermal convection in a Bénard cell, i.e. a fluid-filled tank in which free convective motion is induced by a temperature difference maintained externally on the upper and lower bases [10]. This system is described by the Boussinesq equations and controlled by two parameters, the Prandtl number $Pr = \nu/\chi$ and the Rayleigh number $Ra = \frac{\alpha g \Delta T H^3}{\nu\chi}$. Simulations of two-dimensional convection with $Pr = 7$ have been performed on vector multiprocessor computers, IBM 3090 and ES 9000, using a 128×128 grid, at $Ra = 1.2 \times 10^6$ and $Ra = 2.4 \times 10^6$, in periodic and turbulent motion regime respectively. The physical features revealed by the system were found to be in a complete agreement with previous simulations performed with spectral methods on grids of the same size. The study of the distributions of temperature fluctuations in time and space resulting from the aforementioned simulations shed new light on the subject and allowed for a simple explanation of experimental observations of exponential probability distribution functions of temperature fluctuations as a function of space.

3.3 Compressible flows

Most flows of astrophysical interest are compressible. For instance, compressible turbulence is likely to play an important role in the process of star formation from dense clouds of interstellar gas. Unfortunately, to date, LBE has been directed almost exclusively to incompressible flows. The reason is to be tracked to the "hystorical" link to Lattice Gas. In a Lattice Gas the sound speed is invariably $O(1)$, and since the fluid speed must be much smaller than 1 to avoid spurious numerics, the result is that LG need be operated in the low-Mach regime. This limitation can be removed by a suitable modification of the e.d.f. N_i^{eq}. The basic idea is to "freeze" sound-waves bringing the sound speed as close as possible to zero. Technically, this can be achieved by introducing rest particles (zero speed). For the case of the two-dimensional FHP scheme, a possible e.d.f. leading to the compressible

Navier-Stokes equations in the fluid limit has been recently introduced by Chen et al. [11]: $(u_i \equiv \mathbf{c}_i \cdot \mathbf{u})$.

$$N_i^{eq} = d_0 + \frac{\rho}{7}\delta + \frac{\rho}{3}u_i + \frac{2}{3}\rho(u_i^2 - \frac{(2-\gamma)}{4}u^2)\ i = 1,6 \tag{10}$$

for moving particles, and

$$N_0^{eq} = \rho(1 - \frac{6}{7}\delta) - 6d_0 - \gamma\rho u^2 \tag{11}$$

for rest particles. This represents a three-parameter family which generalizes the e.d.f. resulting from the low-Mach number expansion of the Fermi-Dirac distribution of LG dynamics (corresponding to the choice $d_0 = 0, \gamma = 0, \delta = 1$). In these expressions, γ is a free parameter which can be fixed to the value 1 to overcome unphysical pathologies resulting from LG theory, i.e. breaking of galilean invariance and kinetic-energy dependence of the static pressure. The parameter δ controls the sound speed via the relation

$$c_s^2 = \frac{3}{7}\delta \tag{12}$$

A compressible LBE scheme is therefore obtained by choosing $\delta \ll u^2 \ll 1$. Finally, the parameter d_0 serves the purpose of preserving the positivity of the populations for non-vanishing values of the fluid speed u. Schemes similar to (10-11) have proven capable of tracking coherent non-linearities such as shock waves and propagating fronts. The use of LBE for compressible flows is however still in its infancy; for instance two and three-dimensional applications are (to the author's knowledge) still lacking. Testing the validity of compressible LBE's and their competitivity versus well-established techniques such as PPM [12], Bird's method [13] or Smooth Particle Hydrodynamics [14] is a challenging subject of future research.

4. Performance Considerations

The LBE is basically an explicit finite-difference scheme characterized by a very favourable communication to computation ratio. The computational kernel of LBE codes is the collision routine which requires of the order of b^2 floating point operations per site, as opposed to the b memory references per site required by the streaming routines. As a result, a fairly low communication/computation ratio can be maintained down to the grid-point level, i.e. the finest-grain of the data-parallellization procedure. This is the ultimate reason why LBE lends itself to straightforward and efficient implementations on practically any kind of parallel computing platform.

The earliest performance data pertained to two and three-dimensional implementations on shared-memory architectures such as the IBM 3090 vector multiprocessor. Sustained speeds of about 0.3 Gflops/s could be achieved, with a parallel speed-up of about 5.5 on the full six-way configuration. Recent work on distributed memory multiprocessors based on clusters of IBM RS/6000 superscalar workstations has proved that quasi-linear speed-up can be maintained for configurations of up to 10 workstations connected via a LAN token-ring [15] (see figure 2).Given the

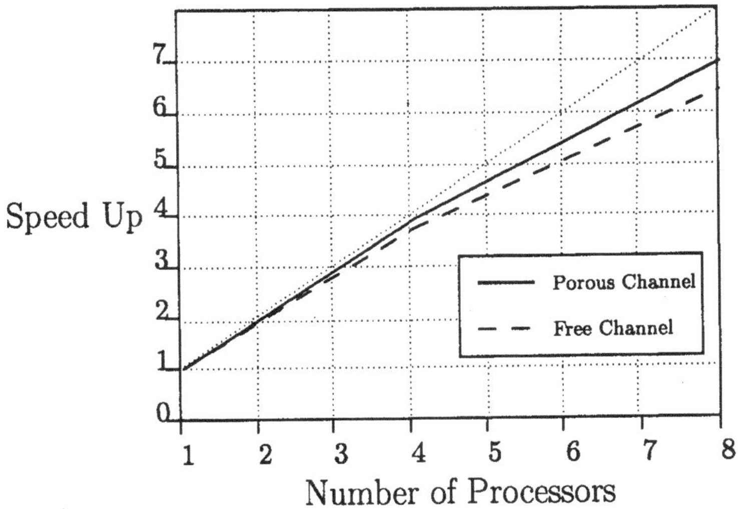

Figure 2: Parallel speed-up on a cluster of IBM RS/6000 workstations

impressive amount of computing power afforded by each single engine (in excess of 100 Mflops for the most recent high-end models) the result is that a cluster of, say, ten such workstations can come close to the Gflop performance, i.e. not far from the top performances (a few Gflops) actually delivered by the most powerful massively parallel machines.

5. Acknowledgements

Prof. A. Lejeune and J. Perdang kind hospitality during this workshop is warmly acknowledged. I'm indebted to W. Benz for a quick but effective introduction to the subject of Smooth Particle Hydrodynamics.

6. References

1. U. Frisch, D. d'Humières, B. Hasslacher, P. Lallemand, Y. Pomeau, J. P. Rivet, *Complex Systems* **1** (1987) 649.
2. *Physica D* Vol. **47** (1991) n. 1-2 special issue.
3. *J. Stat. Phys.***68** 3/4 (Aug. 1992), special issue.
4. U. Frisch, B. Hasslacher, and Y. Pomeau, *Phys. Rev. Lett.* **56** (1986) 15.
5. G. Mc Namara, G. Zanetti, *Phys. Rev. Lett.*, **61** (1988) 2332.
6. F. Higuera and J.Jimenez, *Europhys. Lett.*, **9** (1989) 663.

7. F. Higuera, S. Succi and R. Benzi, *Europhys. Lett.*, **8** (1989) 517.
8. R. Benzi, S. Succi, M. Vergassola, *Phys. Rep.* **222** (3) 1992, and references therein.
9. B. Boghosian, *Computers in Physics*, Nov./Dic. (1991) 585.
10. F. Massaioli, R. Benzi, S. Succi, *Europhys. Lett.*, in press.
11. H. Chen, S. Chen, W. Mattheus, *Phys. Rev. A.*, **63** (1991) 1965.
12. P. Woodward, J. Colella, *J. Comp. Phys.* **54** (1984)174.
13. G. Bird, *Molecular Gas Dynamics*, (Oxford University Press, London,1976).
14. W. Benz, *The numerical modelling of Nonlinear Stellar Pulsations*, ed. J.R. Buchler, (Kluwer Dordrecht, 1990).
15. G. Betello, G. Richelli, S. Succi, *1st Symposium on High Performance Distributed Computing,* (1992) 235.

Note: IBM 3090 and IBM RS/6000 are trademarks of IBM Corporation.

CELLULAR AUTOMATA APPROACH TO REACTION-DIFFUSION SYSTEMS: THEORY AND APPLICATIONS

BASTIEN CHOPARD
Department of Computer Science, University of Geneva
CH-1211 Geneva 4, Switzerland

and

MICHEL DROZ, STEPHEN CORNELL, LAURENT FRACHEBOURG
Department of Theoretical Physics, University of Geneva
CH-1211 Geneva 4, Switzerland

ABSTRACT

This paper reviews the modeling of diffusion and reaction-diffusion phenomena by cellular automata. First, the theoretical issues are discussed (link between the model and the macroscopic description, continuous limit, multiscale expansion). Then, several applications of these models are considered, such as the anomalous kinetics in chemical reactions, the formation of Turing patterns and models of nonequilibrium phase transitions.

1. Introduction

Cellular automata are an interesting tool to investigate complex physical situations, such as systems containing many interacting particles. Lattice gas models for fluids (FHP model, for instance) are famous examples that have been studied intensively. Reaction-diffusion phenomena offer another class of problems for which this approach has proven to be fruitful, as we shall see in this review.

Cellular automata models provide a simplified description of physical systems. Due to their microscopic level of description, these models capture several important aspects that are out of reach of a more traditional approach based on differential equations. Also, they are much simpler than usual molecular dynamics and numerically more efficient. They are appropriate for an implementation on a massively parallel machines or a vector computer.

This review is organized as follows. In section 2, cellular automata models for diffusion and reaction-diffusion processes are introduced. It is explicitely shown how the macroscopic equations can be estabished from the discete lattice dynamics. Models with and without exclusion principle are considered. In section 3, various applications of the above models are discussed. In particular, the problems of anomalous kinetics in chemical reaction, Turing pattern formation and nonequilibrium phase transitions on catalytic surfaces are reviewed.

2. Theoretical concepts

Lattice gas cellular automata models can be thought of as a fully discrete molecular dynamics: particles move in discrete time steps on a regular lattice. Each particle has a velocity of modulus unity pointing along one of the main directions of the lattice. This restriction is known as the lattice gas exclusion principle. Interactions usually take place among particles entering the same site at the same time and result in a new local distribution of particles. The dynamics of such systems is given in terms of boolean variables describing the occupation numbers at each site of the lattice and at each time step. These so-called microdynamic equations are easy to code in a computer and yield a fast and exact implementation of the model. One usually restricts the dynamics so that no more than one particle per direction enters any site at any time.

The physical quantities of interest are macroscopic quantities or average values, such as, for instance, the average density of particles. It is then important to show that our discrete world is, at some appropriate scale of observation, governed by admissible equations: the physical conservation laws are to be present, the symmetry of the space is to be conserved and the discreteness of the lattice should not show up. This connection between the microscopic boolean dynamics and the macroscopic, continuous world has to be established in order to assess the validity of the cellular automata modeling. This step involves several techniques like the Chapman Enskog and multiscale expansions, borrowed from statistical physics.

It is the purpose of this paper to show how this step can be carried out in simple CA models for reaction and diffusion processes. We will first focus on pure diffusion which can be described by a set a linear equations, leading to exact expressions for the continuum limit. Then, we will add a nonlinear term in the dynamics, describing the possibility for the particles to react. We will have to use a multiscale expansion to obtain the appropriate macroscopic equations and we will discuss the validity of the derivation.

Finally, we will briefly introduce new models of reaction diffusion processes that are extensions of the traditional CA approach by not containing an exclusion principle.

2.1. Lattice gas automata for pure diffusion

Diffusion is caused by the random walk of many particles. In a cellular automaton, all the particles move simultaneously, as a result of the synchronous updating rule. The usual sequential algorithm for producing a random walk is to pick a particle and choose at random a neighboring site: if the site is empty, the particle move to it; otherwise the particles stay motionless. Note that the new position of a particle, is not influenced by its previous move, i.e. the particle has no velocity or no inertia.

The mechanism we have just described does not work if all the particles are to choose simultaneously their new position: the same site may appear to be empty to more than one particle and to avoid such collision requires to introduce nonlocal algorithm. Each particle would have to ask its second nearest neighbors if they also want to jump to the same given place. Another way around is to relax the exclusion principle and allow more than one particle sitting at each site. This latter approach will be discussed in a subsequent section.

There is however a simple way to produce a the random walk of many independent particles with a parallel and local rule compatible with the exclusion principle. We assume that each particles have velocity pointing along one of the main direction of the lattice. The result of diffusion is randomly to change these direction of motion at each site and each time step. To be more specific, let us consider the case of a two-dimensional square lattice.

We will use four bits at each site to represent the diffusing particles, each of them corresponding to the absence or the presence of a particle traveling in one of the four possible directions of the lattice (up, right, down or left). In the absence of diffusion, our particles would simply move, at each time step, to a neighboring site, in straight lines, according to their initial velocities. Diffusion is produced as follows: at each time step, the four bits representing the configuration of particles entering at each site undergo a random permutation. For four possible directions of motion, there are $4! = 24$ permutations. On a computer, there is no easy way to generate any of these 24 permutations, at random. It turns out we do not have to consider the complete set of all permutations to simulate a diffusive behavior. A subset of it is sufficient, as can be seen by a direct derivation.[1] It is however important to use a subset of permutations that do not favor any particular direction of motion. In practice, we can simply consider random rotations (of 0, $\pi/2$, π or $-\pi/2$) of the lattice directions (i.e. a random circular shift of the four bits) and have them occur with probability p_0, p_1, p_2 and p_3, respectively. For instance, if two particles enter the same site with respective velocities pointing, say, up and right, and if the diffusion step results in a rotation of that site by π, the two particles will then move down and left. Fig. 1 illustrates this mechanism.

The condition $p_1 = p_3$ is necessary to ensures an isotropic motion. The values of p_0 and p_2 allows us to adjust the diffusion constant and to tune some fine aspect of the dynamics.

Our dynamics is such that it never happens that two or more particles enter simultaneously the same lattice site in the same direction, provided this was true for the initial time. Indeed, a particle that enters a site traveling, say, to the right has to come from the left neighboring site. Since the mechanism yielding diffusion is a permutation of the lattice direction, it may not produce more than one particle traveling to the same site. The exclusion principle is then satisfied and this guarantees that four bits are enough, in two-dimensions, to describe the system at any time (two bits are needed in one-dimension and six in three-dimensions).

In what follows, we will focus on the one-dimensional case because the analytical treatment is quite simple and makes the discussion more pedagogical. Two- and three-dimensional cases works the same but involve more complicated algebra.

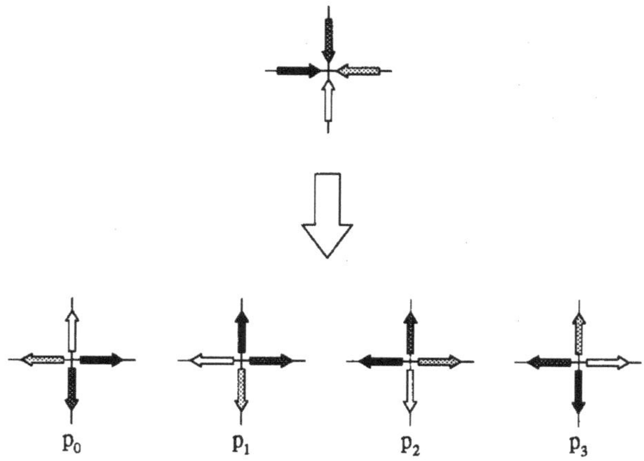

Fig. 1: **How the entering particles are deflected at a typical site, as a result of the diffusion step. The four possible outcomes occur with the preassigned probabilities p_0, $p_1 = p_3$ and p_2. The figure shows four particles but any one of the arrows can be omitted if less particles are present**

The reader is referred to reference[1] for more details. Note that, in one-dimension, the lattice has only two possible directions of motion (left or right) and, therefore, there are only two possible permutations: either these directions are left unchanged after the diffusion step or they are exchanged. It is then possible to work with all the permutations, as opposed to what we said earlier. However, this does not confer any particular properties to the one-dimensional case that are not present in two- and three-dimensions.

In one-dimension, the microdynamics describing our diffusion process can be written in terms of the occupation variables $n_1(r,t)$ and $n_2(r,t)$. If, at time t, a particle enter site r with a velocity pointing along the direction i ($i = 1,2$), then $n_i = 1$. Otherwise, $n_i = 0$. The evolution equation simply reads

$$\begin{aligned} n_1(t+\tau, r+\lambda) &= \mu(t,r)n_1(t,r) + (1-\mu(t,r))n_2(t,r) \\ n_2(t+\tau, r-\lambda) &= (1-\mu(t,r))n_1(t,r) + \mu(t,r)n_2(t,r) \end{aligned}, \tag{1}$$

where τ is the duration of the time step and λ is the lattice spacing. The quantities $\mu(t,r)$ are independant random boolean variables which take the value of 1 with the same probability p for each site and each time step.

To have a better idea of the macroscopic behavior of this system, we first take an average of the n_i

$$N_i(t,r) = < n_i(r,t) > .$$

Strictly speaking, N_i is meant as an average over the ensemble of all possible realizations of the system. In practice, however, this average is assumed to be equivalent to some coarse graining or to a time average.

Since the $\mu(t,r)$ are independant of the $n_i(t,r)$, the average of Eq. (1) simply gives

$$\begin{aligned} N_1(t+\tau, r+\lambda) &= pN_1(t,r) + qN_2(t,r) \\ N_2(t+\tau, r-\lambda) &= qN_1(t,r) + pN_2(t,r) \end{aligned}, \tag{2}$$

where we have defined $q = 1 - p$.

Let us look for solutions of the following form

$$N(t+\tau, r) = A(t)\exp(ikr), \tag{3}$$

where N and A are the column matrices

$$N(t,r) = \begin{pmatrix} N_1(t,r) \\ N_2(t,r) \end{pmatrix} \qquad A(t) = \begin{pmatrix} A_1(t) \\ A_2(t) \end{pmatrix},$$

$A_1(t)$ and $A_2(t)$ are two functions to be determined. With the relation (3), Eq. (2) reads

$$A(t+\tau) = \begin{pmatrix} p\exp(-ik\lambda) & q\exp(-ik\lambda) \\ q\exp(ik\lambda) & p\exp(ik\lambda) \end{pmatrix} A(t). \tag{4}$$

We then write

$$A(t) = \exp(i\omega t)A(0),$$

and one is left with the condition

$$\begin{pmatrix} p\exp(-ik\lambda) - \exp(i\omega\tau) & q\exp(-ik\lambda) \\ q\exp(ik\lambda) & p\exp(ik\lambda) - \exp(i\omega\tau) \end{pmatrix} A(0) = 0. \tag{5}$$

A solution $A(0) \neq 0$ exists provided that the determinant of this matrix is zero, which gives a relation between ω and k (the dispersion relation). This relation is

$$\exp(2i\omega\tau) - 2p\cos(k\lambda)\exp(i\omega\tau) + 2p - 1 = 0, \tag{6}$$

and can be solved for $\exp(i\omega\tau)$

$$\exp(i\omega_\pm \tau) = p\cos(k\lambda) \pm (1-p)\sqrt{1 - \frac{p^2}{(p-1)^2}\sin^2(k\lambda)}. \tag{7}$$

A first observation that can be made from Eq. (7) is that ω may have both an imaginary and a real part. Therefore, our dynamics not only contains dissipation, as is expected for a diffusion process, but also wave propagation. Indeed, suppose that p is very close to 1. In this case $p^2/(p-1)^2$ is much bigger than 1 and the square root can be approximated by $i(p/(p-1)\sin(k\lambda)$. In this so-called ballistic limit ($p = 1$ means that the particles are never deflected), Eq. (2) reduces to the dispersion relation of the wave equation

$$\exp(i\omega_\pm \tau) = p\cos(k\lambda) \pm ip\sin(k\lambda) = p\exp(\pm ik\lambda),$$

or

$$\omega\tau = k\lambda. \tag{8}$$

The dynamics is determined by the two eigen values ω_+ and ω_-. A quantity such as the density

$$\rho(t,r) \equiv N_1(t,r) + N_2(t,r),$$

will evolve according to

$$\begin{aligned}\rho(t,r) &= (\exp(i\omega_+ t)\rho_+ + \exp(i\omega_- t)\rho_-)\exp(ikr)\\ &= (\exp(i\omega_+\tau)^{(t/\tau)}\rho_+ + \exp(i\omega_-\tau)^{(t/\tau)}\rho_-)\exp(ikr)\end{aligned}$$

We are now interested in the continuum limit obtained when the lattice spacing λ and the time step τ go to zero. One has to compute

$$\lim_{\tau\to 0,\lambda\to O}(\exp(i\omega_\pm\tau)^{t/\tau}) = \lim_{\tau\to 0,\lambda\to O}\left[p\cos(k\lambda) \pm (1-p)\sqrt{1-\frac{p^2}{(p-1)^2}\sin^2(k\lambda)}\right]^{t/\tau}.$$

This limit is easy to compute for ω_-, since, when $\lambda \to 0$, the term in the square braket goes to $2p-1$. As $0 < p < 1$, $(2p-1)^{t/\tau}$ goes to zero when $\tau \to 0$. For ω_+, this limit is of the type "0^∞." We take the logarithm and we have

$$\lim_{\tau\to 0,\lambda\to 0}\frac{t}{\tau}\ln(\exp(i\omega_+\tau)) = t\lim_{\tau\to 0,\lambda\to 0}\frac{\partial}{\partial\tau}\ln(\exp(i\omega_+\tau))$$

To carry out this calculation we need to specify how λ and τ are related when they go to zero. There are two natural choices:

(a) $\lambda/\tau \to \text{const} \neq 0$ which reflects a situation where the propagation speed of the particles remains finite. In this case we find

$$[\exp(i\omega_+\tau)]^{t/\tau} \to 1$$

and

$$\rho(t,r) = \exp(ikr)\rho_+,$$

which means that the system is frozen in this limit.

(b) A more interesting case is obtained when $\lambda^2/\tau \to \text{const} \neq 0$. We then get for the limit

$$[\exp(i\omega_+\tau)]^{t/\tau} \to \exp(-tk^2\frac{p}{2(1-p)}\frac{\lambda^2}{\tau}),$$

and consequenlty

$$\rho(t,r) = \exp(-tk^2 D)\exp(ikr)\rho_+, \tag{9}$$

with D the diffusion constant

$$D = \frac{\lambda^2}{\tau}\frac{p}{2(1-p)}.$$

Equation (9) is precisely the solution of the diffusion equation

$$\partial_t \rho = D\partial_x^2 \rho.$$

This result shows that our dynamics properly describes a diffusion process in the limit $\tau \to 0$, $\lambda \to 0$ and $\lambda^2/\tau \to \text{const}$. The diffusion constant is adjustable through the parameter p.

However, in practical applications such as computer simulations λ and τ take some finite values and corrections to the continuous limits are expected. These corrections can be calculated in several explicit situations like, for instance, the stationary diffusion profile in a semi infinite strip with creation or adsorbtion of particles on the various boundaries.[1]

Another simple way to "see" the effect of finite values of λ and τ is to compute the corrections that come up on the differential equation describing the dynamics of our system. For this purpose, we write Eq. (6) as follows

$$\begin{aligned}\cos(k\lambda) &= \frac{\exp(i\omega\tau) + (2p-1)\exp(-i\omega\tau)}{2p} \\ &= \frac{\exp(i\omega\tau) + \exp(-i\omega\tau)}{2} + (1-p)\frac{\exp(i\omega\tau) - \exp(-i\omega\tau)}{2p}.\end{aligned}$$

Thus,

$$i\sin(\omega\tau) + \frac{p}{(1-p)}\cos(\omega\tau) = \frac{p}{(1-p)}\cos(k\lambda). \tag{10}$$

With the correspondance

$$\partial_t \leftrightarrow i\omega \qquad \partial_r^2 \leftrightarrow -k^2$$

we can obtain the corrections we are looking for by expanding Eq. (10) for τ and λ. We find, to second order in $\omega\tau$ and fourth order in $k\lambda$

$$i\omega\tau - \frac{p}{2(1-p)}(\omega\tau)^2 = -\frac{p}{2p(1-p)}(k\lambda)^2 + \frac{p}{24p(1-p)}(k\lambda)^4$$

which translates to

$$\partial_t\rho + \frac{D}{c^2}\partial_t^2\rho = D\partial_r^2\rho + \frac{D}{12}\partial_r^4\rho, \tag{11}$$

where c is defined as $c = \lambda/\tau$. Except for the term containing the fourth order space derivative, this last equation is known as the Telegraphist equation. It describes diffusion processes combined with wave propagation and ensures a finite value c for information propagation speed. In this respect, it is more physical than the pure diffusion equation in which a perturbation is instantaneously transmitted arbitrarly far away from its source. However, the Telegraphist equation admits non-positive solutions which are not acceptable when ρ represents a particle density.

Strictly speaking, the above discussion assumes a systems with periodic boundary conditions, by virtue of (3). Actually, if the system is periodic, any solution of our dynamics is a superposition of solutions given by Eq. (3), with

$$k = \frac{-\pi}{\lambda} + \frac{2n\pi}{L} \qquad n = 0, 1, ..., L/\lambda - 1$$

where L is the system size. However, for non periodic boundary conditions, the differential equations we have derived are expected to be the same due to their local nature. As a matter of fact, without using a discrete Fourier decomposition, one can show[1] that for any site not next to the boundaries the following equation holds exactly

$$\hat{\partial}_t \rho + \frac{d}{c^2} \hat{\partial}_t^2 \rho = D \hat{\partial}_r^2 \rho \tag{12}$$

where $\hat{\partial}_x$ and $\hat{\partial}_x^2$ are the finite difference operator such that

$$\hat{\partial}_t \rho(t,r) = \frac{1}{2\tau} \left[\rho(t+\tau, r) - \rho(t-\tau, r)\right]$$

$$\hat{\partial}_t^2 \rho(t,r) = \frac{1}{\tau^2} \left[\rho(t+\tau, r) - 2\rho(t,r) + \rho(t-\tau, r)\right]$$

$$\hat{\partial}_r^2 \rho(t,r) = \frac{1}{\lambda^2} \left[\rho(t, r+\lambda) - 2\rho(t,r) + \rho(t, r+\lambda)\right]$$

Note that Eq. (12) takes a more complicated form in 2-dimensions[1]. The discrepency between the cellular automata dynamics and the diffusion equation or the Telegraphist equation when λ and τ are finite can be seen graphically by plotting the corresponding dispersion relation. This done in figure 2 and 3, for one- and two-dimensions respectively.

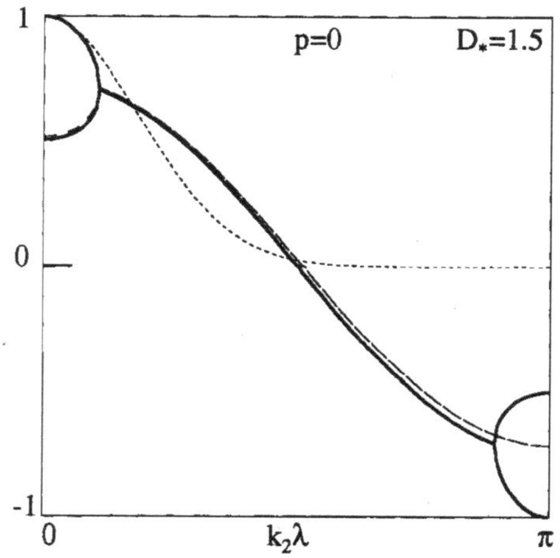

Fig. 2: Comparison between the time evolution of the mode k, according to the CA rule (solid lines), the Telegraphist equation (dashed lines) and the diffusion equation (dotted lines). The situation shown here corresponds to the one-dimensional case. The value of the diffusion constant is $D = 1.5(\lambda^2/\tau)$. Only the real part of the two eigenvalues $\exp(i\omega_\pm \tau)$ is shown.

Figure 2 and Eq. (7) show that there is an eigenvalue -1 in our dynamics. It corresponds to the mode $k\lambda = -\pi$. This mode, which may be present even in

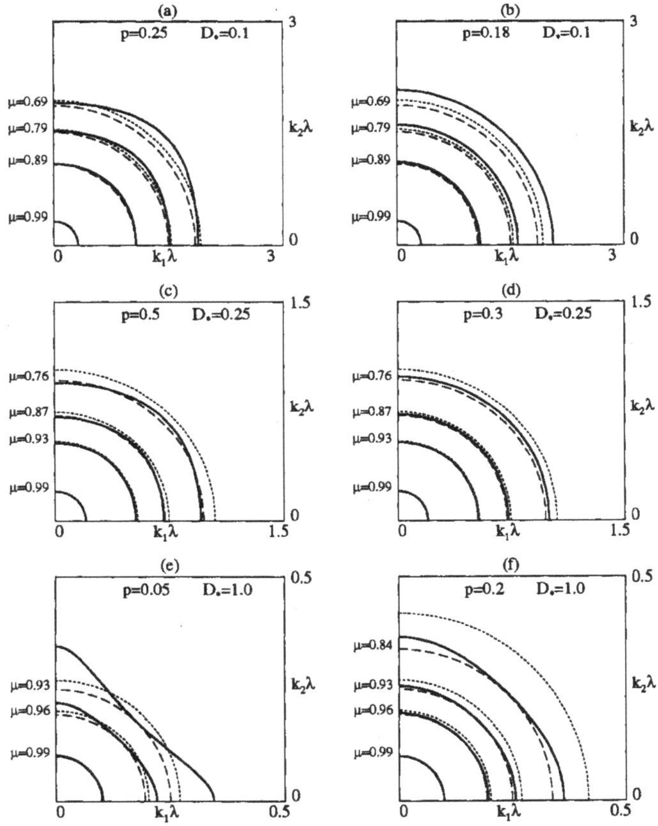

Fig. 3: Contour lines of the diffusive part of the dispersion relation ($\mu = \exp(i\omega\tau)$) for various values of $D_* = (\tau/\lambda^2)D$ and p. The solid lines correspond to the CA model, the dashed ones to the telegraphist equation and the dotted ones to Fick's law of diffusion. For $\mu = 0.99$ the three dynamics are almost indistinguishable. Note that the scales are not the same in all figures.

the continuous limit, will not die out during the time evolution of the system. Therefore, there is, in this model, an additional quatity which is conserved (more precisely that changes sign at each iteration) and that has no physical counterpart. Such quantities, which are frequent in lattice gas automata, are called spurious or staggered invariants. They are usually related to the discrete nature of the lattice.

In our case, the staggered invariant are easy to find from the dispersion relation. There is only one, which can be interpreted by taking the Fourier transform for the mode $k = -\pi/\lambda$

$$A(k = -\frac{\pi}{\lambda}) = \sum_{r \in \text{lattice}} \rho(r,0) \exp(i\pi \frac{r}{\lambda})$$
$$= N_{\text{even}} - N_{\text{odd}},$$

where N_{even} and N_{odd} denote the total number of particles on even and odd lattice sites, respectively. Our result is easily explained by the following fact: all the particles that enter an even site at time t will go to an odd site at time $t+1$, and conversly. Therefore, there is no interaction, at any time, between particles that are not located on the same odd or even sublattice. Particle conservation holds independently for each sublattice and our system actually contains two non interacting subsystems. This feature is known as the checkerboard invariant. In two-dimensions, the dispersion relation yields another staggered invariant when $p_0 = p_2 = 0$. In this case, a particle is always deflected by $90°$ or $-90°$. Thus, all the particles traveling horizontally at time t will move vertically at time $t+1$, leading to another eigenvalue equal to -1 in the dynamics.

2.2. Lattice gas automata for reaction-diffusion processes

In this section, we are going to add a reactive term to our automata model, in order to describe a reaction such as

$$A + B \rightarrow C$$

where A, B and C are different chemical species diffusing on a gel for instance. To account for this kind of reaction, one can consider the following mechanism: at the "microscopic" level of the discrete lattice dynamics, all the three species are first governed by a diffusion rule, as described in the previous section. When a A and a B particle enter the same site at the same time, they disappear and form a C. There are several ways to select the events that will produce a C when more than one A or one B are simultaneously present at a given site. Also, when C's already exist at this site, the exclusion principle may prevent the formation of new ones. A simple choice is to have A and B react only when they perform a head on collision and when no $C's$ are present in the perpendicular directions. Other rules can be considered if we want to enhance the reaction (make it more likely) or to deal with more complex situations ($2A + B \rightarrow C$, for instance). A parameter k can be introduced to tune the reaction rate by controlling the probability for a reaction to take place.

On the microdynamic equations, the reaction term introduces a nonlinearity because it is expressed as products of occupation numbers of the different species. This nonlinearity makes it much more complicated to obtain differential equations describing the behavior of the various particle concentrations. The direct calculation we performed in section 2.1 no longer works and we have to use more sophisticated techniques, similar to that used in statistical mechanics when solving the Boltzmann equation.

In order to illustrate these techiques, we will consider a very simple situation, namely the reaction-diffusion process $A + A \to [inert]$, in one-dimension. The word *inert* means that the reaction produces an inert species which does not play any role in the dynamics. Such a process can be described by the following lattice gas automata dynamics

$$\begin{aligned} a_1(t+\tau, r+\lambda) &= \mu(a_1 - \kappa a_1 a_2) + (1-\mu)(a_2 - \kappa a_1 a_2) \\ a_2(t+\tau, r-\lambda) &= (1-\mu)(a_1 - \kappa a_1 a_2) + \mu(a_2 - \kappa a_1 a_2) \end{aligned}, \tag{13}$$

where the terms in the right hand side of these two equations are computed at position r and time t. The quantity $\kappa(t,r)$ is a random boolean variable similar to $\mu(t,r)$. It is takes the value 1 with probability k, independently for each site and each time step. In order to derive the macroscopic behavior emerging from the dynamics (13), we first take an ensemble average. An important approximation has to be used here to get a closed system of equations. We shall assume that some molecular chaos hypothesis can be made and that $< a_1 a_2 >$ factorizes into $< a_1 >< a_2 >$. This factorization is often referred to as the Boltzmann approximation. It assumes that the correlations can be neglected. This is, of course, not always true and may introduce discrepencies with the actual automata dynamics. With $A_i =< a_i >$, $< \kappa(t,r) >= k$, and $< \mu(t,r) >= p$, Eq. (13) reads

$$A_i(t+\tau, r+\lambda c_i) - A_i(t,r) = \Omega_i(A) - kR_i(A,A), \tag{14}$$

where $c_1 = 1$, $c_2 = -1$,

$$\Omega(A) = \begin{pmatrix} (p-1)A_1 & +(1-p)A_2 \\ (1-p)A_1 & +(p-1)A_2 \end{pmatrix} = (1-p)\begin{pmatrix} -A_1 & +A_2 \\ A_1 & -A_2 \end{pmatrix},$$

and R is the bilinear function

$$R(X,Y) = \begin{pmatrix} pX_1Y_2 & +(1-p)X_2Y_1 \\ pX_2Y_1 & +(1-p)X_1Y_2 \end{pmatrix}.$$

In our case, we simply have

$$R_i = A_1 A_2. \tag{15}$$

After the averaging process, A_i are continuous variables ranging from 0 to 1. The left hand side of Eq. (14) can be expanded in a Taylor serie, up to second order (diffusion is of the order λ^2, as we saw in the previous section)

$$\begin{aligned} A_i(t+\tau, r+\lambda c_i) - A_i(t,r) =& \tau\partial_t A_i \frac{\tau^2}{2}\partial_t^2 A_i + \lambda(c_i\partial_r)A_i \\ &+ \frac{\lambda^2}{2}(c_i\partial_r)^2 A_i + \tau\lambda\partial_t(c_i\partial_r)A_i \end{aligned} \tag{16}$$

Our problem is now to find an equation for the density of A particles $\rho = A_1 + A_2$ from our two equations for A_1 and A_2. In general, there is no simple way to do this. One has to find a projection of the dynamics for A_1 and A_2 (the so-called fast variables) on the slow, macroscopic variable ρ. This program can be achieved by considering an expansion scheme with a small formal parameter ϵ. This method is known as the multiscale expansion since it requires us to introduce several time scale in order to have a correct solution. This method is a standard approach to solving differential equation as a power series[2,3]. In our case, it will also borrow many aspects of the Chapman-Enskog expansion.

The idea of this method is to look for solutions

$$A_i = A_i^{(0)} + \epsilon A_i^{(1)} + \epsilon^2 A_i^{(2)} + \ldots \tag{17}$$

where the superscript ℓ on A_i denotes the function which comes with a coefficient ϵ^ℓ in the expansion. We also write the space and time derivative in term of ϵ by introducing a new variable r_1 such that

$$\partial_r = \epsilon \partial_{r_1}$$

and several time scales $t_1, t_2, \ldots$

$$\partial_t = \epsilon \partial_{t_1} + \epsilon^2 \partial_{t_2} + \ldots$$

In comparison with the usual Boltzmann equations, we have here an additional difficulty in the sense that, strictly spealing, no quantity is conserved in the dynamics (the number of particles goes to zero as time goes on). In order to find a solution to our ϵ-expansion we have to impose that the reaction behaves as a perturbation to the diffusion process, which amount to saying that there is a non zero local equilibrium state in our system. Thus, the reaction rate k is not of order ϵ^0. Actually, we will assume that

$$k = \epsilon^2 k_2. \tag{18}$$

This expression is compatible with the fact that, in the limit of a macroscopic system $k \to 0$ while $k/\tau \to \text{const} \neq 0$.

At the order ϵ^0, Eq. (14) gives

$$\Omega(A^{(0)}) = 0,$$

whose solution is simply $A_1^{(0)} = A_2^{(0)}$. Thus, we still have a free parameter. To find $A_i^{(0)}$ uniquely one imposes that

$$\rho = A_1^{(0)} + A_2^{(0)}. \tag{19}$$

As a consequence we have

$$A_1^{(0)} = A_2^{(0)} = \frac{\rho}{2},$$

and

$$A_1^{(\ell)} + A_2^{(\ell)} = 0, \qquad \ell \geq 1 \tag{20}$$

The next order in ϵ gives

$$\tau \partial_{t_1} A_i^{(0)} + \lambda c_i = \Omega_i(A^{(1)}). \tag{21}$$

This equation allows us to find $A^{(1)}$ since the left hand side is known. To obtain an expression for $\partial_{t_1} A_i^{(0)}$, we observe that the A_i's depend on the times variable only through the the slow variable ρ. We obtain (in our case this step is rather obvious)

$$\partial_{t_1} A_i^{(0)} = \frac{\partial A_i^{(0)}}{\partial \rho}\frac{\partial \rho}{\partial t_1} = \frac{1}{2}\frac{\partial \rho}{\partial t_1}.$$

An equation for $\frac{\partial \rho}{\partial t_1}$ is simply obtained by summing Eq. (21) over i. With the fact that diffusion conserves the number of particle in the system, we have $\Omega_1 + \Omega_2 = 0$ and

$$\frac{\partial \rho}{\partial_{t_1}} = 0.$$

With this condition, Eq. (21) yields

$$A_i^{(1)} = \frac{\lambda c_i}{4(p-1)}\partial_{r_1}\rho. \tag{22}$$

Note that we have $A_1^{(1)} = -A_2^{(1)}$ as required by Eq. (20).

Finally, the second order in ϵ gives

$$\begin{aligned} \tau\partial_{t_1} A_i^{(1)} + \tau\partial_{t_2} A_i^{(0)} + \frac{\tau^2}{2}\partial_{t_1}^2 A_i^{(0)} + \lambda c_i \partial_{r_1} A_i^{(1)} + \frac{\lambda^2}{2}(c_i\partial_{r_1})^2 A_i^{(0)} \\ + \tau\lambda\partial_{t_1}(c_i\partial_{r_1})A_i^{(0)} = \Omega(A^{(2)}) - k_2 R_i(A^{(0)}, A^{(0)}) \end{aligned} \tag{23}$$

Using the results obtained for the order ϵ and summing over i yields

$$\tau\partial_{t_2}\rho + \frac{\lambda^2 p}{2(p-1)}\partial_{r_1}^2\rho = -\frac{k_2}{2}\rho^2. \tag{24}$$

We now collect all the orders in ϵ we found, using the expression of ∂_t, ∂_r and k in terms of t_1, t_2, r_1 and k_2. We get

$$\partial_t \rho = D\partial_r^2 \rho - \frac{k}{2\tau}\rho^2, \tag{25}$$

where $D = (\lambda^2/\tau)(p/2(1-p))$, as before.

Equation (25) is exactly what is expected for the reaction-diffusion process $A + A \to [inert]$. It is however important to remember that its derivation used the assumption that the reaction has small effect compare to diffusion and that the Boltzmann hypothesis can be made. Actually, if the diffusion is efficient enough to mix the particles these two assumption are true. This is however not always the case and our cellular automata dynamics may contains other phenomena that are not present in the differential equation (25).

If, instead of taking the reaction rate k of second order in ϵ, we had taken it of first order, we would have obtained the same equation as Eq. (25), but with an extra term proportional to k only (without a $1/\tau$ factor). In the limit $k \to 0$ this term vanishes and get back Eq. (25). This may be a consistency justification of Eq. (18).

2.3. Models of reaction-diffusion without exclusion principle

Lattice gas automata models take into account all the degree of freedom of a system of many particles as well as fluctuations and correlations of all orders. Also they can be implemeted on a computer in an exact fashion: no numerical error or truncation is involved. Finally, due to their "microscopic" level of description, they may capture some features of a physical situation that are not present in a more traditional representation in terms of partial differential equations.

However, the exclusion principle, which ensures that the dynamics can be described with a finite number of bits is often not convenient. A model for the reaction $3A + 4B \to C$ necessitates that, at least, $3A$ and $4B$ be present at a given site. This is never the case in the one-dimensional model we have discussed so far. Furthermore, the exclusion principle may produce rather artificial terms in the microdynamic equations (e.g. a reaction giving a C type particle only takes place if no C is already present). These terms may alter the reaction in a nonlinear way when the concentration of C is too large. It also makes it difficult to include advection (biased diffusion) in the evolution rule. Last, the exclusion principle, by limiting the total number of particles at each site, leads to poor statistical results when measurments are performed by a computer simulation. Average over many identical experiments are then required, or averages over many time steps and/or over large systems may be necessary.

We have considered several models with no exclusion principle in order to get rid of the above limitation. The idea of these models is to allow for an arbitrary number of particles of each species at each site. The dynamics is still discrete in space and time and consider each particle individually. In this way, all the correlations are preserved, as well as the other advantages of the lattice gas automata mentioned above.

For instance, for a pure diffusion process in one-dimension, we write the microdynamic equation as

$$n(t+\tau, r) = \sum_{\ell=1}^{n(t,r-\lambda)} \alpha_\ell(r-\lambda) + \sum_{\ell=1}^{n(t,r+\lambda)} \alpha_\ell(r+\lambda), \tag{26}$$

where the α_ℓ are boolean variables taking the value 1 with probability $1/2$. Rest particles and reaction terms can be included in a similar way.

It is important to note that $\sum_{\ell=1}^{n} \alpha_\ell$ is a binomial random variable with expectation value of $n/2$ and standard deviation $\sqrt{n}$. Therefore, if n is large enough, one does not have to "loop" over all particles of each site to find how many of them will move to the left and how many to right. A gaussian approximation can be made in order to speed up the computer implementation.

This approach will be used in section 3.2. It is appropriate for massively parallel computers and will be described in great detail in a forthcoming paper.[4]

3. Applications

3.1. Anomalous kinetics in reaction-diffusion systems

Chemical reactions in which several species react are usually described in terms of macroscopic rate equations. These rate equations give the time evolution for the local average concentrations of the different species. They assume that the reaction is completely described by the local average densities, i.e. the reaction introduces no correlations between the reacting species. This is reminiscent of a mean-field approximation in statistical physics, and therefore neglects an important aspect of the problem, namely the microscopic fluctuations.

For homogeneous reaction-diffusion systems, it is well known that the fluctuations play an important rôle. The simplest example is the annihilation reaction $A + A \rightarrow$ [inert]. The solution of the rate equations predicts that the concentration of A will decrease at long time as $a \simeq t^{-1}$ while a calculation taking into account the microscopic fluctuations in the particle density a gives $a(t) \simeq t^{-\alpha}$ where $\alpha = \min[1, \frac{d}{2}]$, and d is the dimensionality of the system[5].

A more complicated class of problem is that for which the initial condition is inhomogeneous. The inhomogeneities of the problem often arise from the presence of sources or sinks of particles. This may lead to the formation of reaction-diffusion fronts. Those fronts play an important rôle in many physical or chemical problems. Examples are given by solid-solid chemical reactions [6], catalysis [7], crystal growth , diffusion limited aggregation[8] or precipitate formation [9].

However, the problem is often too complicated to be approached analytically. Accordingly, it is very useful to have efficient algorithms to simulate the behavior of such systems. As an example, we shall consider a simple model displaying a reaction front which has been analyzed at the level of the rate equations by Gálfi and Rácz [10]. They consider two species A and B diffusing on a two-dimensional substrate and reacting to form a new species C. Provided that the two diffusing reagents A and B are initially separated in space, they will react in a confined region ("reaction front"). An appropriate choice of geometry reduces the problem to one dimension in this approximation. In particular, it is assumed that, at time $t = 0$, the system is uniformly filled with A for $x \leq 0$, and uniformly filled with B for $x > 0$. The study of the properties of this front provides relevant information about the production of C particles. The analytic treatment of Gálfi and Rácz assumes that the reaction rate is expressed in terms of the product of mean particle densities, which we shall henceforth refer to as the "mean field" approximation.

A microscopic description of the problem can be formulated in terms of a master equation for the probability that the sites of an underlying lattice be occupied by the particles of each species at a given time. Although it is easy to write down the equation, it is a formidable task to find its solution. One may hope to find it for one-dimensional systems, for cases in which the problem can be mapped onto an integrable quantum chain [(11)] but it is hopless in higher dimensions.

In terms of the local densities, the dynamics of this reaction-diffusion process can be described by the following partial differential equations

$$\partial_t a = D\nabla^2 a - kR, \qquad \partial_t b = D\nabla^2 b - kR, \tag{27}$$

where a and b are the local concentrations of A and B, k is the reaction constant and D the diffusion constant which, for simplicity, is assumed to be identical for both species. If the diffusion constants are different for the two species, we still expect the same asymptotic scaling behavior, with irrelevant corrections to the scaling function [10]. The reaction term R, is proportional to the joint probability of having one particles of type A and one particles of type B simultaneously present at the same point. Moreover, we are interested in a situation where a reaction front develops, as is the case when A and B are initially separated in space. We assume that $a(x, y, \ldots, t = 0) = a_0$ if $x < 0$ and $a = 0$ otherwise. Similarly, $b(x, y, \ldots, t = 0) = 0$ for $x < 0$ and $b = b_0$ for $x > 0$.

In general, the reaction region (the front) is expected to move through the system as the time goes on. For the sake of simplicity, we shall now consider only the case in which at time $t = 0$ the densities of A and B are the same hence the front is motionless.

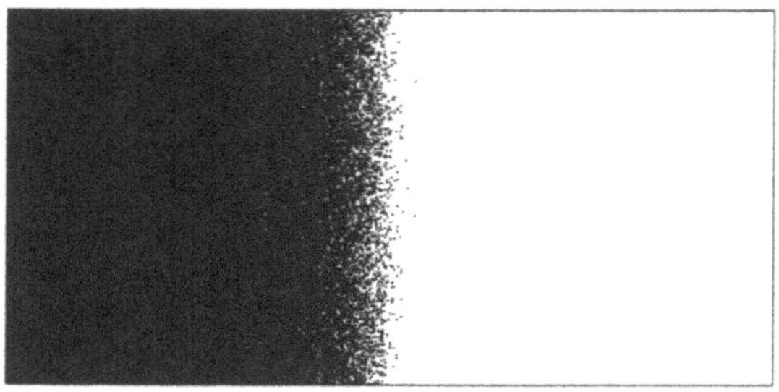

Fig. 4: Snapshot of the A particles as they move in the cellular automata model

Without being explicit on the form of the reaction term, interesting information on the long time behavior of a, b, and R can be obtained from equations (27) by using simple physical arguments and assuming the scaling forms:

$$a(x, t) = t^{-\gamma}\hat{a}(xt^{-\alpha}) \tag{28}$$

$$b(x, t) = t^{-\gamma}\hat{a}(xt^{-\alpha}) \tag{29}$$

$$R(x,t) = t^{-\beta}\hat{R}(xt^{-\alpha}) \tag{30}$$

These scaling forms are only valid for $x \ll L_d$, where L_d is the width of the depletion zone. Relations between these exponents can be derived from two assumptions, [10,13] based on the fact that A and B react only in a localized region:

(i) The reaction zone increases more slowly than the characteristic length of the diffusion process. This means that, in the long time limit, a and b vary over a typical length governed by diffusion, i.e. (grad$a$ $\sim$ grad$b$ $\sim t^{-1/2}$) and thus

$$\gamma + \alpha = 1/2. \tag{31}$$

ii) Due to the gradient concentration of A and B, a flux of particles towards the reaction region is observed. Assuming that the reaction is fed by these particle currents, the reaction-diffusion equations take on a quasi stationary form ($D\nabla^2 a = kR$) and ($D\nabla^2 b = kR$). For the exponents, this amounts to the relation

$$\gamma + 2\alpha = \beta. \tag{32}$$

For consistency with assumptions (i) and (ii), we should also have $\alpha < 1/2$ so that the width of the front grows more slowly than the depletion zone produced by the diffusion. To ensure the quasi stationarity of the solution, we need furthermore that

$$D\nabla^2 a \sim t^{-\gamma-2\alpha} \gg \partial_t a = -t^{-\gamma-1}(\gamma\hat{a} + \alpha x t^{-\alpha}\hat{a}'). \tag{33}$$

From relations (31)) and (32)) one also gets

$$\beta - \alpha = 1/2, \tag{34}$$

which is easy to check numerically since $\int_{volume} R \sim t^{-\beta+\alpha}$. We thus have a scaling theory with one independent exponent.

Relations (31), (32) and (33) are still quite general since no explicit form of the reaction term R has been used. In a mean-field approximation, one neglects the fluctuations and assumes that the reaction term factorizes into a product of the local concentrations of A and B. The mean field reaction rate R_{mf} takes the form

$$R_{mf} = ab. \tag{35}$$

This introduces another relation among our exponents, namely

$$\beta = 2\gamma \tag{36}$$

and we obtain the following the values for the mean-field exponents

$$\alpha_{mf} = \frac{1}{6} \qquad \gamma_{mf} = \frac{1}{3} \tag{37}$$

in agreement with the value of Galfi and Racz.

Numerical simulations have been made using the CA algorithms described in section 2 (see [12,14,15] for more details). In this discussion, we shall restrict ourselves to the one-dimensional case for which the role played by the fluctuations is the strongest. We consider chains of size $2L$. In order to simulate an infinite system, the two extremities of the chain are taken as sources of A and B particles respectively. The chain should be long enough to ensure that, for the observation time considered, the system does not "feel" the boundary conditions, in order to mimic an infinite system. The cellular automata simulations, using the Multi-spin Coding technique, were performed on a chains of length $L = 1024$ and data were collected at ten times between $t = 1000$ and $t = 300\,000$. The values of p_ℓ were chosen as $p_0 = 1/2$, which corresponds to a diffusion constant $D = \lambda^2/2\tau$. In order to measure the profile of the region where C is produced, we monitored the number of C particles created at each time step, as a function of the position. Each measurement was performed over a period of $\Delta T = \frac{\delta T}{10}$ steps, where δT is the time interval between two consecutive measurements. This introduces a systematic error of the order of 1% on each measurement, but is necessary to produce sufficient statistics. The data is further averaged over 1792 independent runs. Several quantities have been measured like the width of the front $W(t)$, and the total reaction rate $N(t)$. The width $W(t)$ of the front is defined as

$$W^2(t) = \frac{\sum_{x=-L}^{L-1}(x - <x>)^2 R(x,t)}{\sum_{x=-L}^{L-1} R(x,t)} \sim t^{2\alpha}, \tag{38}$$

where $<x>= x_f$ is the position of the front defined as the first moment of $R(x,t)$. The time dependence of the width is plotted in figure 5. One finds, over two and a half decades (from t=1000 to 300000) the power law relation

$$W \sim t^{0.292\pm0.005}. \tag{39}$$

This exponent is far from its mean-field value and so fluctuations are clearly important in this case.

It is natural to wonder if there is an upper critical dimensionality d_{up} above which the mean-field predictions are correct. This question has been addressed in [12] and it was shown, based on microscopic arguments that $d_{up} = 2$ for this problem.

Another way to approach this problem is to study the stationary properties of the front formed by the reaction $A + B \to C$ between balanced opposite currents of A and B [16]. Dimensional analysis show that mean-field theory is indeed valid in spatial dimension greater than $d_{up} = 2$. Consistency arguments give exact predictions for the scaling exponents beneath this dimension. These results are confirmed by numerical simulations using the CA algorithms described in section 2 for one, two and three dimensions. The scaling exponents for this problem can be related to the ones for the time dependent problem.

It turns out that this approach is less sensitive to finite-size effects and gives more stable numerical results than the one which consists in following the time dependence of the width of the front. In one dimension, theory and simulations find $\alpha = \frac{1}{4}$, in contradiction to the previously quoted value $\alpha = 0.29$. This means

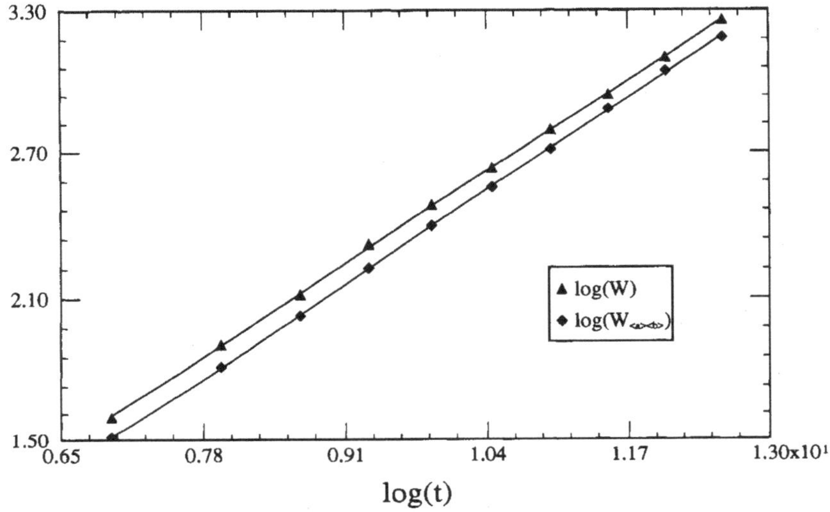

Fig. 5: **Width of the front as a function of the time in a log-log plot**

that slow transients can affect the results obtained by following the time evolution of the front. The more general reactions $nA + mB \rightarrow [\text{inert}]$ can be investigated along the same line [15,16].

3.2 Turing structures

Another interesting application in which CA algorithms are a powerful tool is the formation of Turing structures in reaction-diffusion systems [17]. Turing suggested that, under certain conditions, chemicals can react and diffuse so as to produce steady-state heterogeneous spatial patterns of chemical or morphogen concentrations. Such problems are usually described in terms of rate equations for the evolution of the local average densities of the chemicals [18]. Here, again, this mean-field like approximation neglects the microscopic fluctuations. The investigation of the role played by the fluctuations in such systems has been undertaken only very recently [19]. For the sake of simplicity, we shall only consider here one of the simplest model showing Turing patterns: the Schnackenberg reaction-diffusion model in two dimensions [18,20]. It describes the following autocatalytic reaction:

$$
\begin{array}{cc}
A \xrightarrow{k_1} X & X \xrightarrow{k_2} 0 \\
2X + Y \xrightarrow{k_3} 3X & B \xrightarrow{k_4} Y
\end{array}
\tag{40}
$$

The corresponding macroscopic rate equations for the local average densities x and

y are:

$$\begin{aligned}\partial_t x &= k_1 a - k_2 x + k_3 x^2 y + D_x \nabla^2 x \\ \partial_t y &= k_4 b - k_3 x^2 y + D_y \nabla^2 y\end{aligned} \tag{41}$$

The densities of A and B are kept fixed. The stability of these equations has been studied by Murray.[18] Introducing dimensionless quantities:

$$\begin{gathered} t' = \frac{D_x t}{L^2} \quad r' = \frac{r}{L} \quad d = \frac{D_y}{D_x} \quad \gamma = \frac{L^2 k_2}{D_x} \\ a' = \frac{k_1}{k_2}\left(\frac{k_3}{k_2}\right)^{\frac{1}{2}} a \quad b' = \frac{k_4}{k_2}\left(\frac{k_3}{k_2}\right)^{\frac{1}{2}} b \\ u = \left(\frac{k_3}{k_2}\right)^{\frac{1}{2}} x \quad v = \left(\frac{k_3}{k_2}\right)^{\frac{1}{2}} y \end{gathered} \tag{42}$$

the diffusion-reaction rate equations become:

$$\partial_{t'} u = \gamma(a' - u + u^2 v) + \triangle_{r'} u. \tag{43}$$

$$\partial_{t'} v = \gamma(b' - u^2 v) + d\triangle_{r'} y. \tag{44}$$

A conventional linear stability analysis shows that for some values of the parameters an homogeneous stationary state is unstable towards local density perturbations. Inhomogeneous patterns can evolve by diffusion-driven instabilities providing that the diffusion constants D_x and D_y are not the same, i.e. $d \neq 1$. The region of the parameter space for which homogeneous states of the system are unstable is called the deterministic Turing space (DTS). An example is shown in figure 3 for $d = 31$. The stationary states observed in the Turing space have a hexagonal or a striped structure [21].

All of the above discussion is based on mean-field like equations. It is then legitimate to ask what the effect will be of the fluctuations in such a system. The CA approach is well suited to investigate this question. The idea of our CA algorithms is to simulate as closely as possible the real reaction-diffusion mechanisms at a microscopic scale.

For this purpose we consider a multi-species 2-dimensional lattice-gas with no exclusion principle; i.e. the number of particles of each species at a given site is not restricted. The algorithm proceeds in two steps: the diffusion step and the reaction step [22]:

- Diffusion step: as described in section 2, the particles jump to nearest-neighbor site with an equal probability for each direction. The diffusion coefficient is tuned by varying the number M of consecutive diffusion steps for a given species: $D = M\frac{(\Delta r)^2}{4\Delta t}$, where Δr and Δt are respectively the sizes of the lattice space and time interval.
- Reaction steps: for the reaction

$$nS_1 + mS_2 \xrightarrow{k_j} rS_3$$

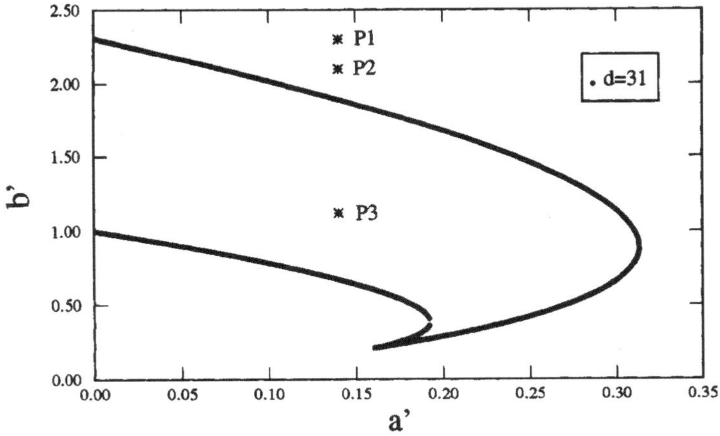

Fig. 6: **Deterministic Turing space for $d = 31$.**

all possible families composed of n S_1 and m S_2 particles have a probability k_j to create r particles S_3.

Note that those rules take into account the fluctuations in a more natural way that the ones used by Dab et al. and Kapral et al.[23] in a similar context. Indeed, the rules of Dab et al. using an exclusion principle, were conceived in a way as to reproduce the rate equations. It is not clear that the fluctuations associated with these rules reproduce the physical fluctuations present in the real reaction-diffusion process.

In the continuous limit and within the Boltzmann approximation ($< n_1 n_2 > = < n_1 >< n_2 >$), our CA model reproduces the diffusion-reaction rate equations[19,23].

The first question we can address is how the Turing space is affected by the fluctuations. Preliminary investigations show that the Turing space is enlarged. Indeed, the solution of the linearized diffusion-reaction rate equations around an unstable homogeneous state and for a square system with periodic boundary conditions of linear size p, takes the form:

$$u(r_1, r_2, t) = \sum_{k_1, k_2} C_{k_1, k_2} \exp(\lambda(\vec{k}^2)t) \cos(\frac{2\pi k_1 r_1}{p}) \cos(\frac{2\pi k_2 r_2}{p}), \tag{45}$$

with $r_1 \in [0, p]$ and $r_2 \in [0, q]$. Inside the Turing space, $\lambda(\vec{k}^2)$ takes a positive value for

$$\vec{k}'^2(a, b, d) \leq \vec{k}^2 \leq \vec{k}''^2(a, b, d). \tag{46}$$

For the CA model, the window $[\vec{k}'^2(a, b, d), \vec{k}''^2(a, b, d)]$ becomes larger. Outside the Turing space, where the homogeneous stationary state is stable within the deterministic theory, all the modes are exponentially decaying with time. This is not true for our CA model. Indeed, for the points $P1$ and $P2$ (see Fig. 6), outside the deterministic Turing zone, some modes are growing. However, for a point well inside the deterministic Turing space ($P3$), the stationary state is not affected by the fluctuations. The structure factor is defined as

$$S^{\alpha}(q_1, q_2, t) = \langle |\tilde{n}^{\alpha}(q_1, q_2, t)|^2 \rangle, \tag{47}$$

where $\tilde{n}^{\alpha}(q_1, q_2, t)$ is the Fourier transform of the density of the species α at time t. This structure factor reaches stationary values which shows that the same pattern is selected both with and without fluctuations. Configuration obtained in the long time regime exhibiting a heagonal geometry is shown in figure 7. The right part corresponds to the solution of the rate equations, while the left part corresponds to the CA simulation.

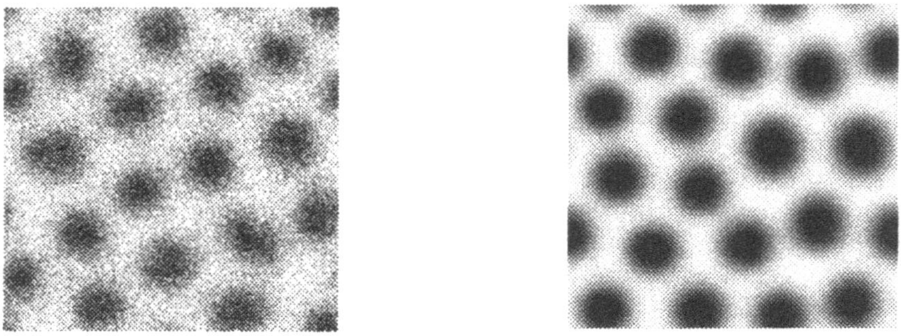

Fig. 7: **Patterns obtained in the Schnackenberg reaction in the long time regime.**

However, the way in which the steady state is reached (starting from an initial state with Poissonian distribution of particles) is very much influenced by the fluctuations. Indeed, the linear stability analysis of the deterministic equations predicts that all the modes with a wave vector belonging to the window defined by (46) have an amplitude which is growing exponentially for short times and that the structure factor behaves as

$$S(\vec{q}, t) \sim \exp(t\lambda(\vec{q}^{\,2})). \tag{48}$$

This is not true in presence of fluctuations has shown for example in Figure 8, where the logarithm of the structure factor $S(2, 0, t)$ is plotted as a function of time.

One sees clearly that the behavior predicted by the rate equations (RE) is drastically different from the results of the CA simulation. For a mode which wave vector is outside the window defined by equation (46), the fluctuations transform an exponentialy decaying mode into a slowly growing mode as shown in Figure 9.

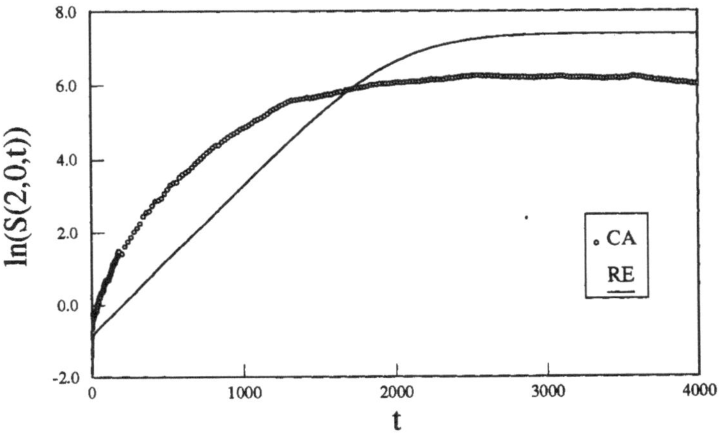

Fig. 8: Time evolution for the component of the stucture factor $S(q_1 = 2, q_2 = 0, t)$ for the deterministic rate equation (RE) and the CA model. The corresponding wave vector is lying inside the window defined by equation (46)

Short time and transient behaviors which can possibly be experimentally observed are drastically affected by fluctuations. A more systematic analysis of the role played by the fluctuations, both in the non-stationary and stationary state, is in progress [23].

3.3 Nonequilibrium phase transitions

Nonequilibrium statistical mechanics is much less understood than equilibrium statistical mechanics. This is true particularly for the problem of phase transitions. In equilibrium, the renormalization group approach [24] provided a framework allowing the determination of the relevant parameters characterizing a phase transition. The critical exponents, describing the behavior of the system at the vicinity of a second order phase transition, fall into universality classes characterized by a small numbers of parameters (dimensionality of the space, number of components of the order parameter, range of the interactions). The situation is not so clear for nonequilibrium phase transitions, and it is useful to study particular models as first steps towards a more global understanding of the problem.

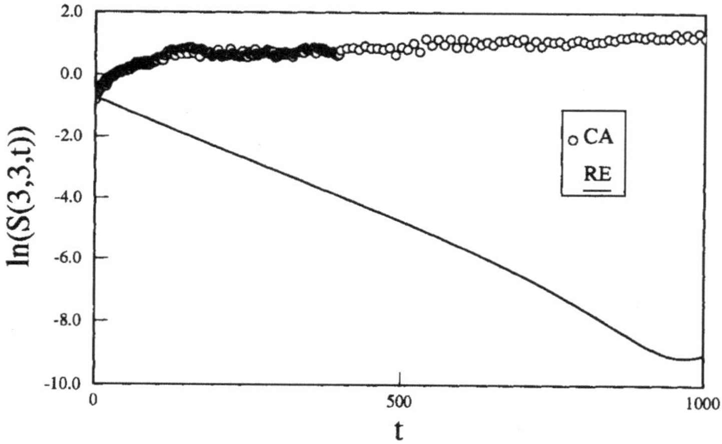

Fig. 9: Time evolution for the component of the stucture factor $S(q_1 = 3, q_2 = 3, t)$ for the deterministic rate equation (RE) and the CA model. The corresponding wave vector is lying outside the window defined by equation (46)

Simple examples of nonequilibrium systems whose steady state displays a phase transition are lattice models where a binary variable $\Psi_i = 0, 1$ is associated with each site and where the evolution rule allows for the existence of an adsorbing state (i.e. $\Psi_i = 1$ everywhere).

Applications range from processes of adsorption-desorption of particles on a surface [25] to models of spreading of a liquid through porous medium [26].

It is now recognized that for such model the character of the phase transition is not affected by the fact that the dynamical rules are implemented in a sequential or parallel manner [27]. However, a parallel implementation, in which all the sites of the system are updated simultaneously, can be more efficient.

As an illustration of what can be achieved, we shall consider extensions of the so-called "A model" which has been introduced by Dickman and Burschka [28] as a simple model describing poisoning transitions similar to the ones observed on catalytic surfaces. One considers a d-dimensional substratum spanned by a regular hypercubic lattice. Each site has two possible states: empty or occupied by a particle A. The first step of the dynamical process is the adsorption. The probability for a vacant site to become occupied during a short time interval δt is $p\delta t$. The second step of the process is desorption. The probability for an occupied site x to become vacant is $q\delta t$, provided that at least one of the nearest neighbors of x is vacant (see figure 10). During the time interval δt, one of the two processes occurs at each site.

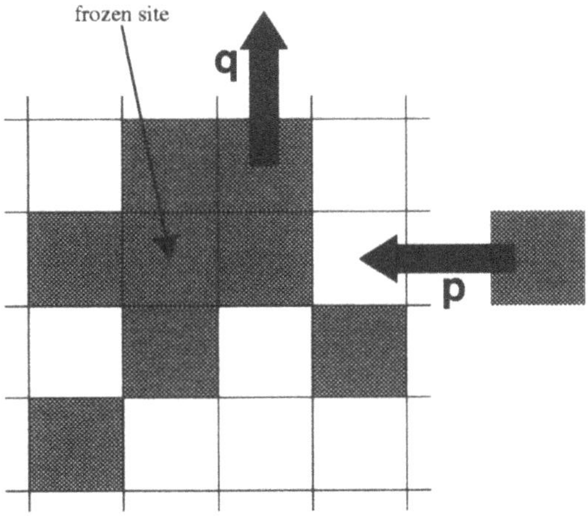

Fig. 10: Schematic representation of the A-model

For simplicity we shall restrict ourselves to the case $q = (1 - p)$.

Qualitatively, one expects that, if p is large enough, an initially empty substratum will be completely covered by A particles, after some time. This is the poisoned phase or the adsorbing state. But, if p is small enough, the desorption mechanism will be efficient enough to prevent such a poisoning. Thus, one may anticipate the existence of a threshold value p_c such that, in the stationary state, the covering fraction of A on the substratum X_A will be 1 for $p \geq p_c$ (poisoned phase) and smaller than 1 for $p < p_c$. If X_A varies continuously across p_c the transition will be of second order and its behavior near the threshold will be described in term of the critical exponent β:

$$1 - X_A(p) \sim (p_c - p)^\beta \tag{49}$$

The cellular automata version of this model is straightforward [27]. One considers a d-dimensional lattice. Each cell of the lattice j has two possible states: $|\Psi_j\rangle = |0\rangle$ or $|A\rangle$. The probabilistic cellular automata adsorption-desorption rules (ADR) are (the abreviation n.n means neareast neighbor):

If $|\Psi_j\rangle(t) = |0\rangle$ then:

$$|\Psi_j\rangle(t+1) = \begin{cases} |0\rangle, & \text{with probability } (1-p) \\ |A\rangle, & \text{with probability } p. \end{cases} \tag{50}$$

If $|\Psi_j\rangle(t) = |A\rangle$ then:

$$|\Psi_j\rangle(t+1) = \begin{cases} |A\rangle, & \text{with probability } p \text{ if } j \text{ has at least one n.n empty} \\ |A\rangle, & \text{with probability 1, if all the n.n are occupied,} \\ |0\rangle, & \text{with probability } (1-p) \text{ if } j \text{ has at least one n.n} \\ & \text{empty.} \end{cases} \quad (51)$$

A new ingredient is now added to the model, namely diffusion. Once the particles have been adsorbed on the surface, they are not necessarily frozen on the surface before a possible desorption occurs; they can diffuse to nearest neighbor empty sites. The diffusion can be simulated by the CA rule (DIFF) described in details in section 1.

The global evolution rule is the following; one applies once the rule ADR and ρ times the rule DIFF. For $\rho = 0$, one recovers the simple A model; for $\rho \to \infty$ the diffusion mixes all the states and one expects the behavior of the system to be mean-field like. Qualitatively, one anticipates that the diffusion will have two effects on the phase diagram. First, the critical probability p_c should increase monotonically with ρ. Second, the values of the critical exponents should vary from those of the pure A model to their mean-field values. How this variation occurs (continuous variation of the exponents as a function of ρ or simple crossover phenomena) is not an easy question and will be discussed later on.

Simulations have been done for a two dimensional system of size $L \times L$, with $L = 128$ and 256 [27]. The stationary quantities were obtained as follows. Starting from an initial state where each cell is empty, one iterates the evolution rules for 1000 steps (per site) which turns out to be sufficient to reach the steady state. The stationary values are then computed by averaging over 200 measurement, each of them separated by 100 iteration steps (per site). Moreover an average was made over 20 samples.

Since it is crucial to have a very precise determination of p_c to obtain the critical exponents, a different approach should be used to find p_c. We applied the method introduced by Grassberger for directed percolation (DP) [29]. We started with a lattice filled with A particles, except for one single site and let the system evolve according to the CA rules. We measured the density of empty sites $n_0(t)$, the average square cluster size $R^2(t)$ and the survival probability $P(t)$ during the simulation. If the system is sufficiently large so that none of the clusters we simulate reaches the boundaries, our data are free of finite size effects.

At the critical probability p_c, we expect power laws behavior for the above quantities, namely:

$$A_i(t) \sim t^{\eta_i} \quad (52)$$

where $A_i(t)$ stands for n_0, R^2 or P. In a log-log plot of the above quantities, upward (downward) curvatures indicates that the value of p is super (sub) critical. The critical probabilities obtained for different values of ρ are quoted in the second column of Table 1.

Having a precise determination of p_c, we can now draw the phase diagrams. In figure 11, the phase diagrams are given for several values of ρ. One notices that the phase boundary behaves differently as a function of ρ as one approaches the critical probability $p_c(\rho)$. As expected, $p_c(\rho)$ is an increasing function of ρ; for large values of ρ, the phase boundary approaches the critical probability in a much more linear way than for smaller value of ρ. This is again in agreement with the interpretation of a more mean-field like behavior, for which $\beta = 1$, as ρ gets large.

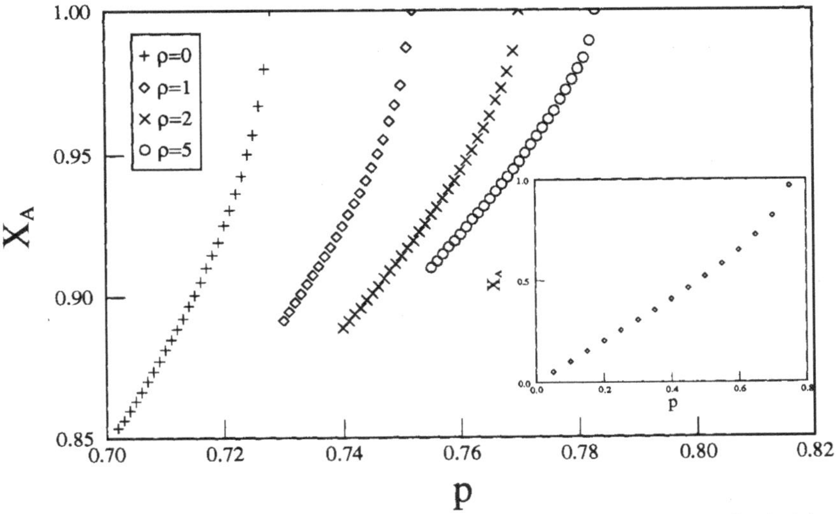

Fig. 11: **Phase diagram for several values of ρ. The density X_A of A particles adsorbed in the stationary state is plotted as a function of p for $\rho = 0,\ 1,\ 2,\ 5$.**

The order parameter critical exponent $\beta(\rho)$ can be extracted from the above data. The values of the critical exponent $\beta(\rho)$ are given in table 1. We note that for $\rho = 0$, the value of β is compatible with the one of directed percolation in $d = 3$, in agreement with a conjecture of Grassberger [30]. One notes that $\beta(\rho)$ increases linearly with ρ for the values of ρ considered.

Table 1: Critical probability p_c and order parameter critical exponent $\beta(\rho)$ for several values of ρ.

ρ	$p_c(\rho)$	β_{eff}
$\rho = 0$	$p_c = 0.7278 \pm 0.0003$	$\beta = 0.56 \pm 0.02$
$\rho = 1$	$p_c = 0.7518 \pm 0.0004$	$\beta = 0.59 \pm 0.03$
$\rho = 2$	$p_c = 0.7702 \pm 0.0003$	$\beta = 0.64 \pm 0.03$
$\rho = 5$	$p_c = 0.7833 \pm 0.0004$	$\beta = 0.70 \pm 0.03$
$\rho = \infty$	$p_c = 0.8$	$\beta = 1.0$

However, the above values of the $\beta(\rho)$ should be interpreted carefully. Indeed, due to finite size of the lattices considered in the simulations one cannot investigate values of p arbitrary close to p_c. At best $(p_c(\rho) - p)/p_c(\rho) \sim 10^{-3}$ can be obtained. It may well be that, as ρ increases, the width of the critical region decreases and that the true critical behavior is not seen in our simulation. The critical exponents quoted in Table 1 would be simply an apparent exponent characterizing how the phase boundary behaves near but not very close to the critical point. However, it is this apparent exponent which is of importance in interpreting the experimental data.

A more satisfactory way to describe the situation is to introduce, by analogy with equilibrium critical phenomena, a so-called "effective critical exponent" defined as follows [31]:

$$\beta_{eff}(\rho, \delta p(\rho)) = \delta p(\rho) \frac{d\ln[1 - X_A(\rho, \delta p(\rho))]}{d\delta p(\rho)} \qquad (53)$$

with $\delta p(\rho) = p_c(\rho) - p$.

This effective exponent has been calculated for several value of ρ. is given in Figure 12.

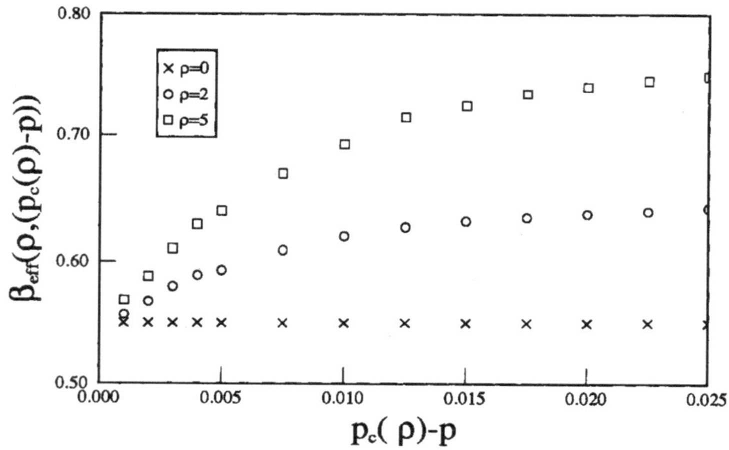

Fig. 12: Effective exponent $\beta_{eff}(\rho, (p_c(\rho) - p))$ as a function of $(p_c(\rho) - p)$ for $\rho = 0,\ 2,\ 5$.

For $\rho = 0$, one sees that the effective exponent does not vary with p in the range investigated $((p_c(\rho)-p) \leq 0.025)$. For $\rho > 0$, the effective exponent is compatible (taking into account errors bars) with the DP exponent when $(p_c(\rho) - p) \to 0$. It increases when $(p_c(\rho) - p)$ increases. The departure from its DP value, at fixed $\delta p(\rho)$, is larger when ρ increases, in agreement with the idea of a crossover towards a mean-field behavior for very large value of ρ.

Several more complicated models aiming to describe homogeneous catalysis on a substrate and using CA algorithms have been investigated in the past years [32]. The first was a CA version a simple model proposed by Ziff et al.[33] of the reaction $A - B_2$ on a catalyst surface [34]. A similar model with different rules taking into account the stochoimetric properties of the reactions has been proposed recently by Ziff et al. [35]. In the long time limit, these models reach stationary states where the catalyst surface may or may not be poisoned by one species or not. These models are too simple to explain several important phenomena observed in experiments such oscillatory behavior associated with spatial structure [36], surface phase transformation [37] and Turing patterns at a molecular scale [38]. Several kinetic mechanisms have to be included to reproduce the observed features and about hundred different models have been proposed in the literature (see Razon et al. [39] for a review). The catalytic processes are complex problems involving intricated interplay between reaction-diffusion processes and surface restructuration. A macroscopic approach is not good enough to fully explain what is observed. On the other hand, a purely microscopic approach is too complicated. The CA approach offers an interesting alternative in terms of a mesoscopic description. This mesoscopic level is fine enough to capture the important phenomena taking place at a small scale but coarsed-grained enough to allow compute efficent computation of the macroscopic properties.

4. Acknowledgements

This work is partly supported by the Swiss National Science Foundation.

5. References

1. B. Chopard and M. Droz, *Jour. Stat. Phys.* **64**, 859 (1991).
2. D. Montgomery, in *Lectures in Theoretical Physics: Boulder Lectures*, **Vol. IX C**, ed. Gordon and Breach (Wesley E. Brittin, 1967) .
3. J.L. Anderson, *Am. J. Phys.*, **60** (1992) 923.
4. B. Chopard, M. Droz and L. Frachebourg, *in preparation*
5. H. Schnörer, I.M. Sokolov and A. Blumen, *Phys.Rev* **A42** (1990) 7075.
6. A.S.Nowick and J.J.Burton, *Diffusion in Solids*, (Academic, New York 1978)
7. R.M. Ziff and K. Fichthorh, *Phys. Rev* **A34** (1986) 2038
8. M. Muthukumar, *Phys. Rev. Lett.* **50** (1984) 839
9. H.K. Henisch, *Periodic Precipitation*, (Pergamon Press, 1991).
10. L. Gálfi and Z. Rácz, *Phys.Rev* **A38**, (1988) 3151
11. M. Droz, M. Henkel and V. Rittenberg, (1993) in preparation.
12. S. Cornell, M. Droz and B. Chopard, *Phys. Rev* **A44** (1991) 4826
13. Z.Jiang and C. Ebner, Phys.Rev. *Phys. Rev* **A42** (1990) 7483
14. B. Chopard and M. Droz , *Euro. Phys. Lett* **15** (1991) 459
15. S. Cornell, M. Droz and B. Chopard, *Physica* **A188** (1992) 322

16. S. Cornell and M. Droz, submitted to *Phys. Rev. Lett* (1993)

17. A. Turing *Phil.Trans.Roy.Soc.Lond* **B237** (1952) 37

18. J.D. Murray *Mathematical Biology* (Springer-Verlag, Berlin, 1990). p 380.

19. M. Droz and L. Frachebourg, *Helv.Phys.Acta* (1993), to appear.

20. J. Schackenberg, *J. Theor. Biol.* **81** (1979) 389.

21. V. Dufiet and J. Boisonnade, *J. Chem.Phys* **96** (1991) 664.

22. L. Frachebourg and M. Droz, (in preparation) (1993).

23. D. Dab, J.-P. Boon and Y.-X. Li, *Phys.Rev.Lett* **66** (1991) 2535; R. Kapral, A. Lawniczak and P. Masiar, *Phys.Rev.Lett* **66** (1991) 2539;

24. S.K. Ma *Modern Theory of Critical Phenomena* (Benjamin-Cummings Publ. Comp. 1976).

25. R. Dickman, *Jour. Stat. Phys* **55**, 997 (1988).

26. S. Cornell, M. Droz, R. Dickman and M.C. Marques, *Jour.Phys*, **A24** (1991) 5605.

27. F. Bagnoli, B. Chopard M. Droz and L. Frachebourg, *J. Phys.* **A25** (1992) 1085.

28. R. Dickman and M. Burschka, *Phys. Lett* **A127** (1987) 132.

29. P. Grassberger, *J. Phys.* **A22** (1989) 3673.

30. P. Grassberger and A. de la Torre, *Ann. Phys.* **122** (1979) 373; P. Grassberger *Z.Phys.* **B47 (1982)** , 365.

31. D.J. Amit *Field theory, the renormalization group and critical phenomena* (McGraw-Hill 1978).

32. X.G. Wu and R. Kapral, *Physica* **A188** (1992) 284.

33. R.M. Ziff, E. Gulari and Y. Barshad, *Phys.Rev.Lett* **56** (1986) 2553.

34. B. Chopard and M. Droz, *Jour. Phys* **A21** (1988) 205.

35. R. Ziff, *preprint* (1992).

36. H.H. Rotermund, S. Jakubith, A. von Oertzen and G. Ertl, *J.Chem.Phys* **91** (1989) 4942.

37. G. Ertl, P.R. Norton and J. Rüstig, *Phys.Rev.Lett* **49** (1982) 49.

38. J. Falta, R. Imbhil and M. Henzler, *Phys.Rev.Lett* **64** (1990) 1409.

39. F. Razon and R.A. Schmitz, *Catal.Rev.-Sci.Eng* **28** (1986) 89.

COMPARISON OF CELLULAR AUTOMATA AND DIFFERENTIAL EQUATION APPROACHES TO REACTION-TRANSPORT PROCESSES

T. Karapiperis
Paul Scherrer Institute, CH-5232 Villigen PSI, Switzerland

Abstract

The transport and chemical reactions of solutes are modelled as a cellular automaton in which molecules of different species perform a random walk on a regular lattice and react according to a local probabilistic rule. Under the assumptions of molecular chaos and a smooth density function, the discrete evolution equations of the automaton go over to the standard reaction-transport equations in the continuum limit. Simulations on one- and two-dimensional lattices show that the discrete model can be used to approximate the solutions of the standard reaction-transport equations. Macroscopic effects resulting from correlations between molecules disappear as the continuum limit is approached. The model is applied to the reactions $a+b \rightleftharpoons c$ and $a+b \rightarrow c$ with homogeneous and inhomogeneous initial conditions as well as to non-linear, chemically reacting systems displaying spontaneous formation of spatial concentration patterns.

1 Introduction

Modelling the transport of solutes in geological media in its full complexity requires a great deal of physical intuition and computational ingenuity. An essential task is to select, on grounds of usefulness and conceptual economy, the suitable dependent variables (e.g. concentrations of aqueous or precipitated species) and the laws (e.g. mass conservation, law of mass action) these variables satisfy. For testing purposes and with practical applications in mind, the model has to be translated subsequently into usable and efficient computer code. The model presented here takes advantage of the fact that the macroscopic behaviour of a large and complex system of reacting molecules is the result of many relatively simple microscopic processes taking place *in parallel* (i.e. simultaneously) and *locally* (i.e. involving molecules within a small spatial neighbourhood). The model is implemented as simple and hence transparent code suitable for *massively parallel computers.* Our approach contrasts with the standard modelling approach which solves systems of non-linear *partial differential equations (PDE)* for macroscopic variables. The latter approach neglects statistical fluctuations by effectively averaging solute properties over a small volume and dealing only with their local mean values; correlations between small numbers of particles are accounted for by introducing a hierarchy of coupled equations for the n-particle density functions and truncating it by some approximation.

Reaction-transport processes are modelled in this work as a *cellular automaton (CA)* [1]. Each species is described by a number of 'particles' residing on the sites of a regular lattice; the particles move randomly between lattice sites in discrete time steps and react with a certain probability upon meeting. The CA evolution rule consists of a combined 'transport' and 'reaction' operation which derives the value of the *occupation number* (i.e. an integer-valued field with a species label, giving the number of particles of a certain species at a site) for the next time step from its present values in a neighbourhood of the site of interest. The ensemble average of the occupation number obeys a discrete evolution equation. The continuum limit of this equation contains the standard advection, diffusion and reaction-rate terms; the probabilities of motion and reaction are chosen so that the desired *advection velocity, diffusion coefficient* and *rate constants* are obtained.

Comparing CA with standard approaches we expect the following gains:

(a) Physicochemical processes are modelled at a *mesoscopic level.* This is to be distinguished both from the *macroscopic level* described by PDE's and from the *microscopic level* of molecular dynamics. In the CA simulation the number of 'particles' is large enough to make statistical concepts meaningful, but is still several orders of magnitude smaller than the real number of molecules involved. The simulation aims at describing macroscopic behaviour which does not depend on the details of the microscopic dynamics, but follows from simple properties of the latter, such as the random nature of collisions. The CA captures the essential aspects of the microscopic dynamics and thus physical insight is implemented at an elementary level. The resulting computer code is simple and allows easy variation of the dynamics and the boundary conditions. Moreover, statistical fluctuations are inherently present in the CA and their macroscopic consequences can be investigated without making any of the simplifying assumptions that lead to the continuous variables (e.g. particle density) of the PDE's. This is important when the rôle of fluctuations becomes significant and the continuous approach is incapable of describing the correct macroscopic behaviour.

(b) By working with *integers*, the CA simulation is free of *round-off errors*, which arise from the projection of real numbers onto finite computer words and lead to possible *instabilities* of the numerical algorithms used to solve PDE's. This is a significant advantage when non-linear reaction terms are present, since there are no universal stability criteria in the case of non-linear PDE's. All quantities in a CA simulation are intrinsically finite and infinities can only arise by extrapolation to some limit (e.g. infinite volume), in very much the way this is done with real systems [2]. The continuity required of the solutions of PDE's is, by contrast, physically counterintuitive, as it presupposes an infinitesimal limiting procedure (e.g. an elementary volume shrinking to a point) which does not correspond to reality beyond a minimal scale.

(c) CA algorithms are *parallel* and can be formulated at *bit level.* This makes them suitable for both massively parallel and vector computers. In particular, the simple nature of physicochemical processes at the mesoscopic level should allow simulation on massively parallel computers with relatively unsophisticated processors (e.g. CM-2 [3]). Supercomputers are also being used to solve PDE's with standard numerical methods and the question of whether CA will provide a more efficient alternative (when fluctuation effects are uninteresting) is still to be answered.

The approach described here is closely related with probabilistic CA models of reaction-diffusion processes [4, 5], except that the latter impose a limitation on the number of particles per lattice site (exclusion principle). This restriction is necessary for simulations on special-purpose computers (e.g.

CAM [6]) and is useful for vectorisation purposes, but we prefer not to impose it, because it limits statistics and causes problems with modelling advection and complex reactions.

In Section 2 the CA model will be introduced and the derivation of the continuum equations will be outlined, emphasising the underlying physical assumptions. The simulation of various reaction-transport processes will be described in Section 3 and the results will be compared with the solutions of differential equations. Section 4 will summarise the findings of this work and will propose future tasks.

2 The Model

For simplicity, we formulate the model on a one-dimensional lattice $\mathcal{L}$ with lattice spacing λ. At time t and position x on the lattice, there are $N_\alpha(x,t)$ particles of species α. The system evolves in steps of time τ. The value of N_α at time $t+\tau$ is given by

$$N_\alpha(x,t+\tau) = \mathcal{C} \circ \mathcal{T} N_\alpha(x,t), \tag{2.1}$$

where the operators $\mathcal{T}$ and $\mathcal{C}$ describe *transport* and *chemical reactions* respectively.

The action of $\mathcal{T}$ amounts to letting each particle move to the right or left or remain stationary with probabilities p_α, q_α and $1-p_\alpha-q_\alpha$ respectively. Formally this is written as

$$\mathcal{T} N_\alpha(x,t) = \sum_{j=-1}^{1} \sum_{n=1}^{\infty} \xi^{(j)}_{x+j\lambda,n}\, \theta(N_\alpha(x+j\lambda,t)-n), \tag{2.2}$$

where the first sum is over the components $\xi^{(-1)}_{x',n}, \xi^{(0)}_{x',n}, \xi^{(+1)}_{x',n}$ of a random triplet $\xi_{x',n}$, $x' \in \mathcal{L}$, taking one of the values $(1,0,0), (0,1,0)$ or $(0,0,1)$ with probabilities p_α, $1-p_\alpha-q_\alpha$ and q_α respectively. The θ-function is defined by $\theta(m)=1$, if $m \geq 0$, and $\theta(m)=0$, otherwise.

$\mathcal{C}$ allows each reaction to occur with a certain probability and, for those that do occur, it subtracts (adds) the appropriate numbers of reactants (products). We write this as

$$\mathcal{C} N_\alpha(x,t) = N_\alpha(x,t) + \sum_{r=1}^{R} \left(\nu^{(f)}_{\alpha r} - \nu^{(i)}_{\alpha r}\right) \eta_{x,r}, \tag{2.3}$$

where $\eta_{x,r}$ is a random Boolean variable and $\nu^{(i)}_{\alpha r}$ and $\nu^{(f)}_{\alpha r}$ are *stoichiometric coefficients* referring to initial and final species of reaction r respectively. The summation runs over all R one-way reactions obtained by separating, in the case of reversible reactions, the forward from the reverse process. Reaction r takes place at x (equivalently, $\eta_{x,r}=1$), with probability $P_r F_r(\{N_\beta(x,t) : \beta \in \mathcal{S}\})$, where the real number P_r is chosen so that the probability lies between 0 and 1, F_r is a function of the occupation numbers and $\mathcal{S}$ is the set of all species. P_r is related to the rate constant and F_r yields the concentration dependence of the reaction rate in the continuum limit.

We obtain a real-valued density by taking an *ensemble average* of $N_\alpha(x,t)$ over several independent simulations: $\rho_\alpha(x,t) \equiv \langle N_\alpha(x,t) \rangle$. Then it follows that [7]

$$\rho_\alpha(x,t+\tau) = \langle \mathcal{T} N_\alpha(x,t) \rangle + \sum_{r=1}^{R} \left(\nu^{(f)}_{\alpha r} - \nu^{(i)}_{\alpha r}\right) P_r \langle F_r(\{\mathcal{T} N_\beta(x,t) : \beta \in \mathcal{S}\}) \rangle \, . \tag{2.4}$$

Using (2.2) we deduce for the first term on the RHS of (2.4)

$$< TN_\alpha(x,t) >= p_\alpha \rho_\alpha(x-\lambda,t) + (1-p_\alpha-q_\alpha)\rho_\alpha(x,t) + q_\alpha \rho_\alpha(x+\lambda,t). \tag{2.5}$$

We consider two reaction rules defined by

$$F_r(\{N_\beta(x,t) : \beta \in S\}) \equiv \prod_\beta \theta\left(N_\beta(x,t) - \nu_{\beta r}^{(i)}\right) \quad \text{(Rule I)}, \tag{2.6a}$$

$$F_r(\{N_\beta(x,t) : \beta \in S\}) \equiv \prod_\beta \prod_{m=1}^{\nu_{\beta r}^{(i)}} (N_\beta(x,t) - m + 1) \quad \text{(Rule II)}. \tag{2.6b}$$

According to Rule I, reaction r takes place once, with probability P_r, if there are sufficient reactants present. In Rule II the probability of reaction is weighted by a function of the occupation numbers designed to give the standard rate law in the continuum limit.

The development so far has preserved all mesoscopic aspects of the simulation. For the purpose of deriving the continuum equations we introduce two assumptions:

(i) The occupation numbers of different species are not correlated (*molecular chaos*).

(ii) N_α does not vary appreciably over several lattice spacings (*smoothness assumption*).

The first assumption allows us to break up $< F_r >$ in (2.4) into a product of expectation values for different species. The second assumption allows us to define ρ_α equivalently as a local average of N_α over neighbouring lattice sites and to express the terms of the above product in terms of species densities. In the next section we shall discuss violations of the first assumption.

Substituting (2.5) and (2.6a) or (2.6b) in (2.4) and taking the continuum limit (λ, τ and $p_\alpha - q_\alpha \longrightarrow 0$, keeping λ^2/τ and $(p_\alpha - q_\alpha)\lambda/\tau$ finite) we arrive at the equations [7]

$$\frac{\partial \rho_\alpha(x,t)}{\partial t} = -V_\alpha \frac{\partial \rho_\alpha(x,t)}{\partial x} + D_\alpha \frac{\partial^2 \rho_\alpha(x,t)}{\partial x^2} + \sum_{r=1}^{R} \left(\nu_{\alpha r}^{(f)} - \nu_{\alpha r}^{(i)}\right) k_r \prod_\beta \sum_{n=\nu_{\beta r}^{(i)}}^{\infty} \frac{\nu_{\beta r}^{(i)}!}{n!} \rho_\beta^n(x,t) e^{-\rho_\beta(x,t)}, \tag{2.7a}$$

$$\frac{\partial \rho_\alpha(x,t)}{\partial t} = -V_\alpha \frac{\partial \rho_\alpha(x,t)}{\partial x} + D_\alpha \frac{\partial^2 \rho_\alpha(x,t)}{\partial x^2} + \sum_{r=1}^{R} \left(\nu_{\alpha r}^{(f)} - \nu_{\alpha r}^{(i)}\right) k_r \prod_\beta \rho_\beta^{\nu_{\beta r}^{(i)}}(x,t), \tag{2.7b}$$

corresponding to reaction rules I and II respectively. We have defined the advection velocity V_α, the diffusion coefficient D_α and the rate constant k_r as follows

$$V_\alpha \equiv (p_\alpha - q_\alpha)\frac{\lambda}{\tau} \ , \quad D_\alpha \equiv (p_\alpha + q_\alpha)\frac{\lambda^2}{2\tau} \ , \quad k_r \equiv \frac{P_r}{\tau} \frac{1}{\prod_\beta \nu_{\beta r}^{(i)}!} \ . \tag{2.8}$$

(2.7b) already has the form of the standard reaction-transport PDE's, whereas (2.7a) acquires approximately that form only for low particle densities ($\rho_\alpha \ll 1$). Since low particle density implies poor statistics, Rule II is more suitable for the simulation of the standard equations. The transition to species *concentrations* is finally made by rescaling densities and rate constants

$$C_\alpha(x,t) \equiv \gamma \rho_\alpha(x,t) \quad , \quad K_r \equiv \frac{k_r}{\prod_\beta \gamma^{\nu_{\beta r}^{(i)}}} \ , \tag{2.9}$$

the first of these equations defining the universal *scaling factor* γ.

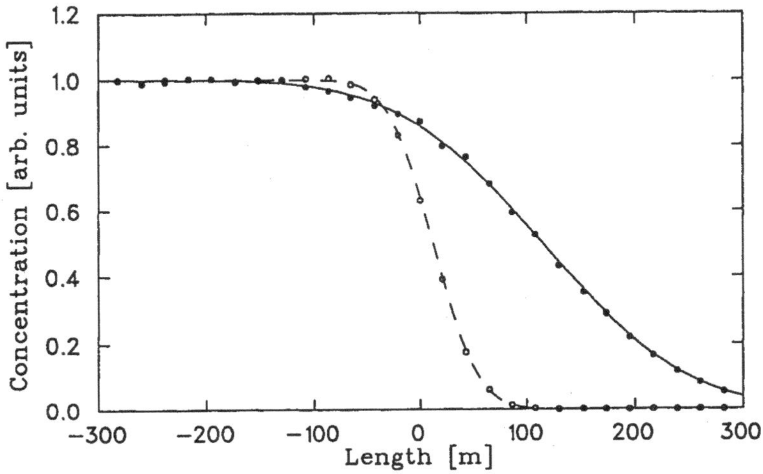

Figure 1: Result of one-dimensional simulation (species 1: solid circles, species 2: open circles) compared with the solution of the transport equation (species 1: solid curve, species 2: dashed curve) at $t = 240y$. Parameters and initial conditions are given in the text.

3 Simulation of reaction-transport processes

The model described Section 2 will now be used to simulate a variety of reaction-transport processes. The reaction-transport equations (2.7a, 2.7b) will serve as reference point: deviations from their solutions will either be related to the discretisation of space and time (making questions of convergence relevant) or will be of a fundamental nature (due to fluctuations or correlations of arbitrarily high order). The PDE's are solved by converting them to finite difference equations (FDE).

Fig. 1 shows the result of a one-dimensional simulation of solute transport without chemical reactions. We simulate simultaneously two solutes with advection velocities $V_1 = 0.5\, m/y$ and $V_2 = V_1/10$. The diffusive term models *mechanical dispersion* in a porous medium with 'diffusion' coefficients $D_1 = a_L V$ ($a_L = 50\, m$ is the *longitudinal dispersion length*) and $D_2 = D_1/10$. The second solute can be thought of as being retarded by a factor 10 due to adsorption on the surface of the solid matrix. Initially both concentrations are equal to 1 (arbitrary units) for $x \leq 0$ and 0 for $x > 0$. The simulation begins with 200 particles of each species per site to the left of $x = 0$ and the concentration is calculated by averaging over cells of 100 lattice sites. Taking $\tau = 10^{-3}\, y$ and $p_1 + q_1 = 1$, we use the second of Eqs. (2.8) to determine λ. The simulation agrees well with the result of the transport equation ((2.7a) or (2.7b) without reaction terms). The fluctuations around the continuous curve are of statistical nature and diminish with increasing number of particles (not shown here).

We introduce chemical reactions by considering first a homogeneous system. We define the particle density $\rho_\alpha(t)$ as the average number of particles per site. If molecular chaos holds, $\rho_\alpha(t)$ satisfies a chemical rate equation obtained from (2.7a) or (2.7b) by dropping all spatial dependence. The system consists of a-, b- and c-particles diffusing and reacting via $a + b \rightleftharpoons c$. We look at the approach to equilibrium by focusing on the time evolution of the *reaction quotient* $\rho_c/\rho_a\rho_b$. We perform the simulation on a two-dimensional lattice of 500 × 500 sites with periodic boundary conditions and begin with a uniform random distribution of 50 000 particles (i.e. 0.2 particles/site) of each species. At each time step every particle moves randomly to one of the four nearest sites. Particles react according to rule I: if there are at least one a- and one b-particle at a site, one a- and one b-particle are removed and a c-particle is added with probability $P_1 = 0.8$, whereas, if there are one or more c-particles, one of them is replaced by an a- and a b-particle with probability $P_2 = 0.2$.

The asymptotic value of the reaction quotient according to the rate equation is obtained by solving

$$\frac{1 - e^{-\rho_c(t)}}{\left(1 - e^{-\rho_a(t)}\right)\left(1 - e^{-\rho_b(t)}\right)} = \frac{P_1}{P_2} \tag{3.1}$$

for any of the densities, subject to the constraints $\rho_a(t) - \rho_b(t) = constant$ and $\rho_a(t) + \rho_c(t) = constant$. The reaction quotient calculated from the simulation converges asymptotically to a higher value. This is due to correlations between a- and b-particles, which are neglected in (3.1). At the rate of one diffusion step per reaction step (d.p.r.), the a- and b-particles originating from the disintegration of a c-particle remain sufficiently close to each other for the probability of their meeting again to be significantly higher than in a uniform random distribution of particles, i.e. if the assumption of molecular chaos were valid. As a result, the forward reaction $a + b \rightarrow c$ is enhanced with respect to the reverse reaction $c \rightarrow a + b$ and relatively more c-particles are produced, resulting in a higher reaction quotient. The conditions of molecular chaos can be systematically approached by performing an increasing number, n_D, of d.p.r. This corresponds to an increasing diffusion constant, but, for a homogeneous system, the rate equation remains the same. The reaction quotient in the asymptotic state is shown in Fig. 2 for $n_D = 1, 2, 3, 5$ and 10. $n_D = \infty$ corresponds to a random redistribution of molecules between reaction steps. We see that, as diffusion increases, the reaction quotient approaches the value obtained from the continuum equations. Similar conclusions concerning correlation effects have been drawn in Ref. [8]. The remaining discrepancy at $m = \infty$ is related to the time discretisation. In fact, for $n_D = 10$ we obtain very good agreement with the solution of the FDE with the same time discretisation, but not with the true solution of the rate equation. This is not surprising, as the discrete evolution equation for the particle density is identical with the FDE. Convergence to the true solution is obtained, for both the simulation and the FDE, by refining the time step.

Moving away from homogeneous systems, we consider a two-dimensional system of a- and b-particles which initially occupy the $x \geq 0$ and $x \leq 0$ halves of the lattice respectively. At later times they diffuse into each other and react via $a + b \rightarrow c$, producing around $x = 0$ a heap of c-particles; the latter diffuse and disintegrate according to $c \rightarrow a + b$. Here and in what follows, reaction rule II is used. Simulations with initially 50 000 particles on each half of the lattice (i.e. 0.4 particles/site) and otherwise the same parameters as before agree, within statistical fluctuations, with the reaction-diffusion equations. In particular, the width of the *reaction zone* (defined as the standard deviation of the product of the a- and b-particle distributions) increases asymptotically with time as $t^{1/2}$, as predicted by the continuum approach. We note that the parameter values we have chosen correspond to a diffusion-limited regime, where chemical equilibrium is attained on a much smaller time scale than

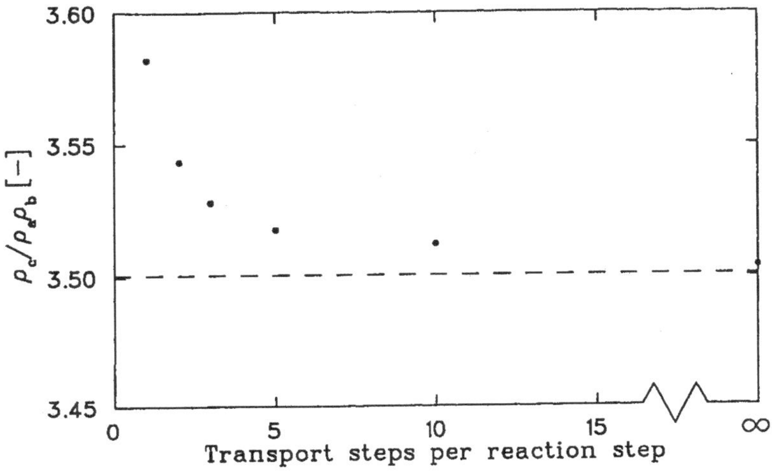

Figure 2: Reaction quotient for a two-dimensional, homogeneous system reacting via $a + b \rightleftharpoons c$ according to rule I, averaged over 700 time steps in the steady state, for different numbers of transport steps per reaction step. The dashed line shows the solution of (3.1).

diffusion. Then particles essentially spread according to the $t^{1/2}$ law expected from pure diffusion, with the relative abundance of different species determined locally by chemical equilibrium.

An interesting situation arises when we turn off the reverse reaction ($P_2 = 0$). Then the a- and b-particles diffuse and annihilate with rate constant $k \equiv P_1/\tau$, while the c-species plays no active rôle. In the case of a two-dimensional system containing initially equal numbers of a- and b-particles uniformly distributed, it is known [9] that the average density of either species behaves asymptotically in time as $t^{-1/2}$ in contrast to the t^{-1} behaviour expected from the rate equation $d\rho_\alpha/dt = -k\rho_\alpha^2$. In the simulation, the asymptotic time dependence is determined by long-range density fluctuations which diffusion has not had the time to destroy; after the bulk has been annihilated, the excess particles within such fluctuations result in regions occupied by one or the other species; then the reaction takes place only along the boundaries of these regions and is slower than under the well-mixed conditions assumed by the rate equation. Simulations using our model reproduce this result. In particular the deviation decreases with the diffusion constant and increases with the rate constant (Fig. 3).

When the a- and b-particles are initially separated, as in the case of the reversible reaction above, then a *depletion zone* develops whose width grows as $t^{1/2}$. It can be shown from the reaction-diffusion equations that the width of the reaction zone, as defined previously, varies asymptotically with time as $t^{1/6}$ [10]. The width of the spatial *distribution of annihilation events* up to time t behaves asymptotically also as $t^{1/6}$. In Fig. 4 we compare the result of the simulation (solid curves) with that of the reaction-diffusion equations: the dotted curves describe the time evolution of the widths of the reaction zone (above) and the distribution of reaction events (below) for 1 d.p.r., while the dashed and dot-dashed curves describe the same quantities for 4 and 10 d.p.s. respectively. The width of the

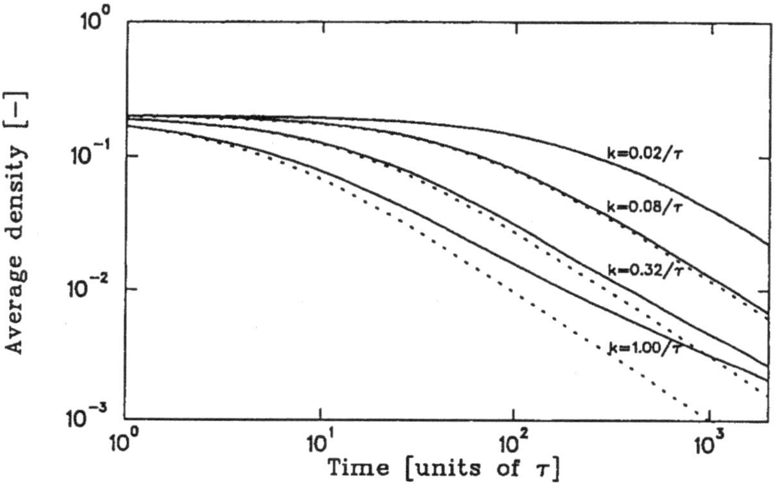

Figure 3: Average density for a two-dimensional, homogeneous system reacting via $a+b \to c$ according to the rule defined by (2.6b), as a function of time, for different values of the rate constant: simulation (solid curves) compared with rate equation (dotted curves).

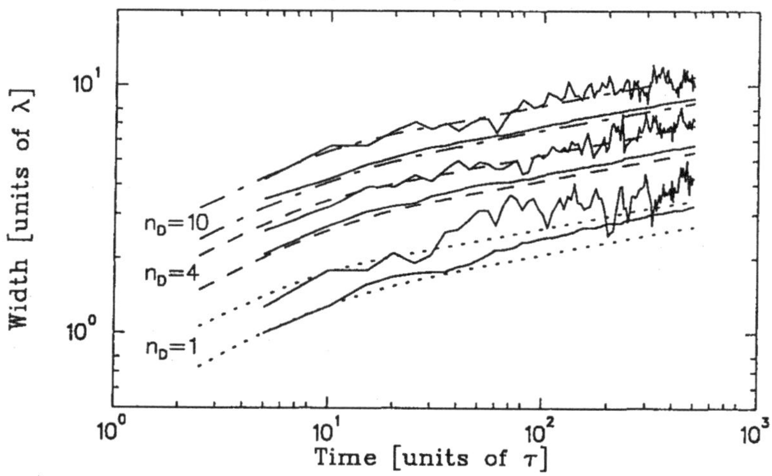

Figure 4: Width of reaction zone and of distribution of annihilation events for a two-dimensional, initially separated system reacting via $a+b \to c$ according to the rule defined by (2.6b), as a function of time: solution of reaction-transport equations (dotted, dashed and dot-dashed curves for 1, 4 and 10 d.p.r. respectively) compared with simulation (solid curves).

distribution of reaction events is clearly less 'noisy' and the comparison is in this case more meaningful. We notice that for 1 d.p.r. the result of the simulation lies above that of the reaction-diffusion equations; moreover, the former displays apparently a higher time exponent than the expected 1/6. This is consistent with Ref. [5]. It is clear that with enhanced diffusion (4 and 10 d.p.r.) the result of the simulation converges systematically to that of the continuum equations. Continuing the simulation to greater times modifies slightly the time exponents, but the following qualitative picture remains: in the absence of sufficient diffusive mixing, a and b annihilate less strongly (probably due to fluctuations of the same nature as in the preceding paragraph) and hence penetrate deeper into each other, resulting in a wider reaction zone.

We now consider reaction-diffusion systems which are significantly more complex than the ones considered so far. These systems are kept away from chemical equilibrium by an external supply of reactants. In such systems, the homogeneous state can become unstable and spatial concentration patterns can develop spontaneously, if appropriate kinetics (e.g. autocatalysis) corroborate with the presence of species with different diffusion coefficients (see e.g. [11, 12]). One such example is the Brusselator [13], defined by the reaction scheme

$$A \xrightarrow{k_1} X, \quad B+X \xrightarrow{k_2} Y+D, \quad 2X+Y \xrightarrow{k_3} 3X, \quad X \xrightarrow{k_4} E. \tag{3.2}$$

Here the concentrations of A and B are kept constant and D and E are continuously removed. Fig. 5 shows the result of a simulation on a one-dimensional lattice of length L (1 010 sites, $\lambda = 10^{-3} L$), with the boundary concentrations of X and Y fixed at the values of the (unstable) uniform steady state. The spatial structure shown is obtained after 800 000 iterations with time step $\tau = 0.833 \times 10^{-4}$ (units of k_4^{-1}). The physical parameters have the values quoted in Fig. 7.13 of Ref. [11]. The concentration is obtained by averaging the particle density over cells of 10 sites and multiplying it by a scaling factor $\gamma \simeq 0.2\,(k_4/k_3)^{1/2}$. Obviously the statistics is still poor for a quantitative comparison with the continuum result, which, incidentally, has converged to a stationary value by the time shown here.

Finally, we show the X-concentration pattern obtained with the reaction scheme

$$A \underset{k_{-1}}{\overset{k_1}{\rightleftharpoons}} X, \quad 2X+Y \xrightarrow{k_2} 3X, \quad B \xrightarrow{k_3} Y \tag{3.3}$$

(Schnackenberg model [14]), starting from a uniform random distribution. Fig. 6 is obtained on a 696×721 lattice of size 1.10×1.14 (arbitrary units) after 4 800 iterations, with $\tau = 0.625 \times 10^{-6}$ (units of k_1^{-1}), $D_Y = 30\,D_X$ and the parameters A and B as in Fig. 2a of Ref. [15]. Time limitations have prevented us from running the simulation up to a time when the pattern becomes stationary. Spatial patterns with the Schnackenberg model have recently been reported by another group [16].

4 Conclusion

The CA defined and simulated in this work is a *simple* and *general* model of reaction-transport processes. Advection and diffusion are obtained as the continuum limit of an ensemble of independent random walks. Chemical reactions are modelled by a local probabilistic rule. In the continuum limit

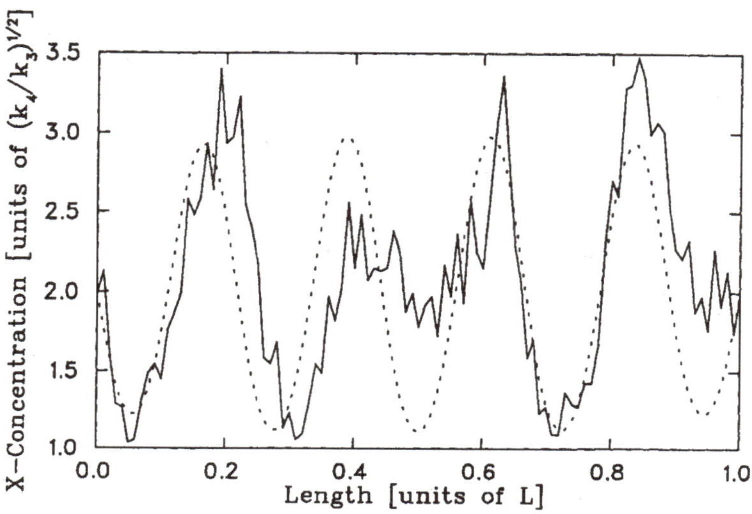

Figure 5: Simulation of one-dimensional Brusselator (solid curve) compared with solution of transport-rate equations (dotted curve).

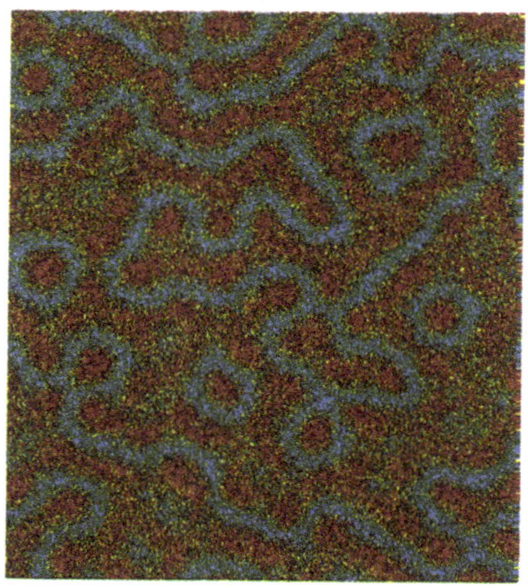

Figure 6: Spatial concentration pattern obtained from simulation of two-dimensional Schnackenberg model; regions of high (low) density are indicated in red (blue).

the model yields the reaction-transport PDE's, if molecular chaos holds and particle density is smooth in space. Since these conditions are not assumed in the simulation, the model accounts for effects due to the underlying microscopic reality and left out by the continuum approach. The full dynamics is transcribed into concise and transparent computer code. This simplicity does not preclude, but, on the contrary, facilitates the application of arbitrarily complex reactions and boundary conditions with little additional effort. Our simulations *demonstrate that the discrete model can be used: (i) to describe a variety of reaction-transport processes, (ii) to approximate systematically the respective PDE's and (iii) to display effects of explicitly microscopic origin.*

Having begun with pure transport and simple homogeneous chemical systems, we concluded by simulating pattern formation in autocatalytic systems, such as the Brusselator and the Schnackenberg model. In the case of $a + b \rightleftharpoons c$, microscopic correlations are found to influence the macroscopic quantities, but the effect disappears as lattice and time discretisation are refined; since the simulation is not performed at the true microscopic level, as explained in the Introduction, these correlation effects are not physically interesting. The asymptotic slowing down of a-b annihilation due to residual long-wavelength fluctuations is, by contrast, a true physical effect, although the time when asymptotic conditions set it depends on details of the discretisation. This is an interesting, albeit physically rare, example of an erroneous large-time prediction by the continuum equations. When a and b are initially separated, then they penetrate deeper into each other, as annihilation becomes weaker, due probably to a similar fluctuation effect (this is consistent with the observation that the effect diminishes with enhanced diffusion). Finally, the simulations of dissipative autocatalytic systems showed qualitatively the expected behaviour (spontaneous formation of spatial concentration patterns), but time limitations prevented us from performing quantitative comparisons.

Mention of 'time limitations' raises the question of computational efficiency. CA simulations of fluid motion and simple reaction-diffusion processes (e.g. [5]) on various supercomputers reach performances of several 10 000 000 site updates per second (u.p.s.). The present model, due to the indefinite number of particles/site it allows, does not vectorise well and will need to be implemented on a massively parallel computer in order to achieve its full potential. In its present form, the best scalar performance is attained on a VAX-9000 and amounts to $\sim$ 1 000 000 u.p.s. for pure transport of one species, $\sim$ 100 000 u.p.s. for $a + b \rightleftharpoons c$ and $\sim$ 20 000 u.p.s. for the Brusselator. On the basis of these figures, we estimate that the simulation of a realistic problem of solute transport in geological media would take at least several hours on a massively parallel computer.

For the purpose of describing reaction-transport processes in porous media, a model of the present general form is indispensable. It is thus clear that a significant improvement in the computational efficiency of the model is a task of first priority. Our aim is to combine optimally the advantages of the present general (but slow) and other fast (but specific) models in order to obtain a *useful* and *efficient* algorithm. This will allow us to address the exciting challenges presented by systems both at the field scale (e.g. solutes of very different concentrations, transport and reaction properties) and at the pore scale (e.g. very complex boundary conditions).

Acknowledgements

A substantial part of the work reported here was performed in collaboration with B. Blankleider. The author has benefited a great deal from discussions with B. Chopard, M. Droz, L. Frachebourg, J. Hadermann, J.-P. Boon and D. Dab. A critical reading of the manuscript by A. Jakob is gratefully acknowledged, as well as partial financial support by NAGRA.

References

[1] S. Wolfram, Rev. Mod. Phys. **55**(1983)601;
D. Farmer, T. Toffoli and S. Wolfram, eds., *Cellular Automata*, Proc. of an Interdisciplinary Workshop, Los Alamos, March 1983, Physica **D 10**(1984), Nos. 1 & 2;
S. Wolfram, ed., *Theory and Applications of Cellular Automata*, Adv. Series on Complex Systems, Vol. 1 (World Scientific, Singapore, 1986).

[2] T. Toffoli, in *Cellular Automata*, ed. D. Farmer *et al.*, Physica **D 10**(1984)117.

[3] B. M. Boghosian, Comp. in Physics **4**(1990)14.

[4] A. Lawniczak, D. Dab, R. Kapral and J.-P. Boon, Physica **D 47**(1991)132.

[5] B. Chopard and M. Droz, Europhys. Lett. **15**(1991)459.

[6] T. Toffoli and N. Margolus, *Cellular Automata Machines: A New Environment for Modeling* (MIT Press, Cambridge, 1987).

[7] T. Karapiperis and B. Blankleider, PSI–Technische Mitteilung TM-41-91-42, Würenlingen and Villigen (1991).

[8] D. Dab and J.-P. Boon, in *Cellular Automata and Modeling of Complex Physical Systems*, ed. P. Manneville *et al.* (Springer, Berlin, 1989), pp. 257–273.

[9] D. Toussaint and F. Wilczek, J. Chem. Phys. **78**(1983)2642.

[10] L. Gálfi and Z. Rácz, Phys. Rev. **A 38**(1988)3151.

[11] G. Nicolis and I. Prigogine, *Self-Organization in Nonequilibrium Systems* (Wiley, N. York, 1977).

[12] J. D. Murray, *Mathematical Biology* (Springer, N. York, 1989).

[13] I. Prigogine and R. Lefever, J. Chem. Phys. **48**(1968)1695.

[14] J. Schnackenberg, J. Theor. Biol. **81**(1979)389.

[15] V. Dufiet and J. Boissonade, J. Chem. Phys. **96**(1992) 664.

[16] M. Droz and L. Frachebourg, Helv. Phys. Acta (1993), to appear.

THE WAVE AUTOMATON

Didier SORNETTE, Patrick SEBBAH and Christian VANNESTE
Laboratoire de Physique de la Matière Condensée, CNRS URA 190
Université de Nice-Sophia Antipolis, Parc Valrose, B.P. 71, 06108 NICE Cedex 2, France

Abstract : We review the definition and properties of a new lattice model recently introduced [Vanneste et al., Europhys.Lett.**17**, 715 (1992)], which is optimized for studying dynamical propagation of waves in large (1024x1024) arbitrary heterogeneous media over long times (several 10^6 inverse band widths). Instead of starting from a wave equation or an hamiltonian which needs to be discretized for numerical implementation, the model is defined by the set of S-matrices, one for each node, describing the interaction of the wave field with the scatterers. We summarize the general method of construction of the S-matrices and discuss the physical meaning of the dynamical S-matrix approach. We give the relation between the wave automaton and the wave equation, the Klein-Gordon equation and Schrödinger equation. We also outline some results obtained on the transport of wave packets in arbitrary random media.

1. Introduction

The wave automaton model [1] is a lattice model of wave propagation in arbitrary media. It is probably the simplest model to capture the essential of time-dependent wave propagation in discrete lattices. Its construction in real space and time makes it ideally designed for studying time-dependent properties of wave propagation in large arbitrary random media at long times. Although this model is analogous, in some range of parameters, to a time dependent tight-binding model with second-nearest neighbor coupling as has been shown in [1a-b], however its formulation is quite different from the usual formulation of tight-binding models. Usually, dealing with wave propagation in heterogenous media implies the resolution of an equation (eg. Schrödinger equation), using an hamiltonian H (eg. $H\Psi_i = V_i\Psi_i + J\Sigma_{|i-j|=1}\Psi_j$ for the tight-binding model). Unfortunately, the time evolution is not given by H directly but by its exponentiated form, the evolution operator $U = e^{iH\tau} = 1 - iH\tau - H^2\tau^2/2 + ...$ Numerical calculations then usually need a scheme to approximate $e^{iH\tau}$ which may lead either to errors which end up accumulating at long times or to complex numerical operations slowing significantly the computation. Again, the specificity of our model is that it is the evolution operator U which is postulated from the start. Besides, because the S-matrix is unitary, U is also perfectly unitary. Except for the calculator precision itself, the problem of energy drifting is eliminated and the number of iterations can be drastically increased, enabling computation of the dynamics over very large times (up to several 10^6 time steps with double precision numbers). The $2L^2$ by $2L^2$ evolution operator U in this model (where L is the size of the 2D-system) consists in a matrix containing $4L^2$ non zero elements in the case where we treat simultaneously the two sublattices and $2L^2$ non zero elements in the case where we consider only one of the sublattice (see [1b] for a discussion of the existence of the two sublattices in a square lattice). U processes the scattering and the conversion of the output field into the new input field. A detailled discussion of more standard numerical methods used to integrated time-dependent equations can be found in [2-4].

The "wave automaton" is in some respect similar in spirit to the cellular automata models [5], in that macroscopic equations are modelled by arrays of variables that follow local interaction rules. Similarly to our guideline in the construction of the "wave automaton", cellular automata have been proposed as alternative to, rather than an approximation of, partial differential equations in modelling physics. In the standard cellular automata models introduced to model fluid dynamics [6], the dynamical variables are typically limited to just a few states (for intance, Boolean variables), hence the name "automaton" in the sense of Von Neumann, and the macroscopic partial differential equation is recovered after a suitable time and space averaging is taken. Since acoustic waves derive from the complete compressible Navier-Stokes equation, a similar approach can be proposed for modeling waves [7]. However, one needs again to average over many lattice sites and time steps to get rid of fluctuations and to recover the macroscopic wave equation, which limits the useful size and the time span of the dynamics. On the contrary, our "wave automaton" uses continuous complex variables $Ae^{i\phi}$, which allows to avoid the averaging procedure since the wave field components must be considered to be discrete realizations of their macroscopic counterparts. Note that we use the term "automaton" in the broader pre - Von Neumann sense of an artificial "creature" (here the complex field) imitating real life (the wave), and not in the restricted meaning of a Boolean or discrete variable.

2. Relation with Wave, Klein-Gordon and Schrödinger equations

It is interesting to note that the scattering processes describing the elementary propagation steps in the wave automaton can be interpreted as a discretized realization of the principle of action-by-proximity embedded in the Huygens'principle using the superposition of secondary wavelets at each point of the system, thus suggesting a possible deep and precise connection with hyperbolic wave equations. In the sequel, we show how to make precise this connection and discuss briefly extensions to treat more general wave equations, such as the hyperbolic wave and Klein-Gordon equation.

The basic idea of the wave automaton is to start from a lattice (square or triangular in 2D, cubic in 3D or others) and consider narrow pulses, characterized by an amplitude and a phase ($Ae^{i\phi}$) propagating along the bonds (see fig.1). The impinging pulses on each node of the lattice are instantaneously scattered in the time domain into out-going pulses which then propagate in a unit time to the neighboring nodes. Each node is carrying a z by z S-matrix (where z is the number of nearest neighbors of a given node or coordinate number), characterizing the scattering at that node. The minimal ingredients that must be kept in order to model "waves" are 1) to deal with complex fields $Ae^{i\phi}$, 2) to obey the condition of energy conservation or equivalently of wave function norm conservation and to satisfy 3) the invariance with respect to time-reversal (in absence of magnetic fields or more generally rotational fields). These conditions imply that the S-matrices must be unitary and symmetric. For the sake of clarity, if we restrict our attention to isotropic media modelled by a square lattice in 2D and cubic lattice in 3D, each scattering process is characterized by three complex parameters t (the transmission amplitude in the direction of the incident wave pulse), r (the amplitude reflected along the incident bond) and d (the amplitude scattered in each channel at $\pi/2$). Using the condition of time reversal invariance (fig.2) and conservation of energy, it can be shown that t, r and d obey the following conditions, |

$$|t|^2 + |r|^2 + (z-2)\,|d|^2 = 1 \quad , \qquad (1a)$$
$$rt^* + r^*t + (z-2)\,|d|^2 = 0 \quad , \qquad (1b)$$
$$(r+t)d^* + (r^*+t^*)d + (z-4)\,|d|^2 = 0 \quad , \qquad (1c)$$

(with z=4 in 2D and z=6 in 3D). In the previous notations [1a-b], the S-matrices are thus given in the 2D square case by

$$S = \begin{pmatrix} t & r & d & d \\ r & t & d & d \\ d & d & t & r \\ d & d & r & t \end{pmatrix}$$

with

$$t = e^{2i\alpha}\,[\cos\delta + e^{2i\rho}]/2 \qquad (2a)$$
$$r = e^{2i\alpha}\,[\cos\delta - e^{2i\rho}]/2 \qquad (2b)$$
$$d = (i/2)\,e^{2i\alpha}\,\sin\delta \qquad (2c)$$

where $0 \le \delta \le \pi$, $0 \le \rho \le \pi$ and $0 \le \alpha \le \pi$ are real parameters. A possible choice [1] consists in imposing the additional condition $\rho=0$ with the motivation to reach the perfect trapping limit ($t=d=0$ and $r=-1$) for $\delta=\pi$. The wave automaton model can then be parametrized by a phase α and a scattering strength controlled by δ and has been shown to be analogous to a time-dependent tight-binding model with second-nearest neighbor coupling. It turns out that the condition $\rho=0$ is fundamentally at the origin of this analogy and a different condition is called for in order to obtain a correspondence with hyperbolic wave equations.

This point is clear by taking $\alpha=-\pi/4$, $\delta=\pi/2$ and $\rho=\pi/4$ which yield $t=d=1/2$ and $r=-1/2$. It turns out that this model corresponds exactly to the Transmission-Line Model (TLM) developed by Johns and Beurle in 1971 [8] to implement the hyperbolic wave equation in free space. In fact, it can be shown exactly that this limit $t=d=1/2$ and $r=-1/2$ yields the finite difference version of the hyperbolic free wave equation in 2D [1e] for the total field on each node, sum of the amplitudes of all the incidents pulses. If one relaxes the constraint $t=d=1/2$ and $r=-1/2$ and just imposes the weaker condition $t=d=a\,e^{i\phi}$, it can be shown exactly [1e] that the wave automaton is then equivalent to the discretized Klein-Gordon equation

$$c^{-2}\delta^2 Y(t)/\delta t^2 \;-\; \Delta Y(t) = m^2\,Y(t) + c^{-2}[1 - e^{2i\phi}]\,Y(t-1) \quad , \qquad (3)$$

where we have noted $\delta/\delta t$ the discretized version of the time derivative $\partial/\partial t$ and Δ the discretized Laplacian operator with

$m^2 = 4\,[1/2 - t]/t$. The condition t=d leads to $\rho=-\delta/2$ and the natural parametrization

$$t = d = a\,e^{i\phi} \qquad (4)$$

and

$$r = -\,e^{i\phi}\,[a \pm i\,(1-4a^2)^{1/2}]\ , \qquad (5)$$

with $0 \le a \le 1/2$ and $0 \le \phi \le \pi$, where

$$a=(\sin\delta)/2 \qquad (6)$$

and

$$\phi=2\alpha+\pi/2\ . \qquad (7)$$

Furthermore, the choice $\phi=0$ suppresses the Y(t-1) term and we then get the finite difference version of the Klein-Gordon equation with a real mass term m^2 and velocity $c=\sqrt{a}$.

This example exhibits the generalization and improvements provided by the wave automaton in comparison to previous TLM. Indeed, taking a coefficient a<1/2 allows to tackle the general problem of the interaction of a wave with a matter field which tends to slow the wave down ($c=\sqrt{a}$). Since the borderline $|t|=a=1/2$ recovers the massless free wave equation, any difference from a=1/2 leading to a slow down of the wave can be interpreted as a wave-matter interaction, well-known to produce an index of refraction different from unity, or in other words a wave velocity smaller than the free wave velocity (in vacuum for light). Physically, the "mass" of the photon propagating in matter comes from the interference between all the scattered waves, i.e., in the language of the "second quantification", the mass stems from the cloud of virtual photons created by the wave-matter interaction. The physical meaning of the Klein-Gordon equation is further elucidated in [1e].

We also would like to point out that the general method underlying the construction of the wave automatoncan be generalized to treat more complex equations, involving several components such as the vectorial Maxwell or elasto-dynamic equations or higher order spatial derivative such as in the thin plate equation. One then needs to consider spinors instead of scalar fields which are incident on each node. Time dependent coupling with heterogeneous matter fields such as in semi-conductor lasers can also be studied with simple extensions of the wave automaton [9]. These developments will be reported elsewhere.

3. Physical content of the Wave Automaton and advantages

The conditions (1) which write the fact that the S-matrix is unitary and symmetric are of course satisfied by the choice of parameters (2). It is clear that the conditions (1) lead to strong constraints on the form of the S-matrix. To illustrate this point physically, let us consider the exemple of a laser beam separated into two beams by a semi-reflective miror (such that only t and d are non-vanishing) and then recombined latter to give two other beams (see fig.3). If we just allow for the conservation of energy for this interaction with the semi-reflective mirors, we could imagine that $t=d=1/\sqrt{2}$ is a suitable choice since the energy $|t|^2+|d|^2=1$ is conserved in this choice. However, upon the recombining process at the other end of the system, we recover a total energy which twice the value that has been put in at the beginning. Note that the recombining process acts in a way similar to the time reverse scattering process shown on fig.2. The solution to this problem is for example found by taking $t=1/\sqrt{2}$ and $d=i/\sqrt{2}$, i.e. allowing for a $\pi/2$ phase shift between the waves transmitted and reflected by the semi-reflective miror. More generally, when dealing with scattering processes in which a wave is incident on several channels, it is all important to choose the S-matrix, i.e. the scattering amplitudes t, r and d such that (1) is verified in the isotropic case.

4. Sub-diffusion and wave localization

We now present our main numerical results obtained on 2D random systems by implementing the wave automaton on a Connection Machine. Fig.4 shows a series of snapshots at different increasing times of the wave intensity field within a random system with $\delta=\pi/2$ and $\Delta\alpha=0.1\,\pi$, after launching at the center of a system (the origin) of size 512 by 512 at $\tau=0$ a narrow Gaussian wave packet of central pulsation $\omega=0.47\,\pi$ and width $\Delta\omega=0.005\,\pi$ (corresponding to a mean free path $l_e \approx 10$). Note that a relatively small number of modes are excited due to the narrowness in frequency of the wave packet. In fig.4 we clearly observe two regimes : the spreading of the wave packets is first diffusive. The ballistic-diffusive cross-over cannot be observed in the case of the wave packet because the source remains active over a time longer than the elastic mean free time. At much longer times of the order of 10^6, the waves finally become localized.

These observations are in qualitative agreement with the standard picture of wave transport in random media [10-12], according to which the wave is ballistic at scales length than l_e, diffusive at scales larger than l_e, until one reaches the localization regime at scales larger than the localization length ξ. In order to exist, the third regime of Anderson localization needs a sufficiently strong disorder [10-12], in dimension larger than two. The present understanding is that waves are always localized in two and less dimensions. In practice however, the Anderson localization will be observed only when the system size is larger than the localization length ξ, which, in our systems, is a function of δ and $\Delta\alpha$.

Fig.5 shows the wave intensity field of a pure eigenmode at frequency f=0.228714 in the case δ=0.5π and $\Delta\alpha$=0.3π, obtained by using a source placed at the origin which oscillates at the pulsation frequency f=0.228714, and whose amplitude as been increased from zero very slowly over a long timespan $\Delta\tau\approx10^5$ in order to minimize the number ($<10^{-5}L^2\approx10$) of excited modes in the frequency range $\Delta f\approx10^{-5}$ around the frequency frequency f=0.228714. Then, keeping the source functioning over a long time $\approx10^6$ at a constant amplitude allows to amplify linearly in time by resonance the mode at f=0.228714, whereas all the other modes which may have been excited have their amplitude remain constant. We check that the obtained wave field corresponds to the single mode at f=0.228714, by Fourier transforming the wave amplitude time dependence at the origin after the source has been finally cut down. One indeed observes the existence of a single peak of intensity $>10^5$ above all others. More directly, we verify that all points of the lattice oscillate with the same frequency f. Although we have not carried a specific study of the spatial properties of the eigenmodes, the spatially intermittent structure of the eigenmodes with the presence of quite many different scales is in qualitative agreement with previous studies [13] which have shown that the localized eigenmodes are fractal and maybe even multifractal up the cut-off length imposed by the finite localization length ξ.

Fig.6 shows the radius of gyration or mean square width $|R^2|=|<\Psi(r)|\,|r|^2\,|\Psi(r)>|^2$ of the spreading wave packet as a function of time up to $\tau=10^6$ launched from an initial Diract impulse at the origin, for δ=0.83π and $\Delta\alpha$=0.1π for two boundary conditions : periodic (upper curve) and open (lower curve). Note that the two curves are indistinguishable except at the largest times at which the tails of the wave packet reach the system border. Here, $|R^2|$ is not averaged over different realizations of the disorder but indeed corresponds to a single specific realization. The observed self-averaging stems from the superposition of the hundred of thousands modes which are excited in this configuration and which overlap on the frequency axis for times less than the inverse average spacing between eigenfrequencies. At longer times, beating becomes significant. For systems with larger disorders so that the number of modes in a localization volume is largely decreased, we observe also significant fluctuations of $|R^2|$ as a function of time.

Instead of the usual diffusive behavior expected at scales $l_e\leq|r|\leq\xi$, we observe a sub-diffusive dynamics in which the mean square width $|R^2|$ of an initial narrow pulse launched at $\tau=0$ scales as $|R^2|\sim\tau^{2\nu}$, with an exponent ν=0.45 significantly less than 1/2 and varying continuously as a function of the disorder. This observation of a sub-diffuse regime, with an exponent $\nu\leq1/2$, allow to refine the standard picture of localization according to which a wave packet diffuses at short times before realizing it is trapped as in an effective cavity of size given by the localization length. The observation of such a diffusive regime has previously been announced in numerical studies in [14] and predicted on the basis of weak localization calculations in [15]. In order to understand better this subdiffusive regime, we report on Fig.7 the dependence of $|R^2|^{1/2}$ as a function of time for different values of the scattering strength (same $\Delta\alpha$=0.4π and increasing δ by 0.1π units from 0.1π to 0.8π from top to bottom. All curves are characterized by three regimes, 1) ballistic at very short times for $|R^2|^{1/2}<l_e$, 2) sub-diffusive at intermediate times for $l_e\leq|R^2|^{1/2}\leq\xi$ and 3) localized at long times. For the smallest δ, the localized regime is not observed since the localization length is larger than the system size. For the largest δ, the subdiffusive regime disappears since localization occurs at the scale of the lattice mesh. The dependence of ν as a function of δ is shown in Fig.8. The exponent ν decreases from its classical diffusive value 1/2 as the scattering strength increases.

This subdiffusion stems from weak localization effects. Indeed, consider the weak localization formula for the diffusion coefficient at large scale L of a wave packet centered on ω :

$$D(L)=D_0\,[1-(2\omega\tau_e)^{-1}\mathrm{Log}(L/l_e)]\ , \tag{8}$$

where $\tau_e(\omega)$ is the characteristic decay time calculated in ref.[1b]. Taking $\tau/\tau_e\sim(L/l_e)^2$, this yields

$$D(\tau)=D_0\,[1-(\omega\tau_e)^{-1}\mathrm{Log}(\tau/\tau_e)]\ . \qquad (9)$$

Using $|R^2|=4D(\tau)\tau$, this recovers precisely the time dependent weak localization perturbation calculations [15]. Using the fact that to first order, $1-(\omega\tau_e)^{-1}\mathrm{Log}(\tau/\tau_e) \approx (\tau/\tau_e)^{-1/\omega\tau_e}$, we thus obtain

$$\nu = 1/2 - (2\omega\tau_e)^{-1} \qquad (10)$$

Eq.(10) explains qualitatively the result of fig.8 that ν decreases as τ_e or l_e decreases. However, no precise quantitative analysis is possible with a Dirac pulse since it does not fulfill the conditions for the weak localization calculations to be valid. Indeed, eq.(10) holds only for narrow wave packets of central pulsation ω and width $\Delta\omega \ll \omega$. A quantitative test of the time dependent weak localization calculations [15] is presented in ref.[1b] where we now consider the simulations of the spreading of wave packets using a set of parameters such that all the conditions required for the weak localization expression (10) to hold are satisfied.

5. Conclusions

We have described in details a new lattice model, the "wave automaton" for the dynamical propagation of waves in arbitrary heterogeneous media, which is efficient for calculations on large systems (1024x1024) over long times (several 10^6 inverse band widths). We have detailed the construction of the wave automaton, and in particular the parametrization of the S-matrix on each node of the lattice. We have used the wave automaton to analyze quantitatively the time evolution of the spreading of a wave packet in a 2D disordered system.

This work was partially supported by DRET under contract 89/271. The computations have been carried out on the Connection Machine of INRIA in Sophia Antipolis and of Institut de Physique du Globe in Paris.

REFERENCES

[1] a) Vanneste C., Sebbah P. and Sornette D., Europhys. Lett. **17** (8), 715-720 (1992); **18**, 567 (1992) (Erratum)
b) Sebbah P., Sornette D. and Vanneste C., "A wave automaton for wave propagation in the time domain : I-Periodic systems", J.Phys.I France (in press)
c) Sebbah P., Sornette D. and Vanneste C., "A wave automaton for wave propagation in the time domain : II-Random systems", J.Phys.I France (in press)
d) Sebbah P., Sornette D. and Vanneste C., "Anomalous diffusion in two-dimensional Anderson localization dynamics", submitted
e) D. Sornette, O. Legrand, F. Mortessagne, P. Sebbah and C. Vanneste, "The Wave Automaton for the time-dependent hyperbolic wave and Klein-Gordon equations in two and three dimensions", submitted

[2] Koonin S.E., Computational Physics (Benjamin/Cummins, Menlo Park, Cal., 1986)

[3] Smith G.D., Numerical solution of Partial Differential Equations : Finite Difference Methods (Clarendon Press, Oxford, 1985)

[4] De Readt H., Europhys.Lett. **3**, 139 (1987); Comp.Phys.Rep. **7**, 1 (1987)
Huyghebaert J. and De Raedt H., J.Phys. A: Math. Gen. **23**, 5777 (1990)

[5] Wolfram S., Theory and Applications of Cellular Automata (World Scientific, Singapore, 1986)

[6] Frisch U., Hasslacher B. and Pomeau Y., Phys.Rev.Lett. **56**, 1505 (1986)
Margolus N., Toffoli T. and Vichniac G., Phys.Rev.Lett. **56**, 1694 (1986)
Toffoli T., Physica **10D**, 117 (1984)

[7] Rothman D.H., Geophys.Res.Lett. **14**, 17 (1987)
Numrich S.K., Krutar R.A. and Squier R.K., J.Acoust.Soc.Am. **89**, 1894 (A) (1991); Numrich S.K., Krutar R.A. and Squier R.K., J.Acoust.Soc.Am. **90**, 2357 (1992)

[8] Johns P.B. and Beurle R.L., IEE Proc. **11B**, pp.1203-1208 (1971)
Johns P.B., The Radio and Electronic Engineer, **44**, N°12, 657-666 (1974)

[9] Lowery A.J., A study of the static and multigigabit dynamic effects of gain spectra carrier dependence in semiconductor lasers using a transmission-line laser model, IEEE

Journal of Quantum Electronics, **24**, N°12, 2376-2385 (1988)
[10] Lee P.A. and T.V.Ramakrishnan, Rev.Mod.phys. **57**, 287 (1985)
[11] Souillard B., Waves and electrons in inhomogeneous media, in Les Houches, Session XLVI, 1986, Chance and Matter, eds. J. Souletie, J. Vannimenus and R. Stora, (North Holland, Amsterdam, 1987)
[12] Sornette D., Acustica **67**, 199 (1989); **67**, 251 (1989); **68**, 15 (1989)
[13] a) Pichard J.L. and Sarma G., Powerlaw decay and fractal character of eigenstates in two-dimensional disordered systems, J.Phys.**C18**, 3457 (1985)
b) Schreiber M., Numerical characterization of electronic states in disordered systems, Inst.Phys.Conf.Ser. **N°108**, Paper presented at the Localization 1990 Conference held at Imperial College, London, 13-15 August 1990, p.65-76 (1991 IOP Publishing Ltd);
Fractal eigenstates in disordered systems, Physica **A167**, 188 (1990)
c) Aoki H., Critical behaviour of extended states in disordered systems, J.Phys.**C16**, L205 (1983)
d) Castellani C. and Peliti L., J.Phys. **A19**, L492 (1986); **19**, L1099 (1986)
e) Aoki H., Fractal dimensionality of wave functions at the mobility edge : Quantum fractal in the Landau levels, Phys.Rev.**B33**, 7310 (1986)
f) Pook W. and Janβen M., Multifractality and scaling in disordered mesoscopic systems, Z.Phys.**B82**, 295 (1991)
g) Schreiber M. and Grussbach H., Phys.Rev.Lett.**67**, 607 (1991)
h) de Vries P., De Raedt H. and Lagendijk A., Localization of waves in Fractals : Spatial behavior, Phys.Rev.Lett.**62**, 2515 (1989)
i) Evangelou S.N., Scaling exponents at the mobility edge, Physica **A167**, 199 (1990)
j) Chalker J.T., Scaling and eigenfunction correlations near the mobility edge, Physica **A167**, 253 (1990)
[14] Scher G.M., J. Non-Cryst.Solids **69**, 33 (1983)
Kaveh M., Phil. Mag. **B52**, 521 (1985)
Haydock R., Phil.Mag.**B53**, 545-559 (1986)
[15] Nakhmedov E.P. and Prigodin, V.N. and Firtov Yu.A., Sov.Phys.JEPT **65**, 1202 (1987)

Figure captions :

Fig.1 : Scattering process at a node of the square lattice.

Fig.2 : Definition of the scattering amplitude in the isotropic case and illustration of the time reversal invariance condition : a) direct scattering event; b) time reverse scattering event.

Fig.3 : Exemple of a laser beam separated into two beams by a semi-reflective miror (such that only t and d are non-vanishing) and then recombined latter to give two other beams. If $t=d=1/\sqrt{2}$, the total out-going energy $|2td|^2 + |t^2+d^2|^2=2$ is twice the injected energy! This paradox is solved by the choice $t=1/\sqrt{2}$ and $d=i/\sqrt{2}$, meaning that the phase difference between t and d is not arbitrary.

Fig.4 : Series of snapshots at different increasing times of the wave intensity field within a random system with $\delta=\pi/2$ and $\Delta\alpha=0.1\ \pi$, after launching at the center of a system (the origin) of size 512 by 512 at $\tau=0$ a narrow Gaussian wave packet of central pulsation $\omega=0.47\ \pi$ and width $\Delta\omega=0.005\ \pi$. The figures a) to d) are closeups which show only a small central part of the system. The last figure d) shows the entire lattice. a) $\tau=500$; b) $\tau=1000$; c) $\tau=10000$; g) $\tau=10000$ showing the entire lattice.

Fig.5 : Wave intensity field of a pure eigenmode of frequency f=0.228714 in a random system with $\delta=0.5\pi$ and $\Delta\alpha=0.3\pi$.

Fig.6 : Radius of gyration $|R^2|=|<\Psi(r)|\ |r|^2\ |\Psi(r)>|^2$ as a function of time up to $\tau=10^6$ (the inverse bandwidth corresponds to $\tau=2$) for $\delta=0.83\pi$ and $\Delta\alpha=0.1\pi$ for two boundary conditions : periodic (upper curve) and open (lower curve).

Fig.7 : Dependence of $|R^2|^{1/2}$ as a function of time for different values of the scattering strength (same $\Delta\alpha=0.4\pi$ and increasing δ by 0.1π units from 0.1π to 0.8π from top to bottom).

Fig.8 : Dependence of ν as a function of δ ($\Delta\alpha$ being kept fixed) in the case where a Dirac impulse is launched at the origin (case represented on fig.7).

Fig. 1

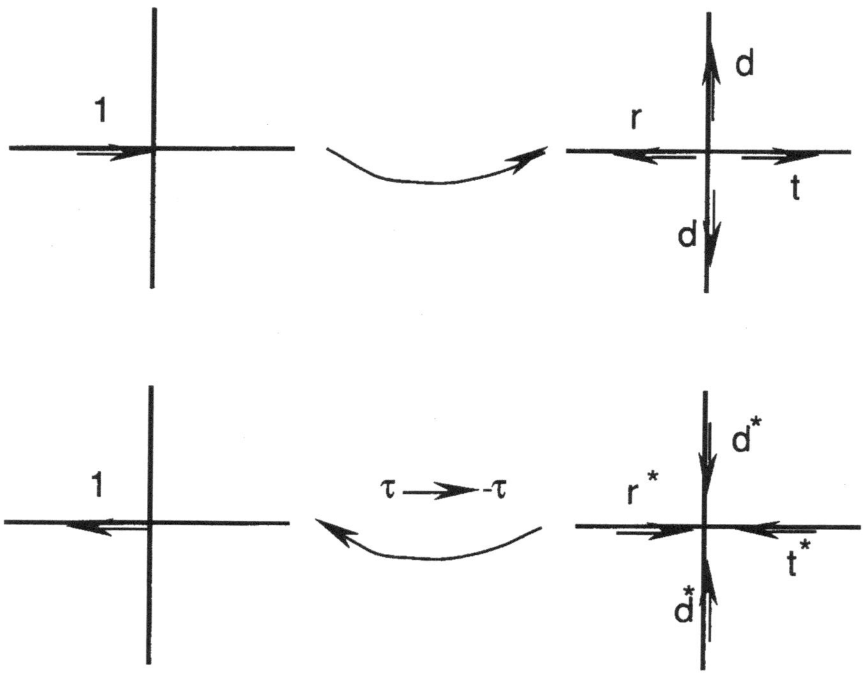

Fig. 2

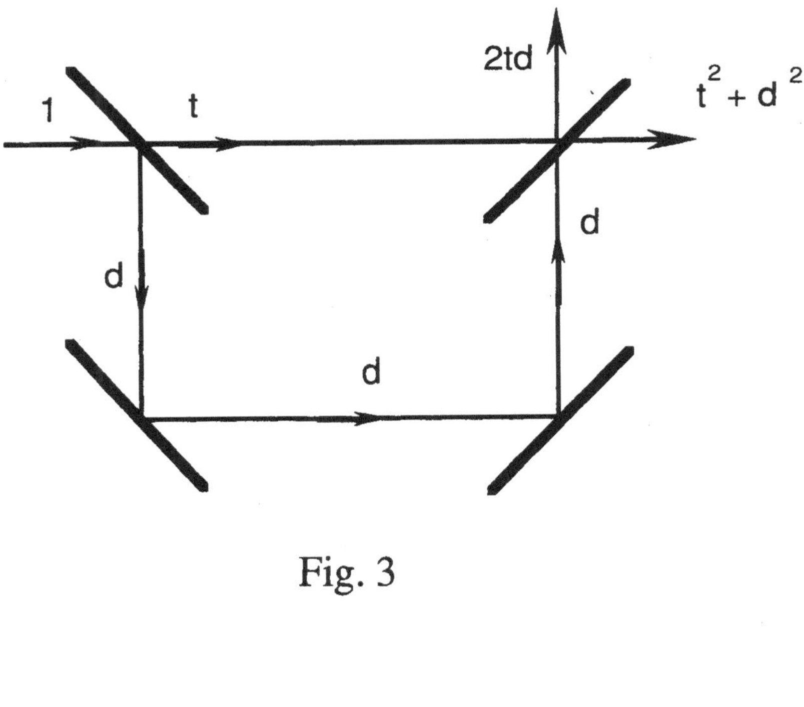

Fig. 3

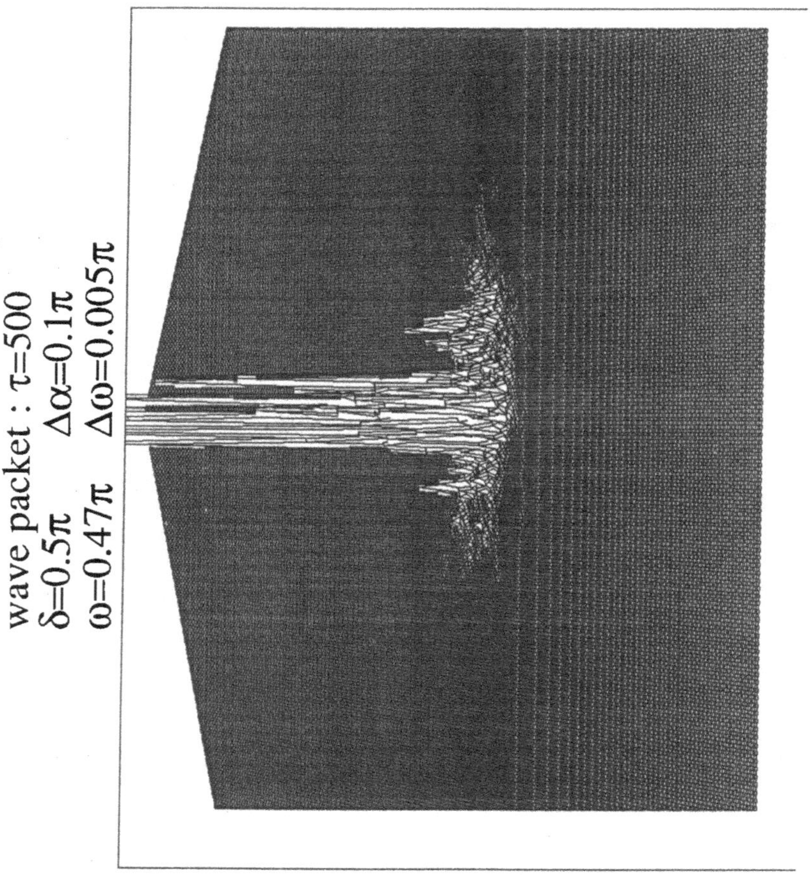

Fig. 4a

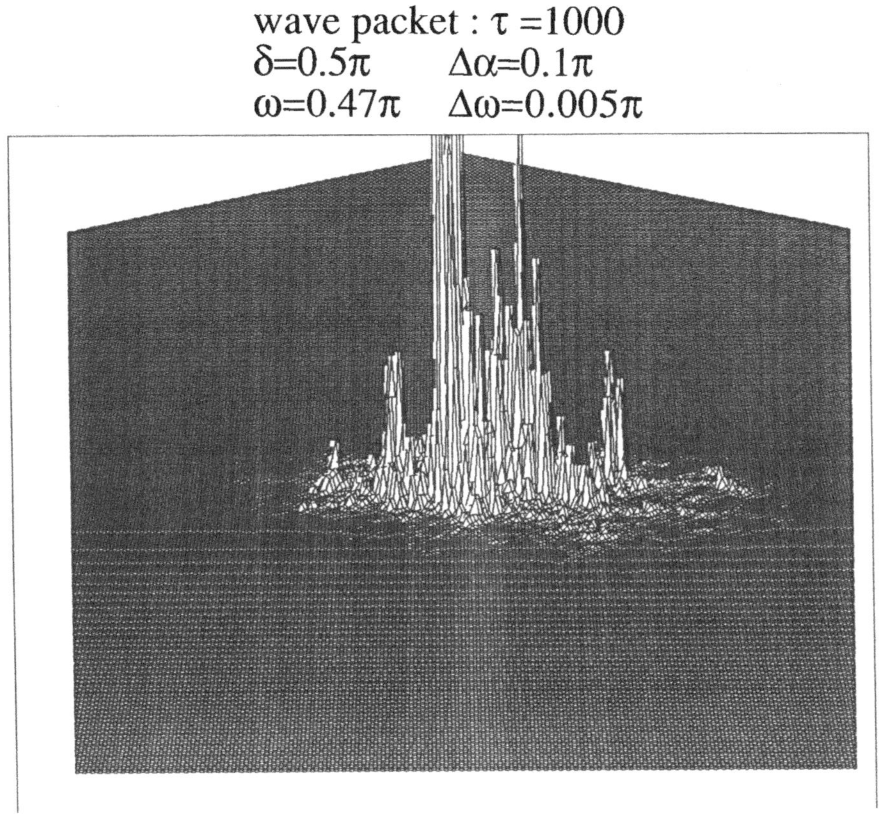

Fig. 4b

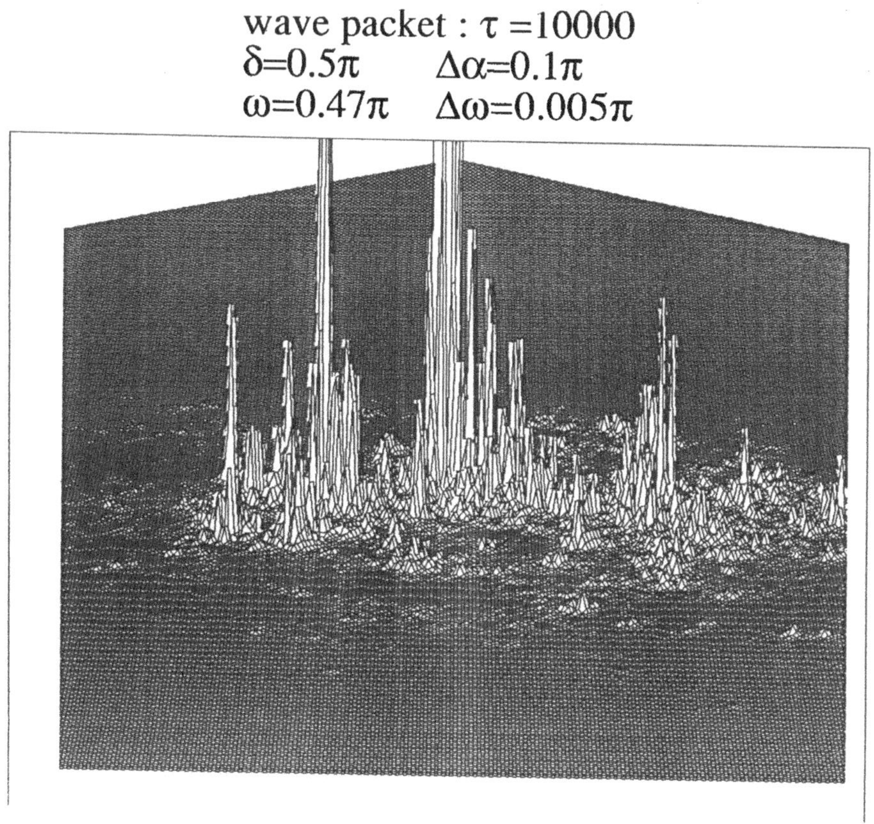
wave packet : τ =10000
δ=0.5π $\Delta\alpha$=0.1π
ω=0.47π $\Delta\omega$=0.005π

Fig. 4c

wave packet : $\tau = 10000$

$\delta = 0.5\pi$ $\Delta\alpha = 0.1\pi$

$\omega = 0.47\pi$ $\Delta\omega = 0.005\pi$

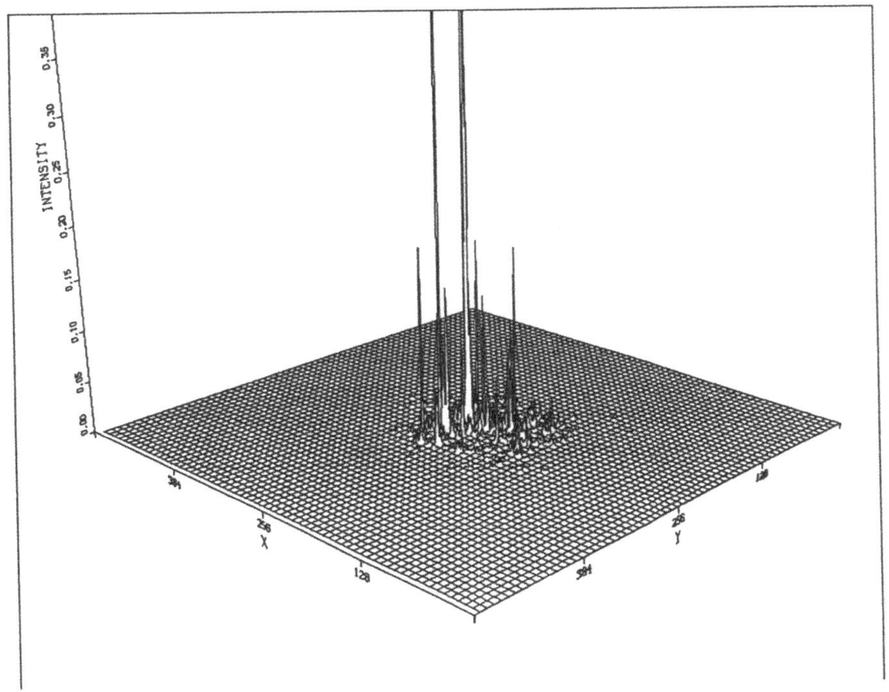

Fig. 4d

Proper mode at f=0.228714
$\delta=0.5\pi$ $\Delta\alpha=0.3\pi$

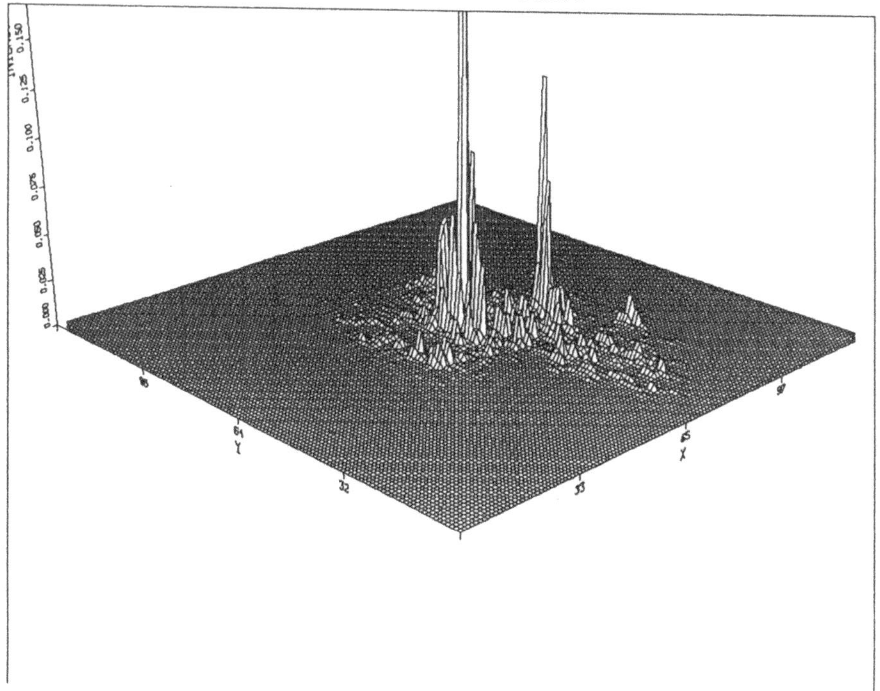

Fig. 5

$\delta=0.83\pi \qquad \Delta\alpha=0.1\pi$

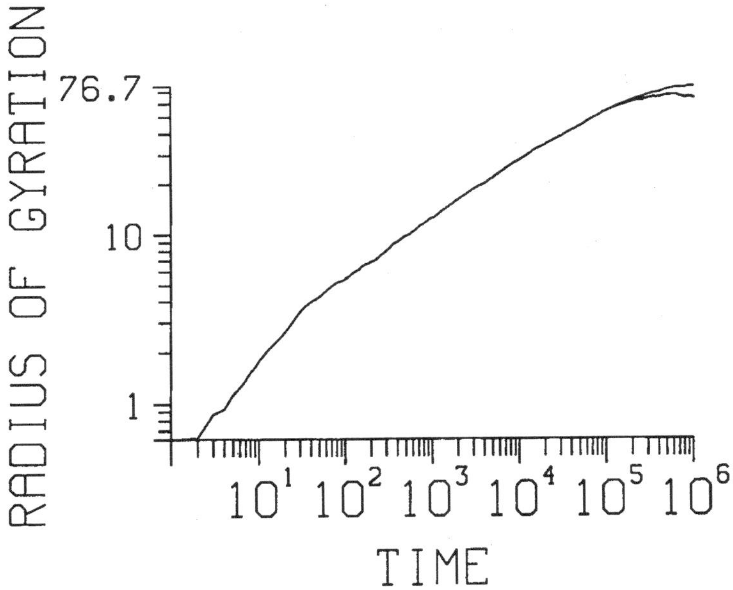

Fig. 6

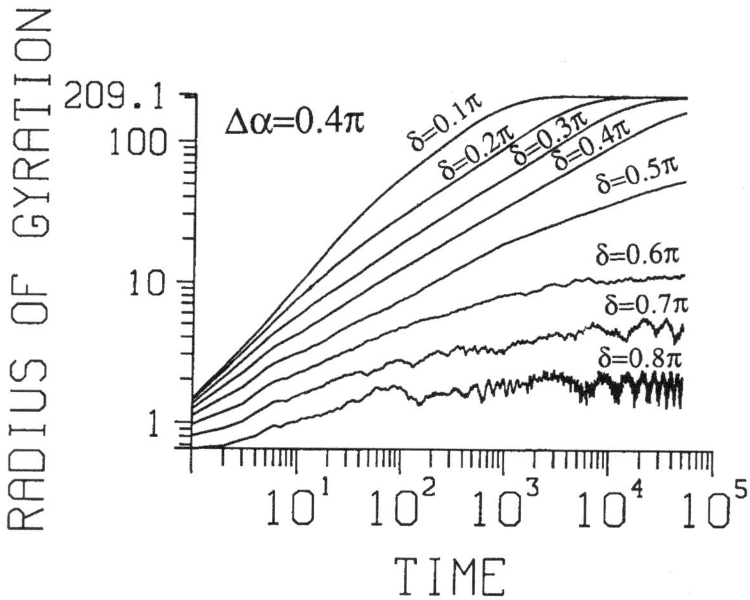
RADIUS OF GYRATION
209.1
100
10
1
Δα=0.4π
δ=0.1π
δ=0.2π
δ=0.3π
δ=0.4π
δ=0.5π
δ=0.6π
δ=0.7π
δ=0.8π
10^1 10^2 10^3 10^4 10^5
TIME

Fig. 7

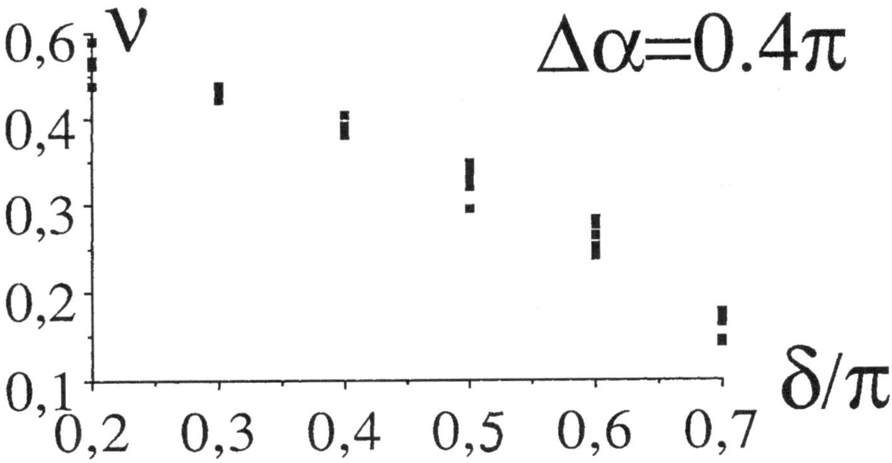

Fig. 8

GENERIC MULTIFRACTAL PHASE TRANSITIONS AND SELF-ORGANIZED CRITICALITY

D. SCHERTZER, S. LOVEJOY†, D. LAVALLÉE*
LMD, BP 99, Université P.& M. Curie, 4 Pl. Jussieu Paris 75252 Cedex 05, France

†Department of Physics, McGill University, 3600 University st., Montréal, Québec, Canada, H3A 2T8

**E. S. R. G., University of California, 5276 Hollister ave. Suite 260, Santa Barbara, Ca. 93111*

ABSTRACT

We discuss statistical mechanisms leading to phase transitions in high dimensional multifractal processes. Second order transitions naturally arise from finite sample sizes, whereas first order transitions are consequences of the scale and dimension of the observations. These transitions are elaborated and their genericity are underscore. We point out the close first order, low temperature phase transitions relation to self-organized criticality and its implications for nonlinear physical systems, in particular in astrophysics.

1. Introduction

Multifractal variability crops up in fields ranging from strange attractors[1,2], turbulence[3-5], statistical physics[6,7], high energy physics[8], astrophysics[9] and geophysics[10]. The basic scaling behavior of these dissipative nonequilibrium systems is determined by exponent functions which have analogues in thermodynamics (for reviews see refs.[11], we will rather follow[12,13]). Two distinct statistical mechanisms lead to phase transitions (discontinuities of the free energy and thermodynamic potential analogues). "Frozen free energy" (second order) transitions[7,8] arise from finite sample sizes[14], whereas much more wild first order transitions are consequences of the scale and dimension of the observations on large samples[3,15]. These are totally different from the high temperature transitions found in (low dimensional) deterministic chaos[16,17] which are basically created by breaks in the scaling symmetry of the probability measure in phase space. In our case these phase transitions are generically produced by the scaling and have direct implications for extreme, catastrophic, events in physical space. They can explain recent results in turbulence[18] as well as the appearance of self-organized critical phenomena[19]. Finally, we give empirical evidence from astrophysical catalogues that the integrated luminosity function does indeed have a first order multifractal phase transition.

2. Multiplicative processes and multifractal codimension formalism

To study these transitions, we rely on multiplicative processes[3,20,21,22] produced by scaling random multiplicative modulations of larger structures by smaller ones which yield highly intermittent space/time fields (see Fig.1a). Each realization corresponds to a finite D-dimensional cut of a process in an infinite dimensional probability space.

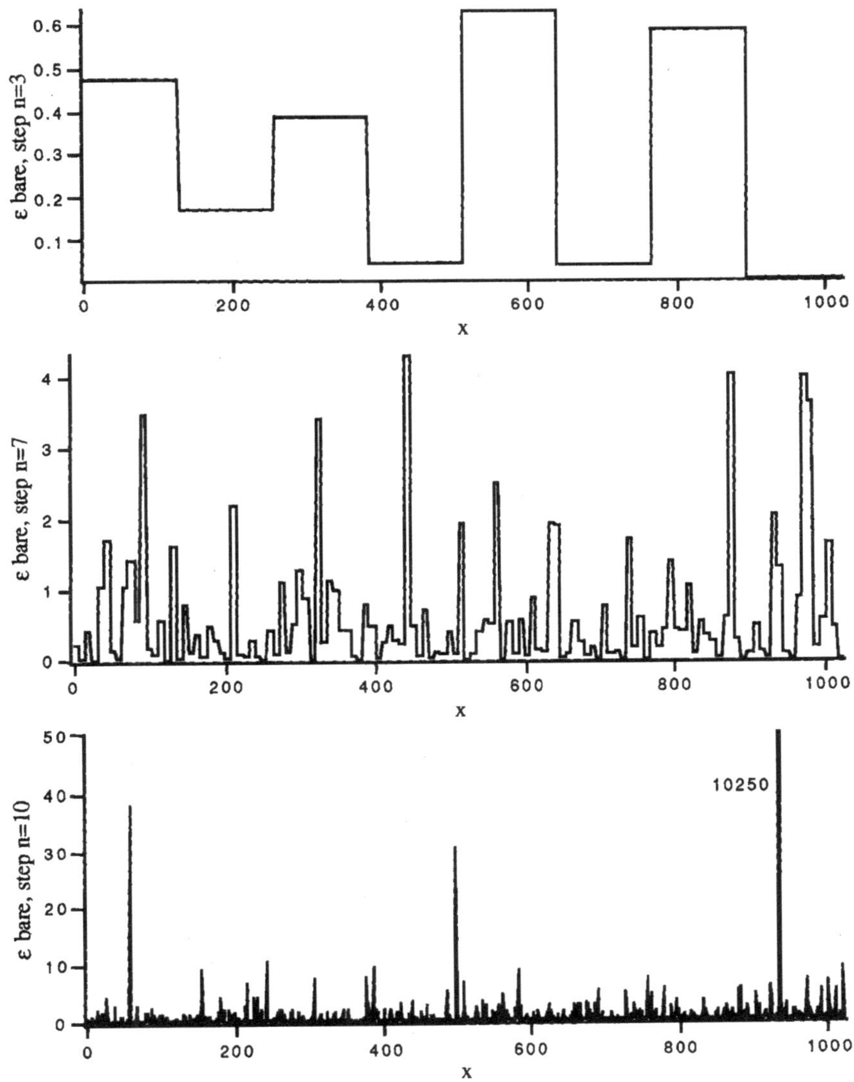

Fig 1a: Construction of a "bare" multifractal field by a multiplicative cascade process ("lognormal" model with C_1=0.9, see discussion below) starting with an initial uniform unit density. At each step the homogeneity scale l is divided by 2 and we display from top to bottom $l = 2^{-3}, 2^{-7}, 2^{-10}$.

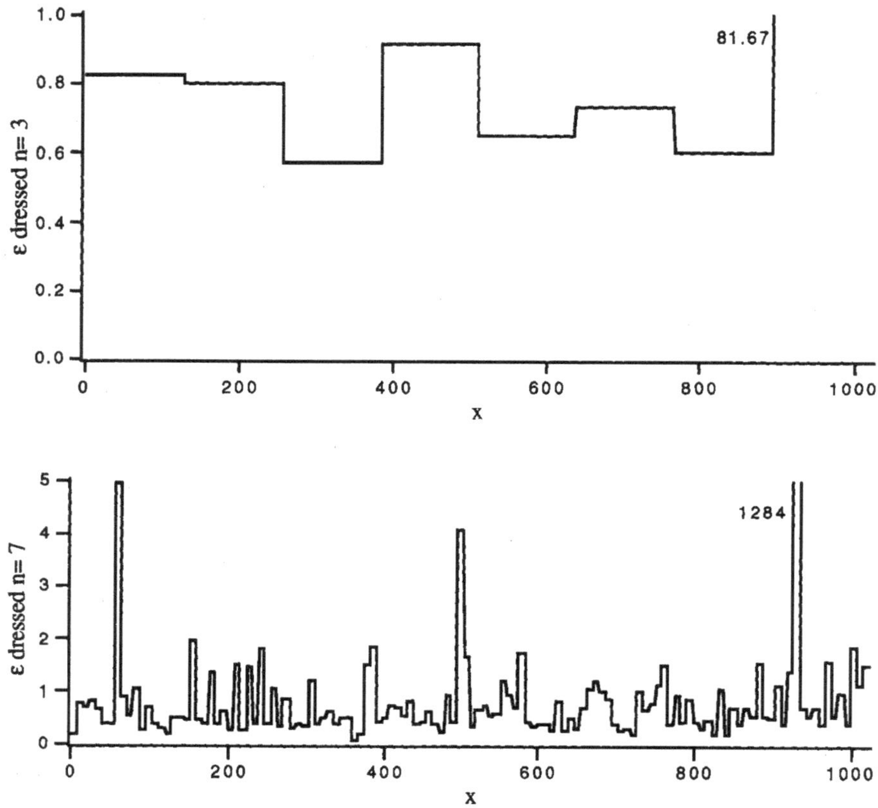

Fig. 1b The "dressed" density, obtained by averaging the "bare" density (Fig. 1a) over the finest resolution scale ($l = 2^{-10}$) up to the scales $l = 2^{-3}\ 2^{-7}$. The sharp contrast between the 1a and 1b arises from smaller scale singularities. As discussed below, among 2^9 realizations, we chose the one displaying the "hardest" singularity.

The multiple scaling behavior of this field ε_λ at scale ratio λ (=L/l the ratio the largest scale L to the scale l), can be either characterized by its probability distribution or by the statistical moments which are obtained via a Laplace transform (here and below the sign ≈ means equality within slowly varying or constant factors):

$$\Pr(\varepsilon_\lambda \geq \lambda^\gamma) \approx \lambda^{-c(\gamma)}; \qquad \langle \varepsilon_\lambda{}^q \rangle \approx \lambda^{K(q)} \approx \int \lambda^{q\gamma} \lambda^{-c(\gamma)} dc(\gamma) \tag{1}$$

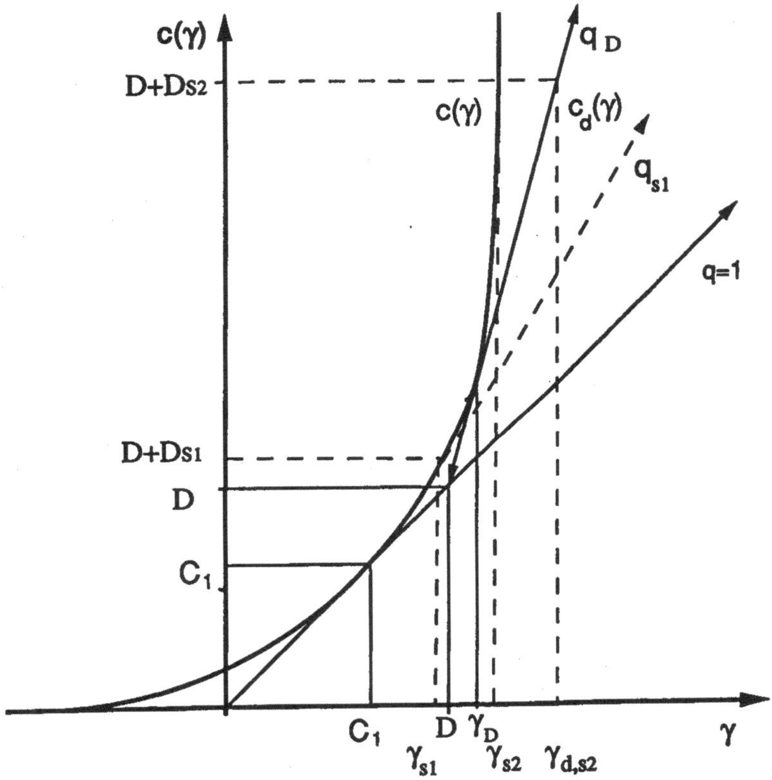

Figs 2a: Schematic diagram of $c(\gamma)$, $c_d(\gamma)$ indicating two sampling dimensions D_{s1}, D_{s2} and their corresponding $\gamma_{s1}<\gamma_D<\gamma_{s2}<\gamma_{d,s2}$; the critical tangent (slope q_D) contains the point (D,D).

the exponent $c(\gamma)$ is[21,22] a statistical codimension since -as discussed below- the probability measures the fraction of the probability space occupied by the singularities exceeding the order γ (Fig.2a). The probability description (γ, $c(\gamma)$) is the multifractal analogue of the (energy, entropy) description of standard thermodynamics, whereas the moment description (q, K(q)) is the analogue of the (inverse temperature, Massieu potential) description. Since a free energy analogue is K(q)/(q-1), discontinuities (phase transition analogues) will be apparent in either the free energy or Massieu potential description. (Entropy, Massieu potential) and (c ,K) are Legendre transform[4] pairs:

$$K(q) = \max_{\gamma} (q\gamma - c(\gamma)) \qquad\qquad c(\gamma) = \max_{q} (q\gamma - K(q)) \qquad (2)$$

these relations establish the one to one correspondence $q = c'(\gamma)$, $\gamma = K'(q)$ (Figs.2a,b).

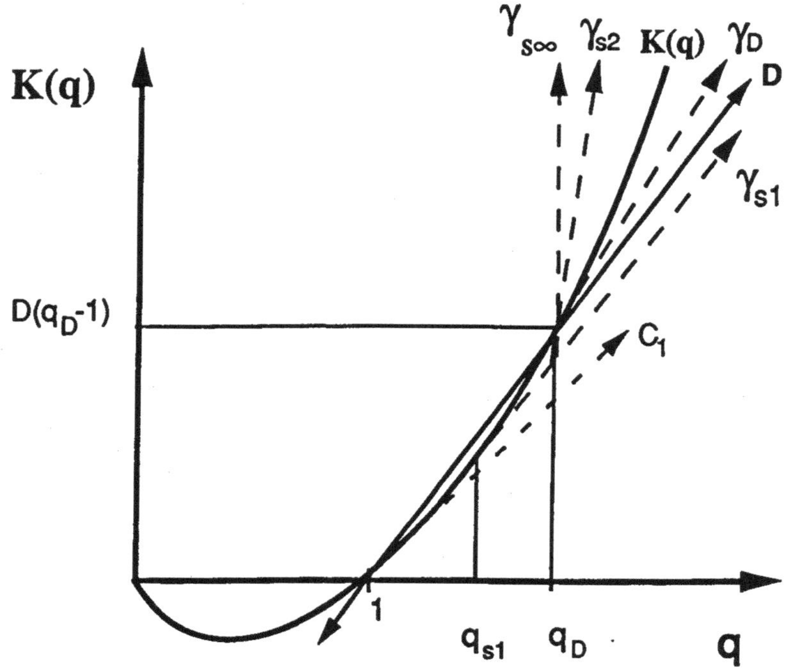

Fig. 2b: Schematic diagram of K(q), with straight lines of slopes $\gamma_{s1}<\gamma_D<\gamma_{d,s2}<\gamma_{d,s\infty}$ ($=\infty$) indicating the behavior for increasing sample size N_s ($N_{s\infty}=\infty$). The line of slope D defining q_D is also shown.

At scale ratio λ, the probability can be estimated as the ratio of the number ($N_\lambda(\gamma)$) of structures with singularities $\geq\gamma$ to the total number of structures (N_λ): $Pr(\varepsilon_\lambda \geq \lambda^\gamma) \approx N_\lambda(\gamma)/N_\lambda$. Whenever $D \geq c(\gamma)$, $c(\gamma)$ also has a geometrical interpretation over a D-dimensional observing set A. In this case, not only $N_\lambda \approx \lambda^D$, but also, on almost any single realization, $N_\lambda(\gamma) \approx \lambda^{D(\gamma)}$, with a positive $D(\gamma)$ which is then a geometric fractal dimension. This restrictive geometric interpretation corresponds to the starting point of ref.[4]. However, the "hard" singularities (see below) which are the most interesting have $c(\gamma)>D$ and are only present in "canonical" multifractals[15,21], which invariants (e.g. turbulent energy flux) are conserved in the "canonical" sense, i.e. on the ensemble average. One may note that recently the need of a codimension formalism has been implicitly acknowledged by Mandelbrot[23].

3. Second order multifractal phase transitions

Consider a sample consisting of N_s independent realizations, each of dimension D, each covering a range of scales λ. Increasing N_s we gradually explore the entire probability space; encountering extreme but rare events that would be almost surely missed on any finite sample (Fig 2a). The sampling dimension[24] D_s quantifies the extent to which the probability space is explored by estimating the highest order singularity (γ_s) we are likely to observe on N_s independent realizations:

$$c(\gamma_s)=D+D_s=\Delta_s \,; \qquad D_s \approx \log N_s/\log \lambda \tag{3}$$

follows from the Eq. 1 and the fact that there are a total of $N \cdot N_s=\lambda^{D+D_s}$ structures in the sample. Δ_s is the corresponding (overall) effective dimension. More extreme singularities would have codimensions greater than this effective dimension ($c \geq \Delta_s$) and are almost surely not present in our sample.

The upper bound γ_s[24] for observable singularities leads to a second order phase transition. Indeed, the Legendre transform of $c(\gamma)$ with $\gamma \leq \gamma_s$ leads to a spurious linear estimate K_s instead of the nonlinear K for $q>q_s$; $q_s=c'(\gamma_s)$ being the maximum moment that can accurately be estimated :

$$K_s(q) = \gamma_s(q-q_s) + K(q_s)\,, \quad q \geq q_s; \qquad K_s(q) = K(q) \quad q \leq q_s \tag{4}$$

hence there is a second order transition associated with a jump in the second derivative of the free energy/Massieu potential:

$$\Delta K''_s = -K''(q_s) \tag{5}$$

and the corresponding free energy $C_s(q)$ rapidly becomes frozen at the value γ_s.

4. First order multifractal phase transitions

We now discuss the more violent first order transitions which may occur for a multifractal process typically observed by spatial and/or temporal averaging on scales $l \gg \eta$ (the inner size of the process) , i.e. with corresponding ratios $\lambda=L/l$, $\Lambda=L/\eta$ with $\Lambda \gg \lambda$. The variability at the observation scale ratio λ may be wilder (Fig. 1b) than the corresponding field obtained by stopping the cascade process at the same scale ratio. Borrowing some renormalization jargon, we may say that the observation is "dressed"[22] by the small scale activity, whereas the process stripped of its small scale activity is called "bare". All the interactions and the resulting fluctuations of the field on scale ratios between λ and Λ are “hidden” from direct observation; nonetheless they are entirely responsible for the much more violent dressed variability.

We have thus to consider the statistical behavior of integrals of multifractals, in particular, the D-dimensional integral over A:

$$\Pi_\lambda(A) = \int_A \varepsilon_\lambda d^D\underline{x}, \tag{6}$$

corresponding to the energy flux (turbulence) or the multifractal probability (strange attractors). Π_λ has multiple scaling behavior corresponding in the usual strange attractor notation to the multifractal measure exponents α, f, τ, but due to the D dimensional

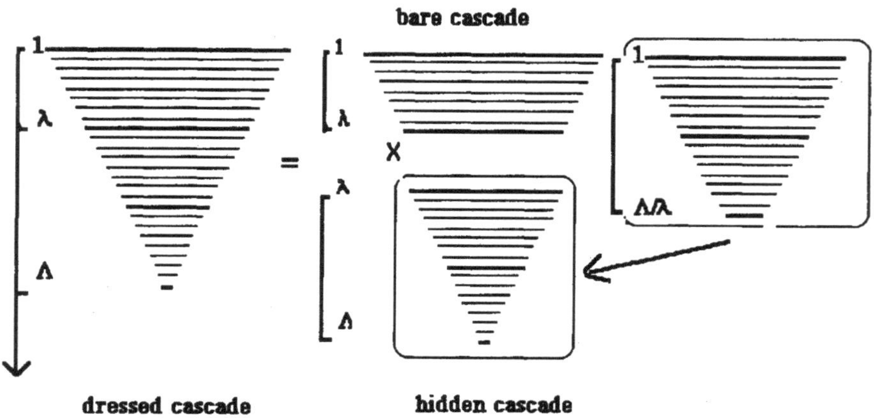

Fig.3: A schematic diagram showing a cascade constructed down to scale ratio Λ, dressed (averaged) up to ratio λ. This is equivalent to a bare cascade constructed over ratio λ, multiplied by a hidden factor obtained by reducing by factor λ a cascade constructed from 1 to Λ/λ.

integration there are extrinsic D dependences[15] (rendered explicit by the subscript "D"):

$$\Pr(\Pi_\lambda(B_\lambda)\geq \lambda^{-\alpha_D}) \approx \lambda^{f_D(\alpha_D)}/\lambda^D;\quad \alpha_D=D-\gamma\,;\, f_D(\alpha_D)= D-c(\gamma) \tag{7}$$

on a ball B_λ of size L/λ. The "trace moments"[22] generalize the partition function used in strange attractors[2] by combining ensemble with spatial averaging ("superaveraging"):

$$\mathrm{Tr}_{A_\lambda}(\varepsilon_\lambda{}^q)\approx <\sum_i \Pi_\lambda(B_{\lambda,i})^q>\approx\lambda^{-\tau_D(q)};\quad \tau_D(q)= (q-1)D -K(q) \tag{8}$$

summing over A at resolution λ, i.e. on a covering of $N_\lambda(A)\approx\lambda^D$ disjoint balls $B_{\lambda,i}$ ($\Pi_\lambda(B_{\lambda,i})\approx\varepsilon_\lambda\lambda^{-D}$). While the intrinsic quantities γ, c(γ), K(q) are independent of the dimension D of the observing space A, unfortunately α_D, $f_D(\alpha_D)$, $\tau_D(q)$ diverge as we increasingly explore the infinite dimensional probability space ($D\to\infty$; contrary to the finite phase space dimension D for strange attractors).

Because of its multiplicative construction, the dressed field (ε_d) factors (see Fig. 3) into a bare ε (large scale) and a hidden ε_h (small scale) component ($\varepsilon_d=\varepsilon\varepsilon_h$). The hidden small scale component may contribute as a highly variable prefactor having occasional avalanche-like effects on the large scale as soon as a singularity of order greater than D occurs (see Fig. 1a-b for an example). This is because the D-dimensional integration cannot smooth it down to the scale of observation; the observation scale (*l*) is no longer effective and the scale of homogeneity η prevails). Since $\Lambda/\lambda>>1$, the dressed singularity (γ_d) computed using the observation scale will be much larger. These events will remain statistically negligible until a critical singularity order γ_D, for $\gamma<\gamma_D$ the dressed codimension (c_d) coincides with the bare codimension (c), but for $\gamma>\gamma_D$ c_d will be determined by maximizing the probability, i.e. minimizing c, *with the only constraint* being the convexity, c_d thus follows the tangent (Fig. 2a):

$$c_d(\gamma_d)= q_D(\gamma_d-\gamma_D) +c(\gamma_D) \quad \gamma_d\geq\gamma_D; \quad c_d(\gamma_d)= c(\gamma) \quad \gamma_d\leq\gamma_D \tag{9}$$

$q_D=c'(\gamma_D)$ is the slope of the algebraic fall-off of the dressed probability distribution, and is the critical order of divergence of statistical moments ($q\geq q_D$, $<\varepsilon_h{}^q>=\infty \Rightarrow <\varepsilon_d{}^q>=\infty$) since a Legendre transform of a linear function diverges ($K_d(q)=\infty$, $q\geq q_D$). It was shown[3,22] that q_D is the solution of:

$$D=q_DD-K(q_D)=q_D(D-\gamma_D)+c(\gamma_D) \tag{10}$$

i.e. the critical tangent (Eq. 9) contains the point (D,D) (Fig. 2a). The latter property can be directly derived by showing that the (statistical) singularity of the trace moment[3,22] density is $q(\gamma-D) -c(\gamma)$ (the opposite of the tangency), and it should reach the critical value D for $\gamma=\gamma_D$, $q=q_D$, hence a simple and direct determination of $c_d(\gamma_d)$:

$$c_d(\gamma_d)= q_D(\gamma_d-D) +D \quad \gamma_d\geq\gamma_D \tag{11}$$

Note that these singular statistics of the hard behavior[15] of the dressed field have been taken as a basic feature of self-organized criticality[19] and are effective as soon as the sample size is large enough ($\Delta_s=D+D_s\geq c(\gamma_D)$). Following the argument for Eqs. 3-4 (see Fig. 2a), the maximum observable dressed singularity ($\gamma_{d,s}$) is given by the solution of $c_d(\gamma_{d,s})=\Delta_s$ and by taking the Legendre transform of c_d with the restriction $\gamma_d<\gamma_{d,s}$ we obtain the finite sample dressed $K_{d,s}(q)$:

$$K_{d,s}(q)= \gamma_{d,s}(q-q_D) + K(q_D) \qquad q>q_D; \quad K_{d,s}(q)=K(q) \quad q<q_D \tag{12}$$

In the limit $N_s\rightarrow\infty$, $\gamma_{d,s}\rightarrow\infty$, and for $q>q_D$, $K_{d,s}(q)\rightarrow K_d(q)=\infty$ as expected. For N_s large but finite, there will be a high q (low temperature) first order phase transition, whereas the scale breaking mechanism proposed for phase transitions in strange attractors[16,17] is fundamentally limited to high and negative temperatures (small or negative q . This transition corresponds to a jump in the first derivative of the K(q):

$$\Delta K'(q_D)\equiv K'_{d,s}(q_D)-K'(q_D) = \gamma_{d,s}-\gamma_D = \frac{\Delta_s-c(\gamma_D)}{q_D} \tag{13}$$

On small samples ($\Delta_s\approx c(\gamma_D)$), this transition will be missed, the free energy simply becoming frozen and we obtain: $K_{d,s}(q)\approx(q-1)D$, which was already discussed with help of some turbulence experiments[3,14].

5. Numerical simulations

The numerical study of the first order transition is particularly demanding since we must not only dress the bare cascade over a large enough range of scales to obtain convergence to the hard behavior ($\Lambda>>1$), but we also require a large enough number of realizations ($\gamma_{d,s}>>\gamma_D$, $D_s>>D$). With these constraints in mind, we chose the "log-normal" multifractal which is the extreme of the family of universal multifractals[15,22,25,26] whose degree of multifractality ($0\leq\alpha\leq2$) is a maximum ($\alpha=2$).

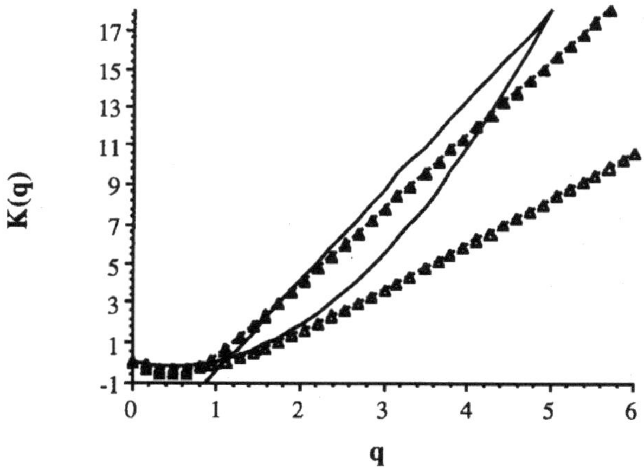

Fig. 4: A numerical simulation of phase transitions: K(q) (theoretical, parabola), $K_s(q)$ (observed second order transition, bare, open triangles), $K_{d,s}(q)$ (observed first order transition, partially dressed[14], closed triangles), $K_{d,s}(q)$ (theoretical first order transition, fully dressed, straight line).

The bare K(q) function is:

$$K(q) = \frac{C_1}{\alpha-1}(q^{\alpha}-q) \quad (14)$$

where $0 \leq C_1 \leq D$ for nondegenerate[22] multifractals. When $\alpha<2$, Eq. 14 is only valid for $q \geq 0$, when $q<0$, both the bare and dressed moments will diverge leading to other (negative temperature) phase transitions to be discussed elsewhere. For simplicity, the cascades were discrete; the unit interval was iteratively divided into lengths 1/2, 1/4, 1/8...1/1024, each interval, and every step being multiplied by appropriately normalized independent lognormal random variables (see Fig. 1a). The individual random variables had unit mean and were statistically independent. The absence of restrictions on the sum of their values allows[15,24] far much larger fluctuations than possible in the usual multifractals and is essential for the first order phase transition.

With $\alpha=2$, $q_D = D/C_1$ and low q_D is obtained with either C_1 large or D small (see bellow). For simplicity, we took D=1 and C_1 large ($=9/10 \Rightarrow q_D=10/9$). Fig. 4 shows the result when $N_s=2^9$. The cascades were constructed over a ratio $\Lambda=2^{10}$, dressed over a ratio 2^7 (Fig. 1 shows the realization giving the highest dressed singularity at this scale), so that the remaining bare range (λ) was only $2^3 \Rightarrow D_s=9/3=3$, $\Delta_s=4$. Finite size effects ($\Lambda<\infty$) must be carefully considered in the statistical estimates which must be performed scale by scale (Fig. 4).

6 Empirical evidence

The existence of multifractal phase transitions may explain long-standing

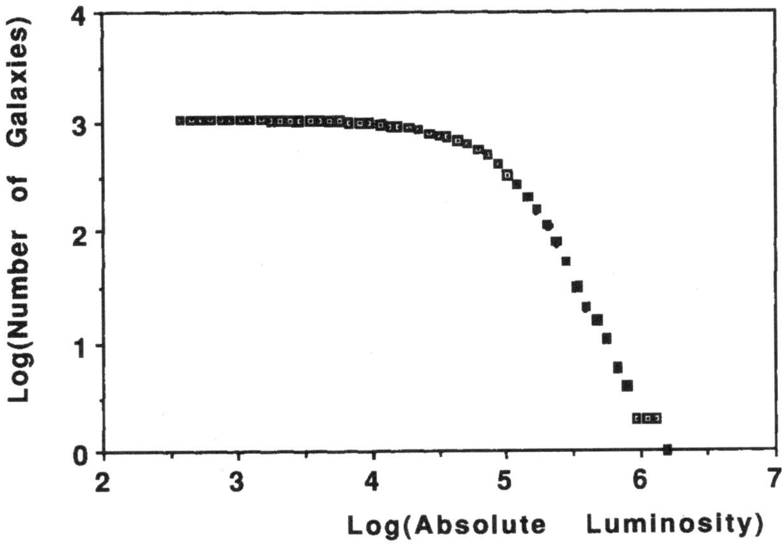

Fig. 5: The data used are from the Center for Astrophysics redshift Catalog[31], apparent luminosity magnitude < 14.5 and velocity<4000km/s. The sample contains 1682 galaxies, of which 436 responsible for the observed q_D. The cutoff at Log(lum)= 4.8 can be physically understood from the missing galaxies due to the r^2 dependence of optical detection of galaxies[30].

geophysical evidence[27] for hard behavior, e.g. the well-known Gutenberg-Richter law in earthquakes [29] as well as more recent evidence in turbulence[18]. Indeed, an alternative method of obtaining a small q_D is having a small D. D<1 may in fact be relevant in turbulence where evidence of hard multifractals has existed for some time. Recall the Kolmogorov relation between the observed velocity fluctuations (Δv_λ), and the energy flux (ε): $\Delta v_\lambda \approx \varepsilon^{1/3}\lambda^{-1/3}$. A simple interpretation of the linear scaling exponent -1/3 is that it represents a fractional integration of order D=1/3 of $\varepsilon^{1/3}$ and as elaborated elsewhere the estimates[16,29] of universal multifractal exponents support a low $q_{\varepsilon D} \approx 2$, consistent with empirical estimates going back ten years[3].

A first order multifractal phase transition mechanism may also explain rather straightforwardly the observed luminosity distribution of galaxies, i.e. the algebraic fall-off of the integrated Luminosity Function. Indeed, although various authors[29] proposed fits to this algebraic fall-off regime of this integrated histogram of observable galaxies exceeding a given level of luminosity, until now no physical interpretation was made. On the contrary, a first order phase multifractal transition with a critical moment order[30] $q_D \approx 2.2$ may explain this regime as shown in Fig. 5. It is likely the straight line part of the graph would extend to much lower luminosities if we had access to all the non-visible galaxies.

7. Conclusion

The fact that nonlinear scale invariant systems generically yield first order multifractal phase transitions and diverging statistics may explain the prevalence of self-organized criticality in nonlinear physical processes: hard cascades with their energy fluxes share not only the same statistics as sand piles and their avalanches, they also share some of the same physics since events of all intensities are generated by the same mechanism. In both cases, the extreme behavior arises as a consequence of canonical conservation of the mean field (energy fluxes and number of sand particles respectively).

8. Acknowledgments

We thank P. Bak, A. Csordas, P. Hubert, C. Hooge, F. Schmitt, T. Tel and Y. Tessier for helpful discussions and comments.

9. References

1. H.G.E., Hentschel, I.. Proccacia., *Physica* **8D** , 435-444 (1983), P., Grassberger, *Phys. Lett*.. **A97**, 227 (1983), P. Grassberger, I. Proccacia,. *Phys. Rev. Lett*., **50**, 346 (1983).
2. T.C Halsey et al., *Phys. Rev. A*, **33**, 1141-1151, (1986).
3. D. Schertzer, S. Lovejoy in *Turbulence and Chaotic Phenomena in Fluids*, T. Tatsumi ed., Elsevier (1984), Schertzer D., S. Lovejoy in *Turbulent Shear Flows 4*, B. Launder ed., Springer (1985).
4. G. Parisi, U. Frisch in *Turbulence and predictability in geophysical fluid dynamics and climate dynamics*, Eds. M. Ghil, R. Benzi, G. Parisi, North-Holland (1985).
5. R.Benzi et al. J. Phys. A, **17**, 3521 (1984). G. Paladin, A. Vulpiani, *Phys. Rev. Lett*., **156**, 147 (1987). K.R. Sreenivasan, C. Meneveau, *Phys. Rev. A*, **38**, 6287 (1988).
6. H.E.P. Stanley, P. Meakin, *Nature*, **6**, 116 (1988). P. Meakin, A. Coniglio, H.E.P Stanley, *Phys. Rev. A*, **34**, 3325 (1986).
7. B. Derrida, J. Phys. 48, 1154 (1986), B. Derrida and E. Gardner, *J. Phys*. **C19**, 5783 (1986), M. Mesard, G. Parisi, M. A. Virascoro, *Spin Glass Theory and Beyond*, World Scientific, Singapore (1987).
8. P. Brax, P. , R. Peschanski, *Phys. Lett. B*, **253**, 225-230 (1991).
9. H. Atmanspacher, H. Scheingraber, G. Wiedenmann, *Phys. Rev. A*, **40**, 3954-3963 (1989).
10. *Non-Linear Variability in Geophysics, Scaling and Fractals*, D. Schertzer, S. Lovejoy eds., Kluwer, Dordrecht-Boston (1991).
11. T. Tel, *Z. Naturforsch*, **43a**, 1154-1174 (1988).
12. H.G. Schuster, *Deterministic Chaos: an introduction*. 2. rev. ed. , New York, VCH, New York (1988).
13. D. Schertzer, S. Lovejoy in ref.10.
14. D. Lavallée., *PhD Thesis*, Physics, McGill U. 142pp (1991).
15. D. Schertzer, S. Lovejoy, *Physica A*, **185**, 187-194 (1992).
16. P. Szépfalusy, T. Tél, A. Csordas, Z. Kovas, *Phys. Rev. A*, **36**, 3525 (1987). A. Csordas, P. Szépfalusy, *Phys. Rev. A*, **39**, 4767 (1989).
17. D. Barkley, A. Cumming, *Phys. Rev. Lett*., **64**, 327 (1990).

18. F. Schmitt et al. Phys. Rev. Lett. **68**, 305 (1992), D. Lavallée et al. *Topological aspects of the dynamics of fluids and plasmas,* Eds. K. Moffat, M. Tabor, G. Zaslavsky, Kluwer 463-478 (1993).
19. P. Bak, C. Tang, and K. Weiessenfeld, *Phys. Rev. Lett.*, 59 (1987), 381.
20. L. Pietronero, A.P. Siebesma, *Phys. Rev. Lett.*, **57**, 1098 (1986).
21. D. Schertzer et al., *Nonlinear Dynamics of Structures.* R. Z. Sagdeev et al, eds., World Scientific, 213-235 (1991).
22. D. Schertzer, S. Lovejoy, *J. Geophys. Res.*, **92**, 9693-9714 (1987).
23. B. Mandelbrot in *Turbulence and Stochastic processes*, Eds. J.C.R. Hunt, O.M. Phillips and D. Williams, The Royal Society (1991).
24. D. Schertzer, S. Lovejoy in *Fractals: Physical Origin and Properties* Ed. L. Pietronero, Plenum, New York (1989). D. Lavallée, D. Schertzer, S. Lovejoy in ref.10.
25. D. Schertzer et al.,*Fractal Aspects of Materials: Disordered Systems*, Eds. D.A. Weitz, L.M. Sander, B.B. Mandelbrot, pp. 267-270, Materials Research Society, Pittsburg, (1988). A.H. Fan, *C. R. Acad. Sci. Paris I*, **308**, 151 (1989).
26. A. Bialas, R., Peschanski, *Nucl. Phys. B*, **B308**, 857 (1988)
27. S. Lovejoy D. Schertzer, *Bulletin of the AMS*, **67**, 21-32 (1986).
28. B. Gutenberg, C. F. Richter, *Bull. Seism. Soc. Am.* **34**, 185-188 (1944)).
29. F. Schmitt et al., *C. R. Acad. Sci. Paris, II*, **308**, 151 (1992), F. Schmitt, D. Schertzer, S. Lovejoy, D. Lavallée, *Phys. of Fluids* (1992, submitted), Chigrinskaia et al.,*Ann. Geophysicae* Supp **11**, 542 (1993), Lazarev et al.,*Ann. Geophysicae* Supp **11**, 542 (1993).
29. G. O. Abell in *Galaxies and the Universe* (1976), P. Schechter,*Ap. J.* **203**, 297 (1976).
30. P. Garrido, *M.Sc. Thesis*, McGill University, Montréal, Québec (1993)
31. Huchra et al , *Ap. J.* Supp, **52**, 89 (1982)

THE SIMULATION OF UNIVERSAL MULTIFRACTALS

S. PECKNOLD, S. LOVEJOY, D. SCHERTZER*
C. HOOGE, J.F. MALOUIN
Department of Physics, McGill University
3600 University St., Montréal, Québéc, H3A 2T8, Canada

ABSTRACT

The existence of multifractal fields in nature is recognized as a widespread phenomenon, one which describes a multitude of different systems. It is therefore necessary to develop an understanding of their behaviour. Simulations play an important role in this understanding. They are used to evaluate new analysis techniques, to determine if results are reproducible according to a theoretically correct model, and particularly to simulate complex interactions in geophysical fields, such as radiative scattering in clouds and the forecasting of weather and earthquakes. They are also necessary for developing sampling, averaging and calibration procedures for remotely sensed geophysical and astrophysical fields. Thus, it is necessary to have a method of producing simulations of all the possible cases of interest. These simulations may be produced for fields and time series with many different parameters, including the sparseness of the actual field, differing probability distributions and spectra, as well as those that exhibit anisotropic scale invariance, e.g. stratified or differentially rotated fields.

1. Introduction

A growing body of geophysical research is devoted to determining the scaling properties of fields that arise in nature as a result of nonlinear processes acting over wide ranges in scale. Geophysical fields including rainfall and cloud fields[1,2], temperature fields[3], topography[4] and the roughness of the ocean surface[5] and wind tunnel turbulence[6] have been analyzed over various time and space scales and have been shown to be multifractal in nature. In addition, such phenomena as earthquakes[7] and sea ice[8], as well as

* *L. M. D./C.N.R.S., BP 99, Université Pierre et Marie Curie*
4 Place Jussieu F-75252 Paris Cedex 05, France

apparent galactic luminosity[9] have been analyzed and shown to have fractal distribution, and further research has demonstrated that all these fields are in fact multifractal in nature.[10,11,12,13] As well, analysis of human speech[14], high energy hadron jets[15] and pollution distribution from the Seveso industrial accident[16] have all demonstrated that multifractal fields arise in many disparate situations. (Universal multifractal parameters are discussed in Section 2 and a fairly complete list of known results is summarized for these fields in Table 1.)

Given this ubiquity of multifractal fields, techniques to model them must be developed. The early scaling models of geophysical fields were based on an assumption of "simple scaling"[17,18], resulting in the description of the field by a single fractal dimension. The physical processes were thus modeled as monofractals, simulations of which were produced by summing many random structures (e.g. the FSP model[18]), by simply filtering white noise (this is how Voss[19] created his impressive monofractal mountainscapes), or by variations on these processes such as "midpoint displacement"[20] or more recently "iterated function systems"[21].

None of these models provide an adequate description of the actual physical nonlinear dynamic process involved. The first step in this direction was the use of multiplicative (nonlinear) processes which emerged from the study of turbulent cascade processes. At first, great attention was paid to the monofractal special case, the "β-model"[22,23,24,25], but this was shown to be unstable to small perturbations (which yielded the multifractal "α-model"[26]. From the point of view of physical modeling, these early multifractal models had several defects: they had discrete rather than continuous spectra (which in addition, gave rise to ugly straight-line structures), but more fundamentally, they involved an infinite number of parameters for their specification (the dimension function). This made it highly unlikely that sufficient theoretical or empirical information would ever be available to enable realistic models to be produced.

It turned out that the solutions to both problems were closely linked. The discrete cascade actually simultaneously involved two limiting processes; the small scale limit and the limit of an infinite number of interactions. By separating the two limits and considering the densification (or more generally nonlinear "mixing" of processes) over a finite range of scales, it was shown[27,28] that due to the existence of stable, attractive "universal" generators of multifractal processes, that many of their dynamical details could be washed out *if sufficiently many nonlinear interactions were available over a finite range of scales*. For universal multifractals, only three basic parameters are necessary to specify the entire process (including the infinite hierarchy of fractal dimensions).

However, important as it was to replace the linear monofractal models by nonlinear multifractal models, such models involved self-similar multifractals in which the small and large scale statistical properties were related by a simple "zoom" or magnification. On the contrary, geophysical fields are seldom (if ever) isotropic, and the operator required to go from large to small scale typically involves reduction coupled with compression, rotation or more complex operations which can vary not only with scale, but also with location. Such anisotropy is associated for example with texture and morphology. The formalism necessary to define the notion of scale and scale change in such an anisotropic framework is Generalized Scale Invariance (GSI)[29,30]. The final step in making physically realistic multifractal models is to therefore to combine continuous cascades (sections 2, 3) with GSI operators (see section 5).

In this paper we bring up to date the techniques of simulating universal multifractals and show for the first time how to combine the use of GSI with the continuous multiplicative simulation technique to provide the most realistic possible model of nonlinear dynamical scale invariant processes. In addition, former problems such as the previous inability to simulate universal multifractals over certain ranges of the parameters have been resolved. It is now possible to simulate universal multifractals over the entire theoretically possible range. Finally, an attempt is made to provide a visual catalogue demonstrating the effects of altering the parameters describing multifractality.

2. Universal Multifractals

From fractals to Multifractals

The transition from a fractal set to a multifractal field is fundamental in terms of the resulting mathematical structure yet is not conceptually difficult to understand. The distinction is that in a (mono-)fractal set we obtain a single value for the dimension, whereas with a multifractal field we can define a series of sets (e.g. using thresholds) each of which will have its own (generally different) dimension. Given a multifractal field, we can look at the sub-sets corresponding to thresholding at each given value. For example, in Fig.1, we look at this type of field, analyzed at scale ratio λ (for example, the ratio between image size and pixel size in an image). The thresholds chosen are λ^{γ_1} and λ^{γ_2}. The points corresponding to field values ε_λ greater than these thresholds form geometric subsets, as long as the corresponding dimension is positive. With each of these geometric sub-sets we can find an associated fractal dimension. If we look at the probability of finding a field value greater than a given (scale dependent) threshold, we find[28] that the probability, $\Pr(\varepsilon_\lambda > \lambda^\gamma)$, can be related to the order of singularity, γ, by :

$$\Pr(\varepsilon_\lambda > \lambda^\gamma) \approx \lambda^{-c(\gamma)} \qquad (1).$$

where ε_λ corresponds to a singularity of order γ and the codimension function, $c(\gamma)$, describes the *sparseness* of the field intensities: when smaller than the dimension of the space in which the process is observed, the codimension is the difference between the dimension of space and the fractal dimension of the geometric subset described by the singularity value. It should be noted that when the codimension is larger than the dimension of support (i.e. the dimension of the embedding space), the meaning is that in any given realization, the corresponding values ε_λ are almost surely not found: $c(\gamma)$ is related to the probability density, and field values with arbitrarily large codimension can appear because the probability space defining the process is infinite-dimensional[31].

Although the codimension function $c(\gamma)$ yields the entire statistical description of the multifractal field, it can only be determined by empirical methods; there is no theoretical form for the function, beyond that determined by the fact that it gives a probability density (which only implies convexity). A more powerful result is what are called *universal multifractals*[31], for which we will be able to write an explicit functional form for the codimension function.

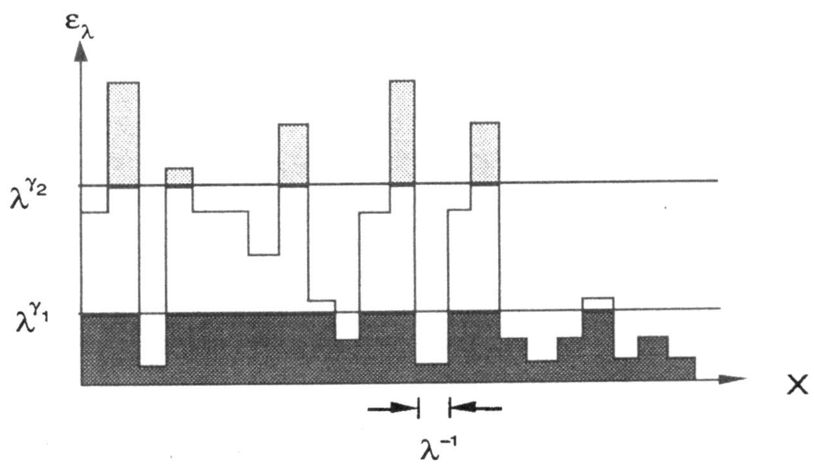

figure 1—A schematic illustration of a multifractal field analyzed over a scale ratio λ, with two scaling thresholds $\lambda^{\gamma_1}$and. λ^{γ_2}, corresponding to two orders of singularity : $\gamma_2>\gamma_1$. (Lovejoy and Schertzer(32))

Cascades

Multifractals can arise in nonlinear systems with both small and large numbers of degrees of freedom. In the former, a few degrees of freedom characterize the probability measure or phase space of the system; these systems are associated with calm, geometric multifractals[33]. On the other hand, turbulent and turbulent-like systems involving many degrees of freedom (e.g. fields) are often associated with stochastic cascade processes and generally lead to wild and hard multifractals[33]. Geophysical and astrophysical applications typically involve fields; we illustrate the cascade process using the example of the turbulent energy flux ε_λ. The energy flux is injected into the system at a large scale cascades down to smaller scales by successive multiplicative modulation down to a very small scale dissipative regime. Figure 2 shows a schematic treatment of the cascade process. In this figure, showing a discrete cascade, independent random increments $\mu\varepsilon$ are generated at each scale, proceeding in factors of two, and the field is multiplicatively modulated with them: the energy is redistributed by boosts and decreases as one moves to smaller and smaller scales. It is termed *isotropic* because the boxes are of the same shape (in this case they are square).

When the redistribution of the cascading quantity ε is uneven between the boxes, it gives rise to what is called intermittency[34,35]. This unevenness could be caused by either a stochastic or deterministic process.

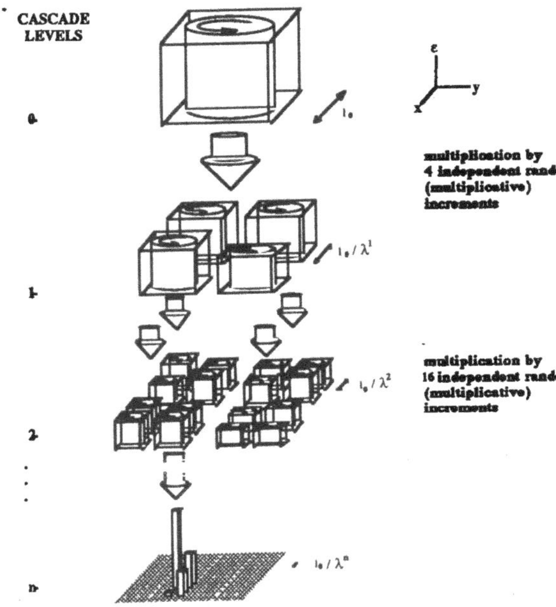

Figure 2—A schematic diagram showing a few steps of a discrete multiplicative cascade process.(30)

Properties of Multifractals

In order to simulate the field of interest, ε, we look at its behaviour at different scales. Using ε_λ as the field at a scale ratio λ, we look at the field at a scale ratio λλ'. Comparing this to the field at a scale ratio λ, we note that for a multiplicative cascade process the field $\varepsilon_{\lambda\lambda'}$ should be statistically the same as taking the field at a scale ratio λ and multiplying each "box" of size 1/λ by the field at a scale ratio λ' which has been reduced by a factor λ. Symbolically, $\varepsilon_{\lambda\lambda'} = T_\lambda(\varepsilon_{\lambda'})\varepsilon_\lambda$. T_λ is a scale changing operator which reduces scales by a factor λ; for the moment we consider isotropic (standard) reductions in scale. In section 5 we indicate how this can be greatly generalized.

Since this is a multiplicative property it is therefore convenient to introduce an (additive) generator of the field, $\Gamma_\lambda = \ln \varepsilon_\lambda$, which satisfies the additive group property[32]:

$$\Gamma_{\lambda\lambda'} = T_\lambda \Gamma_{\lambda'} + \Gamma_\lambda \tag{2}$$

Given that this is the case, we can consider the multiplicative field as being produced down to very small scales by an additive process on the generators.

In multifractal fields, different moments of the field are governed by different power-law relations: i.e. the field is multiscaling. That is, given a multifractal field ε_λ at a scale ratio[28] λ,

$$\langle \varepsilon_\lambda{}^q \rangle = \lambda^{K(q)} \langle \varepsilon_1{}^q \rangle \tag{3}$$

Hence, we have a function K(q) depending on the moment of our field that describes how the statistical properties of each moment behave under isotropic dilations and contractions.

Considering for convenience a normalized field at the largest scale, i.e. $\langle \varepsilon_1^q \rangle = 1$, we can rewrite Eqn.(3) in terms of the generator[28,31] :

$$\langle \varepsilon_\lambda^q \rangle = \langle \exp(q\Gamma_\lambda) \rangle = \exp(K_\lambda(q)) = \lambda^{K(q)} \tag{4}$$

This equation shows that the function $K_\lambda(q) = K(q)\log(\lambda)$ that describes how the statistical moments of the field behave with respect to the scale ratio λ is actually the second Laplace characteristic function corresponding to the generator[31,37] Γ_λ, that is, the logarithm of the Laplace transform of the probability density. Given this, we can now determine the form of the generator.

In order to obtain multiscaling, we wish the generator to be a noise with a possible weighting function, such that it has the following properties[32]:

1) The spectrum of the field must scale as k^{-1}, in order to obtain the scaling behaviour: that is, a $\log\lambda$ divergence of $K_\lambda(q)$.

2) The generator must be band-limited to wave-number between $[1,\lambda]$; this is to ensure that for scales smaller than λ^{-1}, the field will be smooth; λ^{-1} will therefore be the resolution of the field.

3) The probability distribution of the generator must fall off more than exponentially quickly for positive fluctuations. This is to ensure convergence of $K(q)$ for $q>0$.

4) It must be normalized so that $K(1)=0$. This is the condition of conservation of the mean of the field at varying scales; $\langle \varepsilon_\lambda \rangle = 1$.

Having noted these properties, it now remains to determine the type of noise generator to use. As mentioned in the introduction, we know little about the functional form of the codimension function, and thus the $K(q)$ function, beyond the very general restrictions placed upon them until now. The abandonment of the geometric monofractal approach to nonlinear variability for the less restrictive and more accurate multifractal approach was necessary for the proper understanding of actual physical processes. However, in effect, we have traded the description of our nonlinear process via a single parameter (the fractal dimension), for an infinity of parameters describing the hierarchies of dimensions and singularities. This situation is rectified by the introduction of universal multifractals, whereby we can once again describe nonlinear variability in terms of a few fundamental parameters, where these parameters are no longer geometric, but dynamic[31]. We can observe how these parameters, and thus the universality classes they describe, arise by considering the generator of a multiplicative cascade. We therefore examine the properties of the sums of random noises, corresponding to the additive process on the generator of the field, by looking at the problem of random walks[36], which are additive processes.

A random walk in d dimensions is merely the addition of independent random vectors $\mathbf{x}_1, \mathbf{x}_2, \ldots$, with random direction and random length. Let us assume (for now) finite variance of the length. Here we take the expectation or mean of the length = 1. Then the normalized sum of this random walk has its distribution tend toward the normal distribution[36]:

$$S_n = \frac{x_1 + x_2 + \ldots + x_n}{\sqrt{n}}$$

$$c_{ij} = \delta_{ij}/d \tag{5}$$

where the distribution of S_n will tend towards the normal distribution as $n \to \infty$. The covariance matrix of the d components of each vector, c_{ij} (the usual for a normal distribution) will be diagonal with elements 1/d. This is the standard central limit theorem. Although for historical reasons the terminology "universal" is seldom used in this context, this is in fact the most commonly studied universality class for the addition of random variables.

This procedure can be extended to the case where the variances of the vectors are not finite. In this case, the normalized sum of the independent vectors has a distribution that tends toward any member of a class of probability distributions, known as Lévy distributions, which form the complete universality classes of addition of random variables (they are the only stable, attractive classes of random variables under addition). The analogue of Eqn.(5) is[36]

$$S_n = \frac{x_1 + x_2 + \ldots + x_n}{n^{1/\alpha}} \tag{6}$$

where S_n has the property that the distribution $S_n + b_n$ tends toward the distribution of the random variables $\mathbf{x}_i$, characterized by the Lévy parameter α, taking on values $0 < \alpha \le 2$. The term b_n is a recentering term, required for $\alpha > 1$, such that the distribution is centered to zero expectation.

Although the probability density of most members of this class cannot be written in closed form, the characteristic function can be. In fact, the second Laplace characteristic function of this type of random variable, assuming convergence, has the form[37]:

$$\begin{aligned} K(q) &\propto q^\alpha, \ 0 < \alpha \le 2, \ \alpha \ne 1 \\ K(q) &\propto q \ln(q), \qquad \alpha = 1 \end{aligned} \tag{7}$$

where α is the Lévy parameter describing the probability distribution of the noise (in fact α is the exponent of the power-law fall-off of the tail of the probability density[36]).

In our case, we wish to obtain the continuous limit of a multiplicative cascade[31,38]. As a discrete cascade proceeds to the small scale limit, we are in fact taking two limits; an infinite number of random variables and the limit of infinitely small scales. In this case the normalization condition (condition 4) implies that $\langle \Gamma_\lambda \rangle < 0$ which is not in general compatible with the centering and normalization required for the central limit theorem to apply. The result is that discrete cascades will generally not tend to universal limits (a fact that has unfortunately obscured the issue of universality and even lead to claims that no universality classes exist[39]). However, if on the contrary we first keep the total scale ratio λ fixed and finite and proceed by multiplying by independent processes of the same type, then universal behaviour, i.e. that described by Eqn.(7), will be produced[38]. The small scale limit can be taken later; this involves its own difficulties, however these are quite distinct from the question of universality classes. The continuous cascade is built in such a manner by densification, adding more and more intermediate scales in a given multiplicative process.

Thus, since we wish our generator to be stable under addition, we choose it to be a Lévy noise with some parameter α, appropriately weighted and filtered so as to satisfy the other necessary conditions mentioned above. In order to ensure convergence (i.e. to keep K(q) finite for q>0) an extremal asymmetric Lévy variable must be used[27,31,38]. Since the field ε_λ is obtained by exponentiating Γ_λ, which is a sum of these noises (the equivalent of undergoing a multiplicative cascade process), convergence is required.

Multifractals obtained using the Lévy noise generator with parameter α will have the following moment scaling function, K(q), relating the scaling of the moments and the scale:

$$K(q) = \frac{C_1}{\alpha - 1}\left(q^\alpha - q\right) \qquad (8).$$

given that $\alpha \neq 1$, where $0 < \alpha \leq 2$. C_1 is the codimension of the mean of the field, characterizing the sparseness of the mean, and the Lévy index α characterizes the degree of multifractality[31]. As $\alpha \to 0$ K(q) becomes linear and the monofractal β-model[22,23,24,25] is approached; α=2 corresponds to a lognormal model.

We can obtain from this via a Legendre transform the codimension function[31]:

$$c(\gamma) = C_1\left(\frac{\gamma}{C_1\alpha'} + \frac{1}{\alpha}\right)^{\alpha'} \qquad (9)$$

Where $1/\alpha + 1/\alpha' = 1$.

The K(q) has the α-dependence we expect for a Lévy noise, with the linear term arising due to the normalization requirement that $K(1) = 0$. Hence we can determine all the statistical properties of a process using only the two parameters, α and C_1.

For nonconservative fields[28,31], such as density fields f related to ε_λ in a scaling manner ($|\Delta f_\lambda| = \varepsilon_\lambda \lambda^{-H}$), there is a third fundamental parameter, H, which is a measure of the degree of (scale by scale) non-conservation of the field:

$$\langle|\Delta f_\lambda|\rangle = \lambda^{-H} \qquad (10).$$

The value of H can be determined from the scaling of the power spectrum and the moment scaling function. If necessary, a spectral filter (fractional integration) of the form k^H can be used to transform a non-conservative field into a conservative field. This will be described below. For now, we note that since the energy spectrum E(k) is the Fourier transform of the autocorrelation function (a second order moment), we obtain $E(k) \sim k^{-\beta}$, where

$$\begin{aligned} \beta &= 1 - K(2) && \text{(conservative)} \\ \beta &= 1 - K(2) + 2H && \text{(non-conservative)} \end{aligned} \qquad (11)$$

We thus see that the parameter H also specifies the degree of statistical non-stationarity ("non-homogeneity") of the field.

If $H \neq 0$, then the universal forms for K(q) and c(γ) are given by:

$$K(q) \rightarrow K(q) - qH \qquad (12)$$

and

$$c(\gamma - H) = C_1\left(\frac{\gamma}{C_1\alpha'} + \frac{1}{\alpha}\right)^{\alpha'} \qquad (13)$$

(i.e. $\gamma \rightarrow \gamma - H$).

Finally if the parameter α is identically unity then (for all values of H):

$$K(q) - qH = C_1 \cdot q \cdot \log(q)$$

and

$$c(\gamma - H) = C_1 \exp\left(\frac{\gamma}{C_1} - 1\right) \qquad (14)$$

Again as expected. Eq. (14) can be seen to follow from taking the limit as $\alpha \rightarrow 1$ of Eqs.(12) and (13) and applying l'Hôpital's rule.

It is worthwhile at this point to note the correspondence between the formalism of universal multifractals and the strange attractor formalism. In the f(α)-τ(q) notation we have the following[32]:

$$\begin{aligned} f_D(\alpha_D) &= D - c(\gamma),\ \alpha_D = D - \gamma \\ \tau_D(q) &= (q-1)D - K(q) \end{aligned} \qquad (15)$$

Parameters for many multifractal fields have been measured. Some of these are tabulated in Table 1.

Field	α	C_1	H	Range of scales	Refs
Clouds Visible	1.35	0.15	0.3	160m→ 4000km	(2)
Clouds Infra Red	1.35	0.15	0.4	160m→ 4000km	(2)
Clouds Microwave	1.60	0.10	0.35	160m→ 4000km	(1)
Rain Radar	1.35	0.30	0.0	30m→ 64km	(2)
Sea Ice (radar)	1.7	0.05	0.0	50m→25km	(10)
Ocean surface (0.95μm)	1.1	0.25	0.35	1m→50m	(5)
Topography	1.8	0.05	0.5	50m→1000km	(4)
Seismicity	1.35	1.9	0.0	1km→500km	(11)
Pollution (Seveso)	1.2	0.8	-0.2	30m→5km	(16)
Raingauges	1.35	0.20	0.0	50km→ 4000km	(2)
Wind windtunnel	1.30	0.25	1/3	1ms→ 1s	(3)
Wind atmosphere	1.45	0.23	1/3	1ms→ 1s	(6)
Temperature	1.20	0.35	0.30	0.1s→ 1000s	(3)
Low frequency speech	2.0	0.1	-.35	0.1s→1000s	(14)
Hadron Jets	0.7	0.6	–	–	(15)
Apparent Galactic Luminosity	1.0	1.0	–	0.3°→5°	(12)

Table 1: Multifractal parameters determined for a variety of fields with the corresponding ranges of scale. The accuracy of most parameters is ~ ±0.1.

3. Simulation

Discrete Cascades

The simulation of a discrete cascade is straightforward. Given that we wish to simulate universal multifractals, i.e. those whose statistics are described by Eqs.(8)-(9), we specify the C_1 and α desired and use the noise that corresponds to this to obtain the generator. We generate unit Lévy variables with parameter α and modify their amplitude by multiplying by the constant $(C_1(\alpha'/\alpha))^{1/\alpha}$ required to obtain the desired C_1. These noises are generated according to the procedure given by Chambers et al[37], who give an expression for the representation of an extremal Lévy-stable variable of index α, $S(\alpha)$:

$$S(\alpha)=\frac{\sin\alpha(\Phi-\Phi_0)}{(\cos\Phi)^{1/\alpha}}\left(\frac{\cos(\Phi-\alpha(\Phi-\Phi_0))}{W}\right)^{(1-\alpha)/\alpha},\alpha\neq 1 \qquad (16)$$

$$S(1)=\frac{2}{\pi}\left((\frac{\pi}{2}-\Phi)\tan\Phi+\ln(\frac{\pi W\cos\Phi}{\pi-2\Phi})\right) \tag{17}$$

$$\Phi_0=\frac{\pi}{2}(1-|1-\alpha|)/\alpha \tag{18}$$

Here, Φ is a uniform random variable on $(-\pi/2,\pi/2)$, and W is a standard exponential deviate, with W and Φ mutually independent. The constant $(C_1(\alpha'/\alpha))^{1/\alpha}$ is needed as the above procedure assumes that the scale parameter is unity. This technique for generating stable random variables necessitates the production of only two random variables and correspondingly few operations upon them, and is thus superior to the slowly converging algorithm given in Wilson[27] , which uses the property that Lévy variables are stable regions of attraction to generate them by renormalized sums of many hyperbolic variables with exponent α. Although it is stated that the number of terms, n, in the sum must be very large ($n \approx 30$ for $\alpha=1.5$ is suggested), further tests have shown that for accurately reproducing K(q) over a reasonable range (e.g. $q<5$), particularly for α near 2, n is much larger than previously believed[40] .

Given a desired scale ratio λ, at each scale Λ from the largest scale to the smallest scale 2^{Λ} independent noises with the desired parameters are generated These noises, once generated, are then exponentiated to give the multiplicative increments. Each point of the field is then considered to be contained in a specific box at each scale and the field value at that point is given by the product of all the corresponding multiplicative increments. (See Fig.2). The field is then normalized, in keeping with the conservation of the mean condition. To normalize the field, we divide by the expected value of the non-normalized field. This is equal[31] to $\lambda^{C_1/(\alpha-1)}$. It is, in fact, this factor which gives rise to the linear term in q in Eq.(12). This is done rather than dividing by the mean of the specific field produced because we wish conservation only on an ensemble of fields. Conservation holds only on the ensemble average, but will be broken on any particular realization. This is termed canonical conservation.[31,38] Much more restrictive processes involving exact conservation on each realization (called "microcanonical conservation") are also possible, but such processes do not have well-defined universality classes, eliminate the strong singularities, and have properties which depend on the exact details of the construction process including the discrete ratio used to define the microcanonical conservation law.

The disadvantage of using discrete cascades to simulate multifractal fields is that the discrete cascade is nonphysical: energy does not actually cascade in scale steps of factors of two, or of other integer ratio. Also, at least in its usual real-space implementation, it involves straight construction lines which are highly unrealistic: they can result in, for

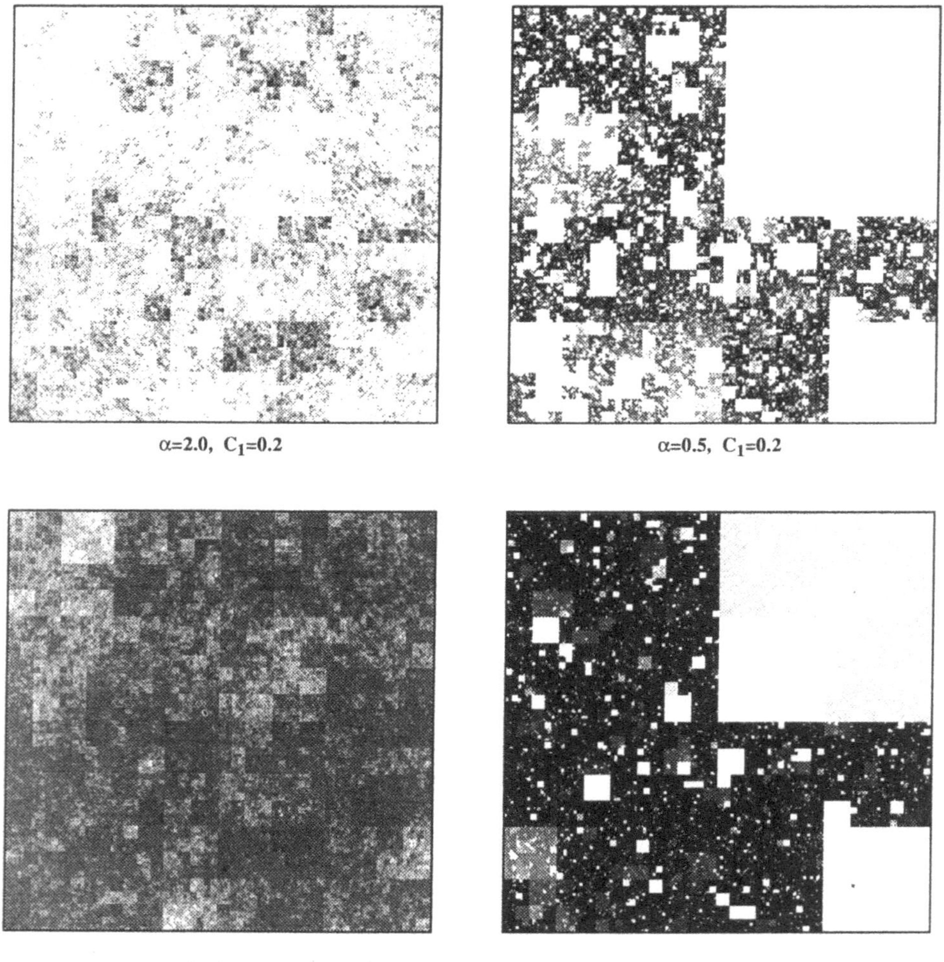

Figure 3. Two–dimensional discrete simulations, 256x256. The seed is constant.

example, square clouds. It is necessary to use continuous cascades, obtained by the densification of discrete cascades, to get an idea of what the actual physical process looks like. Examples of discrete cascades are given in Figure 3.

Continuous Cascades

In order to physically simulate the densification of the cascade process, we use Fourier Transform techniques. We therefore look for the Fourier space generator that will satisfy the four conditions laid out previously. We proceed in the following manner:

Given the desired value for α and C_1, we proceed as for a discrete cascade, producing for each grid point a noise $(C_1/(\alpha-1))^{1/\alpha} S(\alpha)$.

In order to weight the noise to obtain a logarithmic divergence of $K_\lambda(q)$, this noise is then Fourier transformed using an FFT. The distinction between a continuous transform and the discrete FFT is taken into account by multiplying by a constant, $\kappa_d(\lambda)$. This is related to the Euler constant correction described elsewhere[27,28]. Taking into account the scale ratio and dimensionality dependence, this constant is just the ratio of the integral to the sum, in d dimensions, over the total scale ratio:

$$\kappa_d(\lambda) = \frac{\int_1^\lambda k^{-d} d^d k}{\sum_1^\lambda \frac{1}{k^d}} \tag{19}$$

We now wish to obtain the 1/f-noise spectrum, to give multiscaling behaviour. To do this it is necessary to weight the Fourier transform of the Lévy noise. The correct weighting factor can be determined by considering the property of a Lévy random variable[36], $S(\alpha)$:

$$\langle \exp(qS(\alpha)) \rangle = \exp(q^\alpha) \tag{20}$$

This property requires that, to achieve a spectrum of k^{-1}, a weighting factor[31] of $w(\mathbf{x}) \propto |\mathbf{x}|^{-d/\alpha}$ is needed for the real-space noise $S(\alpha,\mathbf{x})$; this corresponds to a factor of $w(\mathbf{k}) \propto |\mathbf{k}|^{-d/\alpha'}$ in k-space. In addition, the noise is band-limited by filtering out all noises except those concentrated in the band with wave-number between 1 and λ, using a filtering function $f(\lambda,\mathbf{k})$. This yields for the generator

$$\Gamma_\lambda = \mathcal{F}^{-1}\left\{ \left(\frac{C_1}{\alpha-1}\right)^{1/\alpha} \tilde{S}(\alpha,\mathbf{k}) |\mathbf{k}|^{-d/\alpha'} f(\lambda,\mathbf{k}) \kappa_d(\lambda) \right\} \tag{21}$$

($\mathcal{F}^{-1}$ represents the inverse Fourier transform). The function $f(\lambda,\mathbf{k})$ is chosen to be one for $|\mathbf{k}| \le \lambda$, and decaying exponentially for $|\mathbf{k}| > \lambda$. This is in order to filter out wave numbers greater than λ (as required by condition 2). Since sharp cutoffs (δ-functions or step functions) in Fourier space cause real-space sine wave "ringing" effects, an exponential decay is used. This adds negligibly (less than 1%) to the constant $\kappa_d(\lambda)$ for $\lambda > 128$. However, if desired, this factor can be taken into account by the appropriate modification of Eqn.(19)

Finally, the weighted noise is transformed back into real space and is exponentiated to give the required field ε_λ. It is also normalized by dividing by the expected value of the field (i.e. $\lambda^{C_1/(\alpha-1)}$) to satisfy condition 4, following the process described in the discrete cascade section.

Examples of continuous cascades in one dimension are given in Figures 4-7 and 12-13. These simulations are over a scale ratio of 256 (extending this to larger ranges rapidly increases the variability of the field, resulting in a field that is difficult to visualize). Computationally, it is not difficult to extend this to much larger fields, depending on the computing power available—on a Silicon Graphics IRIS 4D/35, the generation of 100 one-dimensional simulations with $\lambda=2^{14}$ will take on the order of 8 minutes. For very large sizes the limiting factor is not so much the computation time but the memory requirements of an array of the size of the field.

Fig.4 shows a series of simulations for $C_1=0.9$ and varying values of α. The fields are vertically offset from each other. As α decreases, the high values of the field do not dominate as much as for larger values of α; there are more large deviations from the mean, as we expect for the "thick" tail of the probability distribution. In the case where $C_1=0.01$ (Fig 5), the fields are typically at values near 1, with spikes in both directions. The fields with higher values of α are seen to have a more symmetric distribution of deviations from the mean. Since C_1 is so small, the fields are nearly space-filling. This is seen particularly for $\alpha=1.04$, where the field is nearly uniform in comparison to higher α-values and there are primarily regularities (i.e. downward spikes). This is evidence of the fact that as α decreases, the Lévy noise in fact becomes asymmetric with only negative hyperbolic jumps, i.e. "Lévy holes"[31]. When the field is renormalized, the lowest values are "washed out". This is most obvious when C_1 is small.

Figures 6 and 7 show series of one dimensional simulations with fixed α and varying C_1. It is apparent from them that C_1 is the measure of the sparseness of the field: the higher the C_1, the fewer the field values corresponding to any given singularity. (Because the field is normalized, the spikes appear higher for the fields with higher C_1). It is also again apparent that the higher α fields are dominated by a few large singularities.

Examples of two-dimensional continuous cascade simulations are given in Colour Illustrations 1-3. These fields have been k^{-H}-filtered for easier recognition of the salient features. The three-dimensional representation of some of these fields are given in Figures

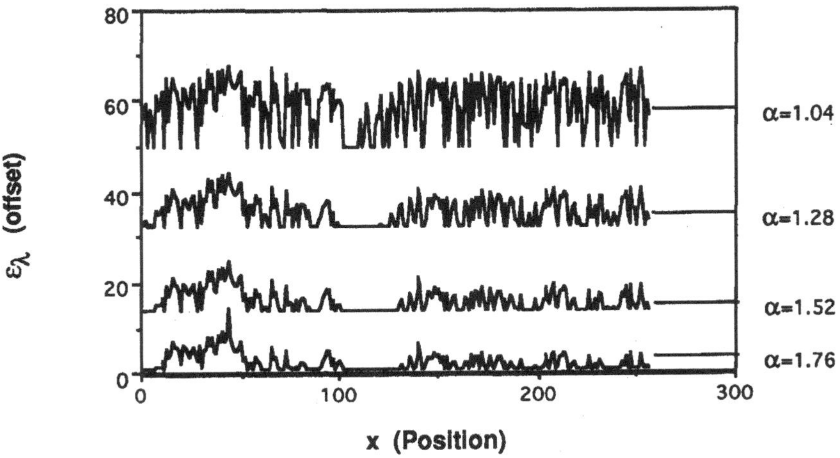

Figure 4. One-dimensional simulation of length 256, with C_1=0.9 and varying α. These simulations have been vertically offset so as not to overlap.

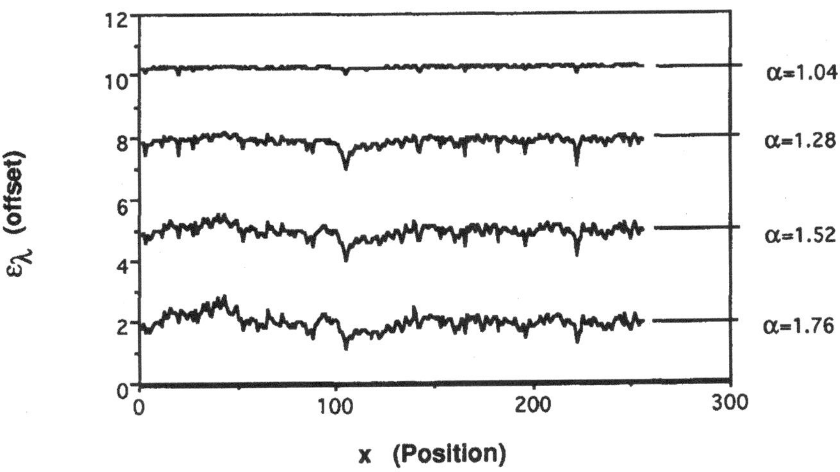

Figure 5. One-dimensional simulation of length 256, with C_1=0.01 and varying α. These simulations have been vertically offset so as not to overlap.

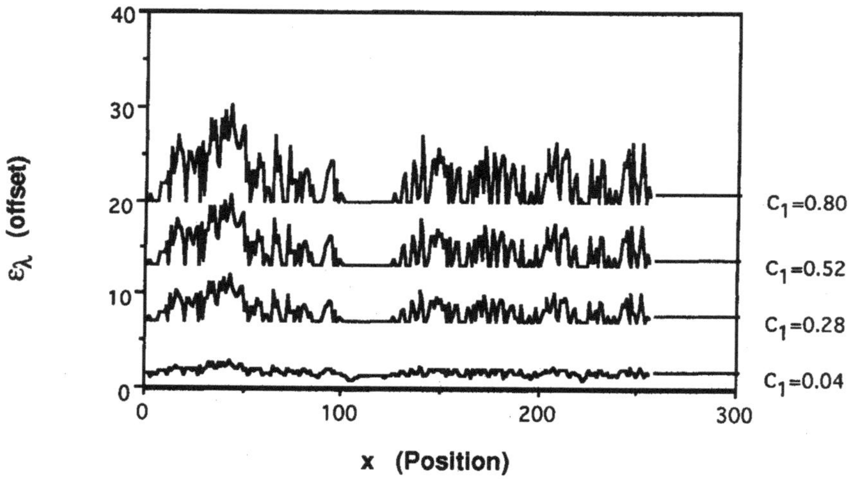

Figure 6. One-dimensional simulation of length 256, with α=1.4 and varying C_1. These simulations have been vertically offset so as not to overlap.

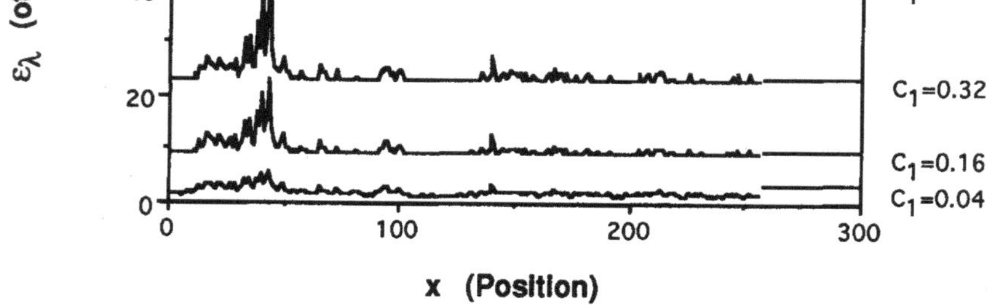

Figure 7. One-dimensional simulation of length 256, with α=2.0 and varying C_1. These simulations have been vertically offset so as not to overlap.

8-11. The same features can be noted for the differing values of α and C_1. The fields with higher values of α are generally dominated by a few areas with upward spikes (i.e. singularities); the lower α-valued fields have more high-valued singularities. The degree of sparseness that C_1 characterizes is fairly obvious from the illustrations. (Although recall that after filtering C_1 is no longer the codimension of the mean of the visualized field, it still provides a description of the sparseness). For higher C_1, the fields are sparsely scattered, with only a few large singularities, while as C_1 is reduced, the field becomes more and more space-filling. Note that in these illustrations the fields are rescaled so that their maximum values are at the same apparent height regardless of value; for the colour illustrations all colours are given by actual field value.

It should also be noted that a modification in the simulation procedure must be made for fields with $\alpha < 1$. In this case, direct use of the factor $w(\mathbf{k}) \propto |\mathbf{k}|^{-d/\alpha'}$ leads to numerical overflow (recall that $\alpha < 1 \Rightarrow \alpha' < 0$); to avoid this problem it suffices to filter by using the numerical transform of the real-space term $w(\mathbf{x}) \propto |\mathbf{x}|^{-d/\alpha}$. The remainder of the simulation program proceeds as above.

4. Fractional Integration and Non–Conservative Quantities

In order to simulate certain physical fields, for example clouds, or galaxies, it is necessary to introduce the parameter H explained above, describing the degree of conservation of the field. To see how this arises in various physical systems consider the example of passive scalar clouds, i.e. cloud fields which are passively advected by the turbulent velocity without themselves affecting the latter; these obey Corrsin-Obukhov scaling[41]. A sizable body of turbulence theory exists for these. From the Navier-Stokes equations and the equation of passive advection we find that the nonlinear terms exactly conserve the following quantities[28]:

$$\varepsilon = -\partial\langle v^2\rangle/\partial t = \text{constant}$$
$$\chi = -\partial\langle \rho^2\rangle/\partial t = \text{constant} \tag{22}$$

where v is the velocity field and ρ the concentration, and ε and χ are the energy and scalar variance fluxes respectively . Dimensional analysis implies that

$$\Delta v(l) \approx \varepsilon^{1/3} l^{1/3}$$
$$\Delta\rho(l) \approx \varphi^{1/3} l^{1/3} \tag{23}$$
$$\text{where } \varphi = \chi^{3/2}\varepsilon^{-1/2}$$

A passive scalar cloud therefore already involves two nonlinearly coupled cascade processes, for ε and χ; in general we will need to use vector cascade processes, using the

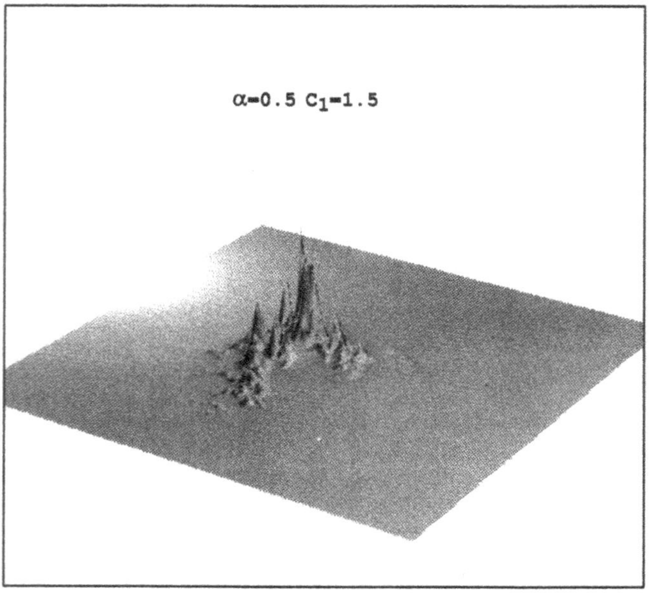

Figure 8. 3–D representations of 256x256 simulated fields.

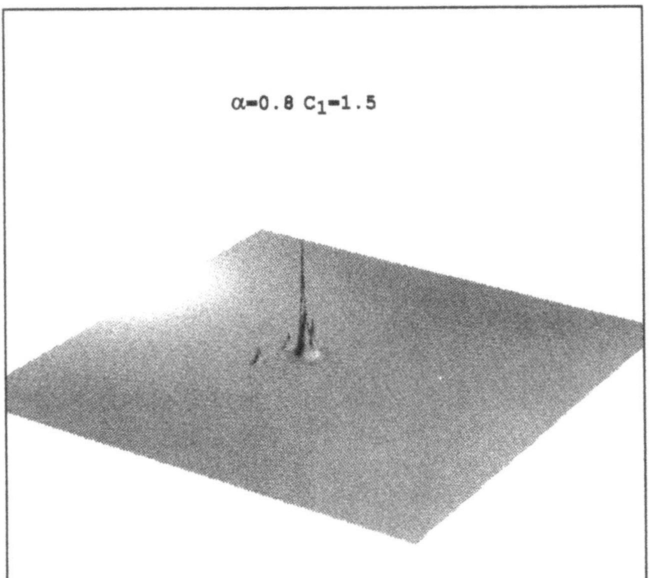

Figure 9. 3–D representations of 256x256 simulated fields.

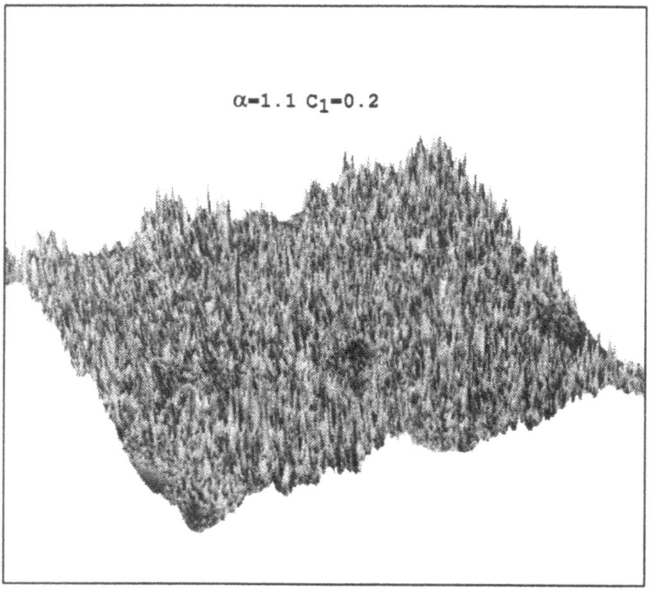

Figure 10. 3–D representations of 256x256 simulated fields.

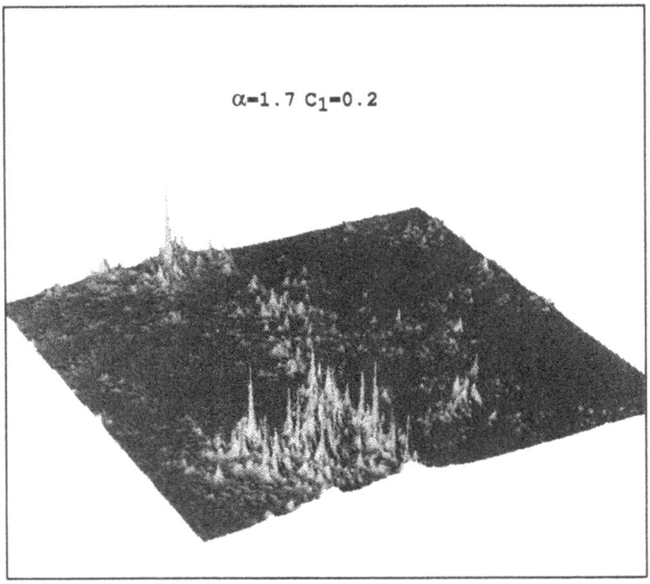

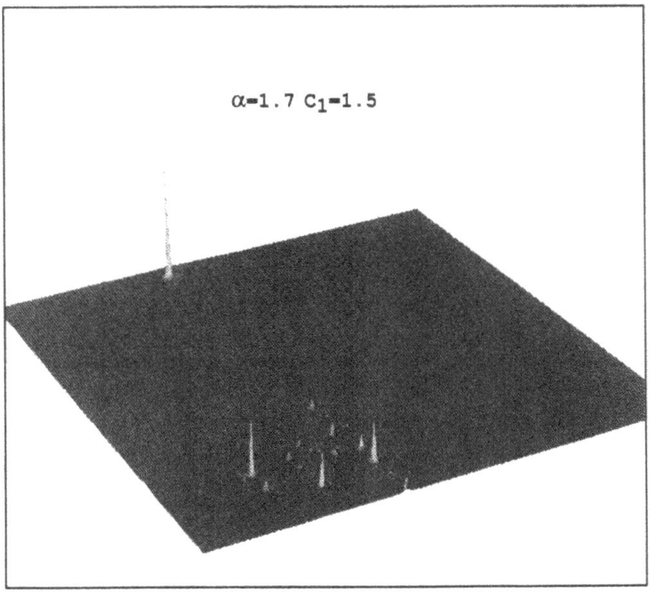

Figure 11. 3–D representations of 256x256 simulated fields.

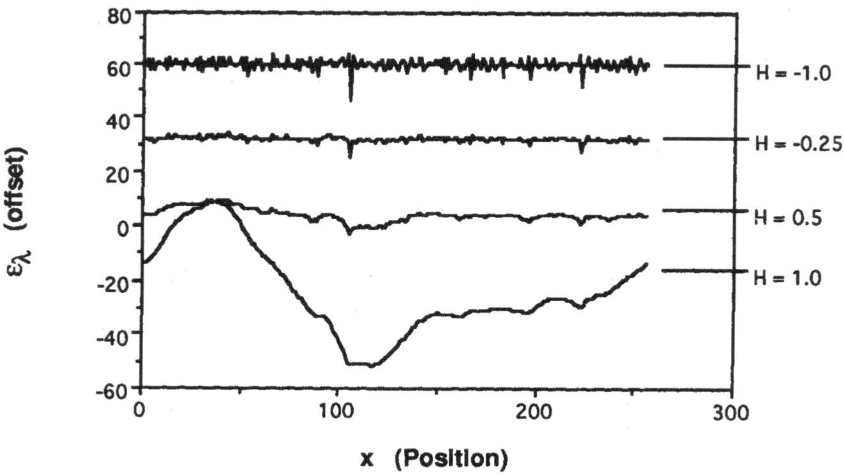

Figure 12. One-dimensional simulation of length 256, with $\alpha=1.8$ and $C_1=0.02$, fractionally integrated with varying H (non-conservation parameter). These simulations have been vertically offset so as not to overlap.

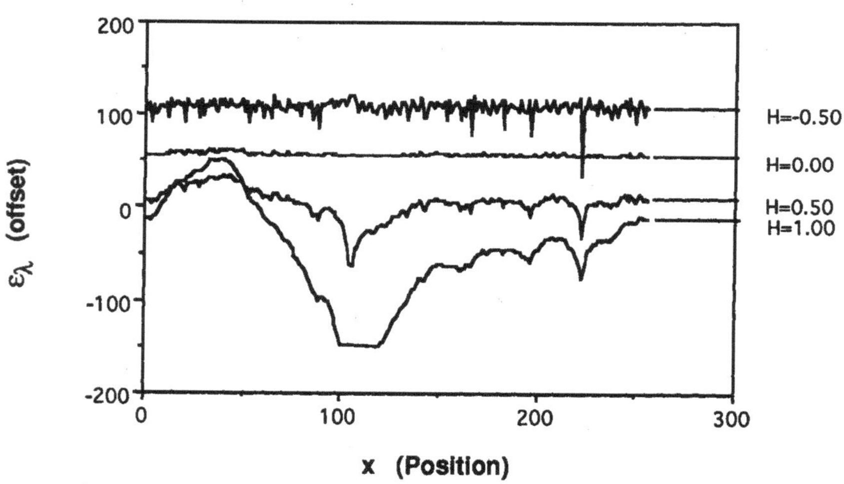

Figure 13. One-dimensional simulation of length 256, with $\alpha=1.5$ and $C_1=0.4$, fractionally integrated with varying H (non-conservation parameter). These simulations have been vertically offset so as not to overlap.

new formalism of "Lie cascades"[42]. Here, we make the simplest assumption, that χ is completely statistically dependent on ε. The field ρ therefore satisfies $\Delta\rho \approx \varepsilon^{1/3} l^{1/3}$. The cube-root of the conserved field is easy to calculate. To introduce the $l^{1/3}$ factor, we transform the field back into Fourier space and multiply it by the factor $k^{-1/3}$. This is in effect a fractional integration of order[31] 1/3. It results in a scale-invariant smoothing of the field. For a general non-conservation parameter H, the simulated conservative field is fractionally integrated by multiplying in k-space by k^{-H}. For H>0, this results in a smoothing of the field; H<0 corresponds to a differentiation and leads to a roughening of the field. It must also be noted that, since the factor k^{-H} will become infinite for positive H as $k \to 0$, the term (corresponding to the mean of the field) cannot be fractionally integrated using Fourier techniques[31]; it must be integrated seperately in real space. The k=0 component $\bar{\varepsilon}$ of a simulation of linear size L is obtained by the following multiplication :

$$\bar{\varepsilon} \to \bar{\varepsilon}\left(L^{H}\right) \tag{24}$$

One-dimensional examples of H-filtering are given in Figs 12 and 13. The effects of H-filtering can be clearly seen. The higher the value of H used, the smoother the field which is obtained (remembering that due to the scale, the H-filtered fields are increased in size relative to the normalized conservative field). On the other hand, negative values of H accentuate the singularities already present in the field. The smoothing and roughening effects are also clearly seen in the 3-dimensional representation of a two dimensional continuous cascade. (See Colour Illustration 4). It can be noted that the smaller spikes, when sufficiently H-filtered, are almost eliminated, while for negative H, the field is extremely spiky. One can also note that negative values of H lead to the production of "regularities", sharp downward spikes, as is expected for a differentiation of the field.

5. Generalized Scale Invariance

A multifractal field is developed from a process that is scale-invariant. In its simplest form, this scale-invariance is isotropic; the resulting multifractals are self-similar, exhibiting no preferred direction in space. However, geophysical fields are seldom, if ever, isotropic. Preferred directions associated with texture and morphology exist whose anisotropy often varies with scale. Isotropic scale-invariance indicates that the (statistical) properties of a field are changed in a power-law way with an isotropic change in scale of space, such as zooming. More generally, we can consider a system to be scale-invariant if this property remains true under more general nonisotropic scale changes. That is, there exists a scale changing operator[31,43] for a given scale ratio λ, T_λ, such that given balls B_λ at this scale ratio,

$$T_\lambda B_1 = B_\lambda \tag{25}$$

T_λ is the rule relating the statistical properties at one scale to another and involves only the scale ratio (there is no characteristic "size"), so that $\mathbf{x}_\lambda = T_\lambda \mathbf{x}_1$, where $\mathbf{x}_\lambda$ and $\mathbf{x}_1$ are vectors on B_λ and B_1 respectively. This implies that T_λ has certain properties. In particular, if and only if $\lambda_1\lambda_2 = \lambda$, then

$$B_\lambda = T_{\lambda_1} B_{\lambda_2} = T_{\lambda_2} B_{\lambda_1}\ , \tag{26}$$

i.e., T_λ has the group properties:

$$T_{\lambda_2} T_{\lambda_1} B_1 = B_\lambda = T_{\lambda_1} T_{\lambda_2} B_1 \tag{27}$$

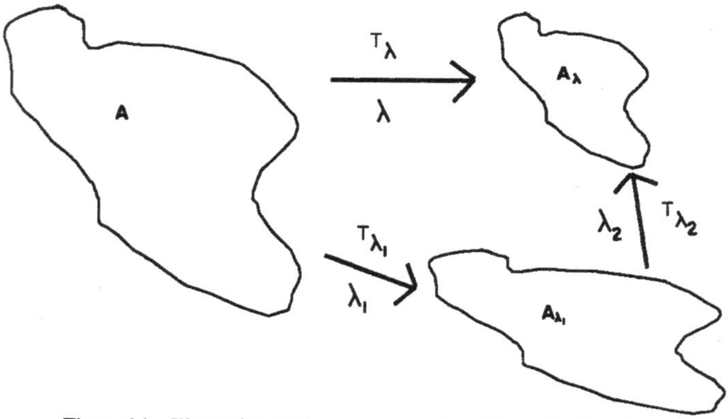

Figure 14—Illustration of the group property of the scale changing operator.(18)

In generalized scale invariance (GSI), T_λ, the scale-changing operator, can be much more general than the standard isotropic dilation. As indicated earlier, in general, a scale invariant system will be one in which the small and large scales are related by a scale changing operation that involves only the scale ratios; there is no characteristic size. Following the development of Schertzer and Lovejoy[43], we note that to be completely defined, GSI needs more than a scale changing operator; it also requires a definition of the unit scale, as well as a definition of how to measure the scale. These three basic elements can be summarized as follows:

i) The unit ball B_1 which defines all the unit vectors. If an isotropic ball (*i.e.*, circle or sphere) exists, we call the corresponding scale the "spheroscale".

ii) The scale changing operator T_λ which transforms the scale of vectors by scale ratio λ. T_λ is a one parameter multiplicative group $\Rightarrow T_\lambda = \lambda^{-G}$, where G is the generator of the group.

If G is an n x n matrix and **x** is an n-dimensional vector then we may use:

$$\lambda^{-G} = \exp\left(-G \cdot \ln(\lambda)\right) = 1 - G \cdot \ln(\lambda) + \frac{1}{2} G^2 \cdot \ln^2(\lambda) - \cdots . \tag{28}$$

This is linear GSI[44]. When G is a more general (*i.e.*, nonlinear) generator, we must define λ^{-G} by differential equations using the fact that G defines infinitesimal transformations.

iii) A measure of scale such as some power of the volume of B_λ ; the exact definition is somewhat a matter of convenience or convention.

The simplest type of scaling is self-similar scaling, as in isotropic cascades. Here, $T_\lambda = \lambda^{-1}\mathbf{1} \Rightarrow G = \mathbf{1}$, where $\mathbf{1}$ is the identity matrix. Taking B_1 as the unit circle, the balls B_λ are circles.

A more interesting case occurs when the scaling is different in two (or more) fixed directions, *e.g.*, coordinate axes. This type of scaling is called "self-affine": G is a diagonal matrix. A simple example is an anisotropic scaling model of atmospheric stratification, which provided the initial impetus for developing GSI[29,30]. Consider a vertical cross-section of the atmosphere (see figure 16), using

$$G = \begin{bmatrix} 1 & 0 \\ 0 & H_z \end{bmatrix} \text{ and } \mathbf{x} = \begin{bmatrix} x \\ z \end{bmatrix} , \tag{29}$$

where x is the horizontal distance and z is the vertical distance. Then the relation $\mathbf{x}_\lambda = T_\lambda \mathbf{x}_1$, where $\mathbf{x}_1$ is the position vector of the unit ball (a circle in this case), becomes

$$\begin{bmatrix} x_\lambda \\ z_\lambda \end{bmatrix} = \lambda^{-\begin{bmatrix} 1 & 0 \\ 0 & H_z \end{bmatrix}} \cdot \begin{bmatrix} x_1 \\ z_1 \end{bmatrix} = \begin{bmatrix} \lambda^{-1} & 0 \\ 0 & \lambda^{-H_z} \end{bmatrix} \cdot \begin{bmatrix} x_1 \\ z_1 \end{bmatrix} = \begin{bmatrix} \lambda^{-1} \cdot x_1 \\ \lambda^{-H_z} \cdot z_1 \end{bmatrix} . \tag{30}$$

When $\lambda \gg 1$, then the cells are strongly vertically aligned. Defining the unit vectors $\mathbf{x}_1$ as circles, this corresponds to structures (*e.g.*, clouds) whose average cross-sections are circular—this is the "spheroscale". When $\lambda \ll 1$, they are horizontally flat.

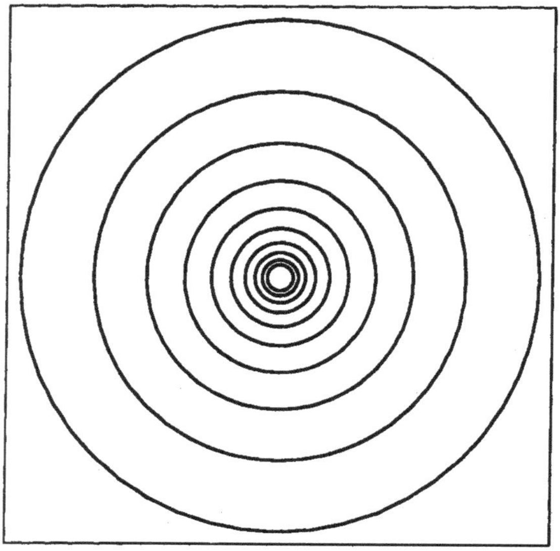

Figure 15—Self-similar scaling.

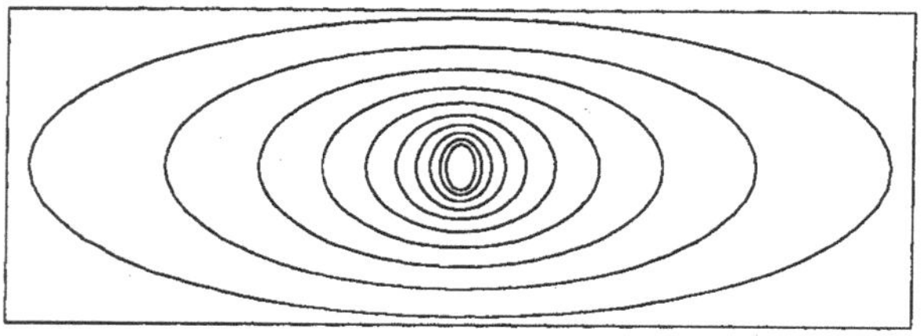

Figure 16—Self-affine scaling with $H_z = 5/9$. The shape of the balls models the average eddy shape; the flattening of the large balls models the fact that the atmosphere is increasingly stratified at large scale, and the small-scale vertically oriented balls correspond to "convective" cells or "rain shafts"

Figure 16 shows an example of a deterministic self-affine fractal set. As expected, the structures which are initially horizontally flat at large scales become more and more vertically flat at small scales.

One might also expect differential rotation (e.g. associated with Coriolis force in the atmosphere for clouds). This can be modeled by a matrix G with off-diagonal elements. We decompose G into quaternion-like elements[29,45] :

$$\mathbf{G} = d\mathbf{1} + e\mathbf{I} + f\mathbf{J} + c\mathbf{K},$$

where

$$\mathbf{1}=\begin{bmatrix}1 & 0\\ 0 & 1\end{bmatrix}, \quad \mathbf{I}=\begin{bmatrix}0 & -1\\ 1 & 0\end{bmatrix}, \quad \mathbf{J}=\begin{bmatrix}0 & 1\\ 1 & 0\end{bmatrix}, \quad \mathbf{K}=\begin{bmatrix}-1 & 0\\ 0 & 1\end{bmatrix}. \tag{31}$$

To obtain $\lambda^{-G} = e^{-G\ln\lambda}$, we use the series expansion of the exponential function with the following identities:

$$(G - d\,\mathbf{1})^{2n} = (e\,\mathbf{I} + f\,\mathbf{J} + c\,\mathbf{K})^{2n} = a^{2n}\mathbf{1}, \tag{32}$$

$$(G - d\,\mathbf{1})^{2n+1} = a^{2n}\,(G - d\,\mathbf{1}), \tag{33}$$

where n is an integer and $a^2 = c^2 + f^2 - e^2$. Writing[29,45] $U = \ln\lambda$:

$$T_\lambda = \lambda^{-G} = \lambda^{-d}\lambda^{(G-d\mathbf{1})} = \lambda^{-d}\left[\mathbf{1}\cosh(aU) - (G - d\mathbf{1})\frac{\sinh(aU)}{a}\right] \tag{34}$$

When $a^2<0$, the above formula holds but with $|a|$ replacing a and ordinary trigonometric functions rather than hyperbolic functions. The case $a^2>0$ corresponds to domination by stratification, whereas $a^2<0$ to domination by rotation.

The application of GSI to simulations is an important one; most geophysical fields include some element of anisotropy, such as gravitation and the Coriolis effect of the earth's rotation in the case of the atmosphere, or, in the case of earthquakes, preferred direction along the faultline. In addition, GSI has been used in the simulation of galaxies[43]; rotation dominance leads to spiral galaxy shapes, while stratification dominance tends to lead toward barred galaxy shapes.

Simulation of GSI

In order to model the effects of GSI, it is necessary to make some changes in the steps explained above. Rather than the usual dimension, we now use the elliptical dimension, given by d_{el}=Tr G, the trace of the GSI matrix. The volume of the balls B_λ varies as $\lambda^{-d_{el}}$, so it is necessary to replace d by d_{el}. In addition, the weighting factor $w(\mathbf{k}) \propto |\mathbf{k}|^{-d/\alpha'}$ that determines the multi-scaling is changed not only by this replacement. It also becomes necessary to replace the modulus $|\mathbf{k}|$ by $\lambda(\mathbf{k})$. All the filters must be appropriately anisotropic.

In order to determine $\lambda(\mathbf{k})$, we consider GSI in Fourier space. Given $\mathbf{x}_\lambda = T_\lambda \mathbf{x}_1$, we wish to determine the Fourier space operator $\tilde{T}_\lambda$ such that $\mathbf{k}_\lambda = \tilde{T}_\lambda \mathbf{k}_1$ (corresponding to T_λ). In fact, by assuming the existence of matrices G and $\tilde{G}$ that generate the scale changing operators (with $T_\lambda = \lambda^{-G}$ and $\tilde{T}_\lambda = \lambda^{-\tilde{G}}$), and requiring invariance of the scalar product under scale changes, it can be shown[44,45] that $\tilde{G}=G^T$. It remains to solve for the scale λ from the equation $\mathbf{k}_\lambda = \tilde{T}_\lambda \mathbf{k}_1$, i.e. to find λ such that $\mathbf{k}_1 = \tilde{T}_\lambda^{-1}\mathbf{k}_\lambda$; when $\mathbf{k}_1$ is on the unit circle, this requires $\left|\tilde{T}_\lambda^{-1}\mathbf{k}_\lambda\right| = 1$. To solve this, we need in addition a definition of the unit ball, i.e. some $U_1=\ln\lambda_1$ that defines the spheroscale. Given this, in two dimensions the scale λ can be found. Eq.(34) is used to determine $\tilde{T}_\lambda$ and the equation $\mathbf{k}_\lambda = \tilde{T}_\lambda \mathbf{k}_1$ is expanded to yield the following transcendental equation[44]:

$$\begin{aligned}
&\ln(P\cosh^2(a(U-U_1)) + Q\sinh^2(a(U-U_1)) - R(\sinh(2a(U-U_1))) = 2U \\
&P = k_x^2 + k_y^2 \\
&Q = \frac{k_x^2(c^2+(f-e)^2) + k_y^2(c^2+(f+e)^2) + 4k_xk_yce}{a^2} \\
&R = \frac{(k_x^2 - k_y^2)c + 2k_xk_yf}{a} \\
&U = \ln\lambda
\end{aligned} \tag{35}$$

This equation is then solved for λ for each value of k_x and k_y, and this value is then used in place of k in the weighting function w(**k**). The same must be done for the k^{-H} filtering if the field is to be fractionally integrated. By examining Colour Illustration 5, the effects of different GSI parameters can be seen. Altering the matrix G to produce rotation- or stratification-dominant fields noticeably changes the texture of the resulting simulation. Stratification dominant fields will have structures smaller than the spheroscale compressed in the direction of stratification, while structures larger than the spheroscale will be stretched in that direction. The total rotation as λ is increased from 1 to ∞ is less than $\pi/2$. Rotation-dominant fields are differentially rotated; the orientation of the large-scale structures is rotated with respect to the smaller-scale ones, an infinite number of rotations as λ is increased from 1 to ∞. Some stratification is useful to make this clearer.

In addition, the size of the spheroscale plays an important role in determining the nature of the resulting field. A very large spheroscale (Fourier space: this is small in real space) combined with stratification, for example, will result in a great deal of "streakiness" at the large scale in real space, since the spheroscale in real space will be at the small scales: these will be isotropic. On the other hand, with a spheroscale midway between the largest and smallest scales, stratification will be seen in opposite directions at the large and small scales in the resulting field.

Simulations can also be made where there is no spheroscale. In this case, a function describing some scale of the desired simulation must be described. Pflug[44] and Lovejoy[46] use the first three symmetric terms of a Fourier expansion to describe $r(\theta)$, i.e.

$r(\theta)=r_0+a_1\sin(2\theta)+a_2\sin(4\theta)+b_1\cos(2\theta)+b_2\cos(4\theta)$. (The Fourier transform of a real field respects the symmetry $F(\mathbf{k})=F^*(-\mathbf{k})$ hence we require $r(\theta)=r(\pi-\theta)$). The scale at which this function is described must be given, and care must be taken that the operation of scale-changing on the given function does not result in an overlapping of scales. (That is, within the range of scales in question the scale-changing operator acting on this $r(\theta)$ function gives a single-valued answer: the definition of scale must be unique). Other methods of defining groups of functions closed under scale-changing are currently being examined. Examples of GSI without spheroscales are shown in Colour Illustration 5. Many different textures are possible with the use of non-spheroscale GSI.

6. Conclusions

Scale invariance seems to be a property common not only to complex geophysical fields, but to aspects of biology and astrophysics as well. In the absence of any symmetry-breaking mechanism, scale invariance will be preserved, and even given the breaking of certain symmetries, anisotropic scale invariance may be preserved. We have examined how quite general assumptions concerning scale invariance lead to the idea of multifractals, and how nonlinear "mixing" (including densification) of multiplicative cascade processes leads to the production of universal multifractal fields. The empirically estimated parameters for many of these fields have been tabulated.

Simulations are an invaluable tool in the comprehension of multifractals, and in their use as physical models. Physical processes on multifractals which have already been explicitly considered with the help of such models include the radiative and diffusive transport on multifractals[47], the electromagnetic scattering properties[48], and the sampling and averaging properties of sparse measures[47]. We have described how to simulate both discrete cascades and the more physically relevant continuous cascade, for a general dimension of space. Several points mentioned in earlier papers have been clarified and the ideas expanded upon. In addition, the production of non-conservative fields by scale-invariant roughening or smoothing has been discussed. This allows generation of fields such as topography and the passive scalar cloud field.

Finally, the use of generalized scale invariance in the production of anisotropic cascades has been discussed. Many physical systems show a preferred direction, and the use of GSI allows the simulation of such effects as differential rotation and stratification in directions which can be functions of scale and position. As well, this framework can be used for the generation of time-series and forecasts, where the space-time transformation is not necessarily an isotropic one.

Further work is necessary in a number of areas to increase the utility of simulations. Higher dimensional GSI is very involved and the simulation of three-dimensional and higher dimensioned fields is not currently possible except for special cases. Additionally, techniques for simulating non-linear GSI, which can only be locally approximated by matrices, are lacking. All of these techniques will also be extended to more general fields of vector quantities, through the use of Lie cascades[42]. Applications of the simulation of universal multifractals are also being developed, including methods for forecasting time evolution of rain and cloud fields as well as prediction of earthquakes, and models of radiative transfer in clouds.

7. References

1. D. Lavallée,D. Jourdan, C. Gautier, C. Hooge, *J. Geophys. Res.* (1993) (in press)
2. Y. Tessier, S. Lovejoy, D. Schertzer, *J. Appl. Meteor.* **32**, (1993) p.223-250.
3. F. Schmitt, D. Lavallée, D. Schertzer, S. Lovejoy, *Phys. Rev. Lett.*, **68** (1992) p305-308.
4. D. Lavallée, in *Fractals in Geography*, eds. L. De Cola and N. Lam (Prentice and Hall) p.171-205.
5. Y. Tessier, S. Lovejoy, D. Schertzer, D. Lavallée, B. Kerman , *Geophys. Res. Lett.* (1993) (in press).
6. F. Schmitt, D. Schertzer, S. Lovejoy, D. Lavallée, *Physics of Fluids* (1993) (submitted 9/92)
7. Y.Y. Kagan, and L. Knopoff, *Geophysical Journal R. Astron. Sci.* **62** (1980) p 303.
8. J. Key and A. S. McLaren, *Geophysical Research Letters*, **18** (1991) p.1437
9. B. T. Jones, V. J. Martinez, E. Saar, J. Einasto *Astrophysical Journal* **332. L1.**(Sept 1988)
10. F. Francis, S. Lovejoy, B. Kerman, D. Schertzer, *Geophysical Research Letters* (submitted).
11. C. Hooge, M.Sc. Thesis, McGill University, Montréal, Québec (1993)
12. P. Garrido, M.Sc. Thesis, McGill University, Montréal, Québec (1993)
13. M. Dogterom and L. Pietronero, *Physica* **A171** (1991) p.239.
14. C. Larnder, N. DeSaulnier-Soucy, S. Lovejoy, D. Schertzer, C. Braun, D. Lavallée,*International Journal of Bifurcation and Chaos*, **2** (1992) p.715
15. S. P. Ratti, G. Salvadori, G. Gianini, S. Lovejoy, D. Schertzer, *Proc. of 21st Int. Symp. on Multiparticle dynamics* (Wuhan,1991)

16. G. Salvadori, S. P. Ratti, G. Belli, S. Lovejoy, D. Schertzer, *J. of Toxicological and Environ. Chem.* (1993) (in press).

17. S. Lovejoy, *20th conf. on radar meteorology* (AMS Boston, 1981) p.476

18. S. Lovejoy, B. Mandelbrot, *Tellus,* **37A** (1985) p.209

19. R. Voss, *Proceedings, Siggraph. Conf.* (Detroit, 1983) p.1

20. A. Fournier, D. Fussel, L. Carpenter, *Comm. of the ACM,* **6** (1982) p.371

21. M. Barnsley, *Fractals Everywhere* (Academic Press, San Diego, 1988)

22. E. A. Novikov, R. Stewart, *Izv. Akad. Nauk. SSSR, Ser. Geofiz.,* **3** (1964) p.408

23. A. M. Yaglom, *Dokl. Akad. Nauk. SSSR,* **2** (1966) p.26

24. B. Mandelbrot, *J. Fluid. Mech.,* **62** (1974) p.331

25. U. Frisch, P. L. Sulem, M. Nelkin, *J. Fluid. Mech.,* **87** (1978) p.719

26. D. Schertzer, S. Lovejoy, *Proceedings, 4th Symp. of Turbulent Shear Flows,* 11.1, 11.8

27. J. Wilson, D. Schertzer, S. Lovejoy, in *Non-Linear Variability in Geophysics: Scaling and Fractals,* eds. D. Schertzer and S. Lovejoy (Kluwer, Norwell, MA, 1991), p.183-207

28. D. Schertzer, S. Lovejoy, *J. Geophys. Res.,* **92** (1987) p.9692

29. S. Lovejoy, D. Schertzer,*Water Resources Research,* **21,8** (1985) p.1233

30. D. Schertzer, S. Lovejoy, *P.C.H. Journal,* **6** (1985) p.623

31. D. Schertzer and S. Lovejoy, in*Non-Linear Variability in Geophysics: Scaling and Fractals,* eds. D. Schertzer and S. Lovejoy (Kluwer, Norwell, MA,1991) p.41

32. S. Lovejoy, D. Schertzer, *Multifractals in Geophysics,* Course Notes (AGU-CGU-MSA Spring Meeting, May 11, 1992)

33. D. Schertzer, S. Lovejoy, *Physica* **A 185** (1992) p.187

34. G. Batchelor, A. Townsend, *Proc. Roy. Meteor. Soc.,* **A199** (1949) p.238

35. D. Schertzer, S. Lovejoy in*Turbulent Shear Flows 4,* ed. L. Bradbury, F. Durst, B. Launder, F. Schmidt, J. Whitelaw (Springer-Verlag, Berlin, 1985) p.7

36. W. Feller, *An Introduction to Probability Theory and its Applications, 2nd Ed.* (John Wiley & Sons, New York, 1971)

37. J. M. Chambers, C. L. Mallows, and B. W. Stuck, *Journal of the American Statistical Association,* **71** (1976) p.340

38. D. Schertzer, S. Lovejoy, D. Lavallée, F. Schmidt in *Nonlinear Dynamics of Structures,* ed. R. Z. Sagdeev, U. Frisch, A. S. Moiseev, A. Erokhin (World Scientific,1991), p.213-235

39. B. Mandelbrot, in *Fractals in Geophysics*, ed. C. H. Scholz, B. Mandelbrot (Birkhäuser Verlag, 1989) p.5

40. Brian Watson, private communication

41. A. Obukhov, *J. Geophys. Res.* **67** (1962) p.3011

42. D. Schertzer, S. Lovejoy, in *Space-time Variability and Interdependence for Various Hydrological Processes*, ed. G. Kovaks (U. Press, Cambridge, 1993) (in press)

43. D. Schertzer and S. Lovejoy, *PAGEOPH*, **130** (1989) p.57

44. K. Pflug, Master's Thesis, McGill University, Montréal, Québec (1991)

45. K. Pflug, S. Lovejoy and D. Schertzer, in *Nonlinear Dynamics of Structures*, ed. R. Z. Sagdeev, U. Frisch, A. S. Moiseev, A. Erokhin (World Scientific,1991), p.72-78

46. S. Lovejoy, D. Schertzer, K. Pflug, *Physica* **A 185** (1992) p.121

47. A. Davis, S. Lovejoy, D. Schertzer, *SPIE* **1558** (1991) p.37

48. M. Duncan, S. Lovejoy, F. Fabry, D. Schertzer, *Proc. 2nd Inter. Symp. on Hydro. Appl. of Weather Radar* (U. of Hanover, Sept. 7-10, 1992)

image #	α	C_1	H	G matrix	u_1,spheroscale	r_0, a_1, b_1, a_2, b_2 (see text)
1	1.7	0.1	0.5	$\begin{pmatrix} 1.20 & 0.05 \\ -0.05 & 0.60 \end{pmatrix}$	$u_1 = \ln 100$	-
2	1.7	0.1	0.5	$\begin{pmatrix} 1.00 & 0.40 \\ -0.30 & 0.80 \end{pmatrix}$	$u_1 = \ln 100$	-
3	1.7	0.1	0.5	$\begin{pmatrix} 1.30 & 0.20 \\ -0.10 & 0.70 \end{pmatrix}$	$u_1 = \ln 50$	-
4	1.7	0.1	0.5	$\begin{pmatrix} 1.30 & 0.20 \\ -0.10 & 0.70 \end{pmatrix}$	$u_1 = \ln 1000$	-
5	1.7	0.1	0.5	$\begin{pmatrix} 1.10 & 0.40 \\ -0.10 & 0.90 \end{pmatrix}$	none	100, 10, -20, 15, 15
6	1.7	0.1	0.5	$\begin{pmatrix} 0.50 & -0.30 \\ -0.20 & 1.10 \end{pmatrix}$	none	100, 0, 20, 0, 20

Table 2: Multifractal and GSI parameters for the images in Colour Illustration 5.

Colour Illustration 1.
2–D simulations , with varying α and C_1.
Seed is constant.
H–filtered with H=0.5
The palette is shown below. White values are high, blue low.

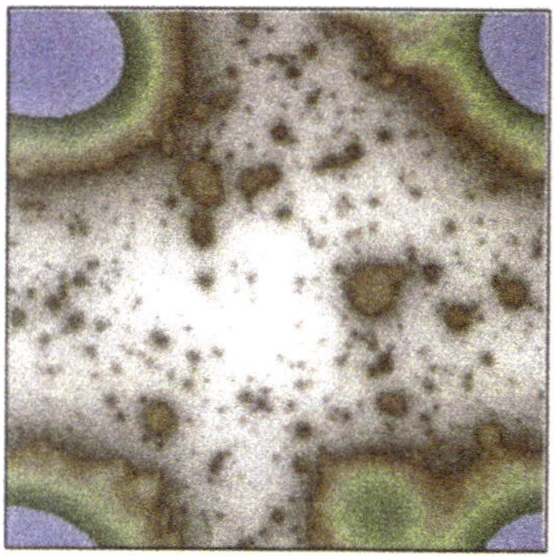

α=0.5 C_1=0.05

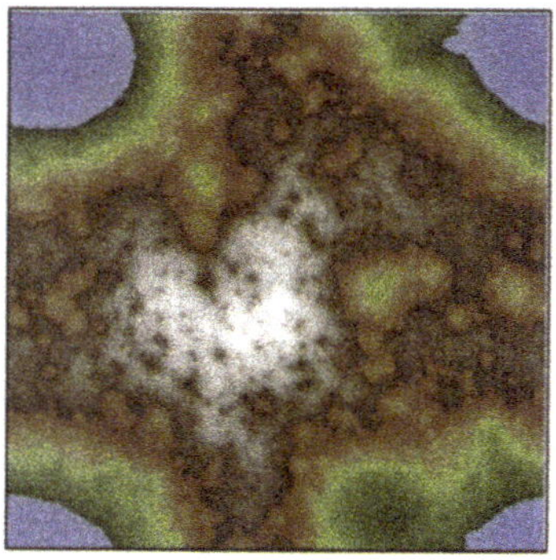

α=0.8 C_1=0.05

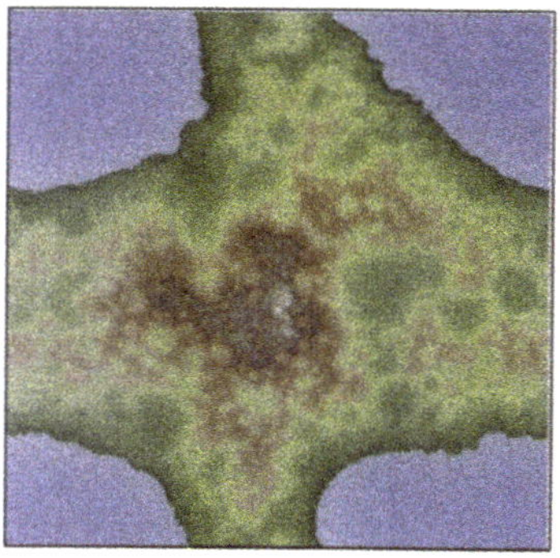

α=0.5 C_1=0.2

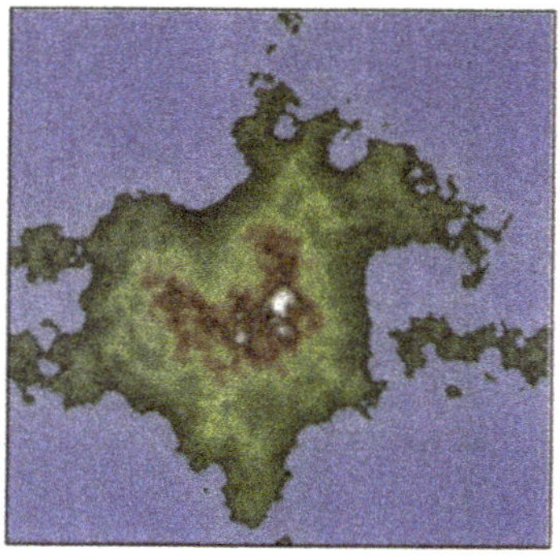

α=0.8 C_1=0.2

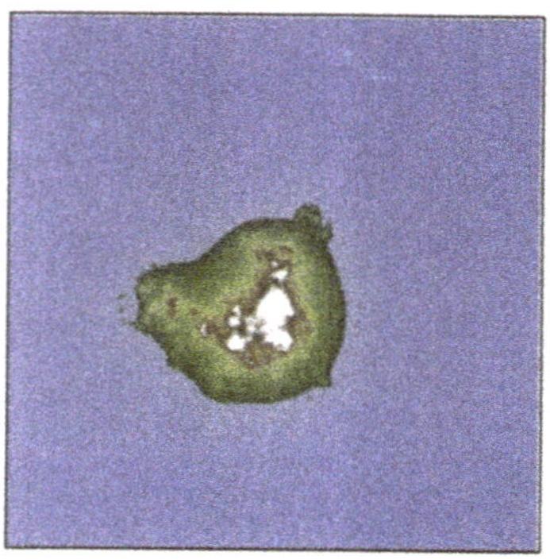

α=0.5 C_1=1.5

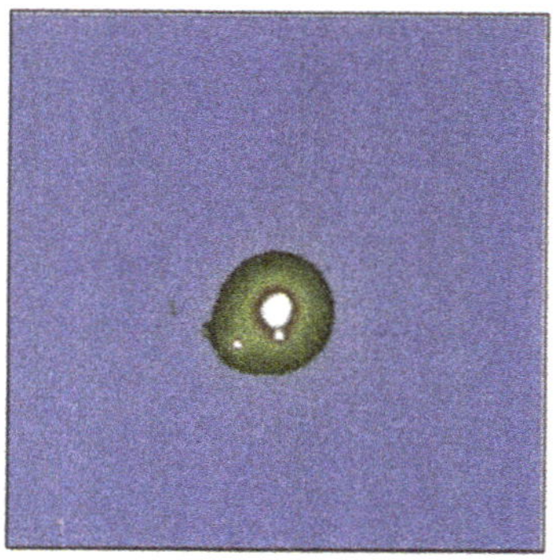

α=0.8 C_1=1.5

Colour Illustration 2. 2–D simulations , with varying α and C_1. Seed is constant. H–filtered with H=0.5
The palette is shown below. White values are high, blue low.

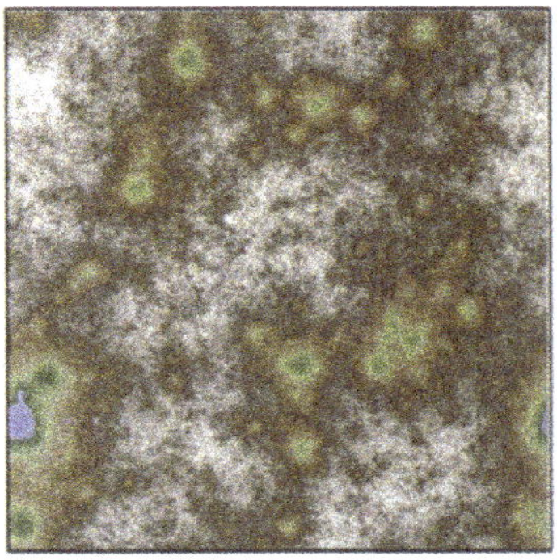

α=1.1 C_1=0.05

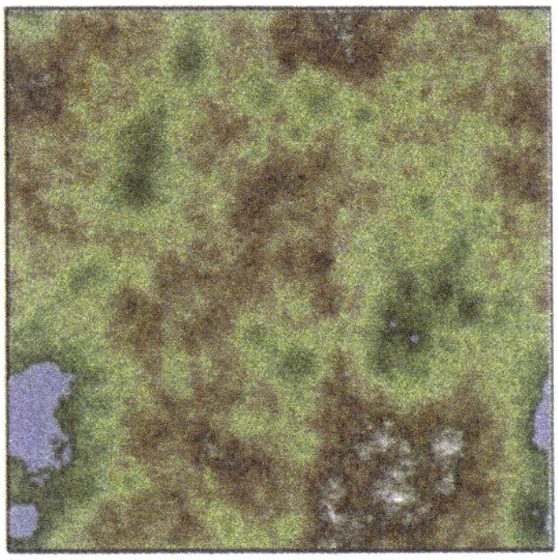

α=1.4 C_1=0.05

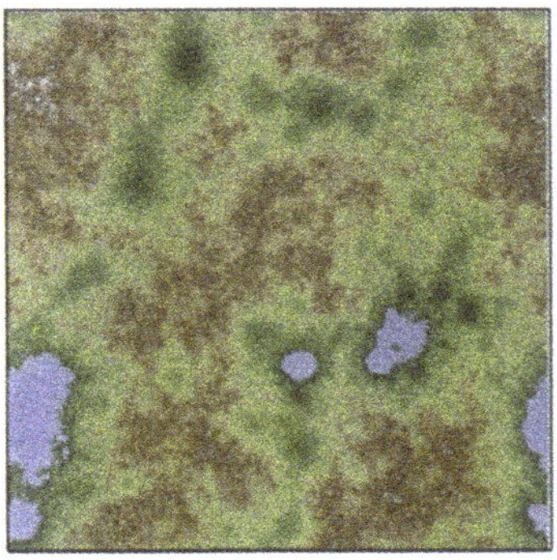

α=1.1 C_1=0.2

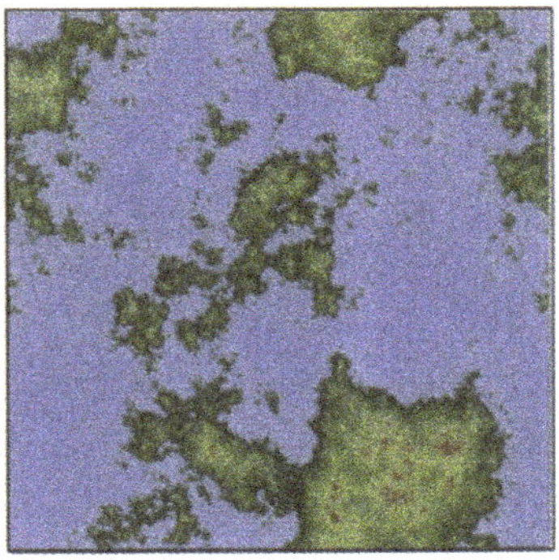

α=1.4 C_1=0.2

α=1.1 C_1=1.5

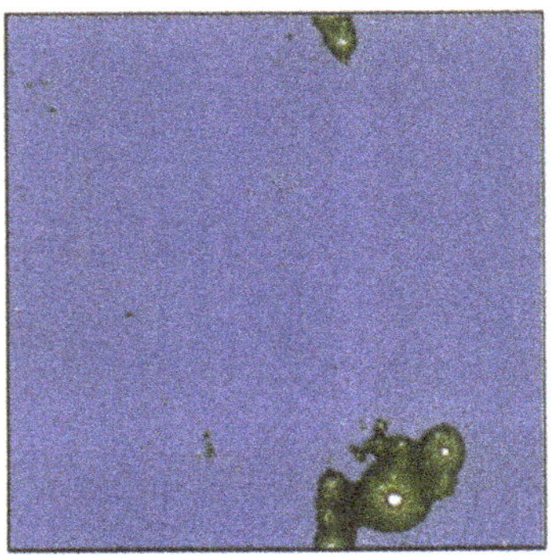

α=1.4 C_1=1.5

Colour Illustration 3.
2–D simulations, with varying α and C_1. Seed is constant. H–filtered with H=0.5
The palette is shown below. White values are high, blue low.

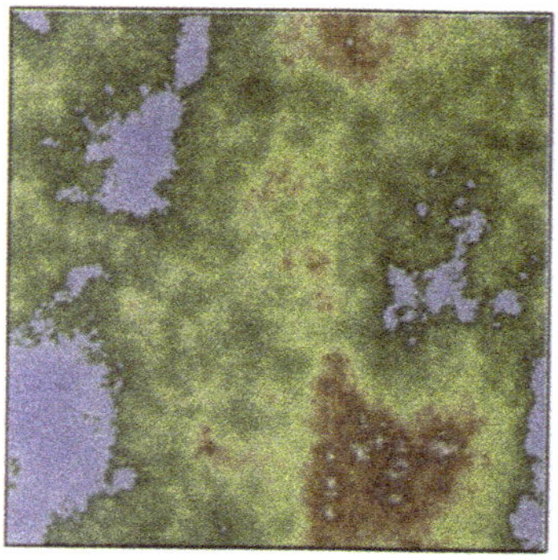

α=1.7 C_1=0.05

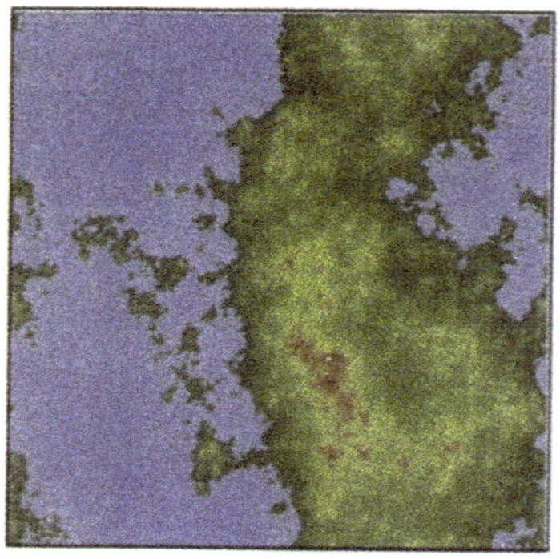

α=2.0 C_1=0.05

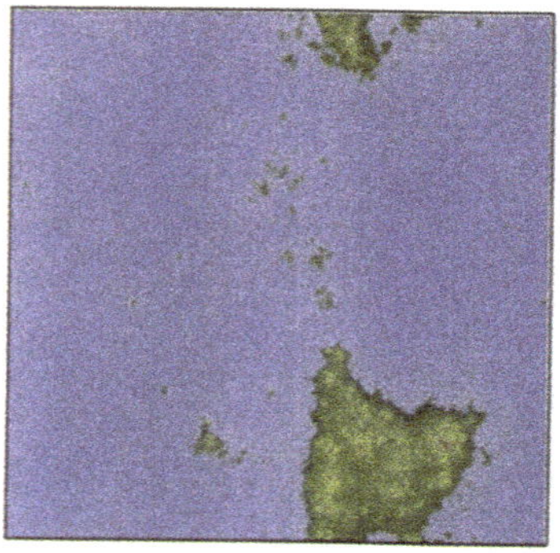

α=1.7 C_1=0.2

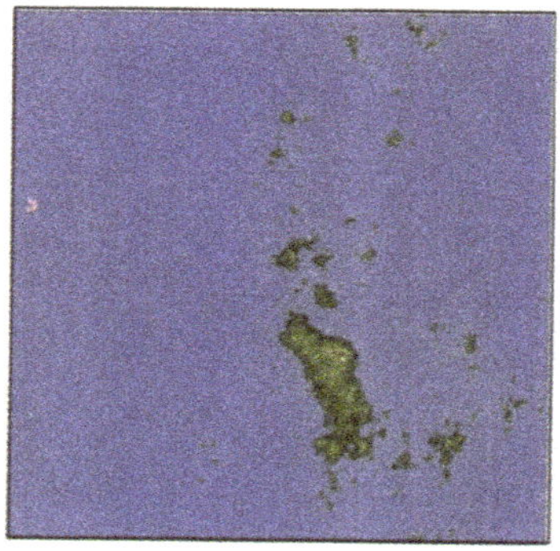

α=2.0 C_1=0.2

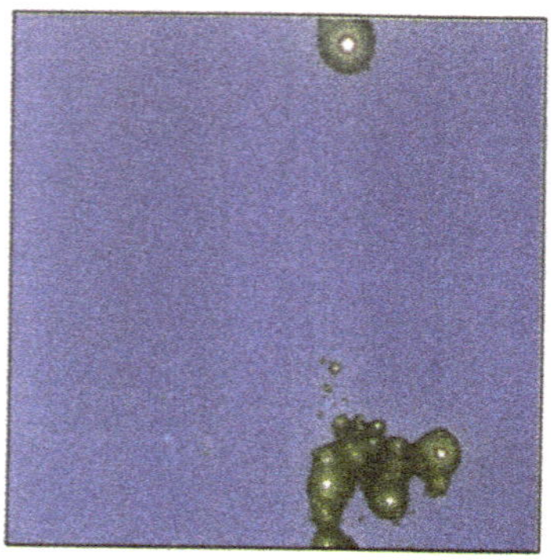

α=1.7 C_1=1.5

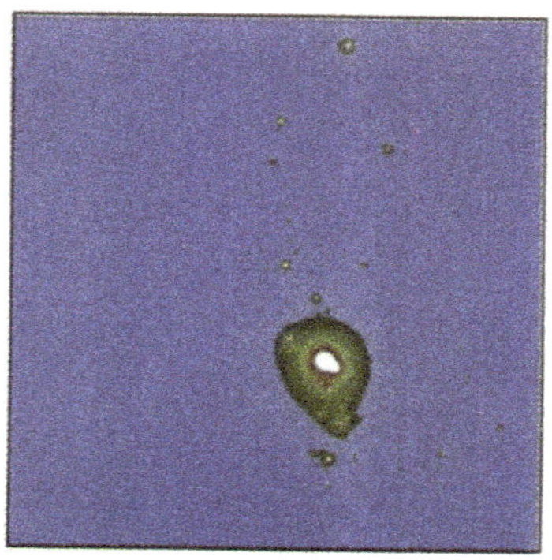

α=2.0 C_1=1.5

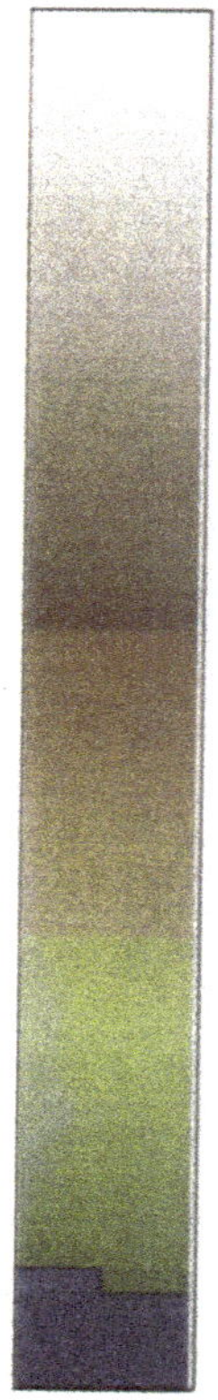

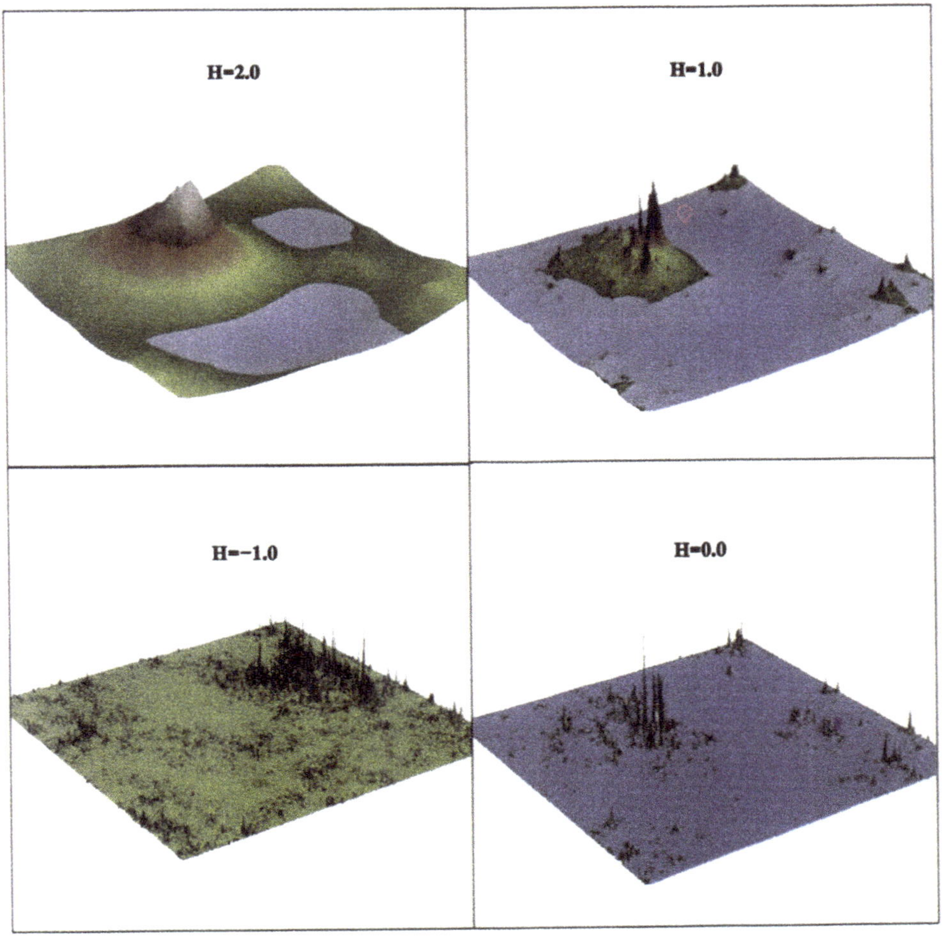

Colour Illustration 4. 3–dimensional representations of fields with α=1.7 and C_1=0.2, and variable H–filtering.

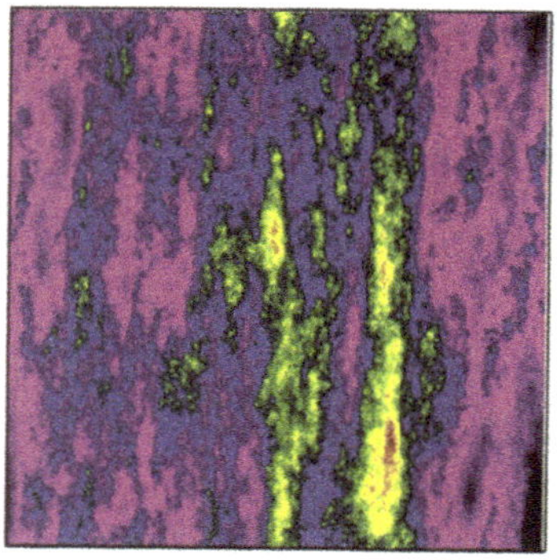

1.Stratification dominant

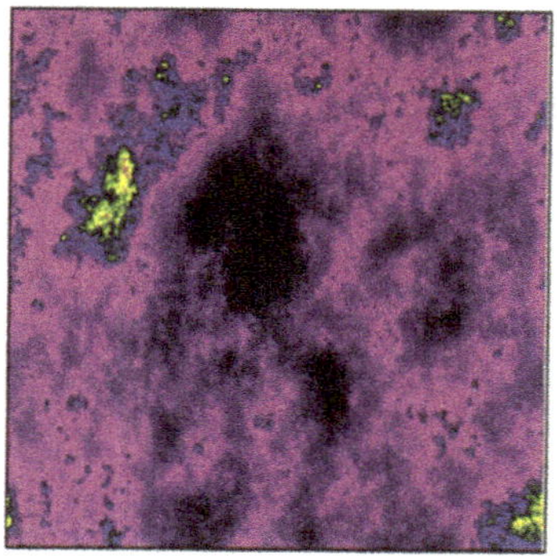

2.Rotation dominant

Colour Illustration 5.
2–D simulations , with varying GSI parameters Seed is constant.
$\alpha=1.7$, $C_1=0.1$.
H–filtered with $H=0.5$
The palette is shown below. Red values are high, black are low.

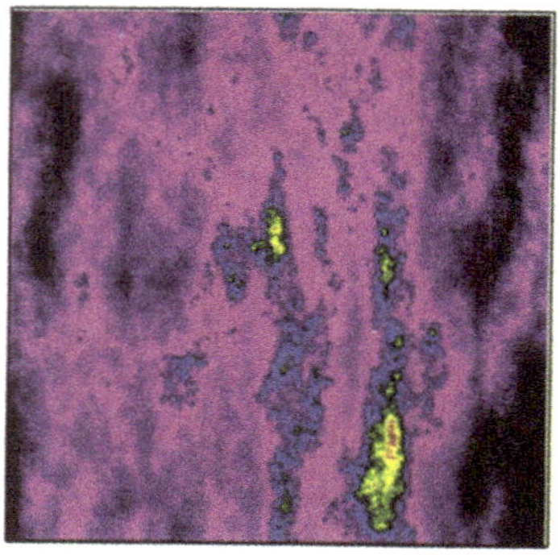

3.$U_1=\ln 50.0$

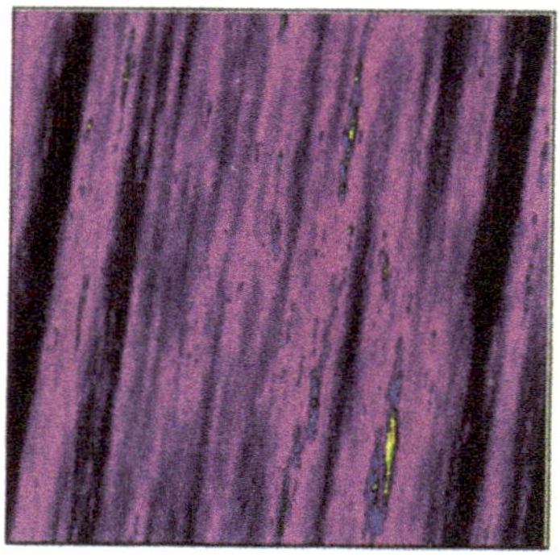

4.$U_1=\ln 1000.0$

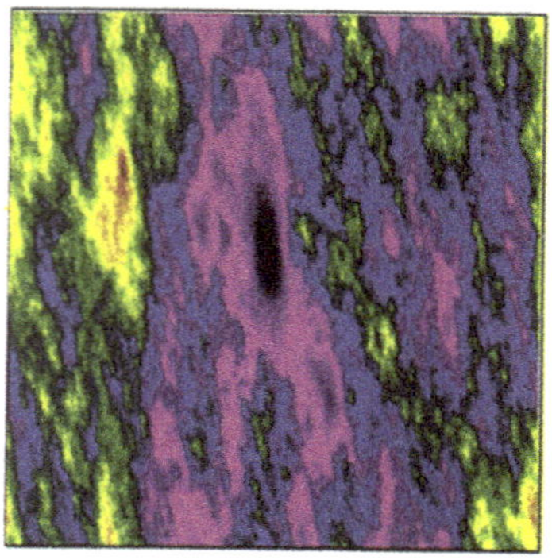

5.Non–spheroscale

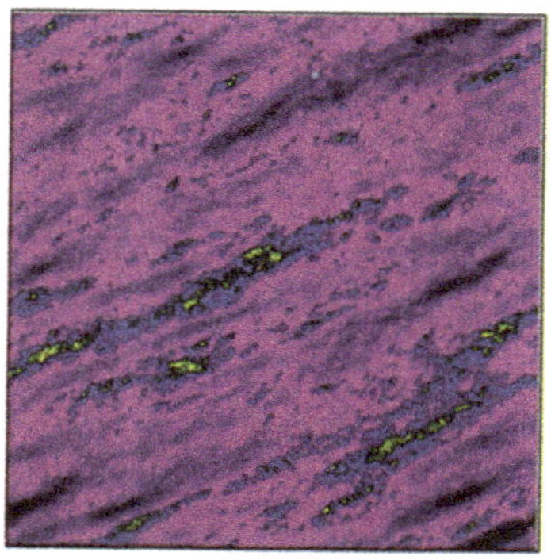

6.Non–spheroscale

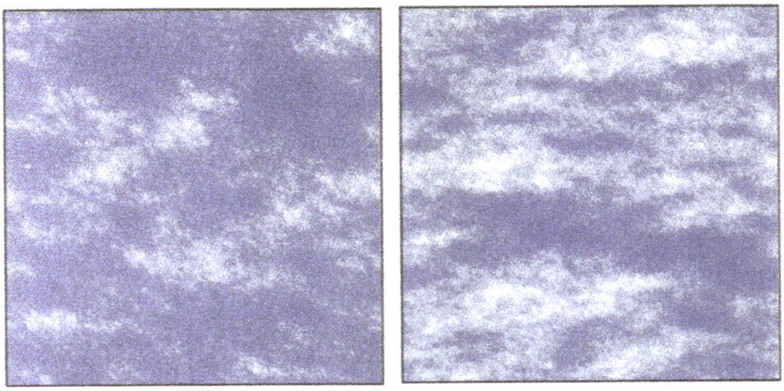

Colour Illustration 6. Simulations using parameters from Table 1. GSI was included. From top, two cloud simulations (visible spectrum),two topography simulations and two simulations of galactic luminosity distribution.

Emergence of Structures from Chaos

Laurent Nottale
CNRS, DAEC, Observatoire de Paris-Meudon, F-92195 Meudon Cedex, France

Abstract. One of the main problems encountered in the study of systems where dynamical chaos occurs is the emergence of structures. We propose a stochastic method for tackling this problem, based on the explicit introduction of the fractal and non-differentiable character of chaotic trajectories at very large time-scales. This method yields a formalism which resembles that of quantum mechanics (but the interpretation is different), and therefore provides us with definite structuration, in terms of peaks of probability density. We exemplify this approach by applying it to the problem of the distribution of planets in the Solar System, and suggest that it may serve as basis for cellular automata models of nondifferentiable processes.

1. Introduction

Chaos (often defined as extreme sensibility to variation of initial conditions) is relevant in a huge variety of natural systems. Although various methods of analysis have been coined to describe the development of chaos (strange attractors, fractal and information dimensions, entropy, characteristic exponents, catastrophe theory...), all of them have up to now struck against the unpassable barrier of unpredictability at large time scales. However, in many systems where chaos arises, spatial and temporal structures are observed experimentally; such structures are in some few cases found or confirmed in numerical simulations, but very rarely understood or predicted from a fundamental theory.

We suggest in the present work a method for tackling this problem: this method is efficient precisely when other methods fail, i.e., on time scales large compared to the "chaos time" (or inverse Lyapunov exponent τ). We shall see that it naturally generates spatial structures. Then we exemplify this approach by its application to the problem of the regularity of the distribution of planets in the Solar System. We hope that the formalism which is presented here may be useful for those who develop cellular automata models, in order to describe chaotic systems, but also for lattice formulations of the Schrödinger equation and its solutions.

2. Formalism

The first step of our approach consists in giving up the concept of well-defined trajectory *at large time scales*, and in introducing families of virtual trajectories (Nottale, 1989, 1992b). The real trajectory is one random realization among the infinite number of trajectories of the family. Such families are now characterized by a probability density ρ. In order to describe them, we assume, once the chaos developed, that the virtual trajectories evolve following a diffusion process, characterized by some diffusion coefficient $\mathcal{D}$. We suggest to describe this diffusion by a Markov-Wiener process $\xi(t)$, as used in Einstein's description of Brownian motion.

Such a Wiener process is non-differentiable, and characterized by a fractal dimension D=2. The simplest way of breaking the differentiability is to assume that, at any point of the trajectory, there are *two* slopes instead of one. An example of continuous curve having such a property is provided by the well-known fractal Koch curve (Mandelbrot, 1983). So we define, following Nelson (1966, 1985), mean forward and backward derivatives, d_+/dt and d_-/dt:

$$\frac{d_\pm}{dt} y(t) = lim_{\Delta t \to 0\pm} \left\langle \frac{y(t+\Delta t) - y(t)}{\Delta t} \right\rangle \tag{1}$$

which, once applied to the position vector $\boldsymbol{x}$, yield *forward and backward mean velocities*, $\frac{d_+}{dt}\boldsymbol{x}(t) = \boldsymbol{b}_+$ and $\frac{d_-}{dt}\boldsymbol{x}(t) = \boldsymbol{b}_-$.

Let us now introduce our main new method. While in every present formulations of Nelson's stochastic mechanics, one writes two systems of equations for the forward and backward processes (or for combinations of them) and eventually combine them in the end into a complex equation, we have suggested (Nottale, 1992b) to work *from the beginning* in terms of complex quantities. So we combine the forward and backward derivatives of Eq. (1) in a complex derivative operator

$$\frac{\mathcal{d}}{dt} = \frac{(d_+ + d_-) - i\,(d_+ - d_-)}{2dt} , \tag{2}$$

which, when applied to the position vector, yields a complex velocity (Nottale, 1992b)

$$\mathcal{V} = \frac{\mathcal{d}}{dt}\boldsymbol{x}(t) = \boldsymbol{V} - i\,\boldsymbol{U} = \frac{\boldsymbol{b}_+ + \boldsymbol{b}_-}{2} - i\,\frac{\boldsymbol{b}_+ - \boldsymbol{b}_-}{2} . \tag{3}$$

The real part $\boldsymbol{V}$ of the complex velocity $\mathcal{V}$ generalizes the classical velocity, while its imaginary part, $\boldsymbol{U}$, is a new quantity arising from non-differentiability. The position vector $\boldsymbol{x}(t)$ is assimilated to a stochastic process which satisfies the following relations (respectively for the forward ($dt > 0$) and backward ($dt < 0$) process)

$$dx(t) = b_+[x(t)]\, dt + d\xi_+(t) = b_-[x(t)]\, dt + d\xi_-(t) \ , \tag{4}$$

where the $d\xi(t)$'s, which characterize the Wiener process, are Gaussian with mean zero, mutually independent and such that

$$<d\xi_{\pm i}\ d\xi_{\pm j}> = \pm 2\, \mathcal{D}\, \delta_{ij}\ dt \ , \tag{5}$$

$\mathcal{D}$ standing for the diffusion coefficient. The non-differentiability of the Wiener process is straighforward from the computation of the velocity, $v \approx d\xi/dt \propto dt^{-1/2}$, which is formally infinite. Note that in terms of the fractal approach (Nottale, 1989, 1992a,b), this expression would read $v \propto \delta t^{(1-D)/D}$, leading to $D = 2$ as expected for a Brownian-like motion.

Equation (5) now allows us to get a general expression for the complex time derivative $\not{d}/dt$. Consider a function $f(x,t)$, and expand its total differential to second order. We get

$$df = \frac{\partial f}{\partial t}\, dt + \nabla f . dx + \frac{1}{2} \frac{\partial^2 f}{\partial x_i\, \partial x_j}\, dx_i\ dx_j \ . \tag{6}$$

We may now compute the forward and backward derivatives of f. In this procedure, the mean value of $<dx_i\ dx_j>$ reduces to $<d\xi_{\pm i}\ d\xi_{\pm j}>$, so that the last term of (6) amounts to a Laplacian thanks to Eq. (5). We obtain

$$d_{\pm} f/dt = (\partial/\partial t + b_{\pm} . \nabla \pm \mathcal{D} \Delta\,) f \ , \tag{7}$$

so that we can finally give the expression for the complex time derivative operator (Nottale, 1992b):

$$\frac{\not{d}}{dt} = \left(\frac{\partial}{\partial t} - i\, \mathcal{D}\, \Delta \right) + \mathcal{V} . \nabla \ . \tag{8}$$

We shall now postulate that the passage from classical (differentiable) mechanics to the new nondifferentiable mechanics that is considered here can be implemented by a unique prescription: Replace the standard time derivative d/dt by the new complex operator $\not{d}/dt$. Let us now give the main steps by which one may generalize classical mechanics using this prescription.

We assume that any mechanical system can be characterized by a Lagrange function $\mathcal{L}(x, \mathcal{V}, t)$, from which an *average* stochastic action $\mathcal{S}$ is defined:

$$\mathcal{S} = \int_{t_1}^{t_2} \langle \mathcal{L}(x, \mathcal{V}, t) \rangle\ dt \ . \tag{9}$$

Our Lagrange function and action are *a priori* complex and are obtained from the classical Lagrange function $L(x, dx/dt, t)$ and action S precisely from applying the above prescription $d/dt \rightarrow \mathcal{d}/dt$. The least-action principle, applied on this new action with both ends of the above integral fixed, leads to generalized Euler-Lagrange equations (Nottale, 1992b)

$$\frac{\mathcal{d}}{dt}\frac{\partial \mathcal{L}}{\partial \mathcal{V}_i} = \frac{\partial \mathcal{L}}{\partial x_i} \ , \qquad (10)$$

which are exactly the equations one would have obtained from applying the correspondence $(d/dt \rightarrow \mathcal{d}/dt)$ to the classical Euler-Lagrange equations themselves. Other fundamental results of classical mechanics are also generalized in the same way. In particular, assuming homogeneity of space *in the mean* leads to defining a complex momentum

$$\mathcal{P} = \frac{\partial \mathcal{L}}{\partial \mathcal{V}} \ . \qquad (11)$$

If one now considers the action as a functional of the upper limit of integration in (9), the variation of the action from a trajectory to another close-by trajectory, when combined with (10), yields a generalization of another well-known result:

$$\mathcal{P} = \nabla \mathcal{S} \ . \qquad (12)$$

We shall now specialize and consider Newtonian mechanics. The Lagrange function of a closed system, $L = \frac{1}{2}m\, v^2 - \mathcal{U}$, is generalized as $\mathcal{L}(x, \mathcal{V}, t) = \frac{1}{2}m\,\mathcal{V}^2 - \mathcal{U}$. Note that its real part becomes $\frac{1}{2}m(V^2 - U^2) - \mathcal{U}$, which is the Lagrangian field proposed by Guerra and Morato (1983). The Euler-Lagrange equations keep the form of Newton's fundamental equation of dynamics

$$-\nabla \mathcal{U} = m\,\frac{\mathcal{d}}{dt}\,\mathcal{V} \ , \qquad (13)$$

which is now written in terms of complex variables and operator. The complex momentum $\mathcal{P}$ now reads $\mathcal{P} = m\mathcal{V}$, so that from Eq. (12) we arrive at the conclusion that, in this case, the *complex velocity* $\mathcal{V}$ *is a gradient*, namely the gradient of the complex action:

$$\mathcal{V} = \nabla \mathcal{S}/m \ . \qquad (14)$$

This is an interesting result owing to the fact that in several derivations of Nelson's stochastic mechanics the fact that the classical velocity V (i.e. the real part of our complex velocity $\mathcal{V}$) is a gradient is *postulated*.

Let us now introduce the probability density ρ. It verifies a forward and a backward Fokker-Planck equation (Nelson, 1966; Welsh, 1970):

$$\partial \rho / \partial t + div(\rho \boldsymbol{b}_{\pm}) = \pm \, \Delta (\mathcal{D} \rho) \; . \tag{15}$$

We shall in what follows consider only the case when the diffusion coefficient is assumed to be a constant, so that we can take it out of the Δ sign. The two Fokker-lanck equations can also be combined in a single complex equation:

$$\partial \rho / \partial t + div(\rho \mathcal{V}) = -i \; \mathcal{D} \Delta \rho \; . \tag{16}$$

The real part of this equation is the equation of continuity,

$$\partial \rho / \partial t + div(\rho \boldsymbol{V}) = 0, \tag{17}$$

while the imaginary part reads $div\{ \rho \; [\boldsymbol{U} \; - \mathcal{D} \nabla \, ln\rho \;] \; \} = \; 0$. It has been demonstrated by Nelson (1966, 1985) that the term inside the *div* operator is zero, so that $\boldsymbol{U}$ is related to the probability density ρ by the relation:

$$\boldsymbol{U} \; = \; \mathcal{D} \, \nabla ln\rho \; , \tag{18}$$

in agreement with our previous result that $\mathcal{V}$ is a gradient. The classical velocity $\boldsymbol{V}$ being also a gradient, we may introduce a dimensionless classical action S,

$$V \; = \; 2\mathcal{D} \, \nabla \mathrm{S}. \tag{19}$$

We find finally that the complex action $\mathcal{S}$ is given in terms of the classical action and of the probability density by the relation:

$$\mathcal{S} \; = \; 2 \, m \, \mathcal{D} \; (\mathrm{S} - i \; ln\rho^{1/2}) \; . \tag{20}$$

This means that the logarithm of the probability density plays the role of the imaginary part of the generalized action.

We may now introduce a complex function ψ from the complex action $\mathcal{S}$,

$$\psi \; = \; e^{i\mathcal{S}} \; , \tag{21}$$

which is related to the complex velocity in the following way:

$$\psi \; = \; \sqrt{\rho} \;\; e^{i\mathrm{S}} \;\; \Rightarrow \;\; \mathcal{V} = -\,2\,i \;\; \mathcal{D} \, \nabla (ln\psi) \; . \tag{22}$$

This is one of the main advantages of our formalism. From this equation and Eqs. (12) and (20), we obtain:

$$\mathcal{P} \psi \; = \; -2i \, m \, \mathcal{D} \, \nabla \psi \; , \tag{23}$$

which is nothing but the *correspondence principle* of quantum mechanics, but demonstrated and written in terms of an exact equation thanks to our introduction of the complex momentum $\mathcal{P}$. We have now at our disposal all the mathematical tools needed to write Newton's equation (13) in terms of the new quantity ψ. It takes the form

$$\nabla \mathcal{U} = 2i\, \mathcal{D} m \frac{d\!\!\!/}{dt} (\nabla ln\psi) \ . \tag{24}$$

Being aware that $d\!\!\!/$ and ∇ do not commute, we replace $d\!\!\!/ /dt$ by its expression (8):

$$\nabla \mathcal{U} = 2i\, \mathcal{D} m \left[\frac{\partial}{\partial t} \nabla ln\psi - i\, \mathcal{D} \Delta(\nabla ln\psi) - 2i\, \mathcal{D}(\nabla ln\psi . \nabla)(\nabla ln\psi) \right] \tag{25}$$

This expression is simplified thanks to the three following identities, which may be established by straightforward calculation:

$$\nabla \Delta = \Delta \nabla \ ; \quad (\nabla f . \nabla)(\nabla f) = \tfrac{1}{2}\, \nabla (\nabla f)^2 \ ; \quad \frac{\Delta f}{f} = \Delta ln f + (\nabla ln f)^2 \ . \tag{26}$$

This implies

$$\tfrac{1}{2} \Delta (\nabla ln\psi) + (\nabla ln\psi . \nabla)(\nabla ln\psi) = \tfrac{1}{2}\, \nabla \frac{\Delta\psi}{\psi} \ , \tag{27}$$

and we obtain

$$\frac{d\!\!\!/}{dt} \mathcal{V} = -\nabla \mathcal{U} / m = -2\, \mathcal{D} \nabla \ \{ i \ \frac{\partial}{\partial t} ln\psi + \mathcal{D} \frac{\Delta\psi}{\psi} \ \} \ . \tag{28}$$

Integrating this equation finally yields

$$\mathcal{D}^2 \Delta\psi + i\, \mathcal{D} \frac{\partial}{\partial t} \psi - \frac{\mathcal{U}}{2m} \psi = 0 \ , \tag{29}$$

up to an arbitrary phase factor $\alpha(t)$ which may be set to zero by a suitable choice of the phase S. Replacing $\mathcal{D}$ by $\hbar/2m$, we get Schrödinger's equation

$$\mathcal{U} = \frac{i\hbar}{\psi} \frac{\partial}{\partial t} \psi + \frac{\hbar^2}{2m} \frac{\Delta\psi}{\psi} \ . \tag{30}$$

The hereabove system becomes equivalent to Nelson's stochastic quantum mechanics (Nottale, 1992b), but with several advantages, such as the new formulation of the correspondence principle, the fact that the real part of Eq. (13) defines an acceleration which is identical to that *postulated* by Nelson (1966, 1985), and more generally the complete symmetry with classical mechanics, including the definition of the Lagrange function.

However we claim that the above formalism has a more general application domain than only quantum mechanics (though clearly with a different interpretation). Assume that one has been able to characterize some chaotic system by a constant diffusion coefficient $\mathcal{D}$ (at least to lowest order approximation), and that the chaos is so developed that, when considered at large time-scale, the trajectories can be approximated by nondifferentiable curves, then (29) is a quasi-quantum equation for such a system, and is expected to yield structures (i.e., peaks of probability) once the boundary conditions are prescribed. We shall briefly exemplify this proposal by an application to the problem of the distribution of planets in the solar system.

3. Application to Celestial Mechanics

The existence of structuration in the Solar System has been recognized for long, at first by Kepler himself. The Titius-Bode "law" (see e.g. Nieto, 1972) is the first empirical attempt at describing these regularities, and was followed by several other proposals (see e.g. Neuhäuser and Feitzinger, 1986; Souriau, 1989). Most attempts at understanding the distribution of planets were based on theories of formation of the Solar System (see e.g. Brahic, 1982). However, there has been increasing evidence, in recent years, that chaos may play a leading role in celestial mechanics (Hénon and Heiles, 1964; Petit and Hénon, 1986; Wisdom, 1987; Sussman and Wisdom, 1988; Laskar,1989, 1990).

Hills (1970) has noticed that it is likely that an epoch of strong dynamical encounters occurred before the planets relaxed into stable orbits. It is also remarkable that numerical simulations (Conway and Elsner, 1988) which have considered various initial conditions for planetary systems, find that systems placed in initially arbitrary distributions are generally chaotic, while when initially placed in Titius-Bode-like laws (increasing planetary separations), they are very stable systems.

With these considerations in mind, we shall investigate the possibility that, at the end of the formation of the Solar System, an epoch of strong chaos occurred (to which the above formalism can be applied), and that the present rather stable state of the system resulted precisely from structuration originating from chaos itself.

Consider a gravitational system described by a Newtonian potential $\mathcal{U}$ such that $\Delta \mathcal{U} = 4\pi G\rho$, and assumed to be subjected to developed chaos: assuming that one can define an average diffusion coefficient, Eq. (29) applies (as a first approximation model) to this problem. In particular, consider a test planet orbiting in the field of the Sun, $\mathcal{U} = -GmM/r$, and subjected to chaotic motion due to the collective field of the other planets. The specialization of (29) to the case of a stationary motion with conservative energy $E \equiv 2i\mathcal{D}m\partial/\partial t$ yields

$$\mathcal{D}\Delta\psi + [\, E + \frac{G\,M}{2\,\mathcal{D}\,r}\,]\,\psi = 0 \quad . \tag{31}$$

The equivalence principle suggests that $\mathcal{D}$ is independent of m. This equation is similar to the Schrödinger equation for the hydrogen atom, up to the substitution $\hbar/2m \rightarrow \mathcal{D}$, $e^2 \rightarrow GmM$, so that the natural unit of length (which corresponds to the Bohr radius) is:

$$a_0 = 4\,\mathcal{D}^2/GM. \tag{32}$$

We thus find that the energies of planets scale as $E_n = -GmM^2/8\mathcal{D}^2n^2$, $n = 1, 2, 3, \ldots$, and that the probability densities of their distances to the Sun are confined to definite regions given by the square of the well-known radial wave functions of the hydrogen atom. We also expect angular momenta to scale as $L = 2m\mathcal{D}\ell$, with $\ell = 0, 1, \ldots, n-1$: this means that, unlike in quantum mechanics, E/m and L/m are quantized rather than E and L. The average distance to the Sun and the eccentricity e are given, in terms of the two quantum numbers n and ℓ, by the following relations (see e.g. Messiah, 1959):

$$a_{n\ell} = \{\tfrac{3}{2}\,n^2 - \tfrac{1}{2}\,\ell(\ell+1)\}\,a_0 \tag{33a}$$

$$e^2 = 1 - \frac{\ell(\ell+1)}{n(n-1)} \tag{33b}$$

Let us now briefly compare these predictions to the observed planets. The difference of physical and chemical composition of the inner and outer solar systems suggests to us that they must be treated as two different systems, i.e., that we expect two different diffusion coefficients for them. The main results are summarized hereafter (see Fig. 1).

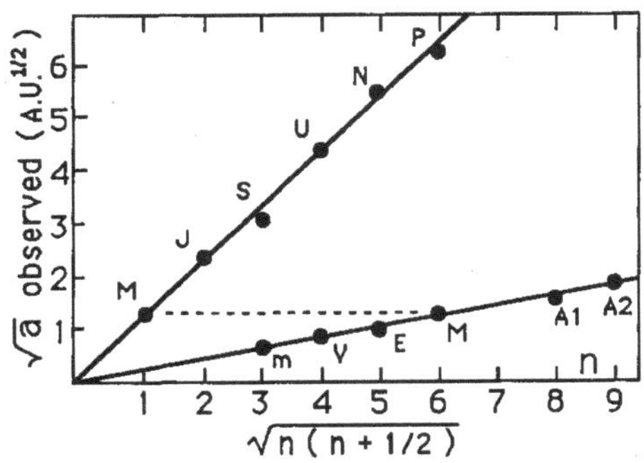

Figure 1. Comparison of the observed average distances of planets to the Sun to our prediction (see text). The abscissa is labelled by the value of n, but is given by $\sqrt{n(n+1/2)}$. A1 and A2 are for the two main peaks in the distribution of asteroids in the asteroid belt.

(i) We may compare the observed values of semi-major axes of the planets to our prediction (Eq. 33a) with $\ell = n-1$, $\sqrt{a} = n\,(1+1/2n)^{1/2}\sqrt{a_0}$, since even the largest eccentricities (Pluto, e^2=0.065; Mercury, e^2=0.042) correspond, to a good approximation, to quasi circular orbits. Note that *the ordinate at origin is predicted to be zero*. This prediction is very well verified for the two systems: we find $a_{int}(0) = 2\;10^{-4}$ A.U. and $a_{ext}(0) = 4\;10^{-3}$ A.U. Mercury, Venus, Earth and Mars take respectively ranks $n = 3, 4, 5$ and 6 in the inner system (n=1 and 2 correspond to distances very close to the Sun, which may explain their emptiness); Jupiter, Saturn, Uranus, Neptune and Pluto ranks $n = 2, 3, 4, 5, 6$ in the outer system (see Fig. 1). The average slopes are $(\sqrt{a_0})_{int} = 0.195 \pm 0.0022$ and $(\sqrt{a_0})_{ext} = 1.014 \pm 0.016$.

(ii) Mars's position is in very good agreement with $n = 1$ of the *external* system also: including it yields an improved slope $(\sqrt{a_0})_{ext} = 1.014 \pm 0.012$. Note also the agreement of Pluto with the outer relation.

(iii) The central peak of the asteroid belt (2.7 A.U.) agrees remarkably well with $n = 8$ of the inner system, and the main peak (3.15 A.U.) with $n = 9$. Including them yields $\sqrt{a_{0int}} = 0.195 \pm 0.0017$. This result may help understand the fact that there is no large planet there: the zone where the belt lies, even though it corresponds to maxima of probability density for the inner system, also corresponds to a *minimum* in the outer system. The region between Mars and Jupiter is where the two systems overlap.

(iv) The quasi-circularity of the orbits may be understood, since the first non circular state, $\ell = n - 2$, is predicted to have eccentricity larger that 0.58 for $n \leq 6$.

(v) Our finding that L/m is quantized rather than L sheds also new light on the distribution of angular momentum among planets (Jupiter 60%, Saturn 25%). The quantum number n remaining small, the distribution of angular momentum is expected to mainly mirror that of mass.

We have not attempted in this contribution to get a theoretical estimate of the diffusion coefficient $\mathcal{D}$, which is our only free parameter. A better understanding of the relation between Lyapunov exponents and diffusion coefficients is needed in this respect (see e.g. Lichtenberg and Lieberman, 1983; Eckmann and Ruelle, 1985). Other questions, such as the maintainance of the flatness of the Solar System, or the low observed dispersion of positions with respect to the predicted ones, will be considered elsewhere.

4. Discussion and Conclusion

As already specified, even though Eq. (29) is a quantum mechanical-like equation, its interpretation must be different from that of quantum mechanics. We have conjectured elsewhere (Nottale, 1989, 1991,1992a,b) that the origin of the quantum properties which are irreducible to the classical behaviour was the nondifferentiability (and fractal structure) of the quantum space-time. On the other hand, the application of the above formalism to classical chaotic systems relies on the *approximation* of nondifferentiability, which is no longer valid on small time-scales, for which one recovers predictable and differentiable trajectories. Then, while this formalism is able to provide us with peaks of probabilities, and then to help us understanding the emergence of structures from chaos, the probabilities nevertheless remain classical. In such a view, one of the (still partly open) problem is to understand how to connect the small time-scale behaviour to the large time-scale one.

Let us conclude by considering the applications to cellular automata models. A first possible application concerns quantum mechanics itself: for example a lattice simulation of the Schrödinger equation could be performed by going down to the elementary process of Nelson's quantum mechanics, namely the forward and backward velocities $\boldsymbol{b}_+$ and $\boldsymbol{b}_-$. This would amount to run two automata instead of one, one for the forward process and the other for the time reversed process, and to combine them in the end (another possibility would be to use the complex formalism itself, and to run independently the real and imaginary parts of the equations). The second application we suggest concerns systems involved in developed chaos, for which the approximation of nondifferentiability, as described by a doubling of the velocity vector, can be considered as valid. This method and its application to various chaotic systems will be developed in forthcoming works.

ACKNOWLEDGEMENTS. I gratefully acknowledge helpful discussions with Drs. S. Bonazzola, J. Heyvaerts, J. Laskar, P.Y. Longaretti and B. Dubrulle.

References

Brahic A. (Ed.), 1982, *Formation of Planetary Systems*, Cepadues Editions, Toulouse.
Conway B.A., & Elsner T.J., 1988, in *Long Term Dynamical Behaviour of Natural and Artificial N-Body Systems*, A.E. Roy (ed.), Kluwer Acad. Pub., p.3.
Eckmann J.P., Ruelle D., 1985, *Rev. Mod. Phys.* **57**, 617.
Guerra F., Morato L.M., 1983, *Phys. Rev.* **D27**, 1774.
Hénon M., Heiles C., 1964, *Astron. J.* **69**, 73.
Hills J.G., 1970, *Nature* **225**, 840.
Laskar J., 1989, *Nature* **338**, 237.
Laskar J., 1990, *Icarus* **88**, 266.
Lichtenberg A.J., Lieberman M.A., 1983, *Regular and Stochastic Motion* (Springer-Verlag).
Nelson E., 1966, *Phys. Rev.* **150**, 1079.
Nelson E., 1985, *Quantum Fluctuations* (Princeton Univ. Press).
Neuhäuser R., Feitzinger J., 1986, *Astr. Astrophys.* **170**, 174.
Nieto M.M., 1972, *The Titius-Bode Law of Planetary Distances: Its History and Theory* (Pergamon Press, Oxford).
Nottale L., 1989, *Int. J. Mod. Phys.* **A4**, 5047.
Nottale L., 1991, in *Applying Fractals in Astronomy*, A. Heck & J. Perdang (eds.), Springer-Verlag, p.181.
Nottale L., 1992a, *Int. J. Mod. Phys.* **A7**, 4899.
Nottale L., 1992b, *Fractal Space-Time and Microphysics*, in the press (World Scientific).
Petit J.M., Hénon M., 1986, *Icarus* **66**, 536.
Souriau J.M., 1989, *Preprint* CPT-89/P.2296.
Sussman G., & Wisdom J., 1988, *Science* **241**, 433.
Welsh, D.J.A., in *Mathematics Applied to Physics* (Springer,1970), p.465.
Wisdom J., 1987, *Icarus* **72**, 241.

Part III :

Astrophysical Applications

SELF-ORGANIZED CRITICALITY IN ASTROPHYSICS

PER BAK
Department of Physics, Brookhaven National Laboratory, Upton, NY 11973, USA

I review recent work in which various astronomical observations are interpreted as evidence of self-organized criticality. These include the light emitted from quasars, pulsar glitches, and emission of hard x-rays from solar flares. The fractal structure of the universe may indicate that the universe operates at the self-organized critical state. The organization into the critical state is illustrated by simple toy "cellular automaton" models.

1. Introduction

"Self-Organized Criticality" (SOC) refers to the tendency of large dissipative dynamical systems to drive themselves to a critical state with a wide range of length and time scales[1-4]. The idea provides a unifying concept for large scale behavior in systems with many degrees of freedom; it complements the concept of "chaos" wherein simple systems with a small number of degrees of freedom display quite complex behavior. It has been looked for in such diverse areas as earthquakes, economics, and biological evolution.

The canonical metaphor is a simple pile of sand. Adding sand slowly to a flat pile will result only in some local rearrangement of particles. The individual grains, or degrees of freedom, do not interact over large distances. Continuing the process will result in the slope increasing to a critical value where an additional grain of sand gives rise to avalanches of any size, from a single grain falling up to the full size of the sand pile. The pile can no longer be described in terms of many local degrees of freedom, but only a holistic description in terms of *one* sandpile will do. The distribution of avalanches follows a power law. There is ample evidence that real sandpiles evolve to the critical state[5-7]. Figure 1 shows the results of recent experiments by Grumbacher et al. They built small heaps on a scale, and monitored the distribution of avalanches of particles falling off the edges. The figure shows log-log plots of the normalized distribution function of avalanches. The experiments were performed using iron spheres (triangles) and glass spheres (circles) of the same size.

SOC provides a natural explanation for the ubiquitous scale free phenomena in nature. The geometry of many objects is best described as fractals with self-similar features on all length scales[8]: Coast lines have features of all sizes, from hundreds of kilometers down to millimeters. River networks consists of streams of all sizes. Turbulent fluids have vortices over a wide range of sizes. Earthquakes occur on structures of faults ranging from thousands of kilometers to centimeters. Fractals are scale-free in the sense that viewing a picture of a part of a fractal one can not deduce its actual size if a yardstick is not shown in the same picture. Actually, the entire universe appears to be a fractal, consisting of clusters of galaxies, organized in clusters of clusters of galaxies and so on[8-10]. The universe has

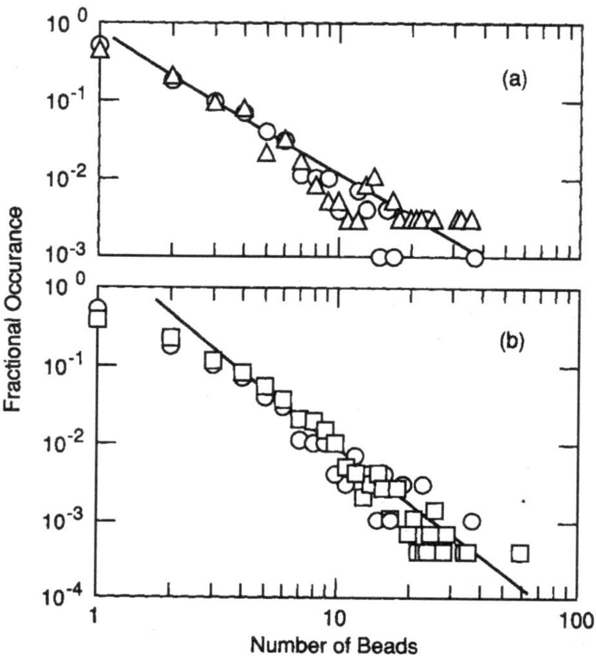

Figure 1. The avalanche distribution from recent experiments on sandpiles by Grumbacher *et al* [7].

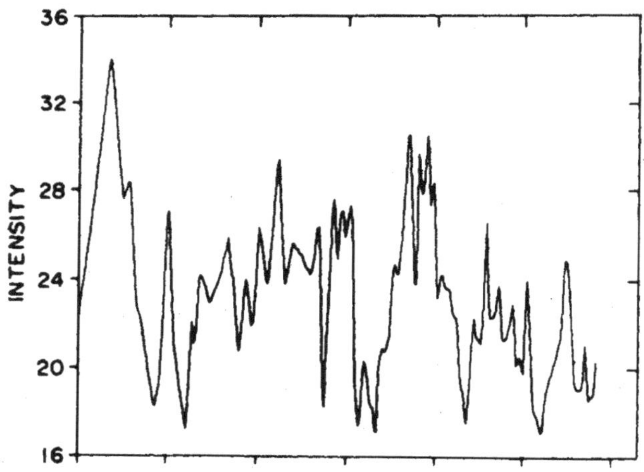

Figure 2. Light emission of the quasar 3C 273 measured over a period of 80 years[12]. The curve is a typical 1/f noise signal, with features on all time-scales.

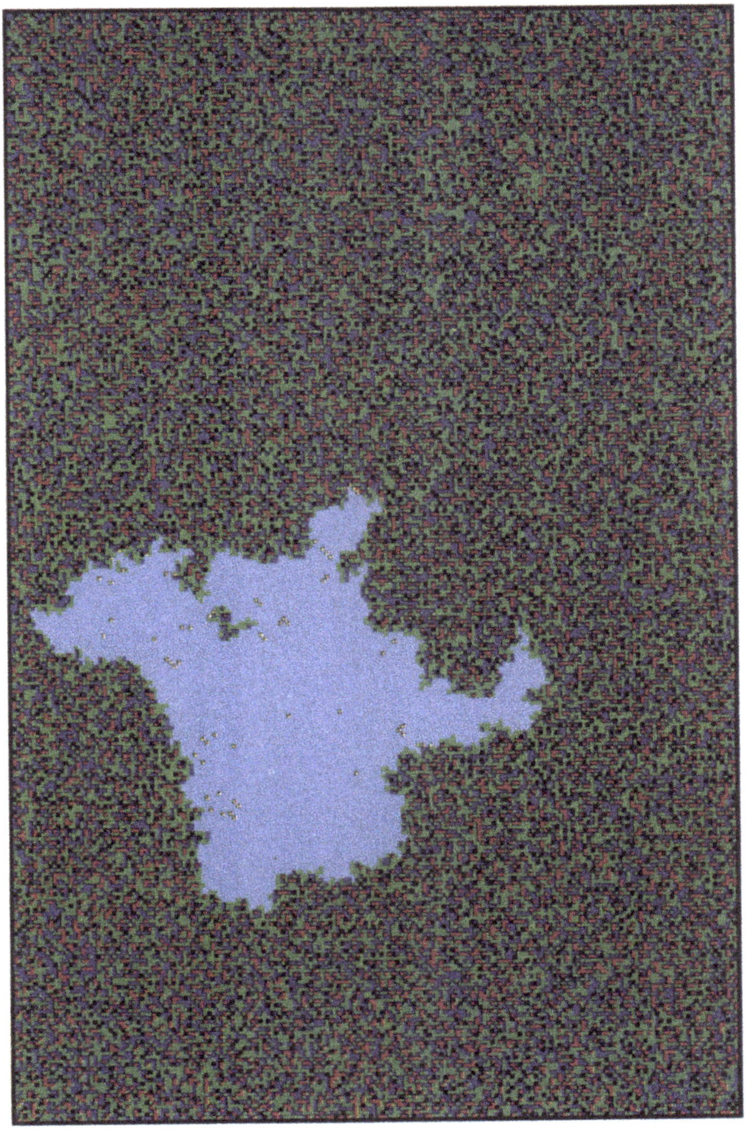

Figure 3. Avalanche propagating in the critical state on a 286 by 184 lattice. The heights z_i from 0 to 3 are represented by black, red, blue, and green, respectively. The still active sites are various shades of yellow, and the tumbled region is cyan. (Bak and Creutz[18])

structures on all scales, including the largest. We may generally think of fractals as snapshots of dynamical systems operating at the self organized critical state[11].

There is another widespread phenomenon in nature which has defied explanation for decades. The temporal evolution of the signal emitted from a variety of sources has a power spectrum decaying as a power law with exponent near unity at low frequencies. This indicates that the signal consists of events of all time scales. Figure 2 shows the light from the quasar 3C 273 measured over a period of 80 years by Fahlman and Ulrych[12]. Indeed, there appears to be fluctuations of all durations, including the total measurement period. This type of behavior is known as 1/f (one-over-f) noise[13]. An 1/f signal can be thought of as a superposition of avalanches whose sizes are distributed according to a power law[2,14]. We boldly interpret the 1/f noise as evidence that quasars and other astrophysical objects exist in the self-organized critical state. In any case, it is hard to imagine that the huge fluctuations in an object of astronomical size are not caused by some kind of concerted cooperative phenomenon extending throughout the object. Random local fluctuations would lead to a Gaussian distribution of variations, with an entirely different $1/f^2$ power-spectrum.

It seems out of the question to construct realistic models of the complex geophysical and astrophysical phenomena that we are dealing with. Thus, we have to restrict ourselves to study simple toy models. These models are mostly cellular automatons. We hope that these models catch the essential physics. The philosophy is very similar to the one used in the study of critical phenomena in equilibrium systems, where universality has permitted us to utilize simple Ising models to calculate power laws for temporal and spatial correlations near the critical point.

2. Earthquakes and Volcanic Activity

2.1. Earthquakes

Before taking off into outer space, let us stay for a while with the planet in the universe for which we have the most reliable formation. The energy distribution of earthquakes is a power law, the famous Gutenberg-Richter law. The idea of self-organized criticality, as applied to earthquakes[15-17], may be visualized as follows. Think of the lithosphere of the earth as a collection of tectonic plates, being squeezed very, very slowly into each other. In the beginning of our geological history, maybe, the stresses were small, and there would be no large ruptures or earthquakes. During millions of years, however, the system evolved into a kind of stationary state where the build-up of stress is balanced in average by the release of stress during earthquakes. Because of the long evolutionary process, the crust has "learned" to respond critically to any initial rupture. The earthquake can be thought of as a critical chain reaction where the process is just barely able to continue.

The more realistic models are continuous deterministic threshold models[15,17], but for illustration the simplest discrete cellular automaton models will do. On a two-dimensional regular lattice of size $L_1 \times L_2$ with open boundaries, integer variables z_i on each site i are used to represent the local stress. The increase of the force (due for instance to tectonic plate motion) is represented by increasing the value of consecutive random sites z_i by unity. When the force somewhere exceeds a critical value z_{cr}, here taken to be 3, there is a rupture, or toppling event wherein 1 unit of stress is transferred from the unstable site to each of the 4 neighbor sites, i.e, the value of z_i is reduced by 4 and the values of z at the 4 neighbor sites are increased by 1. The updating is done concurrently, with all sites updated simultaneously. The initial toppling may initiate a chain reaction, where the total amount of topplings is a measure of the size of an avalanche.

After running the model for a long time, it arrives at a statistically stationary state where the average force remains constant. Fig. 3 shows an avalanche triggered by a single rupture. The lattice here is 286 by 184. The various colors indicate different local stresses, and the blue color covers the sites which have toppled during that particular avalanche. This was a particularly large avalanche, shown for illustrative purposes. In general the final size of the disturbance is unpredictable without actually running the simulation. Some may involve a single tumbling, and others collapse most of the system.

Figure 4 shows a log-log plot of the distribution of the avalanche sizes. The linearity indicates a power law,

$$P(s) \approx s^{-\tau+1}, \ \tau \approx 2.1. \tag{1}$$

This indicates that the model has self-organized into the critical state. We interpret the Gutenberg-Richter law as an indication that the crust of the earth on which we are walking is in a critical state, far from equilibrium.

The power law distribution of earthquakes stems from the fractal nature of the SOC state, with correlated regions ranging over all length scales; those correlated regions, generated by the long term dynamics, are the equivalent of the active faults or fault segments in real earthquakes. The fault structure changes on large geological time-scales. More realistic long-range SOC models produce faults which topologically look much more like a real fractal-like arrangement of two-dimensional faults in a three dimensional matrix[19].

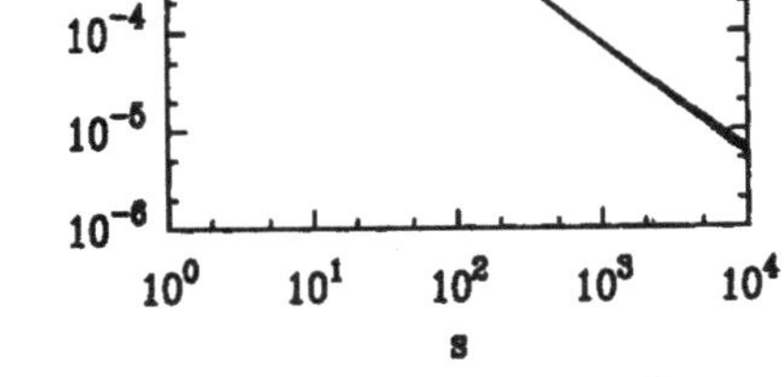

Figure 4. A log-log plot of the distribution of avalanches in the earthquake model.

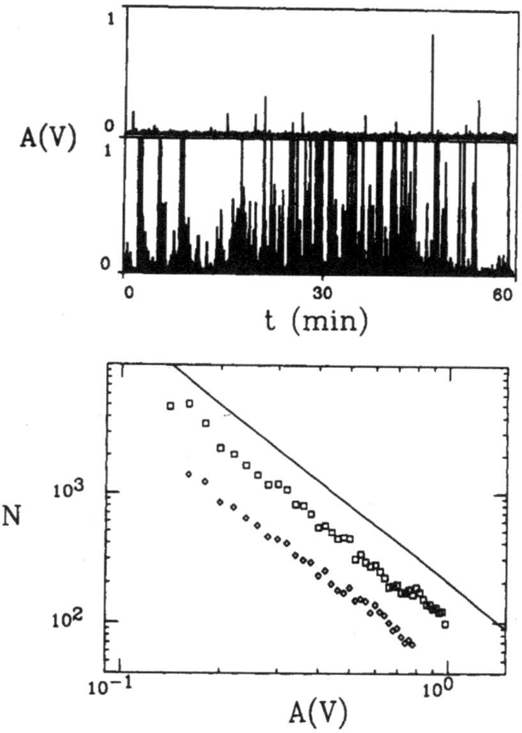

Figure 5. Top: Recorded samples of acoustic volcanic activity at teo different sites near Stromboli. Bottom: Distribution of bursts. (Diodati et al.[20])

The value of the exponent depends on the particular model. In a continuous model studied by Olami, Feder, and Christensen[17] the exponent is a continuous function of a parameter. Thus, one should not look for unique exponents in nature. Indeed different powers have been observed in different geographical areas.

2.1. Volcanic Activity

Volcanic activity appears to be intermittent, with events of all sizes just like earthquakes, with no single periodicity. We sometimes describe the intermittency in terms of "dormant" periods interrupted by bursts of activity. It would be interesting to perform a systematic study of the statistics of volcanic activity through history. Diodati et al.[20] have measured bursts of acoustic emission in the area around Stromboli in Italy. They placed piezoelectric sensors coupled to the free ends of steel rods tightly cemented into rock-drill holes. They observed the distribution function for intervals between the bursts, and the distribution of burst amplitudes, E. Figure 5 shows the latter distribution. Scaling behavior $N(E)=E^{-d}$, with $d\approx 2$ was observed.

3. 1/f Noise and Quasars

The ubiquitous 1/f noise can thought of as a signal formed by superimposing the avalanches occurring in the self-organized critical state. In order to calculate the power spectrum it turns out that one must calculate the expectation value of s^2 vs. duration t, that is

$$\Lambda(t) = \Sigma\, s^2\, P(s,t)$$

where $P(s,t)$ is the probability that an avalanche of size s lasts for t time steps. Christensen et al.[21] have shown that if Λ has the scaling behavior $\Lambda(t) = t^{\mu}$ then the power spectrum $S(f)$ becomes

$$S(f) = f^{-(\mu+1)}\ , \text{ for } -1<\mu<1.$$

In nature, values of the exponent of the 1/f noise in the interval 0.6-2.0 have been reported[13]. The models which we have studied exhibit exponents in the same range. The exponents generally depend on the parameters of the models, so there is less universality than for equilibrium critical phenomena, where exponents usually depend only on the dimension and the symmetry. The connection between the distribution of avalanches and the power spectrum has been explicitly shown for the Olami-Feder-Christensen earthquake model[22]. The spectrum derived from the formula above agrees with a direct Fourier transform of the simulated signal.

Thus, we may speculate that quasars such as 3C 273 emitting 1/f noise exist in a critical state. The intermittent nature of this signal, shown in figure 2, with sudden bursts of large amplitude arises from "starquakes"[23] propagating through the quasar. I have no idea of the nature of these starquakes, since we don't know much about the "engine" of quasars. Vio et al.[24] have analyzed the temporal variability of the luminosity of another quasar, the 3C 245. Indeed, they conclude that the strong bursts manifest the tendency for large scale systems to become organized into a state where they are just at the edge of an instability, and they suggest that the quasar operates at the self-organized critical state.

4. Solar Flares and Pulsar Glitches

4.1. Solar Flares

Observations show that the distribution of solar flare hard x-ray bursts is a power law with exponent 1.8 (Figure 6). The power law spans over five orders of magnitude. Lu and Hamilton[26] have suggested that this power law dependence is a consequence of the coronal magnetic field being in a self-organized critical state, and they constructed a simple model, relying on an earlier physical picture involving re-connection events of magnetic

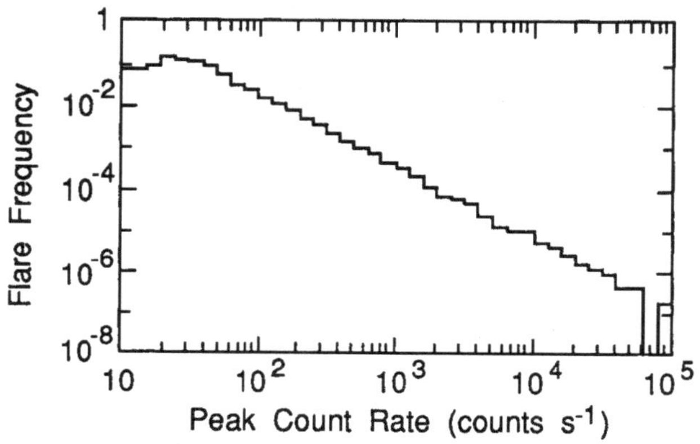

Figure 6. Distribution of solar flare occurrence rate for flares detected with x-ray spectrometer. (Dennis[25])

field lines in the solar corona, driven by convective motion. The model is a cellular automaton:

On a three dimensional grid of points a three-dimensional vector $\boldsymbol{B}_i$, representing the average magnetic field, was specified. The local magnetic gradient $d\boldsymbol{B}_i$ is defined to be the difference between the local magnetic field and the average of its six neighbors. The model is driven by randomly adding driving vectors $\delta\boldsymbol{B}$. The configuration is defined to be unstable to reconnection when the magnitude of the gradient is greater than a critical value $|d\boldsymbol{B}_i| > B_c$. When a reconnection instability occurs, the local magnetic field is cancelled by transporting one-seventh of the gradient vector to each of its six nearest neighbors, so that the local field becomes equal to the average of its neighbors. Each reconnection event releases magnetic energy $(6/7)|d\boldsymbol{B}_i|^2$. The model exhibits self-organized criticality, with power-law distributions for the energy release, peak activity, and duration of x-ray bursts.

The traditionally classification of solar flares into nanoflares, microflares, giant flares and so on is arbitrary, since there is no preferred scale and the fundamental energy release mechanism is the same for flares of all sizes.

4.2. Pulsar Glitches

Pulsar glitches are discontinuities in the period of pulsars. They are caused by increases of angular speed analogous to an ice-skater performing a pirouette and drawing in her hands. In a very interesting paper, Morley and Garcia-Pelayo[23] have analyzed observations from all but the smallest glitches. They translated the changes in frequencies to changes in rotational energy and plotted the cumulative distribution of that energy, which (you guessed it) turns out to be a power law. This was interpreted in terms of a crust cracking model in which the brittle outer crust of the neutron star is under increasing strain due to the decreasing centrifugal force from slowing down. It is expected that the surface of a neutron star is divided into platelets which differ in the amount of magnetic field lines threading their surface (with obvious similarities with the Lu-Hamilton picture of solar flares described above). Strain will cause cracking and relative platelet movement.

Similarities with earthquakes immediately come to mind, and Morley and Garcia-Pelaya call the events "starquakes". The power law correspond to one found by Olami et al.[17] in the limit where the force is conserved during the individual rupture events.

5. Is the Universe Operating at a Critical State?

But not only single astronomical objects, such as the earth, the sun, pulsars, and quasars ppear to be critical. Schulman argues that the dynamics of galaxies is at or near criticality. There is even ample evidence that the entire universe has a fractal distribution of galaxies, with dimension D approximately equal to 1.4[9,10], suggesting some kind of self-organized critical dynamics.

The idea of using simple cellular automata, instead of more realistic, but prohibitively cumbersome, partial differential equations to mimic the dynamics of galaxies and other astrophysical phenomena goes back to Seiden and Schulman[28]. For a more knowledgeable discussion of this and subsequent work I refer to the review article by Schulman in this volume. Kan Chen and I[29] have constructed a simple cellular automaton modelling interacting ballistic particles which evolves to a fractal critical state. We believe that although the model is not a realistic representation of anything, it captures a general scenario for the emergence of scaling behavior.

The simple "toy" model describes an extended system with active propagating particles interacting with other active or passive static particles. The picture that the reader may have in mind is one where objects such as supernovas are exploding, with the emitted material condensing to form new supernovas which may explode, and so on. Active (empty circles) and passive (filled circles) are placed on a 2d lattice of size L by L. The rules for updating the system from time t to $t+1$ are very simple:

(i) Active sites at t "burn out" and become passive at time $t+1$.
(ii) Passive sites are annihilated when they have one, and only one, active neighbor.
(iii) Active sites are created when they have one, and only one, nearest neighbor, which must have a passive site at the opposite position.

Rule (iii) allows configurations •o to propagate along straight lines in free space: the particles have both position and momentum. Active sites are formed next to active sites, unless there is overcrowding causing an instant burnout. Figure 7 shows the evolution of several situations involving collision of particles. The lattice can be thought of as a part of a larger universe in the process of cooling off by emitting particles into our subspace. Open boundary conditions are chosen such that active sites vanish when passing beyond the edges into the environment.

In order to elucidate the correlations for the active sites we consider the following process: starting with a "hot" initial universe with a random distribution of active sites, the system evolves according to the rules (i) - (iii) it comes to rest with no remaining active sites. (For added excitement, one might model the big bang by placing the initial particles in

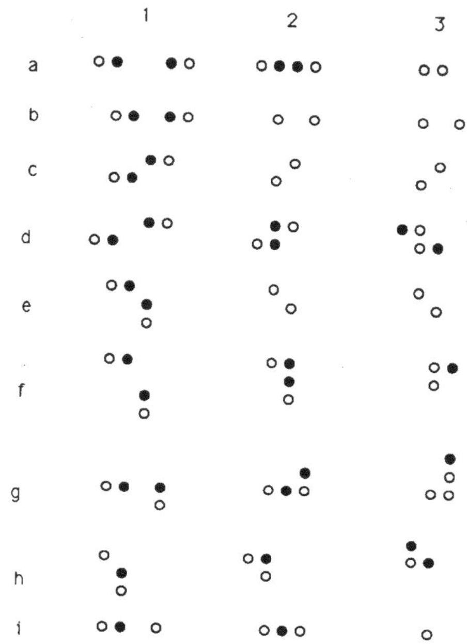

Figure 7. (a-g) collision of propagating active particles. (h) Creation and (i) annihilation of active particles with passive sites. Configurations at three consecutive times, 1, 2, 3, are shown.

a big heap near the center - it makes no difference). The system is then perturbed by adding a single active or a single passive site. As the process is repeated the universe evolves into a statistically stationary state.

In the stationary state, one may measure the total number of sites which becomes active when a single particle is added. The distribution of clusters is a power law, like Eq. 1, with $\tau = 2.4$. The model can be defined also in three dimension where t assumes approximately the same value. The fact that the activity does not decay, or explode, exponentially indicates that the process is a critical chain reaction.

Thus, the background of passive particles has become organized into a complex structure which precisely allows the signal to propagate indefinitely. One might think of the initial state as a "forest fire" which continues until the point where the passive material just cuts communication.

Figure 8 shows a snapshot of the system. The bright sites are grouped together in clusters. In order to investigate whether or not the distribution of active and passive sites is a fractal, we have measured the total number $D(r)$ of active sites at a distance r from another active "visible" sites (Figure 9). The straight line indicates a power law $D(r) = r^{D-1}$, wherethe exponent D is the fractal dimension. We found D=1.6±0.2 in two dimensions, D=1.7±0.2 in three dimensions.

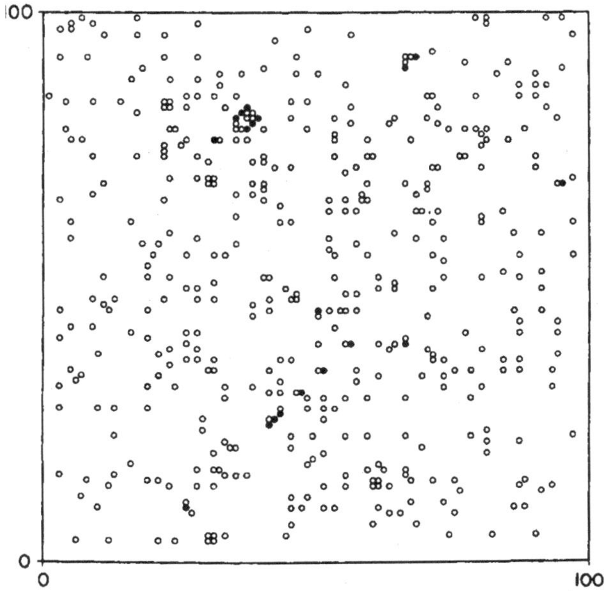

Figure 8. Snapshot of active system. Note the clustering of active sites.

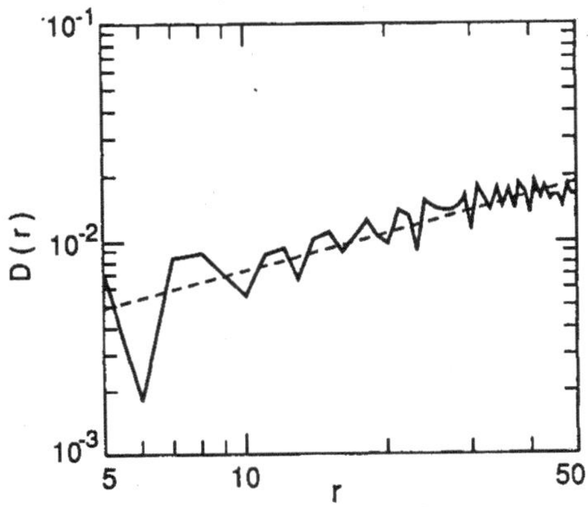

Figure 9. Distribution of the number of active sites at distance r from an active site near the center.

The passive sites, on the other hand, are not fractal but uniformly distributed. Could it be that the bright matter which we observe is fractal, while the total mass distribution is not, and could this explain the problem of the missing mass of the universe?

Acknowledgments

I am grateful to Larry Schulman for discussions on cellular automatons and galaxy formation. Supported by the US department of energy under contract DE-AC02-76-CH00016.

References

1. P. Bak, C. Tang, and K. Wiesenfeld, *Phys. Rev. Lett.* **59** (1987) 381; *Phys. Rev.* **A38** (1988) 3645.
2. C. Tang and P. Bak, *Phys. Rev. Lett.* **60** (1988) 2347.
3. K. Wiesenfeld, P. Bak, and C. Tang, *J. Stat. Phys.* **54** (1989) 1441.
4. P. Bak and K. Chen, *Sci. Am.*, January Issue p 46 (1991)
5. J. Rosendahl and J. E. Rutledge, *Phys. Rev.* **E1**, February (1993).
6. M. Bretz, J. B. Cunningham, P. L. Kurczynski, and F. Nori, *Phys. Rev. Lett.* **69** (1992) 2431.
7. S. K. Grumbacher, K. M. McEwen, D. A. Halvorson, D. T. Jacobs, and J. Lindler, *Am. J. Phys.* January issue (1993).
8. B. B. Mandelbrot, *The Fractal Geometry of Nature* (Freeman, San Francisco, 1982).
9. L. Pietronero, *Physica* **A144** (1987) 257.
10. P. H. Coleman, L. Pietronero, and R. H. Sanders, *Astron. Astrophys.* **200** (1988) L32.
11. P. Bak and Chen, *Physica* **D38** (1989) 5.
12. G. C. Fahlman and T. J. Ulrich, *Ap. J.* **201** (1975) 277.
13. W. H. Press, 1978, *Comments Astrophys.* **7** (1978) 103.
14. K. Christensen, H. C. Fogedby, and H. J. Jensen, *J. Stat. Phys.* **63** (1991) 653.
15. P. Bak and C. Tang, *J. Geophys. Res.* **B94** (1989) 15635.
16. K. Ito and M. Matsuzaki, *J. Geophys. Res.* **B95** (1989) 6853.
17. Z. Olami, H. J. Feder, and K. Christensen, *Phys. Rev. Lett.* **68** (1992) 1244.
18. P. Bak and M. Creutz in *Fractals and Disordered Systems*, ed. A. Bunde and S. Havlin (Springer, Berlin, Heidelberg, New York, 1993).
19. K. Chen, P. Bak, and S. P. Obukhov, *Phys. Rev.* **A43** (1991) 625.
20. P. Diodati, F. Marchesoni, and S. Piazza, *Phys. Rev. Lett.* **67** (1991) 2239.
21. K. Christensen, H. C. Fogedby, and H. J. Jensen, J. *Stat. Phys.* **63** (1991) 653.
22. K. Christensen, Z. Olami, and P. Bak, 1992, *Phys. Rev. Lett.* **68** (1992) 2417.
23. P. D. Morley and R. Garcia-Pelayo, preprint.
24. R. Vio, S. Christiani, O. Lessi, and L. Salvadori, *Astrophys. J.* **380** (1991) 351.

25. B. R. Dennis, *Solar Phys.* **100** (1985) 65.
26. E. T. Lu and R. J. Hamilton, *Astrophys. J.* **380** (1991) L89.
27. L. S. Schulman, these proceedings.
28. P. Seiden and L. S. Schulman, *Science* **233** (1986) 425; Advances in Phys. **39** (1990) 1.
29. K. Chen and P. Bak, Phys. Lett. A **140** (1989) 299.

MODELING GALAXIES: CELLULAR AUTOMATA AND PERCOLATION

L. S. SCHULMAN
Physics Department, Clarkson University
Potsdam, NY 13699-5820, USA

ABSTRACT

We present a cellular automaton model for star formation activity in a galaxy. Mathematically it is equivalent to directed percolation. Its origins are in statistical mechanical studies of Conway's game of Life. The model treats the induction of star formation in one region of a galaxy due to nearby star formation much as one would treat disease transmission. An exposition of a simplified epidemic model ("percolitis") is given to illustrate the technique. Comparison is made with observation, and predictions of the model are given. Feedback effects in the model and the relation to the percolation phase transition suggest a kind of "self organized criticality." We also mention a percolation model for the distribution of galaxies in the universe. Finally, we discuss the way in which short range forces can be promoted to long range structure with the help of a second order phase transitions and the importance of that phenomenon in our rationale for the existence of spiral arms.

1. Introduction

1.1. Background

It is fashionable in biology to analyze everything from courtship to leaf size in terms of the impact on evolutionary fitness. In this spirit, the undirected and apparently purposeless exploratory behavior of juvenile primates ("play") can be viewed as laying the foundation for efficient long term use of their oversize brains. With this perspective, the scientist can be viewed as a case of arrested development: people who never outgrew the exploratory phase in order to get down to the real business of life.

The research I will report in this article began as typically purposeless exploratory behavior, an exploration of the *game* of Life. It became astrophysics, which is what most of this article is about, and may yet become more. I also draw pleasure from observing that support for this work came from IBM, an organization whose farsighted research program is a corporate version of the primate strategy. I hope that in IBM's current economic fitness test it will be found that this policy has been one of the company's strong points.

In this article, I will first describe that exploration of the game of Life. Next I will tell you how this led to the astrophysical application and describe that application in detail. Because the final story on galaxies is far from written, it will also be worth discussing what—to the astrophysicist—our contribution may be.

There are also issues that go beyond the astrophysics. One is the matter of self organized criticality. We don't have it, but we do have a situation in which the system is driven *near* the critical point and in which this proximity is important to the description of Nature. As you will see, the mechanism by which this occurs is a kind of feedback, a phenomenon that is itself robustly present in many systems, so that one is led to wonder whether a softer definition—you don't need to be *at* criticality, only *near*—might also be of value, notwithstanding the ambiguity introduced by such fuzziness. The second of the broad issues touched on is the general problem of directed percolation. Now the jargon-like term "directed percolation" does not suggest a concept of wide impact, but in fact directed percolation encompasses a broad range of phenomena. It is essentially equivalent to a description by a stochastic process. What it brings to the study of stochastic processes is a smooth way to introduce phase transition and criticality concepts. This too will be discussed below.

Finally there is one educational point, particularly directed at astrophysicists, that I hope will emerge from this article. It is this: Short range forces can lead to large scale structure. And there is a corollary: short range details often don't matter and can be ignored in models. As indicated above, our contribution has yet to be fully evaluated; nevertheless, an aspect that has slowed its acceptance is a failure within the astrophysics community to appreciate the point stated above. The emergence of long range order is a part of the fundamental knowledge of the condensed matter physicist and, whatever the ultimate disposition of our own contribution, this fact about Nature should become part of the astrophysics lore as well.

Much of the work reported here is the product of joint endeavor, and in the forthcoming discussion this will be made explicit. Where individual work is involved I will let the personal pronoun carry the message.

1.2. Plan of the article

This article is not a comprehensive review. It is a combination anecdotal survey and report of new results. Several review articles, with varying emphases do exist. One[1] is a general introduction of a relatively non technical nature that appeared in *Science* a few years ago. Two others[2,3] are tilted to the astrophysics. Works more oriented to the statistical mechanics side are Ref. 4 and 5, the former a review article. Other more recent work in this area in which I have been involved will be recounted below.

In §2 I will review the "Game of Life" and describe our experiments and results on this cellular automaton. Then (§3) I will lay out the basics of the stochastic-stimulated-propagation-of-star-formation model. In §4, before presenting the astrophysical consequences of the model, I discuss percolation theory (including the epidemic "percolitis") and show how our model fits into that framework. §5 deals with the astrophysical implications of the model, but is not comprehensive and will mainly point to results already published in review articles, especially the recent one in *Advances in Physics*. In §6 is material that we are still trying to understand and concerns the feedback and self organization—with or without criticality—mentioned above. In §7 is a slight digression in which I recall percolation related work that Seiden and I did on the distribution of galaxies within the universe. It is outside the main thrust of this paper, but recent observational results have revised upward my estimates of its truth value, so it is worth mentioning. Finally in §8 is a summary as well as observations on the power of general notions of collective phenomena.

2. The Game of Life

Over twenty years ago Conway's Game of Life was introduced to the public in the pages of *Scientific American*.[6] I recall discussing this most popular cellular automaton with graduate students at Indiana University. (The rules of this "game" are described in Appendix A.) Many ideas were considered as we looked for ways to deal with a dynamical system without the familiar accouterments: energy, conservation laws, continuous time and space. One could define basins of attraction, compare sizes, contemplate evolution, robustness of patterns and finally of course there was the overwhelming fascination of using newly friendly computers to watch individual patterns move about the two-dimensional universe, sometimes exhibiting the quasi-periodic or reproductive behavior that had suggested a relation between this game and biological life.

In 1973-4 Phil Seiden spent a Sabbatical at the Technion in Israel (I was both there and at Indiana at the time) and he too had become fascinated with this "game." We began to explore together; a vivid memory from that era is that the most challenging part of the problem was to ask good questions. We soon tended toward things with a statistical mechanics flavor. They had (at least) three virtues: They seemed to be a way to address the subject for which this system was supposedly a model, namely, life. They were doable. And perhaps most significantly they took on a life of their own, they introduced their own problems. Since we were in a way pioneers we got to ask the simple questions. Here's the first and simplest: Let the system evolve for a long time. What is the asymptotic density of living squares? Then there were more microscopic questions: surface to volume ratios, average numbers of living neighbors and other correlation related issues. Another class of questions dealt with changing the rules of the game. Increase or decrease the number of living sites needed for survival or birth. Although others subsequently found this a source of endless fascination, what we found most interesting was to introduce an element of randomness. You calculated what the strict Conway rule told you and then you followed or did not follow this with some probability. That probability was parametrized by something we called temperature which had the property that zero temperature gave the Conway rules and infinite temperature gave essentially random birth and death. We soon found ourselves on pleasantly familiar ground: phase transitions, order parameters, entropy, etc. In particular at low temperature the Conway game tended toward zero density (at strictly zero temperature there would be residual patterns). At higher temperatures the system tended to nonzero density. The transition between these patterns was sharp. There was also a geometrical entropy that we defined that seemed to point toward providing one of the desiderata of this subject: a conserved quantity. We all need our familiar props and the absence of conserved quantities like energy or momentum was one of the problems of this early CA. We found that the geometrical entropy divided by the geometrical entropy of a random array at the same density was a (statistical) constant of the motion. This was true over density ranges that spanned an order of magnitude or more. We were measuring the order imposed by the Conway rules. This provided a tool also for studying our phase transition, the aforementioned density discontinuity.

Many of the issues we touched on then[7] warrant further attention. Some have received that attention in the outburst of activity on cellular automata in the intervening years. We continued in a different direction. By the latter stages of our investigation I had become Seiden's guest at IBM. Those who ask why IBM was supporting "research" on the game of Life will further wonder why there was an astrophysicist in the same corridor. Humberto Gerola was interested in our discussions of the Conway game and the framework reminded him of attempts by Mueller and Arnett[8] to model propagating star formation in a lattice-like framework. Their efforts had not produced realistic galaxy images and apparently they did not consider the direction promising enough for follow-up. Gerola's observation quickly led to work on the structure and evolution of galaxies using the techniques we'd developed for the Conway game. I would say that the main things we brought from our modified Life was the randomness and some of the theoretical techniques. The details of this model are the subject of the next section.

3. The Galaxy as a Random Cellular Automaton[5,9–14]

The physics of propagating star formation is not completely understood but it is an observed phenomenon, so that although there are uncertainties in the modeling process I am about to describe, there is confidence that the broad outlines are true. Moreover, as we shall see, one of the virtues of statistical mechanical modeling is that you don't really need full control of the details in order to be able to draw useful conclusions.

Roughly speaking, if you don't disturb the interstellar medium, nothing much happens. A small perturbation away from a uniform distribution is unlikely to intensify (and form dense regions, and ultimately stars) because the gas only heats more and does not reach the point where gravitational forces can overcome the thermal energy (nor do densities reach the point where molecular Hydrogen is shielded against radiative breakup). What it takes to form stars is a shock wave—a large perturbation—passing through the medium, and for the typical shock waves that do exist a star formation episode will produce several hundred stars with masses that can go up to about 30 stellar masses. Where do the shock waves come from? From a previous episode of star formation! The extremely large stars that form live only 5 to 10 million years and end this bright and brief period by going supernova. That supernova can supply the shock needed to form stars in a nearby region. Many details of this process were shrouded in mystery when the stochastic cellular model for galaxies was first introduced and puzzles remain today. Is the star formation episode of which I speak the actual star formation or is it the formation of a molecular cloud from which several apparent episodes of star formation emerge? Then it would be the shock wave from one of these that would initiate the next molecular cloud. I will not go into these issues although my colleagues have certainly attended to them.[15,16]

With this picture, let me describe the progress of events in a spiral galaxy. A supernova goes off. Its shock wave progresses outward. Sometimes it reaches a region ripe for star formation (or formation of a molecular cloud), sometimes it does not. Suppose a nearby region does ignite. Then a few million years later *it* is the source of a shock wave that can continue the process of star formation to yet another region. It is like an disease among a group of susceptible individuals, and the mathematical formalism we develop is suitable for epidemiology as well. Next, let me apply an approach suggested by the framework of the Life game. I will think of the spiral galaxy as a two dimensional array, each "square" of the array a region of the disk. Time is also discretized (a procedure statistical mechanics experience suggests should be harmless). At any given time there is a collection of "living" and "dead" sites, namely sites where star formation is occurring and sites where it is not. On the next time step we allow the neighbors of each living site to come alive, to begin an episode of star formation, with probability p, where p is a given parameter. And so it goes. However, before continuing the description, I must tell you something about the lattice. The "square" is not really a square. It is a region of a galaxy. To best model this we imagine the galaxy divided into concentric rings and each ring cut into little near quadrilaterals, two of whose sides are parts of radii extending from the center of the galaxy and the other two sides are circles centered on the galactic center. See Fig. 1. Furthermore, these rings are not sitting still; they are rotating with the galaxy. This is where we insert a known property of the galaxy. There is a function $v(r)$, the velocity as a function of distance from the center, that for most galaxies is fairly flat, varying relatively little once you get away from the very center. The concentric rings making up the galaxy therefore do *not* rotate rigidly, but rather slip past each other as the inner rings overtake the outer ones. (Only the function $v(r) \equiv \text{const} \cdot r$ would give rigid rotation.)

For the simplest of models these are the only parameters, p and v, and v isn't really a parameter since it is taken from observation. If p is small, but not too small, this procedure gives reasonable looking pictures of galaxies with only one defect to trouble the eye, namely the edge of the galaxy (at its maximum radius) is too sharp. Let me be more specific about this last assertion. As mentioned above, you make a large number of concentric rings, each ring divided into quasirectangular "cells." There are $6N$ of them for the Nth ring, so that all cells have about the same

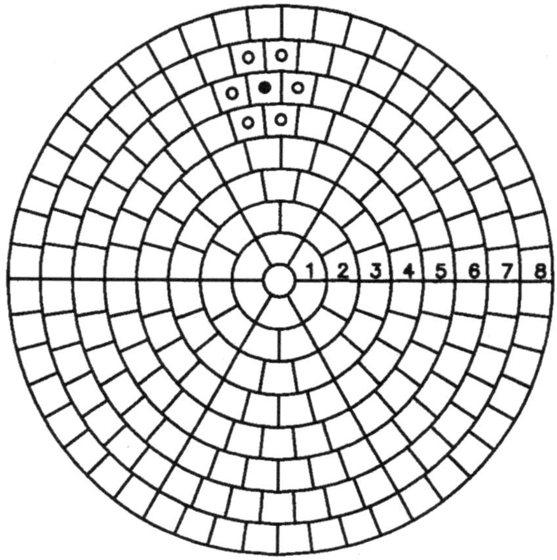

Figure 1. Galaxy geometry for stochastic evolution. The disk galaxy is represented by a sequence of concentric rings, each divided into quasirectangular cells representing potential regions of star formation. A region in which star formation is taking place (solid circle in the figure) can induce star formation in nearby regions (open circles) on the next time step. For a typical simulation, regions are taken to be about 200 pc square and time steps about 10 million years. Rotation of the galaxy is modeled by rotation of the rings with the usual rotation curves [$v(r)$] giving rise to differential rotation in the rings.

size. At any given time step most cells have 6 neighbors (although a few have 5 or 7). Some sites (cells) are alive, possessing star formation. For the next time step, an attempt is made to ignite each one of its neighbors in such a way that the probability of success is p. Successful ignitions give rise to the star formation regions on the next time step. However, when the next occasion for an attempt to transmit star formation arises, the region in question may have different neighbors (because of the differential rotation). An attempt is then made to ignite the new neighbors and the process continues with alternate position and star formation updating. A bit more physics must be put in to relate this to the world of galaxies. Sizes must be assigned to the "cells" and to the time steps. From a statistical mechanics point of view the only thing you want from these sizes is that the conclusions not be sensitive to them. From the point of view of the astrophysicist, there would be a search for significance if some region size turns out to be preferred.

In practice our results were neither wildly sensitive to the choice of sizes nor wholly insensitive. There were good results for cells of 200 pc by 200 pc and time steps of 10^7 yr. Several images are shown in Fig. 2. The pictures show more than merely those cells alive at the given time step, since the usual galaxy photographs show more than only those regions where star formation has taken place in the last 10 million years. Rather, living cells for all of the last 15 time steps are shown, with the more recently living cells depicted using larger symbols. This is a fair representation of the typical photographic sampling of the visual spectrum. A different scheme is used when we produce the infrared image of a galaxy.

The most significant embellishment of the above program involved accounting for the refractory period that resulted from depletion and heating of the gas in a given region due to an episode of star formation. It is reasonable to expect that immediately following star formation the same region could not support another episode, and that the recovery time is an important physical input. At first this was dealt with by reducing the value of the probability for successful ignition in a given region, with that probability gradually recovering to its maximum value on some time scale τ. Later this was replaced by a scheme that took more explicit account of the gas, both because it gave more realistic results and because this allowed comparison with observational results on the gas itself. For each cell, we kept track of its active and inactive gas density, the former presumably related to atomic Hydrogen (HI) and the latter to molecular Hydrogen (H_2). Star formation made all gas inactive, subsequent to which recovery took place on a time scale τ. At all times the probability of star formation in this region (due to a shock wave from a neighboring region) was a certain raw probability p_0 times the square of the *active* gas density (other powers could also be used, but most of our results were not sensitive to this).

The pictures that resulted from this were in my opinion compelling. Spiral arms formed in a robust way. What I mean is that when you saw a movie of the evolution of a galaxy, arms broke up and re-formed. It did not take special conditions to get arms. Other features—less graphic but of a more quantitative, less subjective nature—also emerged and I will discuss them in later sections. Before doing this, however, I would like in the next section to discuss in a more abstract

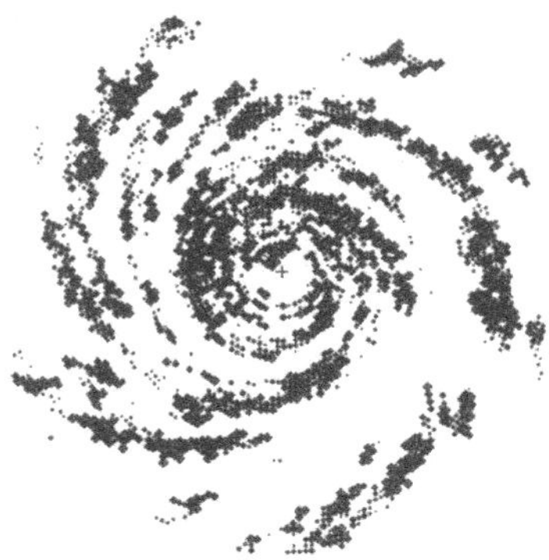

(a)

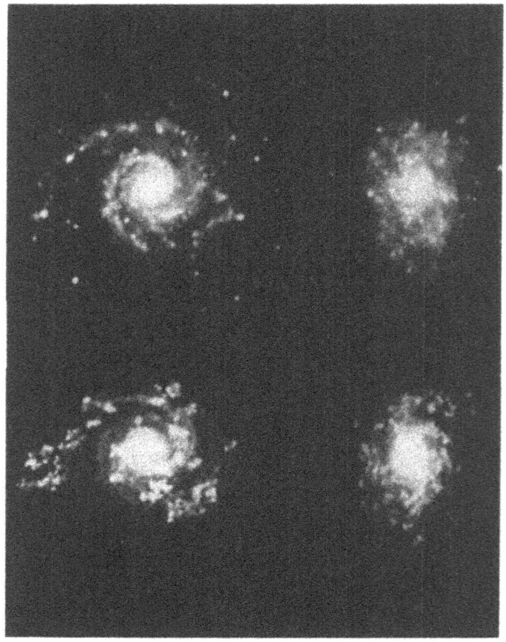

(b)

Figure 2. Images resulting from stochastic evolution. (a.) For the simulation illustrated here we took $p = 0.18$, 49 rings and a rotation curve corresponding to 200 km/sec for the space and time dimensions given in the text. (b.) Left column of images are the galaxies NGC 7793 (upper) and NGC 628. To the right are the results of simulations using the corresponding velocity curves. Galaxy images courtesy D.M. Elmegreen.

way the underlying reasons for the success of this approach.

4. Percolation theory and galactic morphology

Water may or may not percolate through soil and an abstract system that models this process is the basis of *percolation* theory. For the simplest model, you can imagine a gigantic jungle gym made of pipes through which water can pass. Suppose some of the pipes are obstructed so that there is a probability p that a given pipe is open, with the state of individual pipes independent of one another. One then asks whether water pumped in at one end of the lattice (jungle gym) is able to reach the other end. For the analogy with ground water percolation one thinks of the pipes as representing small spaces between particles of the soil, some spaces being large enough to allow water to pass, some not. Rainfall starts at the top and, if percolation is possible, will make its way underground through the soil. What one learns from the model is that for a large lattice there is critical value of the percolation parameter, p_c, such that for $p > p_c$ water will almost always pass and for $p < p_c$ it almost never does. The "almost" has to do with the size of the system and the proximity to p_c: The closer you are and the smaller the system the more there is a possibility of deviation from the most likely behavior.

This model can obviously be applied to more than just water. For the purpose of relating it to galaxies I will use a medical metaphor—this was presaged by our earlier remark that likened propagating star formation to the transmission of disease. It turns out that we are especially interested in something called "directed" percolation, that is, the pipes have one way valves. However, rather than go over to an abstract formulation I will continue with a specific epidemiological model.

The epidemic model concerns a disease called percolitis.[1] As the name suggests, it is not a serious disease. It lasts for one day, beginning and ending at midnight. It conveys no immunity. An individual (healthy or sick) who has come in contact with a sick individual has probability p of contracting the disease the next day, independent of the degree of contact. For simplicity we allow an individual to (re-)infect himself. Suppose we have a town of N individuals and that this is a very social place: Every day, everyone comes in contact with everyone else in town. We are interested in whether, once started, the disease continues to propagate. This will be the *percolation* question: For long times (a large lattice), are there still sick people (water coming out the other end of the jungle gym)?

Let the number of sick people on day t be $n(t)$. Let the fraction sick be $\rho(t) \equiv n(t)/N$. We first derive an "equation of motion" for $\rho(t)$. Consider the situation on day $t+1$. The probability that a randomly selected individual is healthy is (by definition) $1 - \rho(t+1)$. On the other hand, for that individual to be healthy there had to be $n(t)$ *unsuccessful* attempts at transmission (since everyone contacts everyone every day). The probability of a single unsuccessful transmission is $1 - p$ and for $n(t)$ attempts you simply take that power. Thus $1 - \rho(t+1) = (1-p)^{n(t)}$. We next rescale p using N and define the parameter x to be $x \equiv pN$. This gives

$$\rho(t+1) = 1 - \left(1 - \frac{x}{N}\right)^{N\rho(t)} \approx 1 - \exp[-x\rho(t)] \tag{4.1}$$

For present purposes we will look only at the large N approximation in (4.1), although we will lose thereby rare events such as the complete disappearance of the disease through a large fluctuation.[17]

First we consider whether the disease settles down to an equilibrium level. In this case there will be a value $\bar{\rho}$ such that $\rho(t) \to \bar{\rho}$ as $t \to \infty$. Eq. (4.1) becomes

$$\bar{\rho} = 1 - \exp(-x\bar{\rho}) \tag{4.2}$$

It is easy to see (e.g., graphically) that $\bar{\rho} = 0$ always satisfies this equation, and that for $x > 1$ there is a second positive solution. It is also easy to see that for $x < 1$ the iteration converges to $\bar{\rho} = 0$ while for $x > 1$ it is the strictly positive solution to which $\rho(t)$ tends. The implication is that for $x < 1$ the disease dies out, while for $x > 1$ it tends asymptotically to a nonzero level. This conclusion is confirmed by simulations.[1] In fact this system undergoes a percolation phase transition with critical value $x_c = 1$.

This conclusion shows the power of the statistical mechanics methodology. For example, for a large city in which this disease was rampant, our conclusion would be that the disease could be contained by quarantining individuals in non-communicating units of size less than $1/p$. The only property of the disease one needs for this conclusion is the transmission probability p. I don't want to make too much of this in the present case, since for this simple question the mathematical treatment does not go far beyond what one could deduce using words. We will see that for more subtle properties, such as relaxation times, rather stronger conclusions will be drawn.

We turn next to time dependent phenomena. This will lead us to general properties of phase transitions that are of relevance for the galaxy problem and which lie at the heart of the formation of spiral arms in our model. We examine the approach to equilibrium. Let $\delta\rho(t) \equiv \rho(t) - \bar{\rho}$. We expand Eq. (4.1) in $\delta\rho$ to yield

$$\delta\rho(t+1) = (1-\bar{\rho})(1-\exp(-x\delta\rho(t))) \approx (1-\bar{\rho})\left[x\delta\rho(t) - \frac{1}{2}(x\delta\rho(t))^2 + \ldots\right] \quad (4.3)$$

The linearized equation, i.e., dropping quadratic terms in $\delta\rho$, has the form

$$\delta\rho(t+1) = \exp(-1/\xi)\delta\rho(t) \qquad \text{with} \quad \exp(-1/\xi) \equiv (1-\bar{\rho})x \quad (4.4)$$

It is easy to show that whether x is greater than or less than one, it is always true that $\xi > 0$. Thus if the initial deviation is small

$$\delta\rho(t) = \exp(-t/\xi)\delta\rho(0) \quad (4.5)$$

The quantity ξ thus emerges as a relaxation time or, by analogy with problems with a bona fide spatial structure, a correlation length. Let us consider how ξ behaves as $x \to 1$. For the case $x < 1$ we find that

$$\xi = \frac{-1}{\log x} \approx \frac{1}{1-x} \qquad \text{for} \;\; x \nearrow 1 \quad (4.6)$$

The correlation length therefore diverges as x approaches its critical value. This is also true above x_c although the demonstration requires a bit more algebra. For the health officer in our imaginary city this means that if the size of the quarantine zones is near to $1/p$ it will take a long time for things to settle down.

Exactly at criticality one can show [keeping quadratic terms in Eq. (4.3)] that $\delta\rho(t) \approx 2/(t+t_0)$, where t_0 is a constant and t is large. (This is a mean field derivation; because of the proximity to criticality, finite N effects will be more important than for non critical properties.) This transition from exponential to power law equilibration is typical of the critical regime.

We next show how percolitis can fit the "jungle gym" picture used for water percolation. Imagine many copies of this town, one on top of the other. The lowest copy represents the time-zero state of the town; the one atop it the time-one state, and so forth. From every individual in the time-t copy draw a line to every individual in the time-$(t+1)$ copy. Each such line can be considered the pipe of a jungle gym. (There are more pipes now; for the cubic lattice jungle gym direct connections only went to the 6 nearest neighbors.) Suppose that only a fraction p of the pipes are ***not*** blocked. The significance of an unblocked pipe between two individuals is this: If the individual at the earlier time is sick and if the pipe to the other is unblocked, then the individual at the later time gets the disease. So p is the probability of disease transmission. Moreover, allowing passage of water through a path of unblocked pipes and having this continue for a long time corresponds to continued survival of the disease.

The last piece to put in place is the relation of all this to galaxies. Clearly our picture of stochastic propagating star formation corresponds to the random transmission of a "disease." As for the town, we can imagine many copies of our galaxy array, one stacked on top of the other, representing successive steps in our simulation. From each (potential star forming) region pipes go out to its neighbors on the next time step. (This is now more like the jungle gym, in that only a few pipes go out from each cell.) There is also a changing of neighbors, due to the differential rotation in the galaxy. The probability of an unblocked pipe is once again p, the parameter we used for propagating star formation.

One of our earliest observations when doing galaxy simulations was that the best galaxy pictures arose when p was near criticality, just above the phase transition (i.e., above the threshold for percolation). We can now understand why this is so. As for percolitis, there are correlation lengths that grow large near criticality, except that now there is both a temporal relaxation rate and a spatial correlation length, both of which diverge (but at different rates) for $p \to p_c$. It is just through this mechanism that the short range interactions are able to influence global structure.

There is another way to see the role of criticality in our model, one that I like because one need not appeal to anything more difficult than mean field theory in order to see the basic mechanism. By a line of reasoning similar to that used in percolitis we can deduce that the fraction of active sites in the galaxy has the following time dependence:

$$\rho(t+1) = 1 - [1 - p\rho(t)]^6 \tag{4.7}$$

The "6" appears because there are typically 6 neighbors for each site. Eq. (4.7) is derived by looking at a particular cell and assuming that the actual occupation statistics in its neighborhood can be replaced by the average values throughout the galaxy. This is mean field theory. As for percolitis, one finds the equilibrium equation by setting $\rho(t+1) = \rho(t) = \bar{\rho}$, and solving. The resulting curve, $\bar{\rho}$ as a function of p, is shown as the solid line in Fig. 3. The circles represent simulation data. The mean field prediction doesn't look too bad until one realizes that the region of disagreement corresponds to the important parameter range for galaxies, i.e., the range in which the pictures looked best. The error in the mean field theory arises from the neglect of correlations among sites. That is, by replacing the actual neighborhood of a site by an average neighborhood, we implicitly assumed that the probability that two sites are alive is the square of the probability that any one of them is alive. If there is a tendency, e.g., for living sites to be near each other, this assumption would be invalid. We do not have an exact replacement for Eq. (4.7). However, we do have another mean field calculation that throws light on the essential mechanism of the phase transition. Consider a slightly modified density, to be called "local" density. We define it so that it looks only at sites in the galaxy that have at least one living neighbor. To be precise, local density is the probability that a region with at least one living neighbor is itself alive. This can easily be calculated in mean field theory and the results are also shown in Fig. 3. The dashed line is the simplest mean field prediction. The squares are the local density results from simulations. Now the mean field theory is excellent—even as the system passes through its phase transition.

This gives key information on the nature of the phase transition. It says that not much is happening locally. Near any given site that is alive the precise correlations are not important. What is important is another kind of correlation, namely, if there is nothing around you have nothing nearby. That is, the living and dead regions of the galaxy are highly segregated from one another; this is the main effect of criticality. This view is of course consistent with the earlier discussion of diverging correlation lengths, with those lengths corresponding to the characteristic scale for the separation of living and dead regions.

With these insights, we can now describe why our model is successful in producing well articulated spiral arms. First let me tell you what does *not* give spiral arms: differential rotation alone. Imagine you had a bowl of dough (to make a cake) and you dropped into it gobs of chocolate. Now stir. At first you will get nice spiral arms, but as you continue the arms get thinner and thinner until finally you get chocolate cake! Differential rotation gives arms but by itself it also squeezes them out of existence. What is needed is a mechanism for clumping, something to keep the gobs of chocolate together. This is the role of the phase transition. In our galaxy model, a region in which several star formation events occur will be rotated and stretched by the galaxy's overall differential rotation. But the percolation process—in particular when it is close to the critical point—maintains the tendency to clump. Regions of living cells retain their character. Similarly, empty regions have the same property (cumulants of a Boolian random

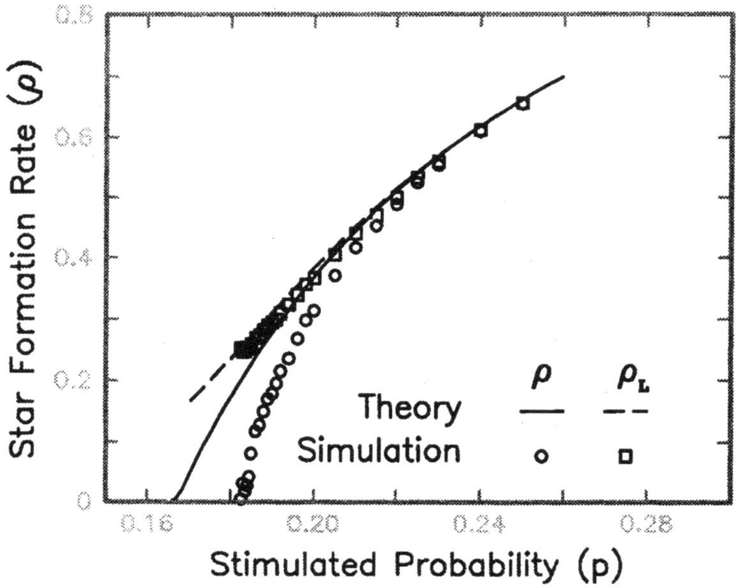

Figure 3. Star formation density (ρ) and local density (ρ_L) versus probability of stimulated star formation. Solid line is the mean field theory prediction for density. Circles are results of simulations. The dashed line (which merges with the solid line at large p) is local density and the squares are the local density taken from simulations.

variable and its complement are the same). Thus the percolation process leads to clumps and big empty regions between them. The differential rotation then spreads these into well articulated arms.

5. Astrophysical implications of the model

5.1. Morphology

The first and most dramatic results of our work were the pretty pictures. Several of these were reproduced in Fig. 2. More dramatic are the moving pictures. In these one can see the formation of spiral arms, their breakup and reformation as stable features. In a moment I will present other consequences of the model, but first I'd like to comment on "pretty pictures." A problem in biology is morphogenesis, and surely one of the difficulties is the lack of precise characterization of the various concepts of form. One of our goals for the Conway game was to approach this problem, and the geometrical entropy mentioned in §2 was a step in that direction. What the sometimes denigrated "pretty pictures" do show is success in getting the *form* of the galaxy. The critical phenomena-related arguments that I gave in §4 therefore represent a mathematical justification for the generation of one kind of form. As such, it is another step toward understanding morphogenesis.[18] What we say is, here is a way to get form: Bring your system near criticality; large correlation lengths will promote short range interactions, interactions that you might have thought inadequate for the creation of global structure. There are non pictorial ways to see that the mechanism works, but *without* the pictures I doubt that anyone would study the system long enough to discover them.

Having made a brief for the worth of addressing morphology, let me return to issues that can be studied with this model that go beyond the recovery of spiral arms. These have been covered in detail in the review papers previously cited, so I will here mention only the highlights.

5.2. Dwarf galaxies

Dwarf galaxies are known to show great variability in their luminosity. Two dwarf galaxies of ostensibly similar properties may differ considerably in their level of star formation activity. The explanation turns out to be simple. 1) When you have small statistics you get large fluctuations, and 2) small objects tend to have large refractory times. The combination of these features leads to bursting behavior, so it is not that one dwarf is bright, another quiescent, rather some are being observed during bursts, others in between. Fig. 4 (taken from Ref. 12) shows simulation results for galaxies whose sizes correspond to those of dwarf galaxies.

5.3. Disk luminosity profile

An observational feature of galaxies is the "exponential disk," a characterization reflecting the fact that the luminosity as a function of radius for galaxies can be reasonably well fit by an exponential. But this exponential is not like other exponentials. The exponential you get, say, in the decay of unstable particles, represents an absence of memory. The amount decaying is proportional to the amount left. In fancier terms, you have time-translation invariance. But for the galactic disk there is no corresponding rationale.

Here is where it was worth putting a bit of physics into our model. In dealing with the refractory time (the time it takes for the star formation capabilities of a

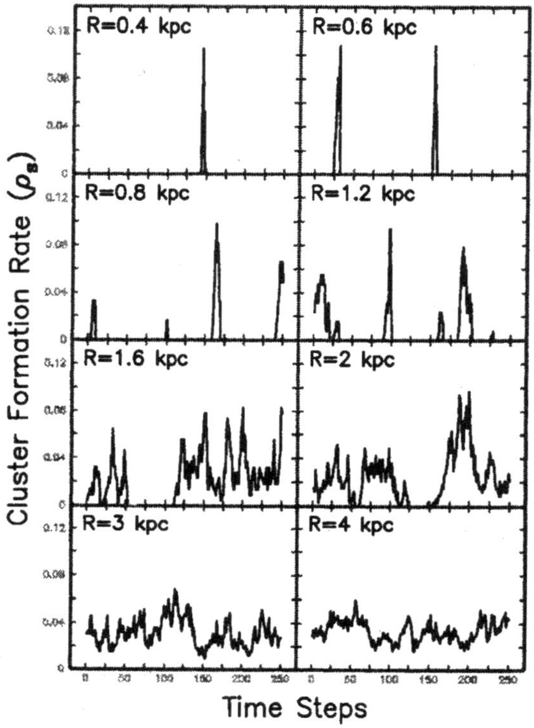

Figure 4. Dwarf galaxy simulations. Star formation as a function of time is shown for galaxies of varying radius.

given cell to recover from a previous episode of star formation) we had introduced an additional variable for each cell, the gas. This was divided into active and inactive gas (HI and H_2, respectively) and the probability of outside stimulation causing star formation was (usually) taken proportional to the square of the active gas. Refractoriness corresponded to changing active to inactive gas after star formation, with gradual return to active status on successive time steps. (In some cases gas was allowed to diffuse from cell to cell, but I will not discuss this here.)

Once the gas became an integral part of the model, it was natural to take the total amount of gas in each cell proportional to the amount of matter one expected to find at the corresponding radial distance from the galactic center. From general considerations, one can deduce this from the function $v(r)$, the velocity at a given radial distance. For the most common behavior, namely v approximately constant, this gives a mass density that drops like $1/r$.

Now our simulations had turned up a systematic feature in the relation of the inactive and active gas. For a wide range of total gas densities (hence, raw stimulated probabilities) the amount of inactive gas was relatively constant. This constancy is a specific consequence of our model and can also be understood analytically. The rest of the gas was active and contributed to luminosity. It follows that for constant-v galaxies the luminosity should drop like $[1/r - \text{const}]$. With this prediction, we reexamined the so-called exponential. It turned out that the fit to an exponential was over a relatively small range and that inverse radius minus a constant gave as good or better a fit—and carried with it an explanation.[13]

5.4. Comparisons with spiral density wave theory

At the time that our model was developed there was already a venerable, if sometimes problematic, theory explaining spiral arms. This is the spiral density wave theory,[19] and it is basically a theory of where the matter is. By traditional fluid dynamics and gravitational theory one begins with the observation that a flat uniform disk is unstable and finds a tendency to get dynamical modes that resemble the galactic arms. (The "problematic" aspect has to do with the stability of these modes.) One then assumes that luminosity follows matter. More recently there have been N-body simulations for galaxy dynamics that independently suggest that spirals do naturally form. This complicates the issue of deciding whether or not our mechanism plays a role in the formation of the arms. It is probably fair to say that at this stage one simply does not have sufficient evidence one way or the other. In their work contrasting grand design and flocculent galaxies Elmegreen and Elmegreen[20] developed an informative display that casts light on this issue. Consider arm/interarm contrast. According to our model this contrast should be considerable in the blue and slight in the infrared. For theories in which the arms represent the matter (with luminosity merely following the matter), both contrasts should be large. In Fig. 5 we show the ratio of arm/interarm contrast in the blue and the infrared versus arm interarm contrast in the infrared. Points are shown for several galaxies and for several positions within the galaxy. For a galaxy for which one already has reason to expect the importance of spiral waves (e.g., grand design galaxies) the points show the anticipated result. For these there is not much difference between the infrared and blue contrasts. Hence the points (mostly the solid symbols in the figure) tend to have values near one. However, other galaxies

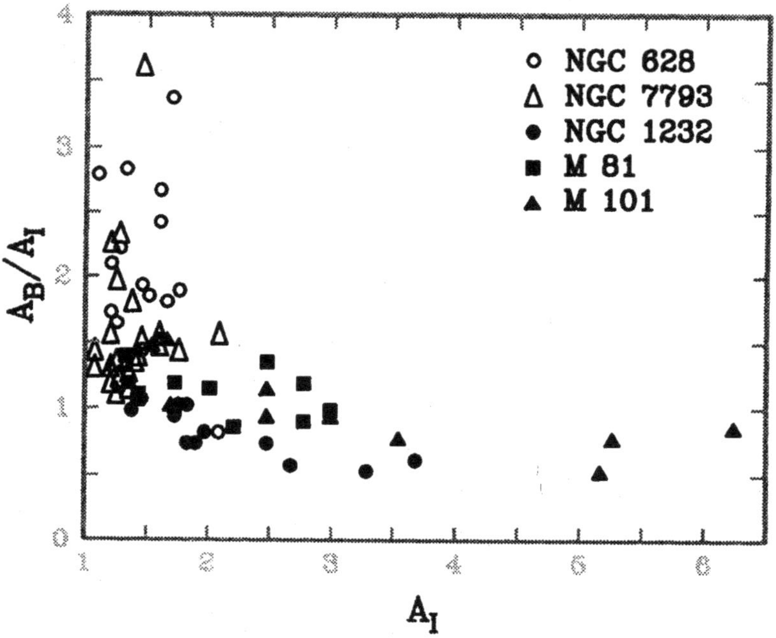

Figure 5. Ratio of blue and infrared arm/interarm contrasts as a function of infrared arm/interarm contrast. Each contrast is itself the ratio of the surface brightness in the arm region to that of an adjacent interarm region. (See Ref. 20.)

show considerable difference between the blue and infrared ratios, and support our model. These are the open symbols in the figure.

My own belief, based on this picture is that there is more than one kind of galaxy and that some are dominated by gravitational and fluid dynamic effects while for others the morphology is a consequence of the statistical mechanics effects described by our model. I believe a real challenge in this field is the development of a common language in which both aspects of galactic dynamics and morphology could be studied in a unified way. There is recent work[21] that suggests this goal may be on its way to being realized.

5.5. Coherent galactic oscillations

For one reason or another there may be galaxies with long refractory times. In that case we predict[14] a new kind of behavior: coherent oscillation. On a time scale of perhaps 300 or 400 million years the galaxy should go bright and dim, continuing this oscillation in a regular way over many periods. The theory of this is not difficult and is partly implicit in the calculations of §6 below. This behavior also leads to a tendency toward the appearance of grand design.

At first it may seem that 400 million years is a long time to watch to see an oscillation, but there are more readily detectable observational consequences. In particular, there should be galaxies rather similar to one another but with markedly different luminosity. In Ref. 14 we suggested that anemic and Markarian galaxies may represent such oscillators. Lately the discovery of low surface brightness galaxies has drawn considerable attention, although our preliminary study of these does not suggest that they are our oscillators. For deciding that a particular galaxy is an oscillator one could compare its morphology to those arising naturally from our simulations (for example a bull's eye pattern) although this may be overly subjective. I would say that if one had an anomalous galaxy (too bright or too dim) and could establish or estimate that it possessed a relatively long refractory period, then it would be a good candidate for one of our oscillators.

6. Feedback and self organized criticality

A recent development in statistical mechanics is the notion of self organized criticality (SOC). In general, critical points are of zero measure in the phase diagram of a system; nevertheless, many systems in Nature are found near criticality, or at least show the fractal structure associated with criticality. SOC purports to explain this by studying ways in which systems, under their own dynamics, drive themselves to their critical points.

In our model we have long recognized the importance of *feedback*. Star formation reduces available gas, lowering the effective value of "p," while quiescence allows buildup of available gas, raising that effective value. Moreover, as emphasized above, the proximity of our system to criticality and the associated long correlation lengths are important parts of our explanation for the stable persistence of spiral arms.

This suggests several questions. Is the galaxy (as we've described it) an example of SOC? If not, what is it, what more general form of SOC could be defined to express the notion that a feedback process can create form? Finally we have found (contrary to our intuitions on scaling) that the feedback seriously modifies the (directed) percolation phase transition, at least for values of p_0 that are as close to the critical point as our simulations could take us.[22]

In this article I will describe a toy model. It exhibits the properties of the full galaxy model and allows more convenient manipulation and perhaps understanding. The basic dynamical law for the evolution of luminosity in the galaxy is

$$\sigma_\alpha(t+1) = 1 - \prod_\beta [1 - p_{eff}\sigma_\beta(t)] \tag{6.1}$$

where $\sigma_\alpha(t)$ takes the value 1 if in cell α at time t there is star formation, zero otherwise. The product runs over cells "β" that are the neighbors of α, typically a set of six cells. The effective value of the probability of stimulated star formation (p_{eff}) in the simplest model is simply a constant. In our more elaborate models it is taken to be a function of the active gas density, hence is time dependent. Furthermore, there is a separate equation of motion for the gas density, somewhat like Eq. (6.1). In the mean field approximation we ignore the correlations between nearby cells and take the expectation of Eq. (6.1) to yield

$$\rho(t+1) = 1 - [1 - p_{eff}(t)\rho(t)]^6 \tag{6.2}$$

where ρ is the expectation of σ, namely the density of living sites. (In Ref. 22 we found that replacing the sixth power in Eq. (6.2) by a fourth or third power took care of much of the effect of correlations by providing the actual number of neighbors usually available for potential stimulation.)

For the toy model I will further simplify the calculation of p_{eff}. Rather than reflecting the immediate neighborhood of a particular site, we will make a mean field assumption for this as well. One form that I found convenient is

$$p_{eff}(t) = p_0 \exp\left[-\sum_{k=1}^{M} \rho(t+1-k)\phi(k)\right] \tag{6.3}$$

The sequence $\phi(k)$ represents the recovery profile. With past star formation, Eq. (6.3) will reduce the probability that a cell comes to life. Our gas program suggests an exponential for this,

$$\phi(k) \equiv \phi_0 \exp(-k/\tau) \tag{6.4}$$

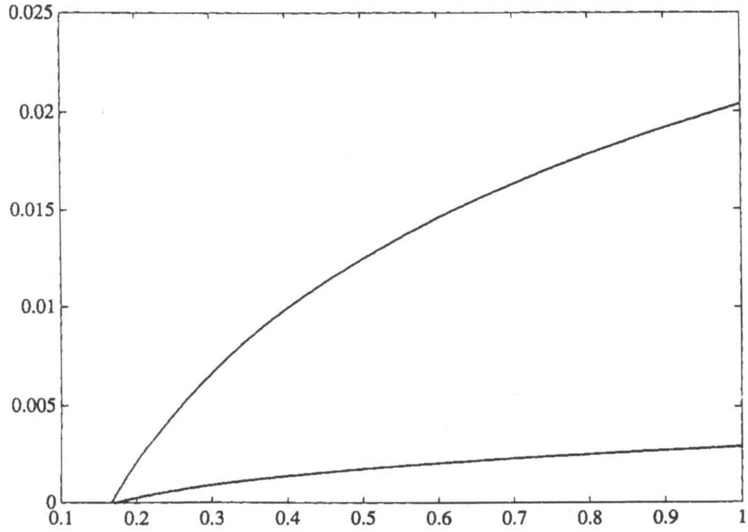

Figure 6. Density and effective probability for star formation stimulation in a mean field feedback model. The abscissa is the raw probability. The upper curve is one half the density and the lower curve is the effective probability minus one sixth (allowing a single ordinate scale). The raw probability ranges from one sixth to 1 while the effective value runs only from from a threshold of one sixth to 0.0029 greater than this.

This specific formulation introduces two additional parameters, M and ϕ_0, whose exact values have no significance for our qualitative conclusions. I also experimented with $\phi(k) \sim 1/k^\gamma$, which I feel would be more likely to show SOC in the true galaxy model; in the calculations I will now report I used the form in (6.4).

In Fig. 6 are shown density and p_{eff} as a function of p_0. (Parameter values were $\tau = 30$, $M = 40$ and $\phi_0 = 2$.) The feedback effect is remarkable. While p_0 varies from one sixth to one, p_{eff} only goes from one sixth to one sixth plus 0.0029. It is interesting that so small change in the effective value of the probability nevertheless brings the density from 0 to about 0.04, the latter well within the reasonable range for galaxies.

The equilibrium phase diagram (Fig. 6) is not the whole story. It also turns out that for p_0 large enough the system oscillates. (Variation of τ causes this as well.) This effect can be analyzed by linearizing about Eq. (6.1). Because we have a difference equation, solution of the linearized equation leads to a polynomial of order M. Just at the point where the appropriate roots of this polynomial move into the right hand plane, a numerical iteration of Eq. (6.1) begins to oscillate. (For the parameter values used in Fig. 6, this occurred at $p_0 \approx 0.5927$.) The frequency is derived from the imaginary part of (the log of) the unstable root and by monitoring the magnitude of the oscillation one can see that one has a standard Hopf bifurcation.

The phenomenon is richer than the logistic equation because the memory makes the iteration effectively involve a large number of variables. For no reasonable parameters, however, did I find further bifurcations. I don't think this should discourage one from looking for multiple bifurcations or chaos in galaxies, since the spatial structure (which is ignored by the mean field calculation) is important. As remarked, I think that power law recovery has a good chance of leading to critical behavior in the galaxy simulation.

From working with many parameter values and forms for the recovery function, I gained the impression that the passage of the system to oscillatory behavior is typical. If this says anything about galaxies, it suggests that behavior of the sort suggested in Ref. 14 (and §5.5 above) is common; moreover, it appears that global time-dependent phenomena—on a scale of hundreds of millions of years—may result from the intrinsic local behavior of the star propagation process.

7. Large scale structure: percolation and the distribution of galaxies

In this section I mention another astrophysical application of our percolation-CA ideas. When Seiden and I learned about the voids and clusters in the distribution of galaxies in the universe—large scale structure—our first thought was that phase transitions and critical phenomena are a natural way to get this. There was a model of galaxy formation that had a percolation flavor, the Ostriker-Cowie model.[23] The picture we used[24] was that galaxy formation involves an enormous explosion (at least for a fraction of the galaxies) and this explosion helps start the galaxy formation process in a neighboring region. Note that we are not wedded to the Ostriker-Cowie model, only to this percolating feature of galaxy formation. In contrast to the propagating *star* formation, the galaxy formation process is concentrated in a relatively brief time interval so that ordinary (rather than directed) percolation is relevant. Depending on the effective percolation parameter (called, as usual, p) one will get limited local formation ($p < p_c$), extensive, homogeneous space filling formation ($p > p_c$), or a fractal distribution ($p \sim p_c$). What we need—and I consider this the physically most delicate part of the problem—is to have $p \sim p_c$. We argued that as the universe expands and gas cools, more or less everywhere at once, the effective value of p increases until it reaches the neighborhood of p_c. (Prior to that, spontaneous galaxy formation would have created limited, small clusters of galaxies.) Once in the neighborhood of p_c a large scale fractal cluster forms. With continued expansion, p again drops—although gases are cooler, distances between newly formed galaxies are larger. We are thus left with one, or perhaps several, fractal clusters of galaxies. I don't think there is much that can be rigorously said about this scenario for obtaining p_c. Perhaps there is a reason for self organized criticality, but without a better idea of the percolation process, I don't think it would be useful to attempt firmer justification.

At this point one runs into a technical problem.[25] It is all very well to appeal to percolation theory to give you voids and clusters, but percolation theory gives you more than that—it tells you about the two-point function. In particular, it drops off like $1/r$. (In three space dimensions it even does this in the crossover region between critical and noncritical behavior.) The two-point correlation function for galaxies was known to drop off like $1/r^\gamma$, with $\gamma \approx 1.7$ or 1.8. Not all scale-free structures are the same!

Before going into this I want to point out a significant prediction of our model. *The fractal structure that we get is in the luminosity distribution, not the matter distribution.* Of course if there is a fractal matter distribution as well, from some dynamical source, the fractal luminosity could sit on top if it,[26] but we do not *require* any nonuniformity in the matter, and at the end of our galaxy formation process, there need not be any significant matter nonuniformity. Where galaxies have formed, there may be a local depletion and the shock waves of formation may have influenced the matter distribution at some distance from the galaxy. But on the larger scale, the scale of the enormous voids, this will be unnoticed.

With regard to the correct power for the two-point function,[27] we quickly understood that there had been significant nonuniform expansion of the universe since the era of galaxy formation. However, detailed calculation, or even simulation, did not seem possible, since our galaxies were only a fraction of the total mass present. Our strategy was to avoid the details and appeal to scale invariance. For dynamics at sizes larger than galaxies there is only gravity, which is scale free. Our luminosity clusters are also scale free. Therefore, as the universe expands, this entire system could only evolve into another scale-free distribution. We thus argued that a power law distribution function evolves to another power law, with only the power changing in time. On a log-log plot, such a distribution function is a straight line. To find this straight line we need two points. If one assumes that for objects on the scale of single galaxy the overall expansion of the universe has little effect while for objects at maximum distance the expansion has its maximum value, then this gives us those two points. We found that if galaxy formation took place when the universe was about 1/40 of its present radius, then the power "γ" would have gone from 1 to 1.8. It follows that with reasonable assumptions for galaxy formation our percolation hypothesis can account for contemporary large scale structure.

Recent observations suggest that it is worth reexamining this model. What would be useful at this point is to develop further measures of the distribution. Fractal dimension is but a single number and different sets of the same fractal dimension may have quite different appearance. By having one or more additional measures one could judge among various proposed mechanism for large scale structure. One promising candidate is "lacunarity"[28] and I would like to see this or other measures evaluated for the many competing suggestions for the galaxy distribution.

8. Discussion

This article presents a model that uses stochastic cellular automata to describe the star formation process in galaxies. Beyond the model's most obvious achievement—the recovery of morphology—we considered more detailed features that showed good agreement with observation. On the issue of *choosing* between this model and more dynamically oriented ones such as the spiral density wave theory, we don't believe that there is a yes-no choice to be made. Both mechanisms operate, in some galaxies one of them may dominate, in others both appear to be important. What seems worthwhile is to have a comprehensive language in which both gravitational and propagating star formation effects are included.

Apart from its astrophysical content, the model carried two broader messages. First there is the value and impact of statistical physics: Short range interactions can lead to global structure; and much of the global structure is independent of the details of the local interactions. This scale enhancement occurs at phase transition critical points, which leads to the second general point. A feedback occurs in the astrophysical model that has the effect of driving the system near to its critical point (when considered as a percolation process). This turned out to be essential for the morphology. Such behavior is characteristic of that predicted by theories[29] of "self organized criticality." From the standpoint of those ideas, our system presents an example in which it is sufficient to be *near* the critical point (but outside the region where critical exponents dominate the behavior).

Finally, we presented a second application of cellular automata (in the guise of percolation theory) to astrophysics. A percolation theory for the formation of large-scale structure recovers the fractal distribution of the luminosity of galaxies while predicting that the matter distribution need not have the same fractal structure and may even be uniform. The detailed physical basis for the model (critically percolating galaxy formation) is speculative, but the results may be of sufficient interest to motivate attempts to justify that physical basis.

Acknowledgements

I thank B. Elmegreen, B. Mandelbrot and P. E. Seiden for stimulating discussions. I am also grateful to the University of Paris VI and Columbia University for their kind hospitality. This work was supported in part by NSF grant PHY 90 15858.

Appendix A. The Rules of Life

The arena is a two dimensional $N \times N$ square lattice which may have periodic or free boundary conditions. One also considers the case $N \to \infty$. At any given time t some of the sites are "alive," others "dead." From the time-t array one generates the time-$(t+1)$ array in the following way: A given lattice site is alive or dead at $t+1$ depending on the number of its living or dead neighbors at time t. Call the site in question α. Let β be an index labelling its neighbors. In the standard Conway rules there are eight neighbors. Thinking of the lattice as a map, there are firstly the four sites in the cardinal directions away from α. In addition we include the four nearest sites located to the northeast, northwest, southeast and southwest. The rule is this: If at t α is alive and if among the eight neighbors either 2 or 3 are alive (and the other 6 or 5 dead), then at $(t+1)$ α is alive. If at t α is dead and if among the eight neighbors exactly 3 are alive, then at $(t+1)$ α is alive. Otherwise, α is dead at $t+1$. These rules can be summarized by

$$\sigma_\alpha(t+1) = \delta(3, \Sigma_\beta \sigma_\beta(t)) + \sigma_\alpha(t)\delta(2, \Sigma_\beta \sigma_\beta(t))$$

where $\sigma_\alpha(t)$ is a zero-one valued variable which is one if site α is alive at time t, zero otherwise. $\delta(n, X)$ is the Kronecker delta and takes the value one if $X = n$, zero otherwise. The sum over β runs over the eight neighbors described above.

The result of this simple rule is the generation of elaborate patterns. You can immediately check that an isolated block of 4 living sites will stay that way. Three living sites in a vertical row (with nothing else nearby) will on the next step become three living sites in a horizontal row. This oscillation continues indefinitely. Then there are "gliders." These are patterns of 5 (or more) sites that over 5 (or more) time steps reproduce their original form but displaced, generally along a diagonal. Far more complicated reproducing patterns have been found. There was for a time a newsletter of aficionados of this system,[30] in which ever more complex "living" creatures were reported. There were also constructions in which certain patterns for this CA were used as elementary computers which by stacking became universal computers.[31]

References

1. L. S. Schulman and P. E. Seiden, Percolation and Galaxies, *Science* **233** (1986) 425.
2. P. E. Seiden and L. S. Schulman, Percolation Model of Galactic Structure, *Adv. Phys.* **39** (1990) 1.
3. P. E. Seiden and H. Gerola, Propagating Star Formation and Structure and Evolution of Galaxies, *Fund. Cosmic Phys.* **7** (1982) 241.
4. L. S. Schulman and P. E. Seiden, Propagating Stochastic Star Formation and Galactic Structure, *Ann. Israel Phys. Soc.* **5** (1983) 251.
5. L. S. Schulman and P. E. Seiden, Percolation Analysis of Stochastic Models of Galactic Evolution, *J. Stat. Phys.* **27** (1982) 83.
6. M. Gardner, Mathematical Games: The fantastic combinations of John Conway's new solitaire game *Life*, *Sci. Amer.* (October 1970) 120. Shorter notes on *Life* appeared in *Scientific American* issues for the next year or so.
7. L. S. Schulman and P. E. Seiden, Statistical Mechanics of a Dynamical System Based on Conway's Game of Life, *J. Stat. Phys.* **19** (1978) 293.
8. M. W. Mueller and W. D. Arnett, *Astrophys. J.* **210** (1976) 670.

9. H. Gerola and P. E. Seiden, Stochastic Star Formation and Spiral Structure of Galaxies, *Astrophys. J.* **223** (1978) 129.
10. P. E. Seiden, L. S. Schulman and H. Gerola, Stochastic Star Formation and the Evolution of Galaxies, *Astrophys. J.* **232** (1979) 702.
11. P. E. Seiden and H. Gerola, Properties of Spiral Galaxies from a Stochastic Star Formation Model, *Astrophys. J.* **233** (1979) 56.
12. H. Gerola, P. E. Seiden and L. S. Schulman, Theory of Dwarf Galaxies, *Astrophys. J.* **242** (1980) 517.
13. P. E. Seiden, L. S. Schulman and B. G. Elmegreen, A Galactic Disk is not a True Exponential, *Astrophys. J.* **282** (1984) 95.
14. P. E. Seiden, L. S. Schulman and J. V. Feitzinger, Coherent Galactic Oscillations, *Astrophys. J.* **253** (1982) 91.
15. P. E. Seiden, Radial Distribution of Atomic and Molecular Hydrogen From Propagating Star Formation, in Kinematics, Dynamics and Structure of the Milky Way, Ed. W. L. Schluter, D. Reidel, 1983.
16. P. E. Seiden, The Role of the Gas in Propagating Star Formation, *Astrophys. J.* **266** (1983) 555.
17. V. Privman and L. S. Schulman, Infinite-Range Mean-Field Percolation: Transfer Matrix Study of Longitudinal Correlation length, *J. Stat. Phys.* **64** (1991) 207; L.S. Schulman and B. Gaveau, Finite-size scaling for mean-field percolation, *J. Stat. Phys.* (1993) to appear.
18. Phase transitions have previously been related to the generation of form through *catastrophe* theory. An example of this is given in L. S. Schulman, Stable Generation of Simple Forms, *J. Theor. Biol.* **57** (1976) 453. In this approach, the morphology is an image of the phase diagram, where one includes in that diagram "limits of metastability."
19. C. C. Lin and Y. Y. Lau, Density wave theory of spiral structure of galaxies, *Stud. App. math.* **60** (1979) 97.
20. B. G. Elmegreen and D. M. Elmegreen, *Astrophys. J. Suppl.* **54**, 61 (1984).
21. W. Benz, in this volume.
22. L. S. Schulman and P. E. Seiden, Feedback Induced Modification of Percolation, 1988, unpublished.
23. J. P. Ostriker and L. L. Cowie, Galaxy formation in an intergalactic medium dominated by explosions, *Ap. J.* **243** (1981) L127.
24. L. S. Schulman and P. E. Seiden, Hierarchical Structure in the Distribution of Galaxies, *Astrophys. J.* **311** (1986) 1.

25. We also ran into publishing problems. A referee report from *Physics Letters A*, where we first sent this in 1984, admonished us against publishing the results of "speculating freely in the bath." We may also have been too free in our prepublication dissemination of these ideas, since we later found them elsewhere in print, unacknowledged.
26. D. Ben-Avraham, S. Havlin and D. Movshovitz, Infinitely ramified fractal lattices and percolation, *Philos. Mag. B* **50** (1984) 297; S. Havlin, D. Ben-Avraham and D. Movshovitz, Percolation on fractal lattices, *Phys. Rev. Lett.* **51** (1983) 2347.
27. The correlation function to use for fractal structures has been provided by L. Pietronero, The fractal structure of the universe: correlations of galaxies and clusters and the average mass density, *Physica* **144A** (1987) 257. However, the power for r that one gets (using Pietronero's function) is the same as that previously calculated using other measures of correlation. See also P. H. Coleman and L. Pietronero, The Fractal Structure of the Universe, *Phys. Rep.* **213** (1992) 311.
28. B. Mandelbrot, private communication.
29. P. Bak, C. Tang and K. Wiesenfeld, Self-Organized Criticality: An Explanation of $1/f$ Noise, *Phys. Rev. Lett.* **59** (1987) 381; C. Tang and P. Bak, Critical Exponents and Scaling Relations for Self-Organized Critical Phenomena, *Phys. Rev. Lett.* **60** (1988) 2347; B. Gaveau and L. S. Schulman, Mean-field self-organized criticality, *J. Phys. A Lett.* **24** (1991) L475.
30. R. T. Wainright, editor "Lifeline," published in the mid 1970s.
31. J. H. Conway and R. W. Gosper Jr., as reported by M. Gardner, *Sci. Am.* Mar. 1971 and R. T. Wainwright: Life Is Universal, preprint, about 1975.

CA SIMULATIONS OF GALACTIC STELLAR DISTRIBUTIONS

A. LEJEUNE
Institut de Physique B.5, University of Liège, Sart Tilman
B-4000 LIEGE 1, Belgium

ABSTRACT

We present a new CA model of galactic stellar distributions. The "decay" processes like the intrinsic evolution of isolated stars and the "catalytic" processes which describe enhanced evolution due to the presence of neighbour stars are described in a probabilistic scheme. Moreover the model takes care of intergalactic matter infall on the galactic disk as well as of the local kinematics and proper motions of individual stars. Characteristics of the galactic system like percolation, fractal pattern of stars and state of self-criticality are studied in detail.

1. Introduction

In the late 70's Seiden and coworkers [5,14] introduced the idea that star formation in a galaxy could be simulated by a propagating forest fire which in turn can be modelled efficiently on a cellular automaton (CA). Although controversial -perhaps mainly because of its elementary nature discarding most of the complex physics- the model, dubbed *Stochastic Self-propagating Star Formation*, has several attractive features. Among the latter the unifying concept that star formation induces a *percolation phase transition* responsible for the morphological types of galaxy appears as particularly relevant.

The model of Seiden and coworkers consists in a 2-D circular galaxy mimicked by a circular lattice, whose cells are obtained by dividing the galaxy into N rings ($r = 1, 2, \cdots, N$) of same width surrounding a central circle ($r = 0$) ; the rth ring is subdivided into 6r cells of same size. This cellular arrangement specifies the geometry of the CA model ; the neighbourhood of each (typical) cell is defined as the set of 6 nearest neighbours. A cell state of this CA simulates the following configurations : (1) a molecular cloud state ; and (2) an OB-association ; the latter are subdivided into two groups, (a) active associations triggering the formation of new active OB-associations from molecular clouds ; and (b) passive OB-associations. The dynamics of the CA is specified by *two local* rules : (1) in the CA neighbourhood of a cell in an active state a new active OB-association is created with probability p per timestep ; (2) the active OB-association remains active during one timestep only. In addition to these local evolution rules, a *global* rule is applied at each timestep to simulate the differential rotation of the galaxy ; this is done by sliding the individual rings in accordance with an observed galactic rotation curve. To run the model all cells are initially in the molecular cloud state. A fraction f of randomly chosen cells is then activated (active OB-associations). In this simplest version the model involves a single intrinsic parameter, namely the rate of induced star formation p , and an external parameter, f specifying the initial conditions[1].

The model exhibits the appearance of spiral structure for $p > p_c \approx 0.18$, p_c being interpreted as a *percolation threshold* (the precise value depending on the rotation law). A detailed theoretical analysis of the neighbourhood of the percolation threshold (correlation length, mean cluster size, etc) can be found in Schulman and Seiden[12].

The strength of the approach by Seiden and coworkers, namely the isolation of a percolation threshold in the parameter p beyond which spiral arms form, is at the same time a weakness : the very definition of a model in a *single* intrinsic coefficient parametrizing a *single* physical mechanism potentially responsible for pattern formation in the galaxy makes it clear that a critical value of this parameter must exist such that beyond this value macroscopic structure will develop via an instability in an originally homogeneous state[2].

Such an argument does not guarantee at the outset that the critical value defines a genuine *percolation threshold.* However, it is geometrically clear that an arm in a spiral galaxy, as opposed to a random matter distribution in an irregular galaxy, can be interpreted as a percolation-like cluster provided that we introduce a suitable definition of a type of percolation adapted to this configuration. With a single physical mechanism included, the schematic model remains too rudimentary to isolate the main physical mechanism responsible for galactic pattern formation.

The purpose of the present analysis is to propose a multi-parameter family of models in which the galactic physics is treated in greater detail. In particular, we shall take account of a schematic, though essentially complete description of stellar evolution. We shall also differ from the approach by Seiden and coworkers by considering a local galactic model adapted to a small part of the galaxy only, of area L^2, say the solar neighbourhood (typically $L \approx 200pc$). This latter region is then simulated by a standard square lattice. In the local approximation differential rotation reduces to a shear flow. Taking again the solar neighbourhood as a reference galactic environment, the radial gradient of the velocity field is $dv/dr \approx 15km/(skpc)$. In addition to rotation we make allowance for proper motions ; the latter are described by a standard *diffusion process* of all of the galactic material.

To the extent that the percolation-like phase transition can be regarded as a singularity of codimension 1 there exists a single basic parameter P , the latter depending on the actual model parameters $p(i), i = 1, 2, \cdots, n$, $P = P[p(1), p(2), ..., p(n)]$. Assuming that the latter dependence is smooth and regular, the standard expressions of the correlation length x , the cluster size s , etc in the basic parameter near the percolation transition, $x \propto 1/ \mid P - P_c \mid^{\nu}$, $s \propto 1/ \mid P - P_c \mid^{\gamma}$, etc, (cf. Stauffer[16]) can be rewritten as $x \propto 1/ \mid p(i) - p(i)_c \mid^{\nu}$, $s \propto 1/ \mid p(i) - p(i)_c \mid^{\gamma}$, etc (same critical exponents), if any of the actual model parameters, $p(i)$, is varied, the remaining

[1]In their model Seiden and coworkers use as a further parameter a refractory time τ over which the probability p is reduced in regions where star formation has just taken place.

[2]If the parameter is zero, no star formation occurs ; if it is large enough coherent structures are expected to form by analogy with the forest fire paradigm

parameters being kept fixed. Accordingly, close to the percolation threshold, ***any*** of the free model parameters induces the phase transition. A percolation analysis does not enable us, therefore, to isolate the main physical variables and mechanisms responsible for pattern formation in a galaxy.

The astrophysical motivation of our CA representation of a local galactic model is threefold. In the first place, a direct macroscopic formalism in terms of partial differential equations to deal with the detailed evolutionary galactic physics is currently lacking. So far only simple global models have been proposed in which the overall evolution of the components of a galaxy have been described by ordinary differential equations assuming spatial homogeneity[15]. The CA framework lends itself to a numerically easy treatment of complex spatial interaction effects which are not readily dealt with by hydrodynamic approaches. In the second place, it is known that *fractal space-time*, or *space patterns* are encountered generically as solutions of CA formulations[10,11]. Suggestions have been made that certain galactic components may indeed be spatially distributed over fractal sets[7] (distribution of supernova remnants). In the third place, it has been conjectured that CA models may capture a class of instabilities which are hard to deal with in continuum dynamics, namely transitions towards dynamical states in which very large numbers of degrees of freedom are excited. Such a transition has been referred to as *self-organized criticality*[2]. Differential models very often involve only small numbers of degrees of freedom, so that they do not naturally lend themselves to model this transition. It seems to us that the transition from a statistically spatially homogeneous state of the galaxy to the structured arm formation is more meaningfully interpreted as a dynamical instability towards a state of self-organized criticality, rather than as a percolation transition : the percolation transition, in its standard acceptance, describes a transition from one static state (absence of a percolation cluster) to another (presence of a percolation cluster) under a change of a control parameter ; the notion of percolation involves no dynamics. The models we are concerned with are strongly time-dependent ; the spatial patterns near the transition point form and dissolve, even though their statistical distributions are essentially time independent (with the exception of the oscillatory solutions). Since the typical power law dependences (distribution of cluster sizes, etc) encountered near self-organized criticality[1,2] are recovered in our model, we propose that the formation of dynamical galactic structures is to some extent linked with self-organized criticality.

As a further introductory remark we observe that in our model the mere occurrence of spirals -reducing to bands in our local model- is just the outcome of *rotation* and the presence of an initial connected "cluster" of empty cells floating in an otherwise uniform gas and dust cloud distribution. Differential rotation stretches this cluster into bands thereby producing an "inter-arm" structure. Star formation occurring within the matter distribution is thus automatically confined to a structured medium. Superposed over this spiral geometry a critical state is found to develop as we change certain model parameters (in our experiment the extent of the initial empty cluster). Accordingly our model is indicative that the critical state is not a necessary condition for the formation of spiral arms, as the work by

Seiden and coworkers seems to suggest. The critical state, amplifying spatial coherence in star formation, makes this structure more conspicuous ; in a nonrotating galaxy randomly oriented percolation-like clusters would develop without showing the macroscopic spatial regularity.

2. Specification of the model

A brief account on a first version of our local galactic model was reported in Ref. 8. We give here full details on our extended model.

2.1. The Geometrical Features

The geometry of our CA is a 2-D square lattice of $L \times L = N$ sites, with periodic boundary conditions (2-torus topology). The sites are identified by integer Cartesian coordinates $(x, y), x, y = 1, 2, \cdots, L$. In our various experiments, $L = 2^n$ with n ranging from 7 up to 10. Each cell has a spatial extension of ≈ 0.2 pc and the biggest lattice we handled (1024×1024) represents a surface of $\approx (200 \times 200) pc^2$.

2.2 The Cell States

The state of site (x, y) specifies the physical nature of the galactic component in cell (x, y) ; each cell is occupied by one and only one among seven components and contains the same mass, i.e. the equivalent of a solar mass $M_\odot$. This point is a weakness of our model due to the inclusion in the described processes of low mass stars ($M \leq M_\odot$) and remnant inert masses. It should be improved in the future.

The seven components are :

- empty sites [E]
- dust and gas [D]
- protostars [P]
- main sequence stars [M]
- stars in an advanced evolutionary stage (red giants) [R]
- remnant inert mass [I]
- low mass stars [L].

The binary variable $N(S; t; x, y) = 1$ if the cell state S = [D], . . . , [L] , 0 if the site (x, y) at time t is empty [E]. We have

$$\sum_{S=[E]}^{[L]} N(S; t; x, y) = 1. \tag{1}$$

We also set

$$N(S; t; y) = \sum_{x=1}^{L} N(S; t; x, y) \tag{2}$$

for the total number of cells of the automaton in state S at time t in the horizontal row located at the vertical coordinate y . The total number of cells of the automaton in state S at time t is given by

$$N(S;t) = \sum_{x,y=1}^{L} N(S;t;x,y) \tag{3}$$

and the total number of cells at time t is given by

$$N = \sum_{S=[E]}^{[L]} N(S;t) \tag{4}$$

The chosen neighbourhood $[x,y]$ of cell (x,y) is the Moore neighbourhood including cells $(x,y),(x-1,y),(x+1,y),(x,y-1),(x;y+1),(x-1,y-1)$, $(x-1,y+1),(x+1,y-1)$ and $(x+1,y+1)$. It will be convenient to use the notation

$$n(S;t;x,y) = \sum_{x',y' \in [x,y]} N(S;t;x',y'), \tag{5}$$

which represents the number of cells in state S in the neighbourhood of cell (x,y).

2.3 The Dynamic Evolution

The evolution rules of the CA are formulated in the form of chemical reactions with the notation

$$[X] + b[B] + c[C] \rightarrow [Y] + b[B] + c[C], \mathcal{P}[XbBcC, YbBcC] \tag{6}$$

to be interpreted as follows : cell (x,y) in state [X] at timestep t transforms into state [Y] at timestep $t+1$ provided that $n(B;t;x,y) = b$ and $n(C;t;x,y) = c$, with probability $\mathcal{P}[XbBcC, YbBcC]$ per timestep. The updating consists in selecting a random number r of uniform distribution in $[0,1]$; if $r \leq \mathcal{P}[XbBcC; YbBcC]$, then the state [X] of cell (x,y) is changed to [Y] .

Some processes are "decay" processes, like the intrinsic evolution of isolated stars (P → M , M → R , R → D , R → I), others are "catalytic reaction" processes which describe enhanced evolution due to the presence of neighbour stars (say stars in binary systems, etc). Moreover the model takes care of intergalactic matter infall on the galactic disk. The processes are summarized in table I.

The physics of the described system like the lifetimes of the different types of star, the respective populations at a given time constraint the various probabilities. The intrinsic evolution rates are fairly well established by stellar evolution theory. The probabilities of the "catalytic reaction" steps as well as the stoechiometric coefficients being largely unknown, these quantities were regarded as parameters which were varied over physically reasonable ranges.

Our model takes account of two protostar formation processes : gravitational condensation of interstellar matter into a protostar, provided that the neighbourhood contains a sufficient amount of dust ; and protostar formation catalyzed by the presence of stars (protostars, main sequence stars and red giants). The latter process is the analogue of the star enhanced star formation process advocated by Seiden and coworkers and parametrized by their coefficient p .

The population of low mass stars [L] will increase with time due to their passive evolution and their long lifetime. Regarding the non-inert stars (P, M, R, L) at each evolutionary stage, we allow for a disintegration reaction back to dust.

The probability associated with the intergalactic matter infall is the only one which is regarded as time dependent for observational reasons.

Table I : Chemical reactions

		Reactions	Probability
Dust	*infall*	$E \rightarrow D$	${}^*P_{DE}$
		$E \rightarrow E$	P_{EE}
	star formation	$D \rightarrow P$	P_{DP}
		$D \rightarrow L$	P_{DL}
		$D \rightarrow D$	P_{DD}
	induced star formation	$D + dD \rightarrow P + dD$	PI_{DPD}
		$D + pP + mM + rR \rightarrow (1+p)P + mM + rR$	PI_{DPS}
Proto	*evolution*	$P \rightarrow P$	P_{PP}
		$P \rightarrow D$	P_{PD}
		$P \rightarrow M$	P_{PM}
	induced evolution	$P + dD \rightarrow (1+d)D$	PI_{PDD}
		$P + mM + rR \rightarrow D + mM + rR$	PI_{PDS}
		$P + mM + rR \rightarrow (1+m)M + rR$	PI_{PMS}
Main	*evolution*	$M \rightarrow M$	P_{MM}
		$M \rightarrow R$	P_{MR}
		$M \rightarrow D$	P_{MD}
	induced evolution	$M + rR \rightarrow (1+r)R$	PI_{MRS}
		$M + mM + rR \rightarrow D + mM + rR$	PI_{MDS}
Red	*evolution*	$R \rightarrow R$	P_{RR}
		$R \rightarrow D$	P_{RD}
		$R \rightarrow I$	P_{RI}
Low	*evolution*	$L \rightarrow I$	P_{LI}
		$L \rightarrow D$	P_{LD}
		$L \rightarrow L$	P_{LL}

${}^*P_{DE}$ taken as time-dependent : $P_{DE} \neq 0$ if $t < T_I$, $P_{DE} = 0$ if $t \geq T_I$

2.4 The Kinematic Evolution

The local galactic kinematics is simulated by a linear shear flow parallel to the x axis and a diffusion of the galactic material. The shear is achieved by shifting the rows (in accordance with a local galactic rotation law) after the chemical updating of the CA at a given timestep has been completed. Since this operation involves just a horizontal shift, the distributions $N(S;t;y)$ are left unaffected. In the absence of diffusion, most of the stellar activity will be confined to bands of horizontal maxima in the initial gas and dust distribution.

The process of diffusion, simulating actual diffusion of interstellar dust and proper motions of individual stars, is modelled by the device of rotating 4-cell blocks as described in Chopard and Droz[4]. This operation is carried out after the rotational updating has been performed. The diffusion algorithm tends to destroy the band structure. Both kinematic effects are turned on with probabilities p_{rot} and p_{mix} per timestep.

2.5 Initial Condition

Since allowance is made for spontaneous protostar formation, we do not need to initiate the star formation process by an initial distribution of active sites, as is done by Seiden and coworkers. In all of our experiments the initial state of the automaton mimics a true original state of a galaxy formed by a diffuse amount of gas only. This state is numerically prepared as follows. First all cells are populated with dust ; then a few "empty pockets", are generated in a n-step random walk starting at randomly chosen initial cells. A fraction $F(D)$ of the lattice is thereby occupied by dust. In this way a lattice configuration at timestep $t = 0$ is obtained, which is generally made up of connected empty zone and connected dust zones at least if $F(D)$ is large enough. The connectivity of each dust zone implies that a signal in one cell of the dust zone can propagate over the whole zone. Although rotation creates an asymmetry between the x and y behaviour, it does not destroy this connectivity ; it may however render y-connections more and more difficult.

2.6 Programming details

The representation of the seven different states is given by the prime numbers $0, 1, 3, 5, 7, 11$ and 13 stored in an integer occupation matrix. We proceed from timestep t to $t+1$ in several phases: stellar evolution (cf. table I) - storage in an intermediate matrix - diffusion - rotation - updating of the original occupation matrix at time $t+1$.

3. Numerical Experiments

Various numerical experiments were performed. Some probabilities should be improved in the future to reproduce the characteristics of typical star forming regions like the Orion cloud. Nevertheless, we believe that the qualitative features of our model remain valid. Table II gives the transition probabilities associated with a timestep $\Delta t = 5.10^6$ years. The time evolution is stopped when the system overshoots the normal lifetime of the galaxy or when the characteristics of the system become unrealistic.

The main physical results the model is able to provide are the time evolution of the different galactic components, the fractal space pattern of stars, the fractal dimension of clusters of stars, the percolation cluster of stellar activity and possible spatio-temporal intermittency. We recall that, by essence, an acceptable model should reproduce the well-known properties of a physical system but also it should forecast unobserved features. This is indeed the case for the proposed model. Moreover, we wanted to answer the question raised in the introduction : is the galactic system in a state of *self-organized criticality* ?

Table II : Numerical data used in the most frequent experiments

Probability		Probability	
P_{DE}	0.0010 (t=0)	P_{PD}	0.8999
P_{EE}	0.9990	P_{PP}	0.0001
P_{DP}	0.00049	P_{PM}	0.1000
P_{DL}	0.00085	PI_{PRD}	0.0005
P_{DD}	0.99866	PI_{PDS}	0.5000
PI_{DPD}	0.9900	PI_{PMS}	0.0500
PI_{DPS}	0.0001	P_{RR}	0.01498
P_{MM}	0.5000	P_{RD}	0.98001
P_{MR}	0.2500	P_{RI}	0.00501
P_{MD}	0.2500	P_{LI}	0.00018
PI_{MRS}	0.7500	P_{LD}	0.000018
PI_{MDS}	0.1000	P_{LL}	0.99980
P_{rot}	0.5000	P_{mix}	0.50000
$F(D)$	72%	Δt	5.10^6 years
Lattice	512 × 512	T_I	1500 Δt

3.1 Temporal Evolution of the Galactic System

For the majority of our experiments the following configuration was adopted : timestep $\Delta t = 5.10^6$ years ; probabilities displayed in table II ; 512×512 lattice ; initial dust percentage of 72%.

Figure 1 shows the time evolution of the mass fractions (taking into account the intergalactic matter infall) of the various galactic components. The spontaneous ignition leads to a drastic reduction of the amount of dust during the first steps and to an equivalent increase of the protostar formation. Later, except for intermittency (see below) or statistical fluctuations, the trend is smooth and regular. The dust percentage keeps decreasing. After an initial bump, the amount of protostars decreases while the amount of main stars is increasing. The red star population as well as the inert components are rather scarce in contradistinction with the low mass star population. Due to the long lifetime of the latter, the low mass fraction increases regularly with time. At time $t = 5.10^9$ years, the system reaches a virtually statistically stationary state. The generation of bands of star formation and evolutionary activity (appearing as galactic arms) is shown in plate 1 at times $t = 500$ and 1000 (in unit 5.10^6 years). We mention once more that in this model the bands themselves are exclusively the result of rotation.

3.2 Clusters and Percolation

We define a *cluster* of stellar activity as a connected collection of cells in any stellar state (P, M, R, I, L), two cells being regarded as connected if they have either a common side or a common vertex. The *percolation cluster* is defined as a cluster extending from the left border to the right border of the lattice network. The search for clusters of activity has been made following the algorithm of Hoshen and Kopelman[6].

Plate 2 shows the two biggest clusters of the system (input data from table II) at time 500 and 1000 (in unit 5.10^6 years), which are percolation clusters.

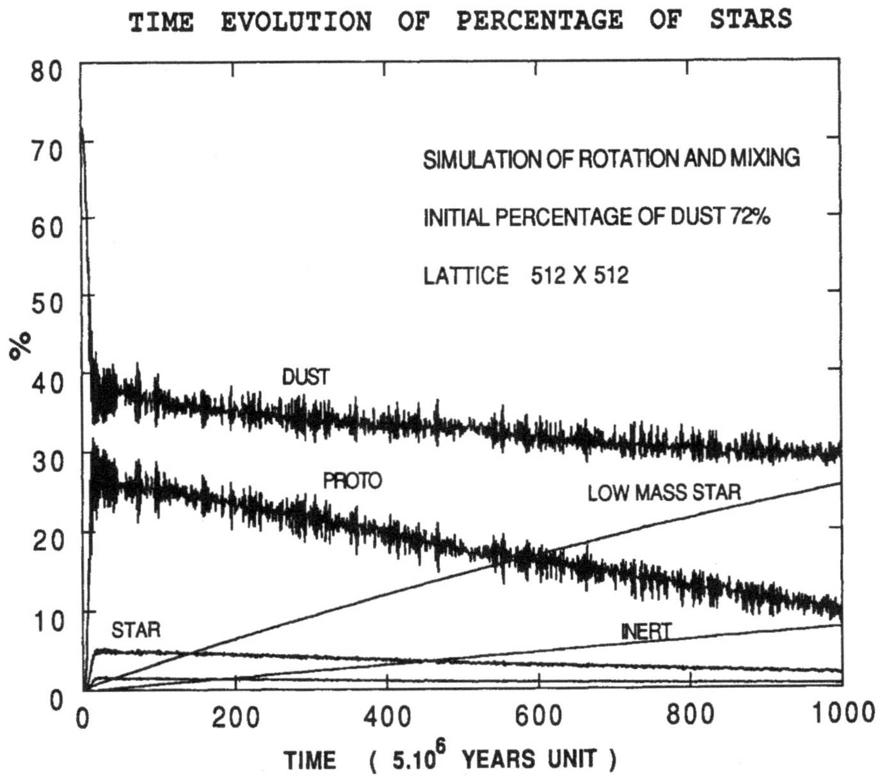

Figure 1 : Total abundance of dust [D], protostars [P], main sequence stars [M], red giants [R], remnant stars [I] and low mass stars [L] as a function of time (5.10^6 years unit). See Table II for numerical data.

A quantity of interest is the fractal dimension of the perimeter of the clusters. Various definitions of the fractal dimension can be used. The equivalent rectangle method gives

$$D_{peri} = \frac{\ln N_{cells}}{\ln \sqrt{l \times L}} \tag{7}$$

where N_{cells} is the total number of cells in the cluster and l and L are the width and the length of the rectangle closely surrounding the cluster. Relation (7) corresponds to a measure of fragmentation of the outer perimeter of the cluster only. The *total* perimeter method gives

$$D^*_{peri} = 1/2 \frac{\ln N_{peri}}{\ln N_{cells}} \tag{8}$$

where N_{peri} represents the total number of "border" cells (outside and inside) in the sense of cells of the cluster having at least one neighbour cell not belonging to the cluster. Eq (8) is a measure of the fragmentation of the outer plus inner perimeters of the cluster, and therefore exceeds the value (7).

Table III displays the results at different times. First the two methods give different result, suggesting that the cluster is a multifractal object. Second, the value of D_{peri} varies smoothly with time showing a more and more homogeneous cluster.

Table III : Fractal dimension of clusters

	t=25	t=500	t=1000
D_{peri}	1.69	1.75	1.80
D^*_{peri}	1.96	2.00	1.98

(time unit : 5.10^6 years)

Detailed analysis of the percolation in close connection with the state of self-criticality will be given in the following section.

3.3 Clusters Distribution

Following Bak and coworkers[2], a state of *self-organized criticality* is characterized by the following relation

$$N(S) = CS^{-\alpha}, \tag{9}$$

where S is the size of a cluster and $N(S)$ the number of clusters of size S . This state corresponds to a system where a large number of degrees of freedom are excited. Such a law is regarded as the fingerprint of criticality.

To test whether our model is consistent with self-organized criticality we plot $\ln N(S)$ against $\ln S$ at time 500 (Fig. 2) (in unit 5.10^6 years). The global shape of the distribution is not a straight line as given by equation (9). The shape is concave.

To investigate the occurrence of percolation, or more precisely, the emergence of a state of self-organized criticality, we have studied the influence of the initial dust and gas fraction $F(D)$, on the spatial pattern of the stellar components.

We recall that the probability of star induced star formation in our model, $\mathcal{P}[DpPmMrR, (1+p)PmMrR]$ is the analogue of the unique free parameter p of enhanced star formation in the model of Seiden and coworkers. Since we know from their work that the latter does produce percolation, we expect percolation to materialize in our model as well provided we vary this parameter over a large enough domain. As we explained in the first section, any one of the free model parameters induces the phase transition (one parameter varies, the others remaining fixed).

To this end, we performed a series of numerical experiments where the initial percentage of dust and gas $F(D)$ was varied from 0 to 100 and we computed the distribution of the *clusters of activity*. Figures 3a-c show the distribution $\ln N(S)$ against $\ln S$ for $F(D) = 54, 66 and 93\%$ when the system reaches a stationary state. The shape of the curve is convex, straight line and concave, respectively.

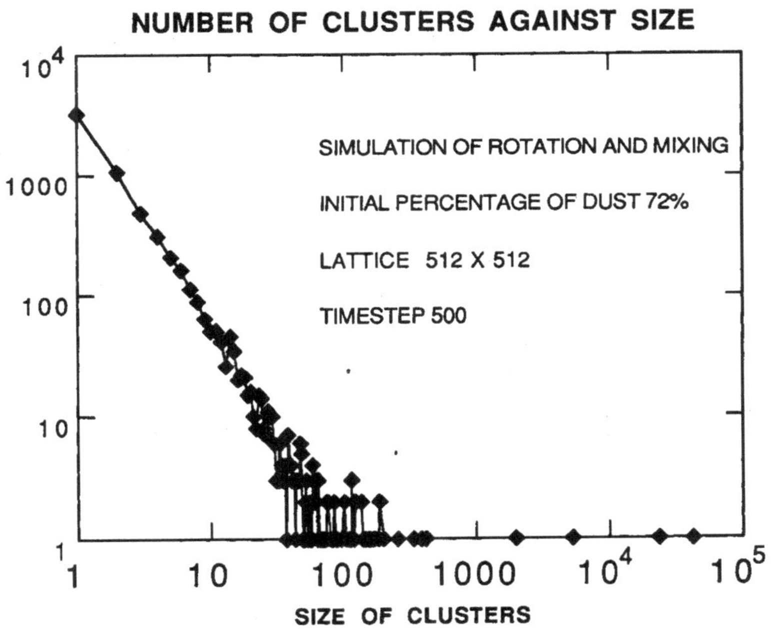

Figure 2 : Log-log plot of cluster size distribution for the numerical experiment corresponding to Table II at evolutionary timestep $t = 500$ (5.10^6 years unit).

We have adopted two methods to estimate the critical value $F_c(D)$. In the first place, we interpolate the observed distribution for different values of $F(D)$ by a quadratic expression

$$\ln N(S) = A - a \ln S + b(\ln S)^2 \tag{10}$$

in which the coefficients are determined by a least squares fit. We then plot $b = b(F(D))$ (figure 4) ; the critical value F_c corresponds to the vanishing curvature $b(F_c(D)) = 0$. In a second method we carry out ten experiments for each value of $F(D)$ and count the fraction $\Delta f(F(D))$ of experiments which produce a percolation cluster (in the sense defined above). The histogram of the frequencies

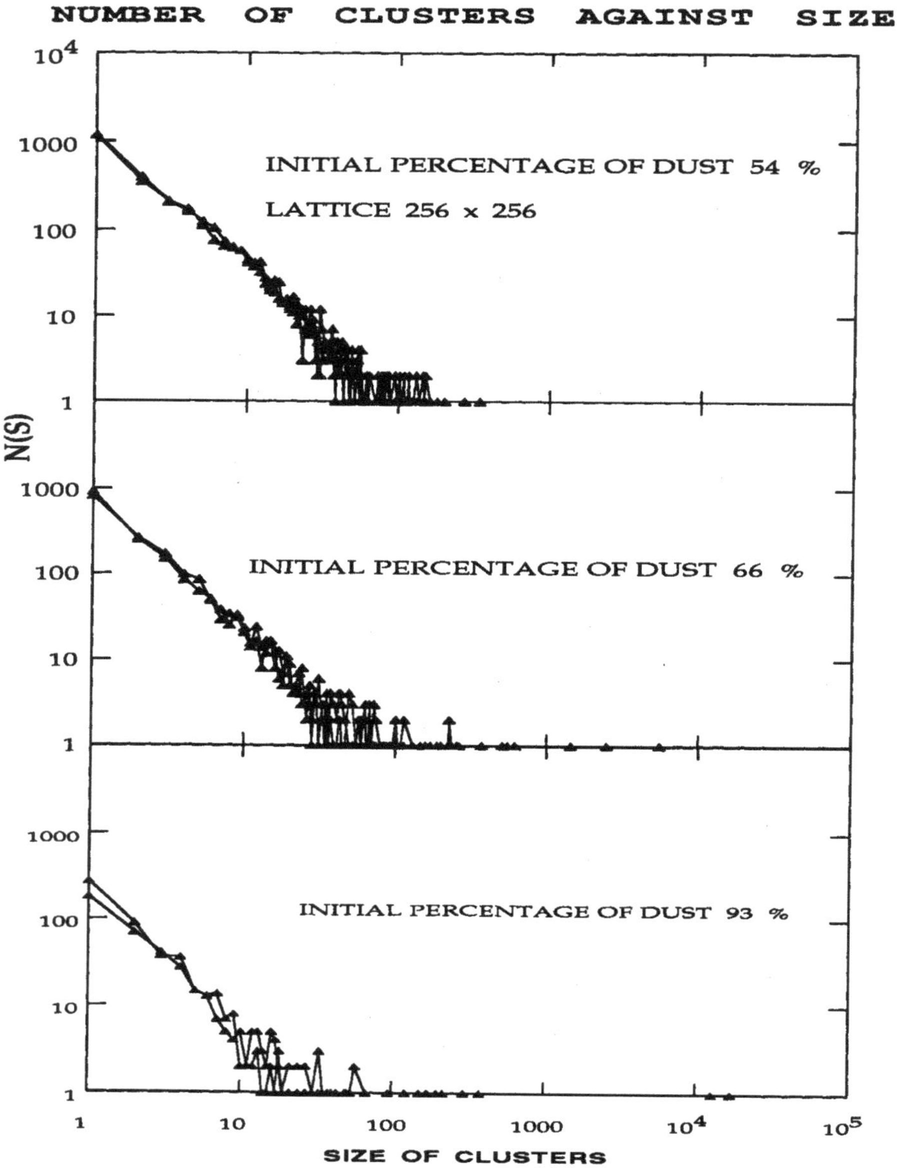

Figures. 3a-c : Log-log plot of cluster size distribution at different values of $F(D)$ illustrating a change of curvature in the tail of the distribution near the critical value.

$\Delta f(F(D))$ then shows a regular transition from 0 to 1 in the neighbourhood of the critical value $F_c(D)$ (figure 4). We read off a critical value $F_c(D) \approx 2/3$. As observed in Stauffer[16] the finite size of the lattice overestimates the true critical value. The exponent of the distribution near the critical point is $\alpha \approx 1.70$.

A visual inspection of the spatial distribution of the different galactic species over the lattice indicates that even for $F(D) \ll F_c(D)$, the structural pattern of horizontal bands of stellar activity remains noticeable ; these bands appear as irregularly interrupted patches ; as $F(D)$ approaches F_c the band structure exhibits a greater continuity and regularity in the horizontal direction.

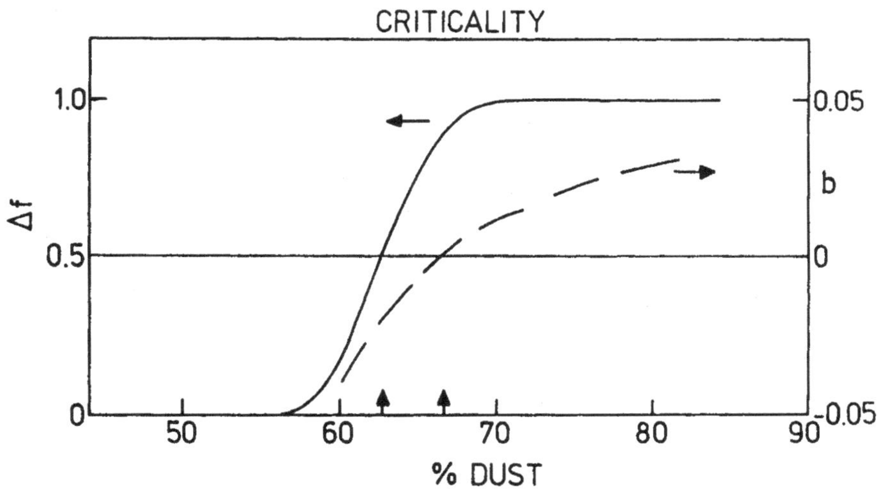

Figure 4 : Critical value F_c estimated from the curvature term b and from the histogram of the frequency of occurrence of percolation against F.

3.4 Fractal Star Pattern

Recently, Ikeuchi and Yoshioka[7], have observed the fractal aspect of various kinds of stars. Their result is, up to now, qualitative only. Using our model we wanted to forecast some quantitative results concerning this fractality.

We make use of two fractal dimensions.

The *mass or capacity dimension* d_m is defined in the following way. If $n_s(r)$ is the number of objects of species S in a disc of radius r of centre coinciding with an object of species S , then we search a representation in the form

$$n_s(r) = Ar^{d_m} \tag{11}$$

with A and $d_m (\leq 2)$ constant. If such a representation holds for r small as compared to the lattice size L , then the exponent d_m defines the fractal mass dimension.

The *correlation dimension* d_c is given in terms of the following function

$$C(r) = \frac{1}{N(S)^2} \sum_{i,j} \Theta(r - | \vec{r_i} - \vec{r_j} |) \tag{12}$$

represented by a power law

$$C(r) = B r^{d_c}, \tag{13}$$

$\vec{r_i} = (x_i, y_i)$ is the position of the ith cell occupied by an object of species [S] . Θ is the Heaviside step function, $N(S)$ is the total number of objects of species [S] at a fixed timestep. If such a law holds, then the exponent d_c is referred to as the correlation dimension. It can be shown that these two fractal dimensions obey the inequality[11]

$$d_c \leq d_m. \tag{14}$$

Table IV shows our results at different timesteps. Inequality (14) is obeyed. The fractal dimensions of the protostar, main and red stars distribution being less than 2 indicates that the distribution of these objects is not spatially homogeneous. Geometrically we can interpret the occurrence of fractional exponents d_m and d_c by saying that these objects distribute over fractal sets of the CA lattice space. Such a result is by no means surprising[10].

Table IV : Fractal dimension of stars

	t=25	t=500	t=1000
Proto	1.8498 1.6438	1.9191 1.9038	1.8644 1.8414
Main	1.8376 1.7899	1.8132 1.7757	1.6916 1.6121
Red	1.6984 1.5664	1.5608 1.4592	1.3893 1.2464
Inert	0.80104 0.70982	1.8906 1.8129	1.9574 1.9155
Low mass	1.5750 1.4622	1.9790 1.9692	1.9756 1.7683
All stars	1.8945 1.6475	1.9634 1.9200	1.9806 1.7944

Time unit : 5.10^6 years First line : d_m , second line : d_c .

The inert stars and the low mass stars have fractal dimension evolving with time toward the value 2 (homogeneous distribution). Therefore, the spatial distribution of *all* stars is close to an homogeneous one. The time dependence of the fractal dimension of the different species of stars, could perhaps be an indication towards measuring the age of a galaxy.

3.5 Spatio-Temporal Intermittency

Figure 5 displays, on an enlarged scale, the 50 first timesteps of figure 1. We observe an oscillatory state in the dust [D], protostar [P] and main stars [M] distributions. The possibility of such oscillations is not unexpected. An elementary differential model of the space averaged evolution of the galactic species has indicated that the feedback mechanism in the star formation process together with the feedback in the enhanced evolution towards red giants can give rise to such oscillations[9].

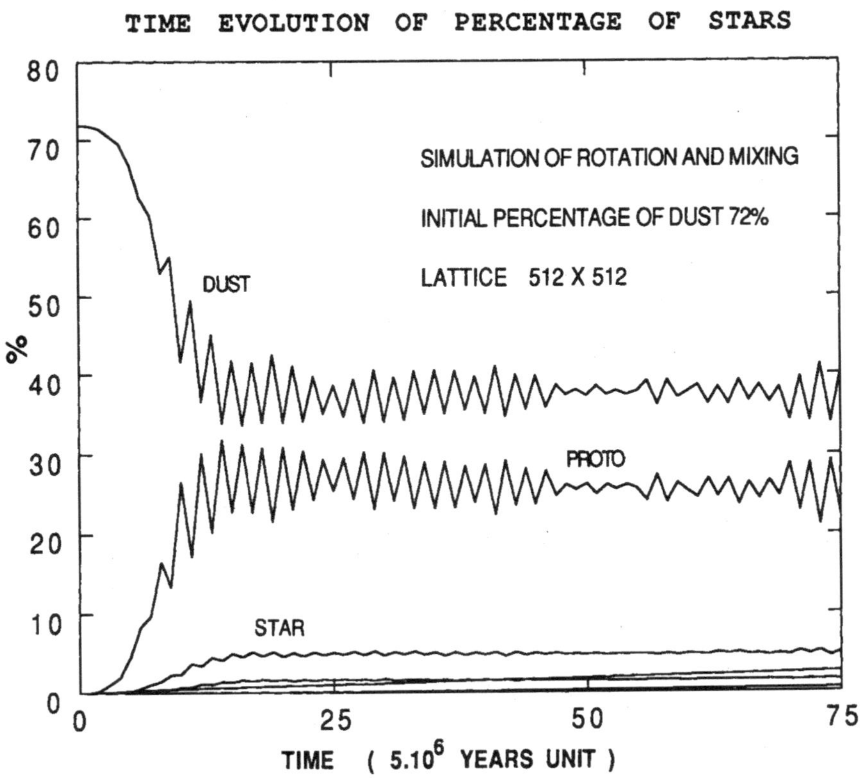

Figure 5 : Same as Figure 1 but with an enlarged timestep scale showing the intermittency phenomena.

The dynamics of these *spatio-temporal intermittency* remains to be investigated. It is not clear whether this intermittent time behaviour can be interpreted as a route to deterministic chaos[3]. In any case, it is a behaviour characteristic of the CA model. Does it correspond to some observations ?

4. Conclusions

The model described in this paper is physically more realistic than previous CA models. If it still remains too rudimentary to be directly comparable with observations, it nevertheless provides a good and detailed description of the galactic evolution. The qualitative conclusion we draw from our experiments is that it is predominantly an effect of differential rotation rather than a pure percolation mechanism that is responsible for the formation of a spiral pattern. The precise fragmented structure of the latter depends however on the closeness to criticality. We have demonstrated that the spatial distribution of the different stellar components is characterized by fractal dimensions. The geometrical pattern of the arms is likewise expected to be captured by fractal dimensions, the latter measuring the degree of fragmentation of the spiral arms.

We plan in the near future to take into account the mass spectrum of the stellar components and to include more detailed physics in the processes of star formation.

5. Acknowledgements

We would like to thank our colleagues J. Perdang and G. Barbaro for their enthusiastic cooperation in this work. Computing support has been provided by IBM Belgium within the ULG-IBM contract "Cooperative Processing".

6.References

1. P. Bak, K. Chen and M. Creutz, *Nature* **342** (1989) 780.
2. P. Bak, C. Tang and K. Wiesenfeld, *Phys.Rev.* **A38** (1988) 364.
3. P. Bergé, Y. Pomeau and C. Vidal, in *Order with Chaos* (Wiley, New York, 1986).
4. B. Chopard and M. Droz, in *Cellular Automata and Modelling Complex Physical Systems* (Springer, Berlin, 1990) **46** p.130.
5. H. Gerola and P.E. Seiden, *The Astrophysical Journal* **223** (1978) 129.
6. J. Hoshen and R. Kopelman, *Phys. Rev.* **14** (1976) 3438.
7. S. Ikeuchi and S. Yoshioka, *Comments in Astrophysics* **13** (1989) 117.
8. A. Lejeune and J. Perdang, in *Proceedings of the Europhysics Conference on Computational Physics* (World Scientific Press, Singapore, 1991) p. 407.

9. J. Perdang, communication presented at the *Contact Group in Astronomy and Astrophysics*, October 1974, Leuven (Belgium).

10. J. Perdang, *Vistas in Astronomy* **33** (1990) 249.

11. J. Perdang, in *Applying Fractals in Astrophysics*, Lectures Notes in Physics (Springer, Berlin, 1991) p. 1.

12. L.S. Schulman and P.E. Seiden, in *Percolation Structures and Processes*, eds. : G. Deutscher, R. Zallen and J. Adler, 1983, p. 251.

13. P.E. Seiden and H. Gerola, *Fundamentals of Cosmic Physics* **7** (1982) 241.

14. P.E. Seiden, L.S. Schulman and H. Gerola, *The Astrophysical Journal* **232** (1979) 702.

15. N.S.Shore in *Birth and Infancy of Stars*, Ecole des Houches, Session 16, eds. : R.Lucas, A.Omont, R.Stora (Elsevier,1985), p. 449.

16. D. Stauffer, in *Introduction to Percolation Theory* (Taylor and Francis, London, 1985).

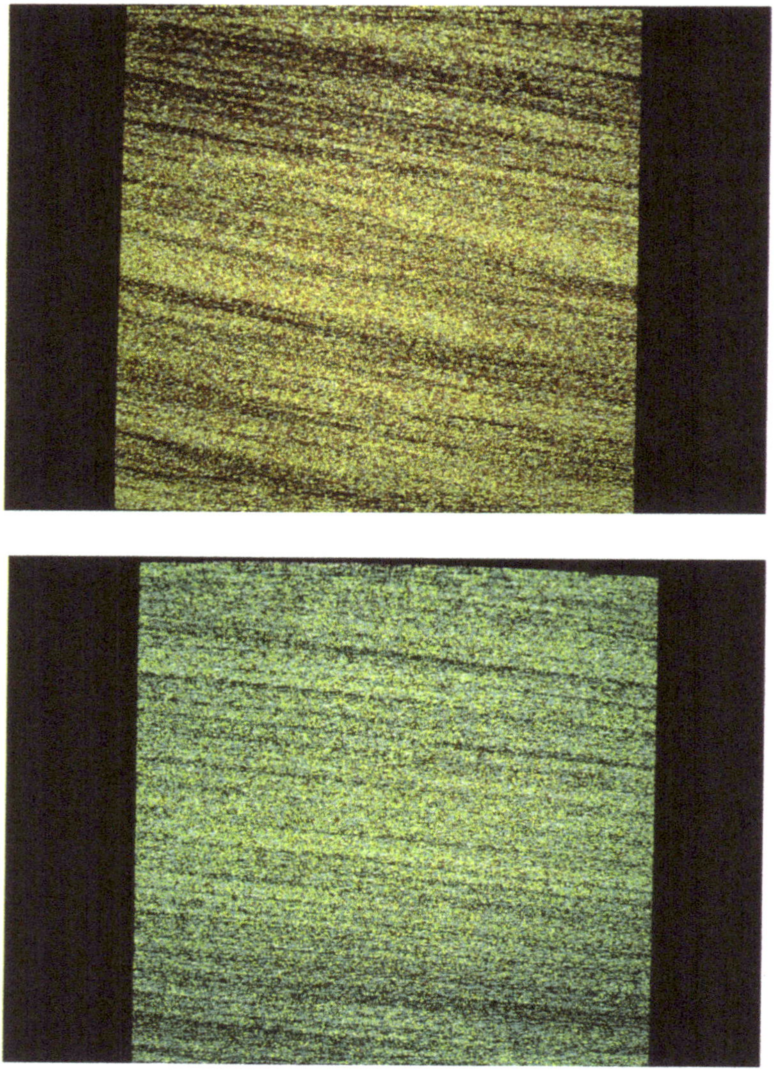

Plate 1 : CA simulation of a local galactic environment at evolutionary timesteps t = 500 and 1000 (5.10^6 years unit). Color scheme : dark blue : empty space ; coral : dust and interstellar gas ; white : protostars ; yellow : main sequence stars ; red : red giants ; cyan : inert mass ; light brown : low mass stars. See Table II for the numerical data.

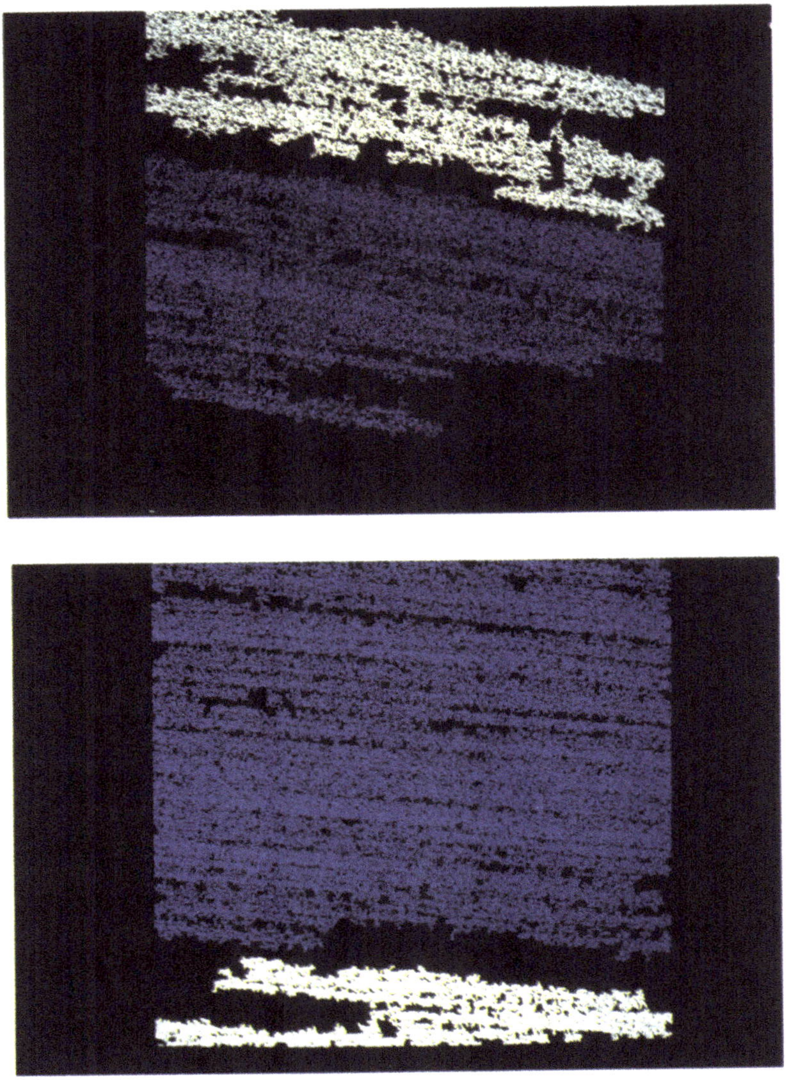

Plate 2 : The two biggest clusters of stars for the same experiment as for plate 1 at timesteps $t = 500$ and 1000 (5.10^6 years unit).

Simulation of Quasi-Radial Stellar Pulsations by Lattice Gas Schemes

J PERDANG

Institute of Astronomy, Madingley Road, Cambridge, CB3 OHA, UK
and
Institut d'Astrophysique, 5 Avenue de Cointe, Liège B-4000, Belgium *

Abstract:
We describe a $2D$ lattice gas scheme devised to deal with approximately radial stellar atmosphere oscillations. Gravitation is treated as an external site-dependent force. The ionisation mechanism responsible for the vibrational instability in real stars is simulated by an excitation and decay process of rest particles. Numerical experiments carried out for an atmosphere of finite height and initially uniform density indicate a violent (oscillatory) relaxation towards a stationary hydrostatic state, followed by a sustained oscillatory behaviour. $3D$ phase space reconstructions of the average radial velocity field of the whole atmosphere indicate a pattern which is not interpretable as a regular oscillation or as low dimensional chaos. On the other hand it does show similarities with phase space reconstructions derived from observational lightcurves of irregular Variables.

Introduction

In this paper we report on work dealing with the application of lattice gas (LG) schemes to stellar *nonlinear nonradial pulsations.*

Nonlinear pulsations of realistic stellar models have been investigated in a strictly *radial* context since the sixties (Christy 1967). In the late seventies it has been established, in the framework of a variety of approximations, that the *radial* stellar hydrodynamic equations support chaotic pulsations of low dimension (*deterministic chaos*; cf Perdang 1990 for details). Integrating the full hydrodynamic equations with a Christy-type code for Population II *Cepheids* over several hundred cycles, Buchler and Kovacs (1987) and Kovacs and Buchler (1988) isolated deterministic chaos in detailed stellar models(attractor dimensions in the range 1.9 to 2.5). Perhaps surprisingly, the fingerprints of this chaotic behaviour (time pattern, return maps, phase space reconstructions) are found to be the same as the signatures of chaotic pulsations of the most elementary one- or two-zone model oscillations.

The status of the observational time series (lightcurves, sometimes radial velocity curves) of the less regular Variables (*Semi-Regulars*, *Mira* stars), the natural candidates for chaotic behaviour, remains puzzling. Typical 'irregular' lightcurves of stellar variability (cf Figs. 9 in Perdang 1985; Figs. 1 in Cannizzo et al 1990; Fig. 2 in Kollath 1990; Fig. 1 in Kollath and Szeidl 1993) show manifest discrepancies with the chaotic signals as found in hydrocode computations (cf Figs. 8 in Kovacs and Buchler 1988) or in idealised one-zone models (for instance Figs. 1,2,3,5 in Tanaka and Takeuti 1988): Observational time series are 'noisy' on small timescales, while theoretical series have a smooth, 'regular' aspect over times short with respect to the cycle length. Over periods long as compared with the cycle, the observations usually show a richer variety of classes of behaviour than the theoretical series, even though the latter may exhibit a catalogued type of intermittency (type I, II or III). Analyses of the observations carried out with the refined techniques of dynamical systems theory by a number of authors (Takeuti 1989, Tanaka and Takeuti 1988, Saijo and Watanabe 1987, Saijo 1988, Cannizzo and Goodings 1988, Cannizzo et al 1990, Kollath 1990, Kollath and Nuspl 1991, Lèbre and Gillet 1992, Kollath and Szeidl 1993) confirm this purely visual impression: From a careful study of *RUCam* Kollath and Szeidl (1993) draw the categorical conclusion that 'the lightcurve of this star cannot be modelled with low dimensional chaos'. It

* Permanent address

seems that this conclusion extends to all well enough documented complex behaviour variables which have so far been analysed with some detail.

On the theoretical side, it has been stressed that under physical conditions representative for the cooler stars in which irregular pulsations are observed, a huge number of ***nonradial*** degrees of freedom are likely to be excited, together with the unstable radial mode(s) (Perdang 1990,1992). An observationally significant theoretical analysis of the oscillatory behaviour of these stars should therefore be based on the full ***nonradial*** equations of fluid dynamics. Unfortunately, at the moment a satisfying integration method capable of dealing with the nonradial pulsation dynamics is still wanting. In fact, due to the presence of a very large (infinite, in the adiabatic limit) number of excited nonradial modes, the standard numerical approaches (catalogued in Boris 1989) cease to be adequate. ***Spectral techniques*** (Peyret and Taylor 1983; Marcus 1986, Canuto et al 1988) ***grid methods***, or ***finite element methods*** (Zienkiewicz 1977; see especially Griffiths 1986) are implicitly predicated on the assumption that a limited number of modes is being excited. The procedure of isolating the nonlinear counterpart of the acoustic modes only (Däppen and Perdang 1984) is not satisfying either in the present situation, since it just ignores the whole spectrum of unstable gravity modes. The application of conventional hydrodynamics to the treatment of 2– and $3D$ Supernova explosions, (Müller 1991, Hachisu 1991), stellar convection, (Porter et al 1990) and astrophysical jets (Norman and Winkler 1986, Todo 1991, Uchida 1991) may be consistent with the assumption of a relatively small number of excited degrees of freedom; for the pulsating cool stars the assumption is not justified.

The cellular automaton (CA) environment is not subject to this criticism. It possesses the attractive property (shared with another unconventional technique, namely smoothed–particle–hydrodynamics; Benz 1990,1991; cf also Eastwood 1986 for general particle methods) of supporting a large number of excitable degrees of freedom: Every lattice cell can be viewed as modelling one degree of freedom. The CA framework is therefore able to deal with huge numbers of instabilities (cf Bak et al 1987,1988).

There are several additional attractive CA features of interest for computing stellar pulsations : (1) The CA model is not subject to numerical instabilities. (2) 'Thermodynamic' noise is generated naturally; this intrinsic noise in turn excites any unstable mode; it also induces bifurcations from one dynamical regime to other (possibly nonradial) nearby regimes. (3) When properly devised, a CA model has a broader range of validity than conventional hydrodynamics; it supplies results on a sub–macroscopic level; and it may remain applicable far from thermodynamic equilibrium.

Noise can in fact modulate the stellar pulsational dynamics considerably. Elementary stochastically disturbed iterations ($2D$ nonlinear Yule schemes) simulating Poincaré maps for real stellar pulsations have traced different complex noise induced intermittencies close to the transition to chaos (Perdang 1990,1991). The effect of stochasticity on stellar pulsations is now also being studied with the help of the Amplitude Formalism (Buchler et al 1993).

The application of the LG techniques to the simulation of stellar pulsations raises a number of new problems. (1) The model must be able to support ***density gradients***. Therefore we have adopted a 'boson'–type LG, rather than a conventional 'fermion' LG. (2) The model must include ***space–dependent gravity***, ideally ***self–gravity***. Since the scheme should also supply information on a submacroscopic scale, we have carried out a detailed analysis of the discrete equation of motion in the presence of a force, for individual LG particles, and for pairs of LG particles in case the finite set of velocities does not allow to satisfy the conventional single–particle equation of motion. We believe that our treatment is more general than the Boltzmann equation approach discussed in Frisch ***et al*** (1987) which requires closeness to the thermodynamic equilibrium state. (3) A CA mechanism simulating the ***stellar vibrational instability*** must be devised. The thermodynamics relevant for an instability is handled by adding two types of rest states to a 1–speed LG, one rest state being a ground state, the other a state of excitation; energy can then be absorbed by the ground state transforming it into a state of excitation; conversely, energy is released from the excited rest state transforming the latter into a ground rest state. (4) A reasonable LG gas must be chosen. We have adopted here a modified form of the very first LG model introduced by Hardy, de Pazzi and Pomeau (HPP, 1976). The major shortcoming of this model, namely the lack of

Mass, radius, core luminosity:

M_o, R_o, L_o ($\leq 1\ M_\odot$, $\geq 1\ R_\odot$, 100-1000 $L_\odot$)

Mass, height of atmosphere :

M_a ($<<M_o$) , H_a ($>R_o$) ; ($\leq$ 0.01 $M_\odot$, 10-100 $R_\odot$)

Observed radial velocities:

$v_r \geq 10$ km/s

Values adopted in experiments :

$M_a \approx 0.01\ M_\odot$; $H_a \approx 100\ R_\odot$; $R_o \approx 0.3\ H_a$; $L_o = 500\ L_\odot$

Table 1.- Stellar parameters representative for RV Tauri, Semi-regular and Long-Period Variables

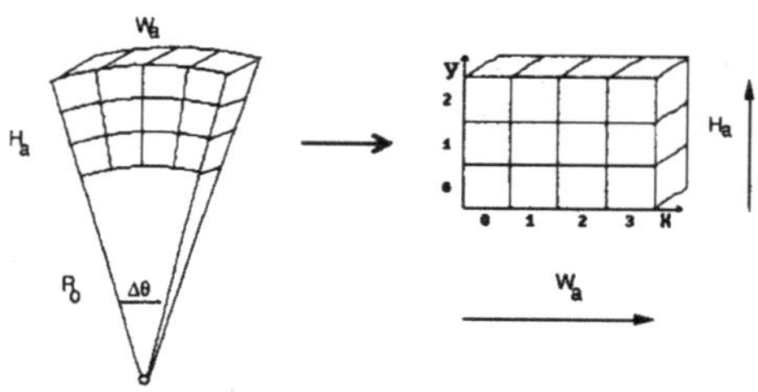

Fig. 1. – Geometry of the atmosphere model

Lattice dimension :

$n_x = n_y = 36$ [1] ; 61 [2] ; 128 [3]

Number of particles in lattice :

$N_p = 800$ [1] ; 1700 [2] ; 8000 [3]

Gravitation at base of atmosphere in CA units :

$g_{CA} = \gamma / n_y$, $\gamma = 2$ in most experiments

Fraction of initially random particles :

$\eta_r = 0.25$

Section of atmosphere modelled by CA

Height of level $j=0,1,\ldots, n_y-1$, and relative height; total width :

$H(j)$ ($H(n_y-1) = H_a$) , $h(j)$ ($h(n_y-1) = 1$) , W_a (= H_a in experiments)

Dimensionless radius of core in units of height of atmosphere:

$r_o = R_o/H_a$ ($\approx$0.3)

Table 2.- Main free simulation parameters

isotropy of the momentum flux tensor, is not thought to be of any serious consequence for the problem we are dealing with: We are concerned here with simulating *oscillations dominated by a radial component* . The nonradial dynamics plays the part of a perturbation effect; the unreal anisotropy may to some extent either exaggerate or underestimate the nonradial contribution; it is unlikely, however, that the character of the global radial behaviour will thereby be modified in any essential way.

The main result of our small–scale numerical experiments is that the simulation of oscillatory stellar atmosphere motions always involves a nonradial contribution of complex time and space behaviour in addition to a radial component of locally complex structure : The spatial pattern traced out by the radial velocity component v_r (lines of constant radial velocity) exhibits in fact a *fractal structure*. Moreover, $3D$ phase space reconstructions of the averaged radial component show a trajectory reminiscent of the observational phase space trajectory of *RU Cam*. There are no reasons to believe that these qualitative features are artifacts of an inappropriate LG model (lack of isotropy), or of an oversimplified physics included in the simulation (probabilistic treatment of force effect, absence of self–gravity, incomplete treatment of radiation). We believe, therefore, that the occurrence of perturbative nonradial motions and fractal structure will survive in any detailed stellar pulsation calculation.

2. The Atmosphere Model

In line with the conventional radial treatment of stellar pulsations, we regard the inner dense core of the star, of radius R_o and mass M_o, as dynamically inert. The core thus preserves its shape of radial equilibrium while the star is pulsating. The pulsations are treated in an external region of the star only, hereinafter dubbed 'atmosphere', of height H_a and mass M_a. The stellar luminosity is produced in the core; its value at the base of the atmosphere is L_o. Typical values of these parameters for the classes of variable stars we are interested in are summarised in Table 1. For the purposes of defining a tractable CA framework, the following approximations are introduced :

(1) As illustrated in Fig. 1, we model a section of the atmosphere by a $2D$ rectangular 'vessel' with base coinciding with the top of the stellar core. This representation is adequate provided that the height H_a is small as compared with the core radius, so that the curvature can be ignored. The $2D$ model further disregards any microscopic motions normal to the plane of the section. In the CA simulation the geometric configuration is thus represented by a square lattice, with n_x and n_y cells in the horizontal (x) and vertical (y) directions respectively.

(2) The boundary conditions for the LG particles we adopt are (a) *periodicity conditions* in the x–direction, and (b) *reflection conditions* at the bottom as well as at the top of the vessel.

(3) Since $M_o \gg M_a$ we disregard self–gravity; the gravitational field in the atmosphere is then given by the radial field due to the core. In the CA model this field is time–independent and directed along the y–axis.

Approximation (1) is not in agreement with the information provided by Table 1. However, at the present exploratory stage we are merely interested in assessing whether noticeable qualitative contributions in the radial oscillations resulting from nonradial effects can be found in CA simulations. The simplest geometry capable of supporting nonradial contributions is therefore adopted. A complication in the geometry will be warranted only for the purposes of precise quantitative results, once it has been shown that new effects are indeed due to appear. The formal width of the atmosphere, W_a, in the CA model is to be taken equal to $(R_o + \frac{1}{2}H_a)\ \Delta\theta$, $\Delta\theta$ being the angle under which the portion of the real atmosphere is seen from the centre of the star (Fig.1). The lateral boundary condition becomes exact if $\Delta\theta = 2\pi$ (i.e. $W_a \approx 3H_a$ for $R_o \ll H_a$); in the majority of our experiments we have taken $W_a = H_a$, implying that we force an unphysical large scale symmetry on the nonradial velocity and density perturbation patterns; there is no reason to believe that this symmetry might have a major influence on the radial (i.e. y) behaviour of the pulsating model. The bottom reflection condition is the simplest condition in line with the dynamical inertia of the core. Regarding the top boundary condition, ideally, the experiment should be devised such as to prevent LG particles from bouncing against the upper layer of cells of the CA; under such conditions no upper boundary condition would be needed, since in the framework of our treatment of the gravitational force all particles undergoing an upward motion are eventually recaptured by gravitation. The practical realisation of such a set up turns out to be too inefficient (a large fraction of the lattice remaining empty). By introducing a fictitious reflection at some maximum height this difficulty is avoided; we should keep in mind, however, that this artificial boundary condition amounts to increasing the gravitational field in the upper layer.

The free CA parameters are summarised in Table 2. The dimensionless gravitational acceleration, g_{CA}, although a free parameter, is subject to the dynamical condition that for instance any individual upmoving LG particle should on average be decelerated once, but not more than twice. The combination of the observational data (Table 1) with the free parameters fix the physical values of the lattice size, λ, the particle mass, μ, and the value of the timestep Δt (Table 3). It transpires from these values that our CA set up is actually capable in principle to support (and to resolve) stellar oscillations of characteristic properties consistent with the observations. For the free parameters selected in our small–scale experiments the timestep remains indeed short as compared to the typical cycle lengths of these stars, and the CA particle speed, $c = \frac{\lambda}{\Delta t}$, is large as compared to the observed radial velocities of the classical Variables.

In the experiments we adopt a system of units (CA units) in which the cellsize λ is 1, the mass μ of an LG particle is 1, and the timestep Δt of the calculations is 1.

3. The Treatment of the Gravitational Force

We sketch our probabilistic treatment of the dynamics of HPP type lattice particles under the effect of the gravitational force field. The full treatment for $2D$ gases with rest states and 1–speed moving states is developed elsewhere (Perdang 1993). The general method adopted consists in establishing a set of equations for the transition probabilities between the different allowed states of motion, such that at every timestep a cluster containing a minimum number of LG particles obeys the classical equations of motion in an appropriate average sense.

To carry out this programme, consider first a $2D$ classical gas (CG) of N_P identical particles, $k = 1, 2, \ldots, N$, of mass μ and positions and momenta $\mathbf{r}_k(t)$, $\mathbf{p}_k(t)$ defined in the $2D$ vessel introduced above ($x \in [0, W_a]$, $y \in [0, H_a]$, boundary conditions as specified in the previous section). The gas is subject to a time independent gravitational field, $\mathbf{g}(\mathbf{r}) = (0, -g(y))$, in Cartesian coordinates.

Our purpose is to simulate the dynamics of this CG by an HPP LG with 2 rest states, defined over the square lattice (cf Fig. 1; cells labelled $\mathbf{q} = (i,j)$, $i = 0, 1, 2, \ldots, n_x$-1, $j = 0, 1, 2, \ldots, n_y$-1), and at discrete times $\tau = 0, 1, 2, \ldots$, in the following sense. If $Q(t; x, y)$ is a dynamic variable of the CG defined at point $\mathbf{r} = (x, y)$ and (continuous) time t, then the corresponding dynamic variable in the LG, $Q(\tau; i, j)$ attached to cell (i, j) covering the point (x, y) is identified with the average value of $Q(t; x, y)$ over the cell (i, j), and the τth time interval $\tau = \mathrm{int}(\frac{t}{\Delta t})$ (int , integer part; $\Delta\tau$, duration of one CA timestep).

For reasons spelled out elsewhere, the LG particles are regarded as bosons, rather than fermions. Therefore a prime number encoding of the CA *field function*, or state of an individual site $\mathbf{q}$ (at discrete time τ), $\Omega(\tau; i, j)$, is adopted in the form

$$\Omega(\tau; i, j) = \prod_a p_a{}^{n(\tau;i,j;a)} \qquad 1$$

the product extends over all 6 particle states, $a = G$ (ground rest state), H (excited rest state), E(ast), W(est), N(orth), S(outh) moving states; the symbols p_a stand for the sequence of *prime numbers*

$$(p_G, p_H, p_E, p_W, p_N, p_S) = (2, 3, 5, 7, 11, 13); \qquad 2$$

the exponent $n(\tau; i, j; a)$ (*occupation number*) gives the number of LG particles in state a and contained in cell (i, j) at timestep τ (cf also Perdang 1993a). The configuration of the automaton at time τ is thus fully given by the matrix $\Omega(\tau)$. For reasons which will become clear in Section 5, the excited rest states will occur in pairs; to indicate this property we shall denote by P an *excited rest pair*, $n(\tau;i,j;P)$ representing the occupation number of these pair states ($n(\tau;i,j;H)$ $= 2\,n(\tau;i,j;P)$).

For convenience we shall use the label a to denote an*arbitrary state*; the label b will be used to identify *rest states*, and c will stand for *states of motion*; a state in some relation (successor state, etc) with a given state a (or b, or c) will be denoted by a', a'', ... (or b', b'', ...; c', c'', ... respectively) regardless of whether b', b'', ... are rest states or not, or c', c'', ... are moving states or not.

In order to translate the *continuity of the motion* of a CG particle into the LG scheme we need to introduce the concept of a *neighbourhood* of an LG state. In the case of the elementary HPP gas with rest states as considered here, the general definition (cf Perdang 1993) collapses to the following statements.

A state c' is a *neighbour of a moving state* c if it is either the ground rest state, or if it is any moving state except the moving state of opposite direction of state c. A state b' is a *neighbour of the ground rest state* $b = G$ if it is *any* moving state, or if it is the excited rest state; the only neighbour b' of the excited rest state $b = H$ is the ground rest state G. Two states are *dynamical neighbours* if they are neighbours and if they are in the same internal state. A moving state c will further be said to be in the direction $\mathbf{d}$ if $angle(\mathbf{d}, \mathbf{v}(c)) < \frac{\pi}{2}$ for an HPP gas; $\mathbf{v}(c)$ denotes the LG velocity (in CA units) of a particle in state c.

We write the change of momentum of an LG particle at time τ , at cell $\mathbf{q} = (i,j)$ and in state a, under the effect of gravitation $\mathbf{g}$ (= $\mathbf{g}$(j)) and the corresponding change in position of the same LG particle at position $\mathbf{q} = (i,j)$ and in state a at timestep τ, as follows

$$\Delta\mathbf{p}(\tau;i,j;a) = \mathbf{p}(\tau+1;i,j;a') - \mathbf{p}(\tau;i,j;a) \equiv \mathbf{v}(a') - \mathbf{v}(a), \quad (a)$$

$$\Delta\mathbf{q}(\tau;a) = \mathbf{q}(\tau+1;a') - \mathbf{q}(\tau;a); \quad (b) \qquad 3$$

a' being the new state of the same particle at time $\tau+1$. Interpret these relations as difference approximations to the equations of a CG particle k at position $\mathbf{r} = (x,y)$ coinciding with site $\mathbf{q} = (i,j)$ of the LG, and of momentum $\mathbf{p}_k(t) \equiv \mathbf{p}(\tau;i,j;a)$ at time $t = \tau\Delta t$. The difference form of the equations of motion for the *classical particle* is given by

$$\Delta\mathbf{p}_k(t) = \mathbf{g}(j), \quad (a)$$

$$\Delta\mathbf{r}_k(\tau) = \frac{1}{2}(\mathbf{p}_k(\tau+1) + \mathbf{p}_k(\tau)); \quad (b) \qquad 4$$

since the gravitational force varies slowly we have not introduced a refined expression for the force (cf Greenspan 1980). Hence, from relations (3.a and b) we should have

$$\mathbf{v}(a') = \mathbf{v}(a) + \mathbf{g}(j), \quad (a)$$

$$\Delta\mathbf{q} = \mathbf{v}(a) + \frac{1}{2}\mathbf{g}(j). \quad (b) \qquad 5$$

The discreteness of the LG positions and dynamic states implies that the latter requirements cannot be satisfied on the individual LG particle level. We therefore adopt a statistical scheme in which Eqs. 5 are satisfied in an average sense.

To this end associate with any local dynamical CA variable, $Q(\tau;i,j)$, a *mean dynamical* CA *variable*, $\langle Q(\tau;i,j)\rangle$, defined by a spatial averaging over the cells of an *averaging neighbourhood* $\mathsf{A}(i,j)$ of cell (i,j),

$$\langle Q(\tau;i,j)\rangle \equiv \frac{1}{Z}\sum_{(I,J)\in A(i,j)} Q(\tau;I,J), \qquad 6$$

(Z, number of cells in $\mathsf{A}(i,j)$); in our numerical experiments this set is a square of $Z = (2h+1) \times (2h+1)$ cells centred around (i,j). For Z small the averaging neighbourhood is a *submacroscopic neighbourhood* of site (i,j), while for Z large the mean dynamical CA variables are identified with hydrodynamic variables. Thus the *number density*, the *linear momentum density* and the *hydrodynamic velocity* are defined by

$$\langle n(\tau;i,j)\rangle = \sum_a \langle n(\tau;i,j;a)\rangle =$$

$$\langle n_o(\tau;i,j)\rangle + \langle n(\tau;i,j;E)\rangle + \langle n(\tau;i,j;W)\rangle + \langle n(\tau;i,j;N)\rangle + \langle n(\tau;i,j;S)\rangle, \qquad 7$$

$$\langle \mathbf{P}(\tau;i,j)\rangle = \sum_a \langle \mathbf{p}(\tau;i,j;a)\rangle \equiv \sum_a \mathbf{v}(a)\langle n(\tau;i,j;a)\rangle, \qquad 8$$

$$\langle \mathbf{v}(\tau;i,j)\rangle = \frac{\langle \mathbf{P}(\tau;i,j)\rangle}{\langle n(\tau;i,j)\rangle}, \qquad 9$$

or

$$\langle P_x(\tau;i,j)\rangle = \langle n(\tau;i,j;E)\rangle - \langle n(\tau;i,j;W)\rangle, \qquad 8.a$$

$$\langle v_x(\tau;i,j)\rangle =$$

$$\frac{\langle n(\tau;i,j;E)\rangle - \langle n(\tau;i,j;W)\rangle}{\langle n_o(\tau;i,j)\rangle + \langle n(\tau;i,j;E)\rangle + \langle n(\tau;i,j;W)\rangle + \langle n(\tau;i,j;N)\rangle + \langle n(\tau;i,j;S)\rangle}, \qquad 9.a$$

and similar expressions for the y–components. Note that $(|\langle \mathbf{v}(\tau;i,j)\rangle| \leq 1)$. We further have the average *total microscopic kinetic energy density* of LG particle motions, $\langle E_K(\tau;i,j)\rangle$, the *excitation energy density*, $\langle E_H(\tau;i,j)\rangle$, the *hydrodynamic kinetic energy density*, $\langle K(\tau;i,j)\rangle$, the *internal kinetic energy density*, $\langle U_K(\tau;i,j)\rangle$, and the *total internal energy density*, $\langle U(\tau;i,j)\rangle$,

$$\langle E_K(\tau;i,j)\rangle = \frac{1}{2}\langle n(\tau;i,j)\rangle - \frac{1}{2}\langle n_o(\tau;i,j)\rangle, \qquad 10$$

$$\langle E_H(\tau;i,j)\rangle = e_H\langle n(\tau;i,j;H)\rangle, \qquad 11$$

$$\langle K(\tau;i,j)\rangle = \frac{1}{2}\frac{\langle \mathbf{P}(\tau;i,j)\rangle^2}{\langle n(\tau;i,j)\rangle}, \qquad 12$$

$$\langle U_K(\tau;i,j)\rangle = \langle E_K(\tau;i,j)\rangle - \langle K(\tau;i,j)\rangle, \qquad 13$$

$$= \frac{1}{2}\langle n(\tau;i,j)\rangle\left(1 - \frac{\langle \mathbf{P}(\tau;i,j)\rangle^2}{\langle n(\tau;i,j)\rangle}\right) - \frac{1}{2}\langle n_o(\tau;i,j)\rangle, \qquad 14$$

$$= \langle n(\tau;i,j)\rangle T(\tau;i,j); \qquad 14.a$$

$$\langle U(\tau;i,j)\rangle = \langle U_K(\tau;i,j)\rangle + \langle E_H(\tau;i,j)\rangle, \qquad 15$$

$$\langle E(\tau;i,j)\rangle = \langle U(\tau;i,j)\rangle + \langle K(\tau;i,j)\rangle = \langle E_K(\tau;i,j)\rangle + \langle E_H(\tau;i,j)\rangle. \qquad 16$$

The quantity

$$\langle n_o(\tau;i,j)\rangle \equiv \langle n(\tau;i,j;G)\rangle + \langle n(\tau;i,j;H)\rangle =$$

$$\big(1-\eta(\tau;i,j)\big)\ \langle n(\tau;i,j)\rangle = \langle n(\tau;i,j)\rangle - \langle n_m(\tau;i,j)\rangle \qquad 10.a$$

is the rest state density, $\eta(\tau;i,j)$ being the fraction of moving particles ($n_m(\tau;i,j)$, number of moving particles); the parameter e_H is the excitation energy of an H rest state (measured from ground state level). In Eq. (14.a) a local *temperature* $T(\tau;i,j)$ ($\leq \frac{1}{2}$) (expressed in energy units) is defined, provided that the lattice gas is locally in thermal equilibrium. Under the latter assumption we also define a *local gas pressure*, $P(\tau;i,j)$, by the usual ideal gas relation for a $2D$ gas

$$P(\tau;i,j) = \langle n(\tau;i,j)\rangle T(\tau;i,j) \equiv \langle U_K(\tau;i,j)\rangle. \qquad 17$$

(η=1 restores the usual results of the HPP gas; Hardy et al 1976; cf also Wolfram 1986).

The change of the average linear momentum in cell (i,j) must obey the ***hydrodynamic momentum equation*** integrated over the cell, or in difference form (unit timestep)

$$\langle \Delta \mathbf{P}(\tau;i,j)\rangle = \langle \mathbf{g}(j)\ n(\tau;i,j)\rangle + \ldots \approx \mathbf{g}(j)\ \langle n(\tau;i,j)\rangle + \ldots. \qquad 18$$

(provided that the averaging neighbourhood is chosen large enough). The dots indicate the contribution to the change of momentum due to the pressure force and due to the flow of matter through the surface of the cell. Expliciting the momentum change $\Delta\mathbf{P}$ we have

$$\frac{1}{Z}\sum_{(I,J)\in A(i,j)}\Big(\sum_a \Delta\mathbf{p}(\tau;I,J;a;\mathbf{g})\Big) = \mathbf{g}(j)\ \langle n(\tau;i,j)\rangle, \qquad 19$$

where the argument $\mathbf{g}$ in $\Delta\mathbf{p}$ indicates that we consider the change of momentum due to gravity only.

Instead of merely imposing that Eq. 19 be satisfied over a macroscopic neighbourhood it is our purpose to select ***as small a neighbourhood as possible***, such that this equation can still be satisfied. In this way our CA model not only captures the hydrodynamic level; it also applies to a submacroscopic level. The microscopic level (individual CG particle motions) cannot be exactly represented by our LG particles.

To this end denote by $Pr(\tau;i,j;a)$ the occupation probability of site (i,j) by a particle in state a at time τ , and by $Tr(\tau;i,j;a\rightarrow a';\mathbf{g})$ the transition probability of state a to state a', at position (i,j), under the effect of the force $\mathbf{g}$, during the time interval τ to $\tau+1$. We have, by definition of an occupation probability

$$n(\tau;i,j;a) = Pr(\tau;i,j;a)\ n(\tau;i,j). \qquad 20$$

To obtain smooth occupation probabilities, we use the relation

$$\langle\langle n(t;i,j;a)\rangle\rangle \approx Pr(\tau;i,j;a)\ \langle\langle n(\tau;i,j)\rangle\rangle, \qquad 20.a$$

where $\langle\langle Q\rangle\rangle$ denotes any averaging of the CA variable Q, over space (with a large enough neighbourhood), over time ($2\Theta+1$ steps), or over both space and time

$$\langle\langle Q(\tau;i,j)\rangle\rangle = \frac{1}{2\Theta+1}\sum_{\theta=\tau-\Theta}^{\tau+\Theta}\langle Q(\theta;i,j)\rangle. \qquad 20.b$$

The occupation probabilities are thus directly recovered from the LG configuration. Likewise the transition probabilities are defined through

$$n(\tau+1;i,j;a') = \sum_a n(\tau;i,j;a)\ Tr(\tau;i,j;a\rightarrow a';\mathbf{g}), \qquad 21$$

with the summation extending over all states a. The probabilistic version of the momentum equation of Eq. 3.a then becomes

$$\Delta\mathbf{p}(\tau;I,J;a;\mathbf{g}) = n(\tau;I,J;a)\ \big(\sum_{a'} Tr(\tau;I,J;a\rightarrow a';\mathbf{g})\ \mathbf{v}(a') - \mathbf{v}(a)\ \big), \qquad 22$$

(state a can transform, a priori, into any state a'). Substitute into Eq. 19, and take account of Eq. 20 to obtain

$$\frac{1}{Z}\sum_{(I,J)\in A(i,j)} n(\tau;I,J)\ \big(\sum_a Pr(\tau;I,J;a)\sum_{a'} Tr(\tau;I,J;a\rightarrow a';\mathbf{g})\ \mathbf{v}(a')\ \big) =$$

$$\frac{1}{Z}\sum_{(I,J)\in A(i,j)} n(\tau;I,J)\sum_a Pr(\tau;I,J;a)\ (\mathbf{v}(a)+\mathbf{g}(j)\ ). \qquad 23$$

This equation operates as a statistical version of the microscopic equation of motion. We observe that if the state of the lattice gas is given at timestep τ , then Eq. 23 contains as only unknowns the transition probabilitites. Eq. 23 thus plays the part of the *defining equation for the transition probabilities.*

The number of unknowns exceeds the number of defining equations (23), so that we need additional constraints to obtain a unique physically acceptable solution. To this end we make use of two natural *selection principles* (which isolate the nonzero transition probabilities):

A1: For a classical LG a force–induced transition occurs between dynamical neighbours only.

A2 : A force–induced transition from state a is to the same state a, or to at most two of its neighbour states, a' and a'', such that two among the states a, a' and a'' are moving states, the third state being a rest state.

(A) If $a \equiv b$ (rest state), then $a', a'' \equiv c^*, c^{**}$, (the two moving states 'in the direction of the force' $\mathbf{g}(j)$).

(B) If $a = c$ (moving state) and $\mathbf{v}(a)$ not parallel to the force, then $a' \equiv c$ and $a'' = c^*$ with c^* in the 'direction of the force', and closest to the latter; if $\mathbf{v}(a)$ parallel to the force direction, then a' $= c$ and $a'' = b$ (rest state).

The latter selection principle translates the CG continuity of motion under the force in the CA environment. It requires that the force expressed in dimensionless CA units is small

$$\max_j |\mathbf{g}(j)| = \epsilon \ll 1. \qquad 24$$

In turn this requirement implies that if a and a' is any pair of dynamical neighbour states, then we have

$$Tr(a \rightarrow a) = 1 - |O(\epsilon)| \quad \text{and} \quad Tr(a \rightarrow a') = |O(\epsilon)| \quad \text{for} \quad a' \neq a. \qquad 24.a$$

4. The HPP Transition Probabilities

System (23) is solved in two steps. In the first place we observe that a formal solution always exists for $A(i,j)$ reducing to the current cell (i,j). This solution is found by solving

$$\sum_{a'} \mathbf{v}(a')\ Tr(\tau;i,j;a \rightarrow a';\mathbf{g}) = \mathbf{g}(j) + \mathbf{v}(a) \equiv \mathbf{V},\ a = G,H,\ldots,S \qquad 25$$

together with the normalisation conditions

$$\sum_{a'} Tr(\tau;i,j;a \rightarrow a';\mathbf{g}) = 1, \quad a = G,H,\ldots,S. \qquad 25.a$$

Eq. 25 is a one–particle microscopic evolution equation: The force destroys state a and creates a new state a'; the associated transition probability $Tr(a \rightarrow a')$ — if it exists — is such as to secure that *statistically* each LG particle state satisfies *individually* the mechanical equation (4.a) of a CG particle. The uniqueness of the formal solution of (25,25.a) is secured by our selection rules.

We discuss the alternatives of a current rest and moving state separately.

(A) If the current state is the *ground rest state*, $b \equiv G$, then Eq. 25 reduces to

$$Tr(b \rightarrow c^*)\ \mathbf{v}(c^*) = \mathbf{g}; \qquad 26$$

where $c^* \equiv S$ is the state of motion in the direction of the gravity field (negative y–axis; there is no state c^{**}) ; hence ($g \equiv |\mathbf{g}|$)

$$Tr(G \rightarrow S) = g;\ \ Tr(G \rightarrow G) = 1 - g;\ \ \text{and}\ \ Tr(G \rightarrow a) \equiv 0\ \ \text{if}\ \ a \neq G, S. \qquad 26.a$$

This formal solution is also a physical solution.

(B) If the current state is a moving state c, then we have 3 alternatives :

(1) If c=N, only the N state is to be kept in Eq. 25

$$Tr(N \to N)\ \mathbf{v}(N) = \mathbf{g} + \mathbf{v}(N); \qquad 27$$

hence

$$Tr(N \to N) = 1 - g; \quad Tr(N \to G) = g; \quad \text{and} \quad Tr(N \to a) \equiv 0 \ \text{ if } \ a \neq N, G. \qquad 27.a$$

which is again a physical solution.

(2) If $c = S$, the counterpart of Eq. 27 yields the unphysical solution

$$Tr(S \to S) = 1 - Q(S), \quad \text{with} \quad Q(S) = -g. \qquad 28$$

(3) If $c = E$ or W, then

$$Tr(E \to E)\ \mathbf{v}(E) + Tr(E \to S)\ \mathbf{v}(S) = \mathbf{g} + \mathbf{v}(E); \qquad 29$$

or

$$Tr(E \to E) = 1; \quad Tr(E \to S) = g; \quad \text{and} \quad Tr(N \to a) \equiv 0 \ \text{ if } \ a \neq E, S. \qquad 29.a$$

which is again an unphysical solution. The same result holds for the W state.

In conclusion, for moving states c a ***physical solution*** is secured only if

$$Tr(c \to c) + Tr(c \to c^*) = 1 + Q(c), \quad \text{with} \quad Q(c) \leq 0, \qquad 30$$

which happens for $c = N$ only. We shall refer to the remaining moving states E,W,S as c^+ states ($Q(c^+) > 0$). Then the previous discussion shows that a c^+ state ***is not individually consistent with the mechanical equation of motion*** (4.a) of a CG particle (in a statistical sense).

For the c^+ states we introduce a multi–state decomposition of Eq. 23 suggested by the following observation. The ***momentum destruction*** process $N \to G$, is described by the transition probability $Tr(N \to G) = -Q(N)(> 0)$. A fictitious negative transition probability, $Tr(c^+ \to G) = -Q(c^+) < 0$, for this process is then rather naturally interpreted as indicating that the reverse process, $G \to c^+$, namely a ***momentum creation*** process, is taking place, with transition probability $Tr(G \to c^+) \equiv -Tr(c^+ \to G)$. Physically the momentum equation for a c^+ state at site (i,j) can then be satisfied if this state is combined with the momentum equation of a ground state G at some site $(i',j') \in \mathsf{A}(i,j)$; the latter G state is then assigned an ***induced transition probability*** by the c^+ state, $Tr_{c^+}(G \to b')$, to transform into a successor state b'.

Suppose, for the sake of simplicity (provisional assumption H) that site (i,j) occupied by a c^+ state, is also occupied by a G state. Then we can again select $\mathsf{A}(i,j) \equiv (i,j)$ in Eq. 23 which becomes

$$\sum_{c^+} Pr(\tau;i,j;c^+)\ [Tr(\tau;i,j;c^+ \to c^+;\mathbf{g})\ \mathbf{v}(c^+) + Tr(\tau;i,j;c^+ \to S;\mathbf{g})\ \mathbf{v}(S)]$$

$$Pr(\tau;i,j;G) \sum_{c^+}\sum_{b'} Tr_{c^+}(\tau;i,j;G \to b';\mathbf{g})\ \mathbf{v}(b') +$$

$$Pr(\tau;i,j;G)\ Tr_G(\tau;i,j;G \to S;\mathbf{g})\ \mathbf{v}(S) =$$

$$\sum_{c^+} S(\tau;i,j;c^+)\ Pr(\tau;i,j;c^+)\ (\mathbf{v}(c^+) + \mathbf{g}(j)) +$$

$$\sum_{c^+} [1 - S(\tau;i,j;c^+)]\, Pr(\tau;i,j;c^+)(\mathbf{v}(c^+) + \mathbf{g}(j)) + Pr(\tau;i,j;G)\ \mathbf{g}(j). \qquad 31$$

The first contribution of the LHS refers to the transitions from c^+ moving states towards the c^+ and S states (summation over all c^+ states). The second contribution refers to the c^+-induced transitions of the ground state into other states b' ($Tr_{c^+}(b \rightarrow b')$). The third contribution gives the 'spontaneous' transition of the associated rest state G into the S state (probability $Tr_G(G \rightarrow S)$). The global transition probability of the rest state towards a successor b' has thus been separated as follows

$$Tr(G \rightarrow b') = Tr_G(G \rightarrow b') + \sum_{c^+} Tr_{c^+}(G \rightarrow b'), \qquad 32$$

with the constraints and normalisation

$$0 \leq Tr_a(b \rightarrow b') \leq 1; \quad \sum_a \sum_{b'} Tr_a(b \rightarrow b') = 1; (a \neq N) \qquad 32.a$$

The RHS of Eq. 23 contains a contribution of the G state, and a contribution of the c^+ states; each individual c^+ contribution, say $C(c^+)$, is formally subdivided into components $S(c^+)\ C(c^+)$ and $[1 - S(c^+)]\ C(c^+)$, where $S(c^+)$ is a free parameter to be determined below. Any N state continues to obey Eq. 27.a.

To solve Eq. 33 we introduce the following identification

$$Tr(\tau; i, j; c^+ \rightarrow c^+; \mathbf{g})\ \mathbf{v}(c^+) + Tr(\tau; i, j; c^+ \rightarrow S; \mathbf{g})\ \mathbf{v}(S) =$$

$$S(\tau; i, j; c^+)\ (\mathbf{v}(c^+) + \mathbf{g}(j)), \qquad 33$$

$$\sum_{b'} Tr_{c^+}(\tau; i, j; G \rightarrow b'; \mathbf{g})\mathbf{v}(b') =$$

$$[1 - S(\tau; i, j; c^+)]\ \frac{Pr(\tau; i, j; c^+)}{Pr(\tau; i, j; G)}\ (\mathbf{v}(c^+) + \mathbf{g}(j)), \qquad 34$$

$$Tr_G(\tau; i, j; G \rightarrow S; \mathbf{g})\ \mathbf{v}(S) = \mathbf{g}(j). \qquad 35$$

We then have a linear system in the transition probabilities whose solution is a physical solution provided that constraints (32.a) are obeyed. We show that this is always the case.

A detailed physical interpretation of these formal operations is discussed elsewhere (Perdang 1993). We just mention that certain 'real' CG particles are regarded as mixtures of two LG particle states, a c^+ state and a G state. If $Pr(t; x, y; \mathbf{p})s(\mathbf{p})$ symbolises a 'real' CG particle existing at time t and at position (x, y) (in lattice cell (i, j)), with a probability $Pr(t; x, y; \mathbf{p})$ in a state symbolised by $s(\mathbf{p})$ (with momentum $\mathbf{p} = |\mathbf{p}|\mathbf{v}(c^+)$, $|\mathbf{p}| < 1$ in CA units, $|\mathbf{v}(c^+)| = 1$), then the associated mixed LG entity is symbolised as

$$Pr(\tau; i, j; c^+)\ p(c^+) + Pr(\tau; i, j; G)\ p(G) \Longleftrightarrow Pr(t; x, y; \mathbf{p})\ s(\mathbf{p}) \qquad 36$$

($Pr(\tau; i, j; a)$, occupation probability; p, prime number encoding of LG state). Under the scaling

$$Tr(c^+ \rightarrow c') = S(c^+)\ Tr'(c^+ \rightarrow c') \qquad 37$$

Eq. 33 collapses into

$$Tr'(c^+ \rightarrow c^+)\ \mathbf{v}(c^+) + Tr'(c^+ \rightarrow S)\ \mathbf{v}(S) = \mathbf{v}(c^+) + \mathbf{g}(j), \qquad 38$$

(cf Eq. 25), with $Tr'(c^+ \rightarrow c^+)$, $Tr'(c^+ \rightarrow S)$ *formal transition probabilities*. From the definition of a c^+ state we see that the normalisation condition for the formal transition coefficients is not satisfied. However, we can now choose the free scaling parameter $S(c^+)$ such that

$$Tr(c^+ \rightarrow c) + Tr(c^+ \rightarrow c^*) = 1. \qquad 39$$

Hence

$$S(c^+) = \frac{1}{1+Q(c^+)}; \quad Tr(c^+ \to c') = \frac{Tr'(c^+ \to c')}{1+Q(c^+)}. \qquad 40$$

With this choice of the scaling factor, Eqs. 33,34,35 reduce to equations already discussed above (cf A and B).

For our HPP–type LG the final solution takes the following simple form.

(A) *Ground state G*

The G self–induced transition probabilities of a G state are :

$$Tr_G(G \to G) = 1-g; \quad Tr_G(G \to S) = g; \quad Tr_G(G \to a) = 0, \quad a \neq G, S; \qquad 41$$

(B) *Moving states* $c^+ = E, W, S$:

For the E state we have

$$Tr(E \to E) = \frac{1}{1+g}; \quad Tr(E \to S) = \frac{g}{1+g}, \qquad 42$$

$$Tr_E(G \to E) = \frac{\langle n(\tau;i,j;E)\rangle}{\langle n(\tau;i,j;G)\rangle} \times \frac{g}{1+g},$$

$$Tr_E(G \to S) = \frac{\langle n(\tau;i,j;E)\rangle}{\langle n(\tau;i,j;G)\rangle} \times \frac{g^2}{1+g}. \qquad 43$$

Note that

$$\frac{Tr(E \to S)}{Tr(E \to E)} = \frac{Tr_E(G \to S)}{Tr_E(G \to E)} = g. \qquad 44$$

For the W state the equations are similar (substitution $E \to W$ in Eqs. 42,43,44). For the S state we have

$$Tr(S \to S) = 1; \quad Tr(S \to E) = 0, \qquad 45$$

$$Tr_S(G \to S) = \frac{\langle n(\tau;i,j;E)\rangle}{\langle n(\tau;i,j;G)\rangle} \times g,$$

$$Tr_S(G \to E) = 0, \qquad 46$$

or

$$\frac{Tr(S \to E)}{Tr(S \to S)} = \frac{Tr_S(G \to E)}{Tr_S(G \to S)} = 0. \qquad 47$$

It is seen from these equations that two different alternatives occur, depending on whether we are dealing with a 'cold' or a 'hot' gas:

(1) If the occupation probability ratio $\frac{Pr(c^+)}{Pr(G)} = \frac{\langle n(\tau;i,j;c^+)\rangle}{\langle n(\tau;i,j;G)\rangle}$, is of the order of the force ('cold gas')

$$\frac{Pr(c^+)}{Pr(G)} = O(\epsilon), \qquad 48$$

then we have

$$Tr(c^+ \to c^+) = O(\epsilon^o); \quad Tr(E \to S) = Tr(W \to S) = O(\epsilon); \qquad 48.a$$

and

$$Tr_{c^+}(G \to c^+) = O(\epsilon^2); \quad Tr_E(G \to S) = Tr_W(G \to S) = O(\epsilon^3). \qquad 48.b$$

For ϵ small enough, (cf Eq. 24), the *induced momentum creation effects* (48.b) *are negligible* if the calculations are performed to order ϵ.

(2) If the probability ratio of the moving state c^+ to the rest state G, $\frac{Pr(c^+)}{Pr(G)}$, is of order 1 ('hot gas')

$$\frac{Pr(c^+)}{Pr(b)} = O(\epsilon^o), \qquad 49$$

then the orders (48.a) continue to hold, while the induced transition terms become

$$Tr_{c^+}(G \to c^+) = O(\epsilon); \quad Tr_E(G \to S) = Tr_W(G \to S) = O(\epsilon^2). \qquad 49.b$$

The correction to the momentum balance due to momentum creation in the process $G \to c^+$ is of same order as the probability of orientational change of momentum $E \to S$, or $W \to S$; the momentum creation in the transition $G \to S$ alone is of a higher order, and therefore negligible if the calculations are performed to order ϵ.

Although our analysis is based on the provisional hypothesis H, the following observation allows us to relax this hypothesis : If cell (i,j) at timestep τ contains no ground state G, we choose a submacroscopic neighbourhood $\mathsf{A}(i,j) = (i,j) \cup (i',j')$, with (i',j') the cell nearest to (i,j) containing a state G. In the symbolism (36) we have

$$Pr(\tau;i,j;c^+)\, p(c^+) + Pr(\tau;i',j';b)\, p(b) \Longleftrightarrow Pr(t;x,y;\mathbf{p})\, s(\mathbf{p}), \qquad 50$$

i.e. a CG particle is modelled by an LG entity made up of a c^+ state at site (i,j) and a G state at a *different* site (i',j'). Since the force obeys $\mathbf{g}(j') \approx \mathbf{g}(j)$ (cf Eq. 18) we have also, for any state a , $Pr(\tau;i'j';a) \approx Pr(\tau;i,j;a)$, if $(i',j') \in \mathsf{A}(i,j)$. Hence Eqs. 50 and 36 are equivalent. The physical interpretation of the symbolic operation is as follows : The a–state occupation probability is independent of the precise microscopic site (i',j') within the averaging submacroscopic neighbourhood; reshuffling the states among the sites of $\mathsf{A}(i,j)$ does not affect the submacroscopic dynamics. In our case a G state is transferred from (i',j') to (i,j); the dynamics then operates on G at site (i,j); eventually we transform back to the initial position (i',j').

The argument introduced here can be applied to avoid other restrictions as well (for instance the constraint of the exclusion principle).

It is finally straightforward, though algebraically tedious to verify that the treatment of the gravitational force given in this section leads indeed to the classical force term $\varrho(t,\mathbf{r})\ \mathbf{g}(\mathbf{r})$ in the RHS of the momentum density equation of the HPP gas (Perdang 1993).

5. The Vibrational Instability Mechanism

To implement an instability mechanism on the CA level we recall that stellar instabilities can be interpreted as phase transitions.

Secular and *dynamical instabilities* appear as *first order phase transitions* (cf the thermodynamic analyses by Thirring 1970, 1972 and by Lynden–Bell and Wood 1968). The bifurcation from the uniformly rotating MacLaurin spheroid towards the Jacobi tri–axial ellipsoid at a critical angular momentum is a *second order phase transition*, (order parameter : excentricity of the equatorial ellipse, or a related parameter, Bertin and Radicati, 1976). The bifurcation towards a *periodic state of motion* (Hopf bifurcation) can also be viewed as a *second order phase transition*; the obvious order parameter is here the oscillation amplitude. On the other hand, the CA formulation lends itself both to theoretical studies (cf the Ising model, Kadanoff 1976) and to convenient numerical simulations of phase transitions.

A simple CA scheme to simulate first–order transitions has been proposed by Chen et al (1989) in a variant of the $2D$ FHP gas (Frisch, Hasslacher and Pomeau 1986). Their idea consists in adding to the standard 6–particle states of the standard FHP fermion gas a 7th excited bound pair state P (again a fermion in their model) whose formation and decay supplies the extra thermodynamic degree of freedom needed for the phase transition. The model incorporates the following new automata rules:

(1) *Inelastic collisions* : 2 opposite momentum states in a cell have a nonzero probability T to transform into a bound pair state P.

(2) *Decay of bound pair* : a bound pair state P has a nonzero probability T' to decay into 2 opposite momentum states.

To model condensation, Chen et al introduce a potential energy between any two bound pairs $P(i,j)$ and $P(i',j')$ at sites (i,j) and (i',j') respectively, given by

$$V\big(P(i,j), P(i',j')\,\big) = -e_o \;\text{ if }\; (i',j') \in \mathrm{N}(i,j), \;\text{ and }\; 0 \;\text{ if }\; (i',j') \in \mathrm{N}(i,j), \qquad 51$$

($\mathrm{N}(i,j)$ = cell (i,j) and its 8 nearest neighbours). The change of potential energy at cell (i,j) due to the formation of a bound pair in (i,j) is then

$$\Delta V = -e_o \sum_{i',j' \in \mathrm{N}(i,j)\setminus(i,j)} n(i',j';P), \qquad 51.a$$

where $n(i,j;P)$ is the number of pair states at site (i,j). The transition probabilities T and T' are written

$$T = \lambda\, \frac{\exp\,(-\beta\Delta V)}{1+\exp\,(-\beta\Delta V)}, \qquad T' = \lambda\,(1-T), \qquad 51.b$$

with λa free parameter $(0\langle \lambda \leq 1)$ and β, the inverse temperature.

This model is found to exhibit a liquid–solid (=bound pair cluster) type phase transition ($\beta \approx 1$, for $e_o = 1$), with two-phase coexistence. We observe further that the numerical simulations indicate that the border of the cluster of bound states seemingly has a fractal structure (Fig. 2 in Chen et al 1989).

In our pulsation model the process of transforming rest pairs into moving particles accompanied by a transfer of non-kinetic energy into kinetic energy, together with its reverse process, has been adapted to simulate the stellar phase transition we are interested in. Indeed, we do believe that the mechanisms proposed by Chen and coworkers supply a computationally adequate CA implementation of the main physics responsible for the vibrational instability in stars, namely the ionisation processes of H, HeI and $HeII$ in the outer stellar layers (cf the systematic study by Cox et al 1979; for a recent exhaustive survey of the vibrational spectrum of $RRLyrae$ models, and of classical Cepheid models, see Glasner and Buchler 1993). The modification we need to make is to interpret the rest pair as an excited state (with associated excitation energy, rather than with positional potential energy). In the framework of the HPP LG, the decay of an excited pair state into two moving states

$$P \rightarrow E + W \quad \text{or} \quad P \rightarrow N + S, \qquad 52$$

then duplicates the dynamics of a genuine ionisation process

$$A \rightarrow A^{+} + e^{-}. \qquad 52.a$$

(The fact that in the CA simulation the outcome of the 'reaction' (52) involves two particles of same mass, instead of two different masses in the real ionisation, has no major qualitative consequences). Physically the simulation process (52), just as the 'real' process (52.a) translates a transformation of internal energy (excitation energy, or binding energy) into microscopic kinetic energy; the latter manifests itself as an increase in pressure. Macroscopically, if the configuration was in a state of hydrostatic equilibrium prior to the operation of reaction (52), then, once the latter becomes operative, the equilibrium is disrupted. The local increase in pressure tends to produce an expansion of the gas. For the reverse processes of (52) or (52.a), on the other hand, if we start with an initial hydrostatic equilibrium, as moving particles are transformed into rest pairs, the local pressure is decreased, thereby triggering a contraction. The microscopic kinetic energy lost in this process is stored on internal degrees of freedom and can be made available again at a later time.

This discussion shows that the conversion mechanisms (52) can be regarded as a simulation of the *gamma mechanism* : These 'chemical equilibria' affect the compressibility of the gas, or equivalently the adiabatic rigidity coefficient γ. We point out at this stage that by modifying γ, caution must be exercised to avoid producing a ***dynamical instability*** in the stellar configuration (in a genuine $3D$ stellar model the latter instability arises for an averaged value of this parameter over the star $(\langle\gamma\rangle < \frac{4}{3})$. To obtain a proper simulation of all relevant factors influencing vibrational instability we start out with an approximate expression of the damping coefficient k of a mode of (adiabatic) frequency

$$k = \{ \ [\ 4\ (\Gamma_3 - 1)\ ] + [\ -\kappa_T\ (\Gamma_3 - 1) - \kappa_\rho] - \frac{4}{3}\ \} \ \times$$

$$\frac{(\Gamma_3 - 1)\ L}{(2\ \omega^2\ I)}\ N, \qquad 53$$

(κ_s , logarithmic derivative of opacity with respect to state variable $s = T$, temperature, or ϱ, density; Γ, generalisation of the adiabatic coefficient γ; L, total luminosity; I moment of inertia; N, dimensionless ratio of weighted averages of squared relative density eigenfunction, $\frac{\delta\rho}{\rho}$, to squared relative radial displacement, ξ, of adiabatic mode ω, $N = \frac{\langle(\frac{\delta\rho}{\rho})^2\rangle}{\langle(\xi)^2\rangle} \ > \ 0$). This expression flows from a perturbation calculation (cf Ledoux and Walraven 1958; for a direct derivation from discrete hydrodynamics, see Perdang 1990; the main physical contributions entering Eq. 53 can be found from an analysis of a one–zone model; Baker and Kippenhahn 1962). Since $\Gamma_3 > 1$, the onset of a vibrational instability depends on two components in the first curly brackets : The expressions $[4(\Gamma_3 - 1)]$ and $[-\kappa_T(\Gamma_3 - 1) - \kappa_\rho]$, isolating the *gamma* and the *kappa effect* respectively. As seen from Eq. 53, the gamma effect on its own does not produce the instability; the instability does require the kappa effect: The necessary condition for the onset of a vibrational instability is that the second expression be negative; by lowering gamma, the net damping coefficient κ can then become negative.

This role of the kappa effect translates a physically natural scenario. Since during a phase of compression of the star the thermodynamic state variables, pressure, density and temperature, increase with respect to their equilibrium value, the opacity increases as well $[\kappa_T(\Gamma_3 - 1) + \kappa_\rho > 0]$; therefore a surplus of radiation energy is absorbed by matter during the contraction stage; this energy is deposited on the internal degrees of freedom of the particles, as excitation energy. Conversely, during expansion, the state variables decrease, producing a lowering of opacity with respect to the equilibrium opacity; accordingly, the surplus of energy stored on the internal degrees of freedom is released; the latter energy may then become available as kinetic energy.

The remaining 'destabilising' factor 4/3 in the first curly brackets results from the 3–dimensional spherical structure of the star (for arbitrary dimensionalities, see the discussion in Perdang 1990).

Being the result of a perturbation calculation, the approximate form (53) for the damping coefficient is quantitatively valid only for $\frac{|k|}{\omega} \ll 1$; for the cool Variable Stars we are interested in this condition is violated (for instance for *Miras*, the ratio $\frac{|k|}{\omega}$ is in the range $0.1 - 1$). However, expression (53) remains essentially correct if we substitute ***nonadiabatic eigenfunctions*** for the adiabatic ones. This operation essentially alters the numerical coefficient N, without changing the form of the relation. The expression continues to serve our purpose of isolating the instability mechanisms.

The full CA implementation of the kappa effect would require a detailed modelling of the matter–radiation interaction mechanisms. In principle such a simulation could be accomplished with the help of a massless LG as a convenient CA model for the radiation field (cf the discussion in Molvig et al 1990). But since the net effect of the kappa mechanism amounts to storing energy on internal degrees of freedom, and releasing again this excitation energy, it suffices, for the purposes of modelling pulsations, to devise a simple CA scheme duplicating these energy exchange processes. In line with our approach of dealing with the gamma effect, we achieve this goal by introducing additional reaction processes. Specifically the following transformations are included :

$$(1) \qquad G + G \rightarrow P; \qquad 54$$

this process simulates the storage of radiation energy on the internal degrees of freedom. It is understood that the two ground state particles occur at the same site (i, j) at time τ. In order to be operative as an instability mechanism, the transition probability for the rest state reactions (54), $Tr(2G \rightarrow P)$, must be a function of the local physics, s, of the star,

$$Tr(2G \rightarrow P) = T_g(s), \quad 54.a$$

such that this probability is large during contraction and small during expansion.

We further take account of the inelastic collisions

$$(2) \qquad E + W \rightarrow P \quad \text{and} \quad N + S \rightarrow P, \quad 55$$

which simulate the the storage of microscopic kinetic energy on the internal degrees of freedom. This process produces the drop in local pressure, as mentioned above. To act as instability mechanisms, the transition probabilities $Tr(E + W \rightarrow P)$, $Tr(N + S \rightarrow P)$, must again be functions of the local physical state of the star,

$$Tr(E + W \rightarrow P) = Tr(N + S \rightarrow P) = T_m(s), \quad 55.a$$

(both probabilities must be equal in a locally isotropic medium); this probability ought to be large during contraction and small during expansion.

As a third process we include the decay reaction (52), which simulates the release of the stored energy in the form of microscopic kinetic energy, accompanied by a local pressure increase. To operate as a kappa instability mechanism the decay probabilities $Tr(P \rightarrow E+W)$, $Tr(P \rightarrow N+S)$, functions of the local physics s

$$Tr(P \rightarrow E + W) = Tr(P \rightarrow N + S) = T_d(s), \quad 52.a$$

are required to be small during contraction and large during expansion.

Notice that our collection of reaction mechanisms devised to handle the kappa effect actually incorporate the gamma effect as well. Accordingly, processes 52, 54 and 55 supply a complete CA simulation of the standard vibrational instability. The reaction rates 52.a,54.a and 55.a depend on the local physics and should be described by expressions of form (51.b). The computationally simplest representation consists in regarding the macroscopic state variable s as a Boolean identifier of local contraction ($s = 0$) or expansion ($s = 1$) only. If, moreover, in order to keep the number of free parameters as small as possible, we assign these probabilities the values 0 or 1 in accordance with the above requirements, we have

$$T_g(s) = T_m(s) = 1 - s = 1 - T_d(s). \quad 56$$

However, the instability mechanisms discussed so far are based on a linear stability analysis only (Eq. 53). It is physically clear that the instability, giving rise to the pulsations, must eventually be killed as the excursions away from the unstable equilibrium of the macroscopic variables become large enough (for instance, in the ionisation process 52.a, if the temperature is sufficiently increased, all of the matter is in the form of A^+ and e^-). This saturation effect can be incorporated in the Boolean state parameter, s, by defining the state $s = 0$ as *weak contraction* or *strong expansion*, and $s = 1$ as *weak expansion* or *strong contraction*.

In our numerical experiments we estimate s by computing the local 'radial' momentum P_y over an 'instability neighbourhood' $\mathsf{N}_{inst}(i, j)$ (cf Eq. 8); contraction [expansion] corresponds to $P_y < 0$ [> 0]; the transition from weak to strong is parametrised ($|P_y| <$ or $> P_{crit}$).

The formalism allows us also to write down a (radial) energy equation supplying the luminosity in every horizontal layer. Denote by $L(\tau; j)$ the luminosity of the horizontal layer j at timestep τ, the radiation input into our atmosphere being provided by the time–independent luminosity

L_o at the bottom of the atmosphere. Denote further by $E_{tot}(\tau; j)$ the total particle energy in the horizontal layer j at time τ, including the microscopic energy (kinetic plus excitation) $E_{tot}(\tau; j)$ (cf Eq. 16, summed over all horizontal cells, but not averaged) and the gravitational energy of the layer. In one timestep the total microscopic energy of each horizontal layer j of the atmosphere is modified by (1) an inflow of radiation coming from the layer below, $L(\tau; j-1)$; (2) an outflow of radiation to the layer above, $L(\tau; j)$; (3) and an exchange of total energy do to the flow of matter. An approximate equation for the luminosity is then (in CA units)

$$L(\tau; j) = L(\tau; j-1)$$

$$+ \alpha \left\{ \left[E_{tot}(\tau-1; j) - E_{tot}(\tau; j) \right] + \frac{1}{2} \left[E_{tot}(\tau-1; j) + E_{tot}(\tau; j) \right] \left[v_y(\tau; j-1) - v_y(\tau; j) \right] \right\}, \quad 57$$

to be solved with the boundary condition $L(\tau; -1) = L_o$, luminosity input from the stellar core (α is a conversion factor depending on the geometry).

6. Numerical Experiments

We report now on a series of numerical results obtained with the 6 state HPP gas, with an external vertical gravitational force field (decreasing with height) and with vibrational instability mechanisms incorporated. The experiments are carried out on a 61 × 61 lattice. Submacroscopic averages are computed over 7 × 7 or 11 × 11 cells (Eq. 6). Although this lattice size may appear as prohibitively small to allow us to draw any strong conclusions, a few testruns performed on a 128 × 128 lattice (with A Lejeune) show in fact the same type of characteristic behaviour. Therefore, we do feel confident that the new qualitative features traced in our experiments are indeed real, rather than artifacts of small number fluctuations.

Initial conditions

The following initial conditions are adopted in all of our experiments:

(1) The inital *particle distribution* of N_P particles is *statistically uniform*; the available CA space is filled up to a relative height h ($N_P = 1700$; $h = 0.6$ in most experiments).

(2) The initial distribution is at a *uniform temperature* (fraction η_o of number of particles in random motion; $\eta_o = 0.25$ in most experiments).

(3) All *rest particles* are in the G state.

Conversions from CA units to the physical parameters of stars are listed in Table 3.

These initial conditions thus specify a configuration which (statistically) depends on the vertical coordinate y only (cf the first frame in Fig. 2). Accordingly we simulate a macroscopic stellar configuration of *initial radial symmetry*. This initial configuration is *far from a hydrostatic equilibrium* state; moreover, the surface layers of the initial configuration are *far from local thermodynamic equilibrium*.

Two evolutionary phases can be distinguished in all of our runs.

(1) *Relaxation towards a hydrostatic state*

The first several hundred timesteps (at least 200–400, sometimes 1000) show a *transient evolution* depending strongly on the initial conditions. If we start with $\eta_o = 0$ (zero temperature gas), the atmosphere initially undergoes a global collapse. If $\eta_o \neq 0$, initially only an inner collapse is observed (contraction of deeper atmosphere), while at the surface an 'evaporation' of particles is taking place; for η_o small the instantaneous density pattern of the evaporated matter violates radial symmetry (instantaneous number of N state surface particles in averaging box dominated by statistical fluctuations). Under both alternatives an eventual roughly radial stratification is built up; the space averaged density increases gently downward over about 80% of the depth; in the bottom layers it jumps up to about 3 particles per cell. Also, under both alternatives, the initial collapse is followed by a general expansion which in turn transforms again into a collapse.

For $\eta_o = 0.25$ we have exhibited the details of the relaxation process in the sequence of frames displayed in Figs. 2 and 3.

Table 3.- Conversion to physical units

Cellsize in physical units:

$\lambda = H_a/n_y = 1.93\ 10^{11}$ [1] ; $1.14\ 10^{11}$ [2] ; $5.45\ 10^{10}$ [3] cm

typical order of $\lambda \sim 1\ R_\Theta$

Number of CA cells; equivalent number of cells in total atmosphere

$N_{CA}=n_x\ n_y = 1.296\ 10^3$ [1] ; $3.721\ 10^3$ [2] ; $1.6384\ 10^4$ [3]

$N_a = \lambda^{-3}\ (4/3)\ \pi\ [(R_o+H_a)^3 - R_o^3] = (4/3)\ \pi\ n_y^3\ [(1+r_o)^3-(r_o)^3]$

$\rightarrow (4/3)\ \pi\ n_y^3\ (1+3R_o/H_a)$ if $H_a >> R_o$

$\rightarrow 4\ \pi\ n_y^3\ (R_o/H_a)^2$ if $H_a << R_o$

$= 4.24\ 10^5$ [1] ; $2.06\ 10^6$ [2] ; $1.906\ 10^7$ [3]

Fraction of volume of atmosphere simulated by CA

$f_a = n_x\ n_y/N_a$

$= 3.06\ 10^{-3}$ [1] ; $1.81\ 10^{-3}$ [2] ; $8.60\ 10^{-4}$ [3]

Mass of one LG particle:

$\mu = f_a\ M_a/N_P$

$= 7.62\ 10^{25}$ [1] ; $2.12\ 10^{25}$ [2] ; $2.14\ 10^{24}$ [3]

typical order of $\mu \sim 10^{-8}\ M_\Theta$

Gravitational acceleration at level j

$|g(i,j)| = GM_o/[R_o+H(j)]^2 = [GM_o/R_o^2]\ [r_o/(r_o+h(j))]^2$

Gravitational acceleration at base of atmosphere

$g_o = GM_o/R_o^2 = 3.03\ 10^{-1}\ cm/s^2 = g_{CA}\ \lambda/\Delta t^2$

Timestep :

$\Delta t = [\lambda\ g_{CA}/g_o]^{1/2} = 1.88\ 10^5$ [1] ; $1.11\ 10^5$ [2] ; $5.30\ 10^4$ [3] sec

typical order of $\Delta t \sim 1$ day ($\sim 8.64\ 10^4$ s) $\sim$ 1/100 cycle length of long-period variables

Kinetic energy of moving LG particle :

$(1/2)\ \mu\ \lambda^2/\Delta t^2$

$= 4.02\ 10^{37}$ [1] ; $1.12\ 10^{37}$ [2] ; $1.13\ 10^{36}$ [3] erg

Area of sphere at half height of atmosphere :

$A_a = 4\pi\ (R_o + H_a/2)^2$

$\rightarrow \pi\ H_a^2\ (1 + 4\ R_o/H_a)$ if $R_o << H_a$

$\rightarrow 4\ \pi\ R_o^2\ (1 + H_a/R_o)$ if $R_o >> H_a$

$= 3.90\ 10^{26}\ cm^2$

Horizontal area of simulated part of atmosphere :

$A'_a = \lambda\ W_a = H_a^2/n_y$

$= 1.35\ 10^{24}$ [1] ; $7.94\ 10^{23}$ [2] ; $3.78\ 10^{23}$ [3] cm^2

Fractional area intercepted by simulated atmosphere :

$\phi_a = A'_a/A_a\ (\sim f_a)$

$= 3.45\ 10^{-3}$ [1] ; $2.04\ 10^{-3}$ [2] ; $9.72\ 10^{-4}$ [3]

Luminosity traversing simulated part of atmosphere
(CA luminosity in physical units) :

$L'_a = \phi_a\ L_o$

$= 6.66\ 10^{33}$ [1] ; $3.94\ 10^{33}$ [2] ; $1.87\ 10^{33}$ [3]

(of order of L_Θ)

CA luminosity in dimensionless CA units :

$L_{CA} = L'_a/[\mu\ \lambda^2/\Delta t^3]$

15.6 [1] ; 19.5 [2] ; 44.1 [3]

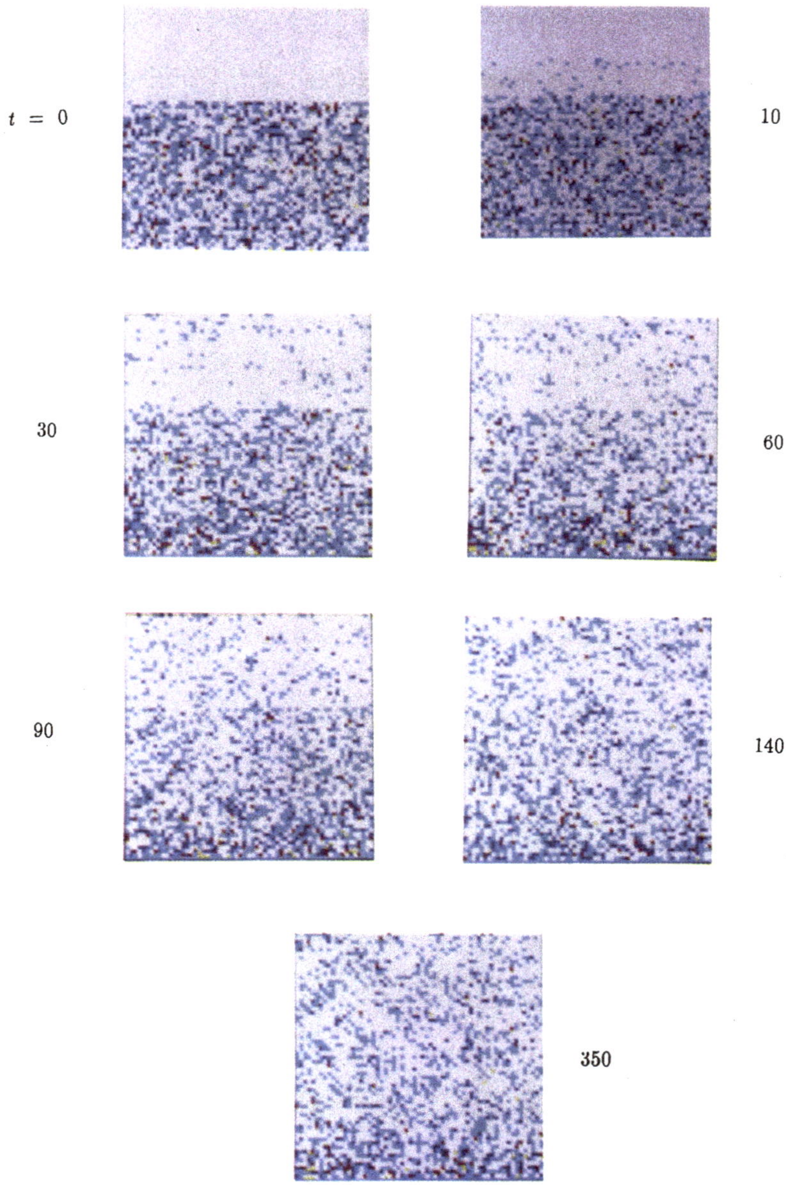

Fig. 2. – Microscopic cell occupation during relaxation phase; the colour refers to the total number of particles per cell (cyan: 1 particle/cell; blue : 2; red : 3; yellow: > 3)

Fig. 3. – Transient evolution phase. The frames show the instantaneous density profile at low (increment $0.5 \times n_o$ between successive colours) (left), and high resolution (increment $0.2 \times n_o$) (middle), as well as the local instantaneous velocity field; red and blue indicate expansion ($v_r > 0$) and contraction ($v_r < 0$) respectively, dark red and dark blue correspond to 'large speeds' $|v_r| > 0.1$

The instantaneous microscopic cell occupation at selected timesteps is illustrated in Fig. 2. The frames indicate in particular that the scar of the initial surface density discontinuity remains visible at least up to timestep 120. A high bottom density in the deepest layers of the atmosphere has developed at step 60.

Fig. 3 displays the (submacroscopic) averages of the 'radial' velocity component, as well as of the particle density. The initial evaporation process appears as patchy (lack of radial symmetry), as is shown by the sporadic occurrence of blobs of matter above the initial atmosphere. Perhaps more interesting, during the expansion phase following the initial contraction (after step 100), pockets of upstreaming surface matter are observed to develop which reach the top of the available CA space. This scenario seems to suggest a *massloss* phenomenon which, in the framework of our experiment is killed off by the reflecting boundary conditions.

It is seen from the radial velocity frames that the outer expansion zone, visible at steps 10, 20 and 30, rapidly shrinks, and at the same time an expansion at the bottom tends to develop. The latter eventually 'percolates' all through the whole atmosphere around step 80. Note that the boundary between expansion and contraction is strongly irregular, thus violating radial symmetry: tongues of expanding matter are intruding into the originally dominant contracting velocity field. During the subsequent expansion phase (after timestep 80) small irregular aggregates of collapsing matter can be observed in an expanding sea (cf timestep 150).

Regarding the density pattern, the isopycnics are deformed into wiggles, reminiscent of the incipient structures traced in simulations of supernova explosions (direct $2D$ hydrocode integrations, Hachisu 1991; or smoothed-particle simulations, Benz 1993). The density patterns in the supernova experiments develop tongues transforming into filamentary patterns (cf Hachisu 1991), chiefly as a result of the violent dynamical instability at the origin of the explosive motion (no saturation effect); in contrast, our experiments are controlled by a milder instability; any developing density tongue is eventually destroyed, as a consequence of the saturation effect blocking the vibrational instability. In any event, the analogies of our density profiles with the incipient $2D$ density structure in supernova explosions give us confidence in the reality of the breaking of the radial symmetry as suggested by our experiments.

To the extent that these experimental results are not to be dismissed as small number noise, they lead to the conjecture that the alternations between contraction and expansion show a behaviour reminiscent of a *percolation* mechanism: At the submacroscopic level, and prior to the pulsation, the medium contains coexisting patches (aggregates) of positive and negative radial velocities; if a growth is observed in one type of aggregates, say the negative velocity aggregates, leading to a negative velocity percolation cluster, then the medium eventually exhibits a macroscopic contraction; at the next stage, among the surviving positive velocity aggregates, the growth of one cluster developing into a positive percolation cluster manifests itself macroscopically as an expansion. The alternation between contraction and expansion would appear here as resulting from *alternative percolation phase transitions*, the global changes in the medium brought about in one transition playing the part of the order parameter. The confirmation, or dismissal, of this conjecture requires further analysis.

(2) *Stationary Pulsation Phase*

Fig. 4 refers to the results of a 5000 timestep run. It shows in the first place the global space average of the radial velocity field over the whole atmosphere (Eq. 6, with averaging neighbourhood = whole CA space), and smoothed over time (Eq. 20.b, with $\Theta \approx 15$). Barring the initial relaxation phase (which extends here at least up to 500 as follows from a closer examination of the results), the subsequent radial velocity curve exhibits well defined *cycles* of *variable shape* and *length* P; the average cycle length $\langle P \rangle$ is about 160 timesteps (in rough agreement with a direct estimate of the period of the fundamental radial mode in the atmosphere); individual periods may show relative deviations of more than 30%; this order of magnitude in the scatter of individual cycles is actually observed in the Long Period Variables. The oscillation is thus *not monoperiodic*, although it seemingly does preserve a characteristic feature of monoperiodicity, namely a cycle length P of the order of a true period of the configuration.

The oscillatory structure cannot be interpreted as mere random noise due to the small scale of our

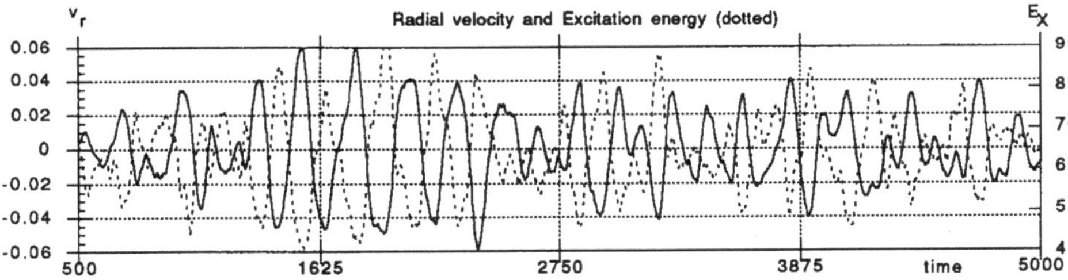

Fig. 4.– Averaged radial velocity v_r (in CA units) and global excitation energy E_X shown over 5000 timesteps. Initial relaxation phase ≈ 500 steps not shown.

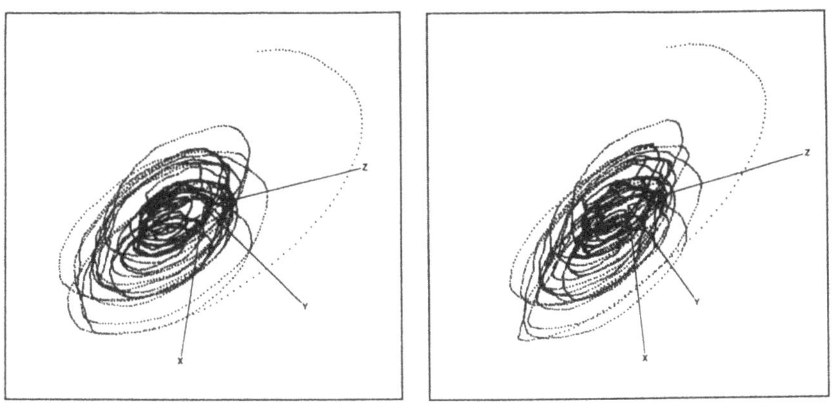

Fig. 5. – Stereoscopic view of $3D$ Phase Space Reconstruction (trajectory (R, v_r, a)) corresponding to run shown in Fig. 4 .

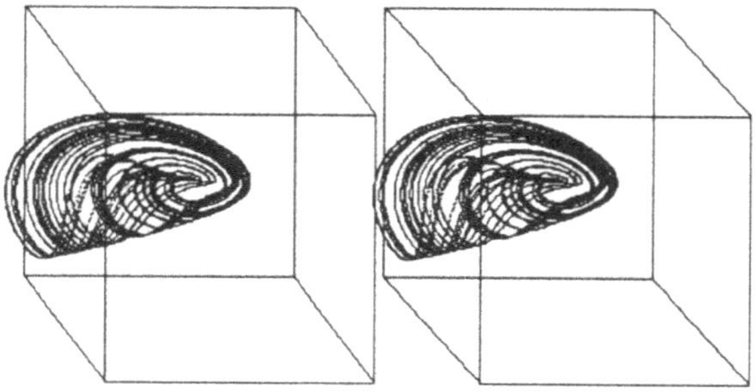

Fig. 6. – Comparison with the $3D$ phase space trajectory of a schematic one–zone stellar model ($\dot{x} = y$; $\dot{y} = -0.4\,x + 0.6\,y + z$; $\dot{z} = -3\,z - y + 1.8\,yz$)

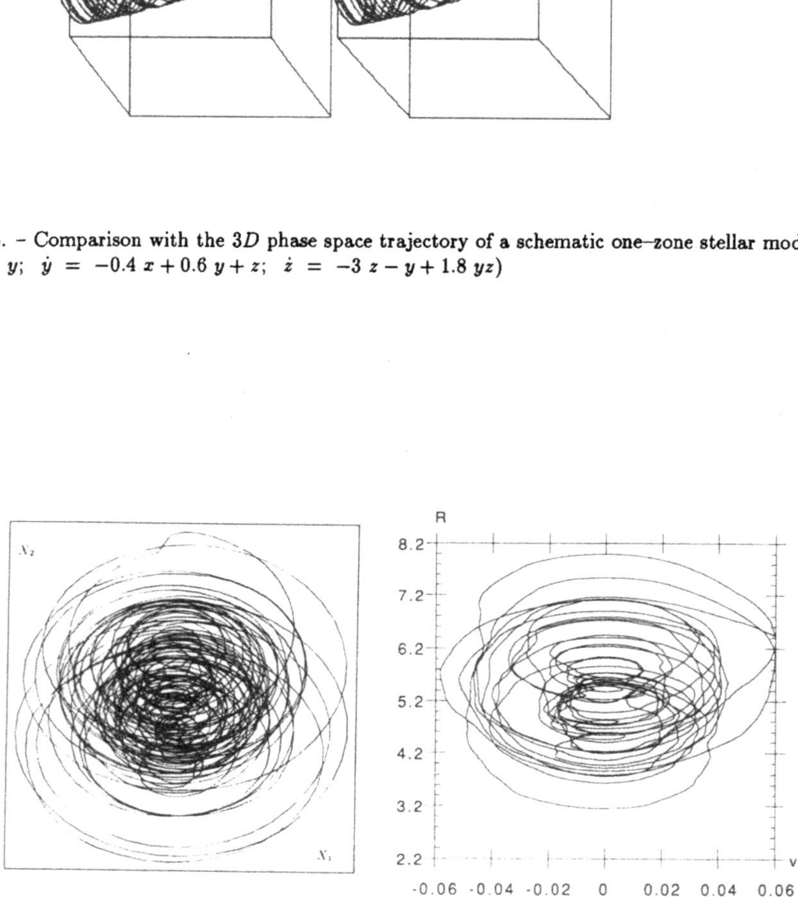

Fig. 7. – Projection of the trial phase space trajectory of the observational lightcurve $s(t)$ of $RU Cam$ in the plane $X_1(t) = s(t)$, $X_2(t) = s(t+\tau)$ (Kollath and Szeidl 1993) (a); comparison with the projection of the LG trajectory of Fig. 5 into the the plane $X_1(t) = v_r(t)$, $X_2(t) = R(t)$ of the LG simulation (b)

experiment. In fact, as transpires from Fig. 4, the excitation energy (Eq. 11 averaged over the atmosphere) varies practically in phase opposition with the velocity field ($\Delta E_H \propto - \Delta v_r$); such a variation is precisely expected if the pulsations are driven by the alternations of an internal energy storage and an energy release.

For the purpose of a closer comparison of our LG experiments with the results of elementary theoretical models (one–zone models) or full radial hydrocode integrations of detailed models, or with genuine observations of variable stars, we have constructed a $3D$ trial phase space representation from our velocity signal. Since the conventional delay method suffers from a lack of direct physical interpretability of the phase space coordinates — besides the technical problem related to the arbitrary choice of the delay time — we adopt the following procedure : From the time series $v_r(t)$ we compute numerically an associated *acceleration*, $a(t) = \frac{d}{dt}\, v_r(t)$ and a *radius* $R(t) = \int dt (\, v_r(t) \; - \; \nu \,)$. Care must be exercised in choosing a convenient averaging to obtain a stable numerical derivative, and in selecting the value for the constant ν in the integration to eliminate a secular variation in R). A stereoscopic plot of the orbit $\mathbf{X}(t) = (\; R(t) \;,\; v_r(t) \;,\; a(t) \;)$ in the trial $3D$ phase space is represented in Fig. 5. Although an animation of the $3D$ representation is required to convey a proper information on the structure of the carrier of this trajectory, the illustration is sufficient to indicate that the orbit is in fact carried by a fuzzy plane of the form $a = KR + Q$, K, Q constants. We mention at this stage that the existence of a fuzzy plane carrier of the orbit traced out in the second order return maps for *o Ceti* was reported in the literature (Perdang 1991).

For comparison we also show a corresponding *chaotic orbit* in a $3D$ phase space (Fig. 6) which we have computed from a representative Normal Form (Perdang 1992) or equivalently, from a parametrised one–zone model (of the form devised in Tanaka and Takeuti, 1988); we have adjusted the parameters to give rise to the type of chaotic behaviour as observed in full radial hydrocodes (Kovacs and Buchler 1988). A mere visual inspection of both figures makes it clear that the CA result does not exhibit the typical self–similar pattern characteristic of low–dimensional chaos. The complex pulsation behaviour of our LG is thus *not interpretable as low dimensional chaos.*

Finally, in Fig. 7a we reproduce a result from a trial phase space analysis due to Kollath and Szeidl (1993), and obtained from the observations of the *W Virginis* Variable *RU Cam*, namely the projection of the orbit in the plane $X_1(t) = s(t)$, $X_2(t) = s(t + \tau)$, with $s(t)$ the smoothed observed signal (delay $\tau \approx \frac{1}{4}$ cycle length). This observational orbit is compared with the projection into the plane $X_1(t) = v_r(t)$, $X_2(t) = R(t)$ of our LG orbit (Fig. 7b). The structural similarity of both figures is undeniable. For other Variable Stars (*Miras*) we have likewise found similarities between the return maps of first and second order of observational data on the one hand and our LG simulations on the other (Perdang 1993).

The $3D$ phase space reconstruction (Fig. 5) suggests a simple interpretation of the structure of the theoretical signal $v_r(t)$. The projection of the orbit in the $a(t) \; - \; R(t)$ plane (directly detectable in the stereoscopic plot) is, in a rough approximation, a straight line segment of negative slope, or, with straightforward notations,

$$a(t) = -[K + \Delta K_o(t)]\Delta R(t) \; + \; \Delta K_1(t), \quad K + \Delta K_o(t) \; > \; 0; \qquad 58$$

where $\Delta K_o(t)$, $\Delta K_1(t)$, are time–dependent zero–mean fluctuating terms. If we disregard the latter, the relation collapses into the equation of motion of the simplest oscillatory time behaviour, namely a *monoperiodic oscillation*

$$\Delta R(t) = \Delta R_o \cos(\frac{2\,\pi}{P} t + \phi) \;, \quad K = (\frac{2\,\pi}{P})^2; \qquad 59$$

(ΔR_o, amplitude; P, period; ϕ, phase). Although we have already noted that our LG (radial) motions cannot be regarded as being monoperiodic, the fact that relation (58) holds, indicates that we can interpret these motions as being regulated by a main radial mechanism (responsible for the order of magnitude of the cycle length). The latter is disturbed by nonradial motions whose existence is directly observed in the numerical experiments. The nonradial contributions in a real star are set up either as a result of direct instabilities of the equilibrium against nonradial disturbances, or as instabilities of the already established radial motions against nonradial disturbances. In our CA simulation there may be a further effect, namely 'thermal' noise. Although qualitatively correct, the effect of this noise could be quantitatively spurious, since, as a result of the small scale of our experiments, thermal noise is indeed strongly amplified with respect to real thermal noise in genuine stars.

7. Outlook

Although, at this stage, the work summarised in this paper manifestly remains of an exploratory nature, we are entitled to draw one strong conclusion. The fact that our admittedly rudimentary numerical simulations show a closer qualitative agreement with the observations of the cooler variable stars than the results of the currently most elaborate radial hydrocode integrations (cf Figs. 7) lends support to the view that nonradial contributions cannot be disregarded in the numerical treatment of the pulsations of these stars. On the quantitative side, it is clear, however, that in the context of CA simulations only very large scale LG experiments can really decide whether the nonradial effects are sufficient to account for the current discrepancies between observations and conventional radial stellar fluid dynamics.

At a quantitative stage, satisfactory CA experiments should be based on an LG model satisfying the isotropy requirement (extensions of the FHP gas); they also should include an improved treatment of the geometry of the star (curvature effect), and they should take account of self-gravitation. To provide for a less constrained intensity of the gravitational field, a higher level CA formulation (Perdang 1993a) of the force effect seems to be required. Details of the matter–radiation interaction (precise opacity table) are needed to obtain an observationally significant equation for the surface luminosity.

On a more fundamental level, we point out that hydrodynamics possesses specific topological properties (invariants) related to the dimensionality of the embedding space. For instance, as is already obvious intuitively, incompressible fluid flow in 2 dimensional carrier spaces is subject to $2D$ integral constraints, differing essentally from the conservation laws constraining 3–dimensional flow (general results are given in Arnold and Khesin 1992). Therefore, even qualitatively a $2D$ treatment cannot be regarded as a really satisfying framework for genuine stellar pulsation dynamics. A full $3D$ formulation seems absolutely necessary for a meaningful comparison of real observational data with theory.

References

Arnold V.I., Khesin B.A. 1992 *Annual Review of Fluid Mechanics* **24** 146

Bak P., Tang C., Wiesenfeld K. 1987 *Physical Review Letters* **59** 381

Bak P., Tang C., Wiesenfeld K. 1988 *Physical Review* **A38** 364

Baker N., Kippenhahn R. 1962 *Zeitschrift für Astrophysik* **54** 114

Benz W. 1990 in *The Numerical Modelling of Nonlinear Stellar Pulsations*, NATO ASI **C 302** 269

Benz W. 1991 in *Proceedings of the Second European Predoctoral Astrophysical School on Theory and Observation of Late Stages of Stellar Evolution, and Computational Methods in Astrophysical Fluid Dynamics*, editor: C. de Loore, Springer, Berlin

Bertin G., Radicati L.A. 1976 *The Astrophysical Journal* 206 815

Boris J.P. 1989 *Annual Review of Fluid Mechanics* **21** 345

Buchler J.R., Goupil M.J., Kovacs G. 1993 *Astronomy and Astrophysics* (submitted)

Buchler J.R., Kovacs G. 1987 *The Astrophysical Journal (Letters)* **320** L57

Cannizzo J., Goodings D. 1988 *The Astrophysical Journal* **334** L31

Cannizzo J., Goodings D., Mattei J.A. 1990 *The Astrophysical Journal* **357** 235

Canuto C., Hussaini M.Y., Quarteroni A., Zang T.A. 1988, *Spectral Methods in Fluid Dynamics*, Springer, Berlin

Chen H., Chen S., Doolen G.D., Lee Y.C., Rose H. 1989 *Physical Review* **A40** 2850

Chen S., Lee M., Zhao K.H., Doolen G.D. 1989 *Physica* **D37** 42

Cox A.N., King D.S., Dodson S.W. 1979 *The Astrophysical Journal* **228** 201

Däppen W., Perdang J. 1984 *Memorie della Società Astronomica Italiana* **55** 299

Eastwood J.W. 1986 in *Astrophysical Radiation Hydrodynamics*, editors: Winkler K.H., Norman L., 415, Reidel, Dordrecht

Frisch U., Hasslacher B., Pomeau Y. 1986 *Physical Review Letters* **56** 1505

Frisch U., d'Humières D., Hasslacher B., Lallemand P., Pomeau Y., Rivet J.P. 1987 *Complex Systems* **1** 649

Glasner A., Buchler R. 1993 *Astronomy and Astrophysics* (in press)

Greenspan D. 1980 *Arithmetic Applied Mathematics*, chapt 1, 2 , Pergamon, Oxford

Griffiths D.F. 1986 in *Astrophysical Radiation Hydrodynamics*, editors: Winkler K.H., Norman L., 327, Reidel, Dordrecht

Hachisu I. Y. 1991 *Numerical Astrophysics (Japan)* **2** 43

Hardy J., de Pazzis O., Pomeau Y. 1976 *Physical Review* **A13** 1949

Kadanoff L. 1976 in *Phase Transitions and Critical Phenomena* **5a** , editors: Domb C., Green M.S., 1 , Academic Press, London

Kollath Z. 1990 *Monthly Notices of the Royal Astronomical Society* **247** 377

Kollath M., Nuspl J. 1991 in *Rapid Variability of OB Stars: Nature and Diagnostic Value*, ESO Conference and Workshop Proceedings **36** 363

Kollath Z., Szeidl B. 1993 *Astronomy and Astrophysics* (in press)

Kovacs G., Buchler J.R. 1988 *The Astrophysical Journal* **334** 971

Lèbre A., Gillet D. 1992 *Astronomy and Astrophysics* **255** 221

Ledoux P., Walraven Th. 1958 *Handbuch der Physik* **51** 353, Springer, Berlin

Lynden-Bell D., Wood R. 1968 *Monthly Notices of the Royal Astronomical Society* **138** 496

Marcus P.S. 1986 in *Astrophysical Radiation Hydrodynamics*, editors: Winkler K.H., Norman L., 359 and 387, Reidel, Dordrecht

Molvig K., Donis P., Miller R., Myczkowski J., Vichniac G. 1990 in *Cellullar Automata and Modelling of Complex Physical Systems* **206**, Springer, Berlin

Müller E. 1991 in *Supernovae*, Les Houches Summer School, Session **LIV**, editors: Audouze J., Bludman S., Mochkovitch R., Zinn-Justin J., Elsevier

Norman M.L., Winkler K.H. 1986 in *Astrophysical Radiation Hydrodynamics*, editors: Winkler K.H., Norman L., 187, Reidel, Dordrecht

Perdang J. 1985 in *Chaos in Astrophysics*, NATO ARW **C161** 11

Perdang J. 1990 in *Cours de Structure Interne, Aussois, 2e Session*, editors: Hubert A.M., Schatzman E., 311, Observatoire de Paris

Perdang J. 1991 in *Rapid Variability of OB Stars: Nature and Diagnostic Value*, ESO Conference and Workshop Proceedings **36** 349

Perdang J. 1992 in *Nonlinear Phenomena in Stellar Variability*, Proceedings of the IAU Colloquium **134** (in press)

Perdang J. 1993 *Astrophysics and Space Science* (submitted)

Perdang J. 1993a (these Proceedings)

Peyret R., Taylor T.D. 1983 *Computational Methods For Fluid Flow*, Springer, New York

Porter D.H., Woodward P.R., Yang W., Qi Mei 1990 *Annals of the New York Academy of Sciences* **617** 234

Saijo K. 1988 *Bulletin of the National Science Museum (Japan)* **E11** 1
Saijo K., Watanabe M. 1987 *Bulletin of the National Science Museum (Japan)* **E10** 1
Takeuti M. 1989 in *In the Numerical Modelling of Stellar Pulsations*, NATO ARW **C302** 121
Tanaka Y., Takeuti M. 1988 *Astrophysics and Space Science* **148** 227
Thirring W. 1970 *Zeitschrift für Physik* **235** 339
Thirring W. 1972 *Essays in Physics* **4** 125
Todo Y. 1991 *Numerical Astrophysics (Japan)* **2** 193
Uchida Y. 1991 *Numerical Astrophysics (Japan)* **2** 219
Wolfram S. 1986 *Journal of Statistical Physics* **45** 471
Zienkiewicz O.C. 1977 *The Finite Element Method*, McGraw–Hill, London

SMOOTH PARTICLE HYDRODYNAMICS AND SUPERNOVA EXPLOSIONS

Willy Benz
Steward Observatory, University of Arizona
Tucson, AZ 85721, USA

and

Marc Herant
Astronomy Board/Lick Observatory, University of California
Santa Cruz, CA 95064, USA

ABSTRACT

We present a review of the Smooth Particle Hydrodynamics (SPH) method in which we show how to derive the appropriate algebraic expressions corresponding to the mass, momentum and energy conservation equations describing the dynamics of fluids. We also discuss briefly a number of technical aspects of the method. In order to illustrate the capabilities of this technique we present calculations of two types of hydrodynamical instabilities taking place in type II supernova explosions. The first instability is a classical Rayleigh-Taylor instability occurring behind the outward moving shock and leading to an extensive chemical mixing of the ejecta. The second instability consists of a neutrino driven convection which is identified as a primary ingredient for a successful explosion.

1. Introduction

It has been now about fifteen years since the first paper by Lucy[1] describing the Smooth Particle Hydrodynamics (SPH) technique appeared in the literature. During these years SPH has grown substantially, becoming a mature hydrodynamic method able to compete, sometimes successfully, with most other numerical schemes. Much of the development of SPH is due to the work of Gingold and Monaghan[2,3], Monaghan and Gingold[4] and Monaghan[5,6].

SPH is a fully Lagrangian scheme to solve the usual conservation equations of fluid dynamics. Its main advantage reside in its ability to evaluate spatial derivatives without the use of an underlying grid. Hence, the method avoids the usual problem associated with multi-dimensional Lagrangian schemes, namely the rapid distortion of the mesh, without having to deal with the additional complexity introduced by the advection terms associated with an Eulerian approach. In addition, its gridless nature allows the method to follow arbitrary deformations without having to constrain the flow to remain within a computational box. Whereas spatial derivatives are computed in a novel manner, time integration proceeds in a usual way by using either a leap-frog algorithm or a low order Runge-Kutta integrator.

SPH is essentially a method of interpolation and approximation rather than of finite differencing. As with any interpolation method, the local value of a quantity is determined by its value at points in the direct vicinity. Points at which the characteristics of the flow are known are called particles, hence local gradients will be estimated from the knowledge of the flow variables at neighboring particles. One of the main technical difficulty of SPH lies precisely in determining efficiently the subset of all particles which contribute to the local properties. A naive approach leads to a scheme for which the computer time requirement scales like N^2 limiting its use to only a few thousand particles. With the introduction of hierarchical tree methods[7,8,9] originally designed for N-body calculations, SPH has experienced a real explosion in usage. These methods can be easily adapted for use in SPH codes[10,11] speeding up tremendously the calculation of the gravitational forces and allowing easily for a variable smoothing length, the SPH analog of an adaptive grid.

This paper is divided in two parts. We start by reviewing the fundamentals of the method (section 2). We follow by deriving the relevant algebraic expressions describing the conservation of mass, momentum and energy (section 3) and by discussing various technical aspects (section 4). As it is impossible to describe all the variants of SPH that have been proposed, we limit ourselves to what we could call the "standard" version. As an illustration of the capabilities of the method, we present in the second part calculations related to hydrodynamical instabilities occurring at various stages during the explosion of a supernova.

2. Kernel Estimation

As mentioned above, SPH can be understood in the framework of interpolation theory. A complete derivation is beyond the scope of this paper, but we present some of the basic features which are necessary to understand the method. Let us start by considering the following function :

$$\langle f(\vec{r})\rangle = \int W(\vec{r}-\vec{r}\,',h)f(\vec{r}\,')d\vec{r}\,' \tag{1}$$

where W is a function that we call the kernel and h is a measure of the width of the kernel. We further impose a normalization condition on W:

$$\int W(\vec{r},h)d\vec{r} = 1 \tag{2}$$

From Eqs. (1) and (2) it follows that:

$$\langle f(\vec{r})\rangle \xrightarrow{h\to 0} f(\vec{r}) \tag{3}$$

Eq. (1) therefore defines the kernel estimate $\langle f(\vec{r})\rangle$ of $f(\vec{r})$. In particular if $W(\vec{r}-\vec{r}\,',h) = \delta(\vec{r}-\vec{r}\,')$ than, obviously, we have the equality $\langle f(\vec{r})\rangle = f(\vec{r})$.

By considering the fact that W is a function strongly peaked at $\vec{r}\,' = \vec{r}$ one is allowed to expand $f(\vec{r})$ in a Taylor series in Eq. (1) about $\vec{r}$. If W is chosen to be an even function of r only (i.e. spherically geometric kernel), that is $W(\vec{r},h) = W(|\vec{r}|,h)$, than one can show that the following relation applies:

$$\langle f(\vec{r})\rangle = f(\vec{r}) + c\ (\vec{\nabla}^2 f)\ h^2 + \ O(h^3) \tag{4}$$

where c is independent of h. Hence, approximating $f(\vec{r})$ by $\langle f(\vec{r})\rangle$ is said to be second order accurate in h. Notice that since the term in h^2 is multiplied by the gradient of f, this implies perfect equality for constant or linear functions. Notice also that for the term in h^2 to be small, the function f should not be discontinuous or having too large gradients over the size of W. In fact, if this were the case, the Taylor expansion, used above to derive the expression, would be unjustified to begin with.

If we now assume that the flow variables are known on a finite number of discrete points only, distributed in space according to the number density distribution:

$$n(\vec{r}) = \sum_{j=1}^{N} \delta(\vec{r} - \vec{r}_j) \tag{5}$$

we can use Eq. (1) to derive the following expression:

$$\langle f(\vec{r})\rangle = \sum_{j=1}^{N} \frac{m_j}{\rho(\vec{r}_j)} f(\vec{r}_j) W(|\vec{r} - \vec{r}_j|, h) \tag{6}$$

where we have written the number density at $\vec{r}_j$, $\langle n(\vec{r}_j)\rangle$, in the form:

$$\langle n(\vec{r}_j)\rangle = \rho(\vec{r}_j)/m_j \tag{7}$$

Notice that this procedure actually introduces the concept of particles of mass m_j.

If now, for example, we choose in Eq. (6) $f(\vec{r}) \equiv \rho(\vec{r})$ we get the following expression for the fluid density:

$$\langle \rho(\vec{r})\rangle = \sum_{j=1}^{N} m_j W(|\vec{r} - \vec{r}\,'|, h) \tag{8}$$

This expression can be interpreted as follows. Every particle of mass m_j is smoothed in space according to W which can be regarded as its density distribution in space. The density at any point in space is then obtained by summing up the contributions from all particles at that point. In fact, the name of the method, Smooth Particle Hydrodynamics, actually derives from this interpretation.

Notice that to compute the fluid density only the particle locations and their masses are needed. The continuity equation is satisfied automatically provided the masses of the particles are constant and none are lost. Furthermore, since the density depends on the locations of neighboring particles this means that not only edge effects are inevitable but also that setting up initial conditions with arbitrary density distribution is not a totally trivial matter.

3. Hydrodynamical Equations

In this section we recall the derivation of the equations of SPH starting from the hydrodynamic conservation equations. As mentioned above, we limit ourselves to the description of the minimum set of terms necessary to model astrophysical self-gravitating fluids. Many modifications can be (and have been) made to these basic equations in order to include more physics or address particular problems. However, the following equations can be considered as forming the real backbone of the method and except for self-gravity they will be present in one form or another in all implementations of SPH.

3.1 Momentum Conservation

In this section, we briefly recall how the momentum conservation equation is cast in the SPH formalism. For more details the reader is referred to the existing literature[12,13]. We start with Euler's equation

$$\frac{\partial \vec{v}}{\partial t} + (\vec{v}\cdot\vec{\nabla})\vec{v} = -\frac{\vec{\nabla}P}{\rho} \tag{9}$$

Multiplying all terms by the kernel and integrating over the solution space V yields:

$$\int_V \left(\frac{\partial \vec{v}}{\partial t} + (\vec{v}\cdot\vec{\nabla})\vec{v}\right) W(|\vec{r}-\vec{r}\,'|,h)d\vec{r}\,' = -\int_V \frac{\vec{\nabla}P}{\rho} W(|\vec{r}-\vec{r}\,'|,h)d\vec{r}\,' \tag{10}$$

where now all variables are taken to be functions of $\vec{r}\,'$.

For the sake of simplicity let us now assume that the smoothing length h is constant in both space and time. Notice that the ideal case would be to allow h to vary in both space and time but this introduces complicated correction terms that will be not be discussed here. With this assumption we rewrite the left-hand-side of Eq. (10) in the following way:

$$\int_V \frac{\partial \vec{v}}{\partial t} W(|\vec{r}-\vec{r}\,'|,h)d\vec{r}\,' + \int_V (\vec{v}\cdot\vec{\nabla})\vec{v}W(|\vec{r}-\vec{r}\,'|,h)d\vec{r}\,' = \frac{\partial\langle\vec{v}\rangle}{\partial t} + (\langle\vec{v}\rangle\cdot\vec{\nabla})\langle\vec{v}\rangle = \frac{d\langle\vec{v}\rangle}{dt} \tag{11}$$

where any quantity (scalar or vector) $\langle f(\vec{r})\rangle$ is defined by Eq. (1). The expression for the advection term above has been obtained by a Taylor expansion of the integrand. The right hand-side of Eq. (10) can be estimated integrating the expression by parts followed by a similar Taylor expansion of the integrand. Assuming that the boundaries of the computational volume are taken at infinity or that pressure vanishes on the boundary we obtain the following expression

$$\frac{d}{dt}\langle\vec{v}\rangle = -\vec{\nabla}_r\left\langle\frac{P}{\rho}\right\rangle - \left(\frac{P}{\rho^2}\right)\vec{\nabla}_r\langle\rho\rangle \tag{12}$$

We now apply the results of section 2 to rewrite Eq. (12) in terms of sums. The acceleration of particle i is given after regrouping all terms by:

$$\frac{d\vec{v}_i}{dt} = -\sum_{j=1}^{N} m_j \left(\frac{P_i}{\rho_i^2} + \frac{P_j}{\rho_j^2} \right) \vec{\nabla}_i W(|\vec{r}_i - \vec{r}_j|, h) \tag{13}$$

where $\vec{v}_i \equiv \langle \vec{v} \rangle_i$. Eq. (13) describes momentum conservation in the SPH formalism.

It is easy to show that Eq. (13) conserves momentum exactly. We notice that forces are evaluated in pairs, so that the force due to particle j on particle i is given by:

$$m_i \left(\frac{d\vec{v}_i}{dt} \right)_j = m_i m_j \left(\frac{P_i}{\rho_i^2} + \frac{P_j}{\rho_j^2} \right) \vec{\nabla}_i W(|\vec{r}_i - \vec{r}_j|, h) \tag{14}$$

but from the definition of W we have $\vec{\nabla}_i W(|\vec{r}_i - \vec{r}_j|, h) = -\vec{\nabla}_j W(|\vec{r}_i - \vec{r}_j|, h)$ so that Eq. (14) can be rewritten as:

$$m_i \left(\frac{d\vec{v}_i}{dt} \right)_j = -m_j m_i \left(\frac{P_i}{\rho_i^2} + \frac{P_j}{\rho_j^2} \right) \vec{\nabla}_j W(|\vec{r}_i - \vec{r}_j|, h) = -m_j \left(\frac{d\vec{v}_j}{dt} \right)_i \tag{15}$$

which ensures total momentum conservation. Notice also that angular momentum is conserved locally as well since all forces are always directed along a line joining the center of the particles.

Finally, the sum in Eq. (13), although formally extending on all particles, is in practice determined only by particles for which $\vec{\nabla}_i W(|\vec{r}_i - \vec{r}_j|, h)$ has a non-negligible contribution which is in turn determined by the choice of the kernel. This reflects the fact that pressure forces are indeed short range forces and that only neighboring particles will contribute to the local pressure gradients. Because of this, the efficiency of an SPH code will be determined by the speed at which all particles within a fixed radius around any given particle can be found.

3.2 Energy Conservation

Energy conservation is achieved using the first law of thermodynamics which for the adiabatic case is written

$$\frac{\partial u}{\partial t} + (\vec{v} \cdot \vec{\nabla}) u = -\frac{P}{\rho} \vec{\nabla} \cdot \vec{v} \tag{16}$$

where u is the specific internal energy of the fluid.

In proceeding the same way as for the momentum equation, that is by multiplying by W and integrating over the solution domain whose boundaries are taken to be at infinity and transforming integrals in sums, we get the following equation:

$$\frac{du_i}{dt} = \frac{P_i}{\rho_i^2} \sum_{j=1}^{N} m_j (\vec{v}_i - \vec{v}_j) \vec{\nabla}_i W(|\vec{r}_i - \vec{r}_j|, h) \tag{17}$$

where as before we define $u_i \equiv \langle u \rangle_i$. The right-hand side is nothing else, of course, than the SPH representation of the PdV/dt term. One also notes for future use that:

$$\rho\vec{\nabla}\cdot\vec{v} = -\sum_{j=1}^{N} m_j(\vec{v}_i - \vec{v}_j)\vec{\nabla}_i W(|\vec{r}_i - \vec{r}_j|, h). \tag{18}$$

As with the momentum conservation equation, one easily shows that total energy is indeed conserved in the following way. The variation of the total internal energy is given by:

$$\frac{d}{dt}\sum_{i=1}^{N} m_i u_i = \sum_{j=1}^{N} m_i \frac{du_i}{dt} = \sum_{i=1}^{N}\sum_{j=1}^{N} m_i m_j \frac{P_i}{\rho_i^2}(\vec{v}_i - \vec{v}_j)\vec{\nabla}_i W(|\vec{r}_i - \vec{r}_j|, h)$$

by expanding the right-hand-side term, interchanging indexes and making use of the identity $\vec{\nabla}_i W(|\vec{r}_i - \vec{r}_j|, h) = -\vec{\nabla}_j W(|\vec{r}_i - \vec{r}_j|, h)$ we obtain after some rearranging of terms:

$$\frac{d}{dt}\sum_{i=1}^{N} m_i u_i = \sum_{i=1}^{N}\sum_{j=1}^{N} m_i m_j \left(\frac{P_i}{\rho_i^2} + \frac{P_j}{\rho_j^2}\right)\vec{v}_i \vec{\nabla}_i W(|\vec{r}_i - \vec{r}_j|, h)$$

by comparing this expression to Eq. (15) one gets:

$$\frac{d}{dt}\sum_{i=1}^{N} m_i u_i = \frac{d}{dt}\left(\frac{1}{2}\sum_{i=1}^{N} m_i v_i^2\right) \tag{19}$$

which shows that the work done by pressure forces is at the expense of internal energy and thus ensures the conservation of total energy.

3.2 Dissipation

So far we have only described the inviscid equations. It is well known that these equations are unable to describe shocks and that large, unphysical, oscillations in the field variables will result if one still dares to try. It transpires that there is no mechanism built in so far that allows for the dissipation of kinetic energy into heat. Such a mechanism is always present in nature due to the intrinsic viscosity of the fluid. No matter how small the viscosity is, it will always be sufficient to dissipate kinetic energy in a shock since the gradient of the variables across the shock will adjust itself so as to allow for it. Since in most astrophysical situations the molecular viscosity of the fluid is extremely small, dissipation should, ideally, only occur in shocks. In SPH, this has been done by introducing the standard artificial viscosity term. As usual, this method consists in assigning a completely artificial viscosity to the fluid and modifying the equations of momentum and energy conservation accordingly. Two different types of viscosity are usually introduced that both translate, when introduced in the momentum equation, into an artificial viscous pressure term of the form:

$$\text{bulk viscosity} \quad \Pi_l = -\alpha\rho l c_s \vec{\nabla}\cdot\vec{v} \tag{20}$$

$$\text{von Neumann} - \text{Richtmyer} \quad \Pi_q = \beta\rho l^2(\vec{\nabla}\cdot\vec{v})^2 \tag{21}$$

where α, β are free parameters, l a typical length scale over which the shock is spread and c_s the sound speed. Note that both artificial viscous stresses include the term $\vec{\nabla}\cdot\vec{v}$. We could use Eq. (18) to express $\vec{\nabla}\cdot\vec{v}$ but it has been shown by various authors[4] that this still leaves unacceptably strong, shortwavelength postshock oscillations in the field variables. To avoid this we follow[4] and define first an estimate of the velocity divergence at particle i due to the presence of particle j by:

$$\mu_{ij} = \frac{h(\vec{v}_i - \vec{v}_j)\cdot(\vec{r}_i - \vec{r}_j)}{|\vec{r}_i - \vec{r}_j|^2 + \epsilon h^2} \tag{22}$$

where the term ϵh^2 has been added to avoid divergence for small $|\vec{r}_i - \vec{r}_j|$. The corresponding pressure due to both artificial viscosities (bulk and von Neumann-Richtmyer) will be written as:

$$\Pi_{ij} = \begin{cases} \frac{-\alpha c_{ij}\mu_{ij} + \beta\mu_{ij}^2}{\rho_{ij}}, & \text{if } (\vec{v}_i - \vec{v}_j)\cdot(\vec{r}_i - \vec{r}_j) \le 0.; \\ 0. & \text{otherwise.} \end{cases} \tag{23}$$

where we have defined $c_{ij} = 0.5(c_i + c_j)$ and $\rho_{ij} = 0.5(\rho_i + \rho_j)$. This ensures that this term remains symmetric in i and j hence ensures momentum conservation. The condition that Π_{ij} be zero for receding particles makes sure that the dissipation modeled by this artificial viscosity only adds entropy to the system.

The final conservation equations can now be written by introducing the artificial viscous pressure term in the equation of momentum conservation Eq. (13)

$$\frac{d\vec{v}_i}{dt} = -\sum_{j=1}^{N} m_j \left(\frac{P_i}{\rho_i^2} + \frac{P_j}{\rho_j^2} + \Pi_{ij}\right)\vec{\nabla}_i W(|\vec{r}_i - \vec{r}_j|, h) \tag{24}$$

and the corresponding contribution to the energy equation Eq. (17)

$$\frac{du_i}{dt} = \frac{P_i}{\rho_i^2}\sum_{j=1}^{N} m_j(\vec{v}_i - \vec{v}_j)\vec{\nabla}_i W(|\vec{r}_i - \vec{r}_j|, h) + \frac{1}{2}\sum_{j=1}^{N} m_j \Pi_{ij}(\vec{v}_i - \vec{v}_j)\cdot\vec{\nabla}_i W(|\vec{r}_i - \vec{r}_j|, h) \tag{25}$$

3.3 Self-gravity

In many astrophysical situations self-gravity can not be neglected. We derive in this section the appropriate modifications to the momentum conservation equation due to the introduction of gravitational forces. First, it is necessary to recall that interactions between particles are always computed in pairs, that W can be interpreted as an individual particle's density distribution and finally that particles are taken to be spherically symmetric. Keeping these points in mind, Poisson's equation can be solved to obtain the force exerted on particle i from particle j

$$-\left(\vec{\nabla}\phi_i\right)_j = -G\frac{M(|\vec{r}_i - \vec{r}_j|)}{|\vec{r}_i - \vec{r}_j|^2}\frac{(\vec{r}_i - \vec{r}_j)}{|\vec{r}_i - \vec{r}_j|} \tag{26}$$

where $M(|\vec{r}_i - \vec{r}_j|)$ is defined by the following expression

$$M(|\vec{r}_i - \vec{r}_j|) = 4\pi \int_0^{|\vec{r}_i - \vec{r}_j|} u^2 \rho(u) du = 4\pi \int_0^{|\vec{r}_i - \vec{r}_j|} u^2 W(u, h) du \qquad (27)$$

This definition introduces a new constraint on the kernel, namely

$$\int_0^{\infty} r^2 W(r, h) dr < \infty$$

in order to avoid an infinite mass. The total gravitational forces felt by particle i can now be written as the sum of all the individual forces

$$-\vec{\nabla}\phi_i = -G \sum_{j=1}^{N} \frac{M(|\vec{r}_i - \vec{r}_j|)}{|\vec{r}_i - \vec{r}_j|^2} \frac{(\vec{r}_i - \vec{r}_j)}{|\vec{r}_i - \vec{r}_j|} \qquad (28)$$

This term can now be added to the momentum conservation Eq. (24) so that body forces are now taken into account properly. Notice that, contrary to all previous expressions, the sum in Eq. (28) applies to all particles as the terms do not vanish with increasing distance. This reflects the fact that gravity is a long range force. In practice, if this summation is not done in an efficient way, the method will scale like N^2 and therefore lose all its potential applicability because of the tremendous computer time that will be required even for a modest number of particles. The use of hierarchical tree methods is probably the most efficient way to compute the gravitational forces since they scale like $N log N$ and allow to rapidly find all neighboring particles at the same time.

4. Technicalities

In this section we briefly review a number of technical points associated with the method. As it is often the case, it is precisely those apparently small technicalities that end up making a significant difference in the performance of a particular code. Again, this section has not the pretension to be comprehensive but merely to point out some interesting technical aspects of the method.

4.1 Variable Smoothing Length

The equations discussed so far have been derived with the assumption that h is constant in both space and time. To make an analogy with finite difference codes, this is the equivalent of a fixed grid. It is clear that this will lead to resolution problems if the typical length scale of the problem changes by large amounts. These limits on the dynamical range achievable by the method are actually quite severe. For example, a lower limit in density is reached when a particles has lost all its neighbor and is given by

$$\langle \rho(\vec{r})_{i,\ min} \rangle = m_i W(0, h)$$

It is important to realize that by introducing a smoothing length dependent on position (and therefore time) new terms are introduced in the equations since:

$$\vec{\nabla}_{r'} W(\vec{u}; h(\vec{r}, \vec{r}')) = -\frac{1}{u}\frac{\partial W}{\partial u}\vec{u} + \frac{\partial W}{\partial h}\vec{\nabla}_{r'} h \tag{4.1}$$

where we haved used $\vec{u} = \vec{r} - \vec{r}'$. The second term on the right-hand-side of this equation is proportional to the gradient of the smoothing length. A similar term also appears when time derivatives are taken. These new terms should ideally be incorporated in all SPH equations derived above. In practice, these terms are difficult to estimate and have only been discussed very little[14]. The general approach is to simply neglect them which is probably correct as long as the smoothing length varries little among particles contributing to the local average.

The actual value of h can be derived in a natural way by asking that the number of neighbors to any given particle remains about constant. This requirement can be expressed (in 3D)

$$h = h_0 \left(\frac{\rho_0}{\rho}\right)^{\frac{1}{3}} \tag{29}$$

where ρ_0 and h_0 are constants. In practice, this scaling poses the problem that h is needed to compute density. Hence, Eq. (29) is actually an implicit equation. In order to circumvent these difficulties, we take the time derivative of Eq. (29) and obtain using continuity equation

$$\frac{dh}{dt} = \frac{1}{3} h \vec{\nabla} \cdot \vec{v} \tag{30}$$

This differential equation is added to the system and is solved along with the hydrodynamics equation. Notice that this requires almost no additional computer time since the divergence of the velocity is already computed in order to solve the energy equation. Alternate schemes have been proposed to vary h, see for example [11,15]

4.2 The Kernel

So far we have discussed the SPH formalism without specifying the kernel W. Along the way we added however, a number of constraints that any kernel should satisfy. The most used kernels satisfying all these conditions and published in the literature are listed below.

Exponential kernel[16]

$$W(r,h) = \frac{1}{8\pi h^3} exp(-v) \tag{31}$$

Gaussian[2]

$$W(r,h) = \frac{1}{\pi^{3/2} h^3} exp(-v^2) \tag{32}$$

Spline[17]

$$W(r,h) = \frac{1}{\pi h^3} \begin{cases} 1 - \frac{3}{2}v^2 + \frac{3}{4}v^3, & \text{if } 0 \le v \le 1; \\ \frac{1}{4}(2-v)^3, & \text{if } 1 \le v \le 2; \\ 0. & \text{otherwise.} \end{cases} \tag{33}$$

where we have defined $v = \frac{r}{h}$.

The following brief comments can be made. All three kernels interpolate to second order in h, and all three are always positive in the range of interest. The advantage of Eq. (33) is that it is defined on compact support. This provides a clear limit to the number of neighbors to be considered since beyond $v = 2$ all contributions are identically zero. In the use of Eq. (31) or Eq. (32) an artificial cut-off in v has to be introduced to avoid an N^2 calculation.

4.3 Time Integration

As mentioned already in the introduction, the novelty and the strength of SPH resides in the way the spatial derivatives in the hydrodynamics equations are computed. Time stepping in SPH codes on the other hand is done in the exact same way it is done in all explicit hydrodynamics codes, namely using a standard leap-frog algorithm[11] or a predictor-corrector method[18]. Both integration methods lead to second order accuracy in time integration although only one force evaluation per time step is required. The size of the time step is determined from the standard CFL [19] condition to guarantee the stability of the scheme. A further very interesting variation has been introduced[11] in which particles are integrated according to their own time step. This method, derived from standard practice in N-body methods, is clearly very advantageous if multiple, largely different time scales are present in the same problem.

An alternative, somewhat unusual in numerical hydrodynamics, consists in using a second order Runge-Kutta integrator with adaptive time step[20]. A tolerance is given to which the integration has to be performed and the integrator adapts the time step to satisfy this tolerance. Although this approach requires two force evaluation per time step, the size of the time step is no longer directly determined by the CFL conditions but rather by the time derivatives of the forces. In many cases, we have found that this requirements allowed timestep significantly larger than would otherwise be allowed by the CFL condition thus making up for the extra computational requirements. Yet another advantage resides in the fact that all variables are defined at the same time and not in a staggered fashion as required by the leap-frog scheme. This makes the introduction of additional physics much easier as proper time centering is no longer an issue.

4.4 Finding Nearest Neighbors

As already mentioned, the usefulness of SPH is directly related to the speed at which the list of locally contributing neighbors can be found. All the efficient methods are too complex to be described in details here so we will limit ourselves to briefly mentioning them.

In the case of a constant smoothing length or in simulations in which the ratio between the maximum and minimum smoothing length remains small ($\leq 4-5$), the most advantageous method remains based on linked lists[21]. By overlying a grid over the particles with a mesh size exactly equal to the interaction length of the kernel, any locally contributing particle will be located in the same cell or one of the 26 (in 3D) adjacent cells. Linked lists can be built that allow to locate these particles with high efficiency. Besides the limitation to almost constant (in space) smoothing length, this method also suffers from the presence of the grid. Because of storage limitations an upper limit to grid sizes will be around 256^3. Since the mesh size of the grid is given by the interaction length, the amount of physical space covered by the grid is limited.

A more costly but more general way to determine the list of locally contributing neighbors relies on hierarchical trees[10,11]. These methods, by suitably organizing the data structure allow a fast (order $N \log N$) search for neighbors. Using these tree structures naturally fits into the general philosophy of SPH namely a gridless approach to hydrodynamics. There are no spatial limitations of any kind introduced by these techniques. The only real penalty is yet an additional level of complexity.

5. Hydrodynamical Instabilities in Supernova Explosions

The fortunate advent of supernova 1987A (SN 1987A) in the relatively nearby Large Magellanic Cloud, the first naked eye supernova in nearly five centuries, has radically changed our views on how massive stars explode and die[22]. The wealth of data collected during the months following the explosion of SN 1987A has given us a unique opportunity to refine our current understanding of Type II supernovae (and SNae Ib) by directly confronting theory with observations. SN 1987A has permanently shattered the simplified picture of an SNII as a spherically symmetric expanding "onion skin" envelope, made up of distinct, well-separated shells of different chemical composition, which leave behind a neutron star in the center. The general message conveyed by the observations is that the neat spherical arrangement of concentric shells (corresponding to heavier and heavier elements as one goes deeper in the star and the nuclear burning has progressed further) of the pre-supernova star is destroyed by extensive mixing occurring in the explosion. This has naturally given a preeminent role to multidimensional hydrodynamics in attempts to interpret the observations.

Early evidence for mixing in SN 1987A was provided by attempts to model the light-curve (the time evolution of the luminosity) using one-dimensional radiation transport codes. Such attempts showed that it is impossible to reproduce the observed light-curve if chemical elements remain segregated during the explosion. In order to successfully explain the observations, it was found that it is necessary to assume a certain amount of chemical mixing across shells[23]. Additionally, the early detection at 200 days of γ-rays from the decay of ^{56}Co[24] is conclusive evidence of either holes in the ejecta or outward mixing of the very inner layers of the ejecta where ^{56}Ni (parent of ^{56}Co) originates through explosive nucleosynthesis. In an unmixed, spherically symmetric model, γ-rays would have been expected to start emerging from the ejecta after at least 500 days after the explosion[25]. This extensive mixing was later further substantiated by the detections of high velocity (> 3000 km $s^{-}-1$) wings in the infrared line profiles of ^{56}Fe[26,27] indicating the presence of iron at large radii in the expanding ejecta (note that after the shock has emerged from the star, the debris are coasting and the velocity distribution through the ejecta is proportional to the radius, so that position can be simply related to velocity: higher velocity means further out in the ejecta). Finally polarization measurements[28], speckle imaging[29] as well as systematic redshifts in infrared and x-ray line profiles all indicate significant, large scale, asymmetries in the ejecta.

Out of the work of numerous groups (for reviews see[22,30]) in investigating the multi-dimensional hydrodynamical behavior of the explosion and its connection to the observations, the following picture is beginning to emerge. During the initiation of the explosion by the "delayed mechanism", neutrinos deposit some of the energy liberated by the collapse of the core of the progenitor at the bottom of the stellar envelope. This leads to large scale, violent convective motions which are crucial to maintain the energy and matter supply to the outgoing shock and prevent a failure of the explosion. Minutes to hours after the beginning of the explosion, as the blast-wave propagates through the envelope, Rayleigh-Taylor (RT) instabilities develop at the interfaces separating the hydrogen, helium, and carbon-oxygen shells. These instabilities generate sizable extrusions of heavy elements into the outer shells. Finally, in the days which follow, the nickel produced by explosive nucleosynthesis in the first seconds of the explosion decays to cobalt and iron releasing mostly locally its radioactive energy in the form of heat. As a result, a huge, low-density cavity forms in the center of the expanding remnant.

In this part of the paper, we first describe our work in modeling the hydrodynamics of SN 1987A during and after the propagation of the shock through the stellar envelope, as RT instabilities and radioactive decay modify the homologous expansion. We show that although simulations do indeed show significant mixing, it is not sufficient by itself to satisfy observational constraints, and points to "premixing" during the early times of the explosion. Motivated by these results, we go on to present simulations of the post-collapse stage, at the time when the explosion is initiated, and show that indeed, a successful explosion will de facto be accompanied by violent convective motions. Most of the simulations presented here were performed with a two dimensional, cylindrical geometry, version of SPH, which is straightforward to obtain from the standard SPH formalism[31]. Finally we conclude with a discussion of the special advantages, disadvantages and future prospects faced by SPH methods in modeling the Type II Supernova problem.

5.1 Instabilities Associated with the Shock Propagation

The idea that instabilities might arise in the wake of a supernova blast wave is not new[32,33], but received a new impetus after the observations of SN 1987A showed that mixing had to have occurred in the ejecta. It was a happy coincidence that at the same time, both computational capabilities and techniques were coming to the stage where they could deal with such a challenge, at least in two dimensions. Early simulations[34,35] quickly showed that indeed, RT instabilities develop at the interfaces between chemical shells in a few hours as predicted by 1D stability analysis[36]. The question remained, though, that these simulations did not mix the nickel produced in the explosion far enough to account for the observations. However the simulations were not pursued long enough, in part because of limitations due to their grid-size, to model the effects of the decay of the nickel produced in the supernova explosion. It was then conjectured that nickel decay, by locally depositing energy in the form of heat, might boost the mixing appreciably in the weeks following the explosion. Fortunately, SPH is not limited by the extent of any grid, and we were able to perform simulations[37,31] of the explosion of a potential progenitor model of SN 1987A from five minutes to several months. Our main conclusion is that the radioactive decays do indeed modify the structure of the ejecta, but do not accelerate the nickel outward sufficiently to explain the observations.

Like other groups, we initiate the explosion by artificially depositing $1-2\times10^{51}$ ergs ($1-2$ foes) at the center of a one dimensional model of the progenitor star. We evolve the problem in 1D for 300 seconds with a classic grid-base lagrangian code, until the expanding shock wave has just penetrated in the hydrogen envelope. We then map the result to a 2D SPH representation matching the 1D density, energy, and velocity profile at 300 seconds. Initial random perturbations of amplitude $\pm10\%$ and with characteristic wavelength $\approx 3^{\circ}$ are introduced behind the shock to seed the instabilities. The SPH computational domain consists of a wedge centered on the equator with opening 60° and periodic boundaries. We do not run in the full 180° because of adverse numerical effects from the pathological z axis boundary. Note that since the angular size of the instability is much smaller than the opening angle, this restriction should not be a problem.

We display in Figure 1 two snapshots illustrating the results of one of our simulations, the first snapshot (Fig 1a) is taken 400 minutes after the begining of the explosion, whereas the second one (Fig. 1b) is taken 90 days after the explosion. Particles are colored according to chemical composition with red being ^{56}Ni, yellow representing the rest of the metals, green shows helium and finally blue particles trace hydrogen. Clearly, the RT instabilities affect primarily the interfaces between carbon/oxygen, helium, and hydogen regions (see Fig. 1a). In the original progenitor, such interfaces are the site of entropy (and therefore density) jumps resulting from different nuclear burning histories. These sudden density drops are still represented in the post-shock density profile as areas of strong outward decrease of density. As the shock begins to plow through the hydrogen envelope, it begins to slow down considerably and and communicates this decceleration to the postshock material via a pressure gradient, with pressure decreasing inward. These crossed gradients of pressure and density at the interfaces are what cause the growth of the RT instability. As the shock breaks out of the star, the instability ceases, and the RT mushrooms soon stop to grow.

As time goes by, ^{56}Ni decays into ^{56}Co which in turns decays into ^{56}Fe with a timescale of 77 days. This decay is followed by a release of energy that heats the material locally as long as the latter is optically thick, a reasonable approximation for the first 50 days or so following the explosion. This energy release turns out to be insufficient to dramatically affect the overall dynamics of the debris. However, it is enough to affect the internal structure of the ejecta (see Fig. 1b). As a result of the radioactive heating, nickel has swollen, filling the entire central cavity and crushing some of the RT fingers. Hence, despite its small mass (about 0.075 $M_\odot$) the nickel turns out to occupy a significant fraction of the volume surrounding small "nuggets" made of metals, helium or hydrogen.

However impressive the RT growth and subsequent "nickel bubble" may look, they are insufficient by far to explain the extent of nickel mixing required by the observations. As can be seen from Fig. 1a, if the nickel is assumed to be at the bottom of the ejecta 300s after the explosion, it is essentially still at the bottom of the ejecta after 400 minutes. Located too far away from the main driving regions (chemical interfaces) the nickel gets only slightly mixed by the otherwise large scale instabilities. Note that observations would require that a small fraction of the nickel be located close to the tip of the mushrooms. Given that this result is confirmed by two other groups using totally different numerical techniques[31,35] it is most probably physical. This has led us to postulate that an early mechanism, probably connected with the explosion mechanism itself which we did not model, might premix the nickel far enough out in radius that it would be able to participate in the RT instabilities and thus be mixed to the extent required by the observations. Notice that this premixing has to be quite large. From Fig. 1b we see that a small fraction of metals has been transported to the tip of the fingers where we would like to actually see some nickel. Hence, we conclude that close to the entire metal core has to be mixed *after* the nucleosynthesis of nickel.

Finally, there is another motivation linked to the RT instabilities to demand earlier instabilities. We have found, in agreement with other groups[31,35] that in order to show fully developed growth, the initial perturbations need to be at least of order ±5%, a rather large amount. Further, as is evident from the figures, the instability does not saturate, fingers are formed and as the driving force of instability shuts off (shortly after the shock emerges from the envelope), they coast to their ultimate extension without further increase of complexity. The dependence on initial perturbations and the non-saturation does not originate from the numerical method. In cases where the driving force of the RT does not shut off early, numerical experiments do show growth which is nearly independent of initial perturbations and which saturates. The conclusion then is that in the case of SN 1987A, the RT instability does not allow many e-folding times of growth. This leads us to claim that in order to yield the appropriate nickel mixing, the RT instability needs to be seeded by early instabilities during the initial phase of the explosion.

5.2 Instabilities Associated with the Explosion Mechanism

We have shown in the previous section how observations of SN 1987A seem to require an early break of the spherical symmetry of the explosion. For this reason, it seems natural to suspect the explosion mechanism itself, one of the thorniest problem of astrophysics through the last 30 years. We first give a brief description of the sequence of events which is generally assumed to lead to a SN II explosion, and then present some aspects of our work on the multi-dimensional aspects of the hydrodynamics.

As a massive star runs out of fuel at its center because the products of nuclear burning have reached the maximum nuclear binding energy allowed by nature (iron), the core contracts under its own gravity. Soon, the reaction $e^- + p \rightarrow \nu_e + n$ becomes important and the core both loses energy through escaping neutrinos and pressure support through the disappearance of degenerate electrons. The collapse runs away and nearly reaches the free-fall rate. As the density increases, neutrinos are trapped in the core, but since they are relativistic, $\gamma = 4/3$ and this is insufficient to halt the collapse which proceeds until nuclear densities are reached, at which time neutron degeneracy sets in. This sudden stop of the collapse at the center creates a bounce shock in the outer layer of the core as it crashes against what essentially is a hard wall. Subsequently the bounce shock moves out to a few hundred kilometers before being stalled by infalling matter. However, meanwhile neutrinos trapped in the proto-neutron star formed at the center start to leak out with a time scale of a second. Additionally, more neutrinos are produced by hot matter accreting onto the proto neutron star surface. A small fraction of these neutrinos heat the matter between the shock and the proto-neutron star surface; they potentially revive the stalled shock and produce a successful "delayed" explosion.

The potential for hydrodynamical instabilities in such a scenario is quite obvious. Neutrino energy deposition is a decreasing function of radius, hence it is mainly the matter in the close vicinity of the neutron star which is heated the most. This "heating from below" naturally induces convective motions which, if they are of large scale, radically change the one-dimensional picture outlined above.

To investigate this, we have performed in collaboration with S.A. Colgate a number of numerical simulations of the very early phases of the explosion of SN1987A. We obtained our initial conditions by using very simplified model for the core collapse of the progenitor. In this model, we assumed that at $t = 0$ a sound wave (rarefaction wave) propagates from the center and signals to every shell of matter it encounters to start falling inward at a fraction (2/3) of the free fall velocity[39] . We arbitrarily decide on a proto-neutron star mass, and stop the collapse when the proper amount of matter has reached the center. As for the RT problem, we build an SPH representation of the density, internal energy, and velocity profiles of the infalling envelope in two dimensions. We assume that the proto-neutron star surface is a solid wall, and accrete particles onto it only if they reach a density larger than 10^{11} g cm^{-3}. A time dependent neutrino flux is emerging from this inner boundary representing the cooling of the neutron star[40]. This neutrino flux heats the matter, mainly via absorptions by free nucleons (such as $\nu_e + n \rightarrow p + e^-$) and may eventually drive a successful explosion. At the same time as the matter is heated, it also cools by emitting neutrinos, mainly through the inverse of the process given above, namely the capture of electrons and positrons by protons and neutrons. However, for simplicity, we did not include these additional neutrinos in our heating term.

In all these neutrino processes, the energy transfer between (massive) nucleons and (light) leptons is negligible, so that we have $\epsilon_e \sim \epsilon_\nu$ neglecting the mass difference between proton and neutron. The cross-section for each process is $\propto \epsilon_e^2$ or $\propto \epsilon_\nu^2$. This energy dependence of the cooling rate means that the hotter the material, the higher the average electron energy, and therefore, the better the cooling.

However, since electrons are fermions, cooling will also depend on the degree of degeneracy of the electron gas. In particular, in a hydrodynamical situation, net cooling is compensated by compression to restore pressure equilibrium; if degeneracy sets in, the electron gas can be driven to higher and higher chemical potential and fermi energy as the cooling and the compression rate both runaway. Such behavior is illustrated by the fact that variations in the amount of time during which the infall proceeds (and thus variations of the proto-neutron star mass) before "hardening" the proto-neutron star surface and releasing the trapped neutrinos critically determines the failure or success of the explosion in our simulations. This timing is important because the infalling envelope gradually thins out as time passes, and thus the probability of a degenerate cooling runaway is reduced.

Figure 2a shows a snapshot at 32 ms of a simulation where the mass of the proto-neutron star was taken to be 1.5 $M_\odot$, and which yields a successful explosion. The colors have been made proportional to specific internal energy with explicit values (in units of $1.91\ 10^{15}$ ergs/g) given in the color bar. The inner boundary located at 50 km at this time represent the surface of the neutron star through which a time dependent neutrino flux is leaking.

As a result of neutrino energy deposition at the bottom of the accreted material, a very strong convection develops whereas the rest of the envelope is still infalling, forming an accretion shock at about 200 km. As material passes through the shock, matter is dissociated into alpha particles and free nucleons, and continues its infall towards the proto-neutron star. When it arrives in the direct vicinity of the proto-neutron star, it absorbs significant amounts of energy from the neutrino flux and, since the entropy increases, it becomes buoyant. The heated material rises back up, thus maintaining energy supply to the shock. Convection is therefore seen to be an essential element of the explosion as it allows the transport of low entropy matter towards the neutron star where it is heated by neutrinos, and the outward transport of high entropy matter which is able to drive the shock. Further, convection expels high temperature material which then cools through adiabatic expansion and thereby, the neutrino losses which peak at high temperatures are reduced.

The convection depends both on the strength of the gravitational force and the entropy contrast. As the explosion blast wave moves out in radius, the post-shock entropy increases, and the strength of gravity decreases thus progressively slowing down the convection. At a time of two seconds, which is as far as we took these simulations, the nickel produced in the explosion has overtaken approximately its own mass. This premixing is still insufficient compared to what is demanded by the RT simulations but it will increase in simulations which are pursued longer. Also, the amount of energy in the exploding star is found to be too small by about a factor of 3. However, given the number of assumptions and simplifications that went into these simulations, we consider these results as very encouraging.

5.3 SPH and Type II Supernova Explosions

In the case of the modeling of supernova explosions, SPH is especially well-suited for two reasons. The first is that a supernova explosion involves many successive changes of scales, e.g. in the RT / nickel decay simulations the typical scale changes by a factor of 10,000. SPH automatically adjusts to these changes while other methods require the use of complicated expedients such as moving grids and/or embedded grids of different resolution. The second advantage is that being a Lagrangian method, SPH allows for perfectly straightforward, diffusion-free, tracking of chemical species. This is particularly relevant to supernova studies where the degree of chemical inhomogeneity is one of the most stringent quantitative information brought in by the observations.

One additional point, which is not specific to supernova explosions, is that SPH often allows the inclusion of sophisticated physics in a relatively painless manner. For instance, the explosion mechanism problem described above requires complicated equations of state, determining chemical compositions, heating and cooling rates from the interaction of neutrinos with the matter, etc. These have all been straightforward to integrate in the SPH code.

Despite all these positive features, working with SPH is not without difficulties. The main one is probably the initial setup. Unlike for grid code, the density cannot be arbitrarily set since it is a weighted sum of neighboring masses. Therefore, matching a given non-uniform density profile has to be achieved by a combination of varying the masses and the spatial distribution of the particles. Such adjustments are often difficult. Large changes in mass of neighboring particles have to be avoided at all cost; if they occur, a "cannon ball among marbles" type of behavior is seen and numerical artefacts appear. In this respect, the supernova problem is especially difficult to setup because the large changes in density (up to ten orders of magnitude) forces one to vary particle masses by large factors while at the same time, the mixing that goes on during the simulations may bring particles of different initial radii (and therefore very different masses) near one another. Only experience built on many trials and errors has allowed us to get around these problems.

Finally, it is worth noting that all the simulations which are presented here were run on DEC workstations sitting on the authors' desks, in a time never exceeding a few days, which is the case for very few classical multi-dimensional Eulerian simulation of similar scale. While it is true that the era of global networking has come, we can say from experience that local computing offers a flexibility which has served us well. The countless iterations necessary to evolve a code into a tool capable of dealing with a specific physical problem such as the explosion of SN 1987A are best dealt with at home, rather than in remote computer systems.

6. Acknowledgments

We would like to thank F.K. Thielemann, P. Pinto, R. Hix, A.G.W. Cameron for many enlightning discussions and K. Nomoto for providing us with a SN1987A progenitor model. Part of this work has been supported by NSF grant AST 9206378 at the University of Arizona.

6. References

1. L. Lucy, *Astron. J.*, **82** (1977) 1013.
2. R. A. Gingold and J. J. Monaghan, *Mon. Not. R. astr. Soc.*, **181** (1977) 375.
3. R. A. Gingold and J. J. Monaghan, *J. Comput. Phys.*, **46**, (1982) 429.
4. J. J. Monaghan and R. A. Gingold, *J. Comput. Phys.*, **52**, (1983) 374.
5. J. J. Monaghan, *SIAM J. Sci. Stat. Comput.*, **3** (1982) 422.
6. J. J. Monaghan, *Comput. Phys. Rep.*, **3** (1985) 71.
7. A. W. Appel, *S.I.A.M.*, **6** (1985) 85.
8. J. G. Jernigan, in *Dynamics of Star Clusters*, I.A.U. Symp. 113, eds. J. Goodman and P. Hut (1985) 275.
9. J. Barnes and P. Hut, in *The Use of Supercomputers in Stellar Dynamics*, eds. P. Hut and S. L. W. McMillan, Springer Lectures Notes in Physics (1986) 267.
10. W. Benz, R. L. Bowers, A. G. W. Cameron and W. H. Press, *Astrophys. J.*, **348** (1989) 647.
11. L. Hernquist and N. Katz, *Astrophys. J. Suppl.*, **70** (1989) 419.
12. W. Benz, in *The Numerical Modeling of Nonlinear Stellar Pulsation: Problems and Prospects*, ed. J. R. Buchler (Kluwer Academic Publishers, Dordrecht 1990)
13. J. J. Monaghan, *Ann. Rev. Astron. Astrophys.*, **X** (1992) XXX.
14. A. E. Evrard, *Mon. Not. R. astr. Soc.*, **235** (1988) 911.
15. M. Steinmetz and E. Müller, *Astron. Astrophys.*, preprint.
16. D. Wood, *Mon. Not. R. astr. Soc.*, **194** (1981) 201.
17. J. J. Monaghan and J. C. Lattanzio, *Astron. Astrophys.*, **149** (1985) 135.
18. J. C. Lattanzio, J. J. Monaghan, H. Pongracic and M. P. Schwarz, *SIAM J. Sci. Stat. Comput.*, **7** (1986) 591.
19. R. Courant, K. O. Friedrichs and H. Lewy, *Math. Ann.*, **100** (1928) 32.
20. E. Fehlberg, *NASA T.R.*, **287.** (1968).
21. R. W. Hockney and J. W. Eastwood, in *Computer Simulation using Particles* (McGraw-Hill, New York) (1981)
22. D. Arnett, J.N. Bahcall, R.P. Kirshner and S.E. Woosley, *Ann. Rev. Astron. Astrophys.*, **27** (1989) 629.
22. T. Dotani et al., *Nature*, **330** (1987) 230.
23. P. Pinto and S.E. Woosley, *Astrophys. J.*, **329**, (1988) 820.
24. S. M. Matz et al., *Nature*, **331** (1988) 416.
25. D. Arnett and A. Fu, *Astrophys. J.*, **340**, (1989) 396.
26. J. Spyromillio, W.P.S. Meikle and D.A. Allen, *Mon. Not. R. astr. Soc.*, **242** (1990) 669.

27. M.R. Haas, S.W. Colgan, E.F. Erikson, S.D. Lord, M.G. Burton and D.J. Hollenbach, *Astrophys. J.*, **360** (1990) 257.
28. D.J. Jeffery, *Astrophys. J.*, **375** (1991) 264.
29. C. Papaliolios, M. Karovska, L. Koechlin, P. Nisenson, C. Stanley and S. Heatcote, *Nature*, **338** (1989) 565.
30. H. Bethe, *Rev. Mod. Phys.*, **62** (1990) 801.
31. M. Herant and W. Benz, *Astrophys. J.*, **387** (1992) 294.
32. R.A. Chevalier, *Astrophys. J.*, **207**, (1976) 867.
33. R. Bandiera, *Astron. Astrophys.*, **139**, (1984) 368.
34. D. Arnett, B. Fryxell, E. Müller, *Astrophys. J. Lett.*, **341** (1989) L63.
35. I. Hachisu, T. Matsuda, T. Shigeyama, K. Nomoto, K. *Astrophys. J. Lett.*, **358** (1990) L57.
36. W. Benz and F.-K. Thielemann, *Astrophys. J. Lett.*, **348** (1990) L17.
37. M. Herant and W. Benz, *Astrophys. J. Lett.*, **345** (1991) L412.
38. B. Fryxell, D. Arnett and E. Müller, *Astrophys. J.*, **367** (1991) 619.
39. M. Herant, W. Benz and S.A. Colgate, *Astrophys. J.*, **395** (1992) 642.
40. A. Burrows, *Astrophys. J.*, **334** (1988) 891.

Figure Captions

Figure 1: Rayleigh-Taylor instabilities associated with the propagation of the shock through the progenitor. Plotted are individual particles at t=400 minutes (1a) and 90 days (1b). Colors represent chemical composition (red=nickel, yellow=metals, green=helium, blue=hydrogen). Notice the swelling of the nickel due to radioactive energy release.

Figure 2: Neutrino driven convection developing close to the neutron star. Colors represent internal energy in units of $1.91\ 10^{15}$ ergs/g. Notice the large scale nature of the convection which manages to deform the accretion shock.

FIGURE 1

(a) (b)

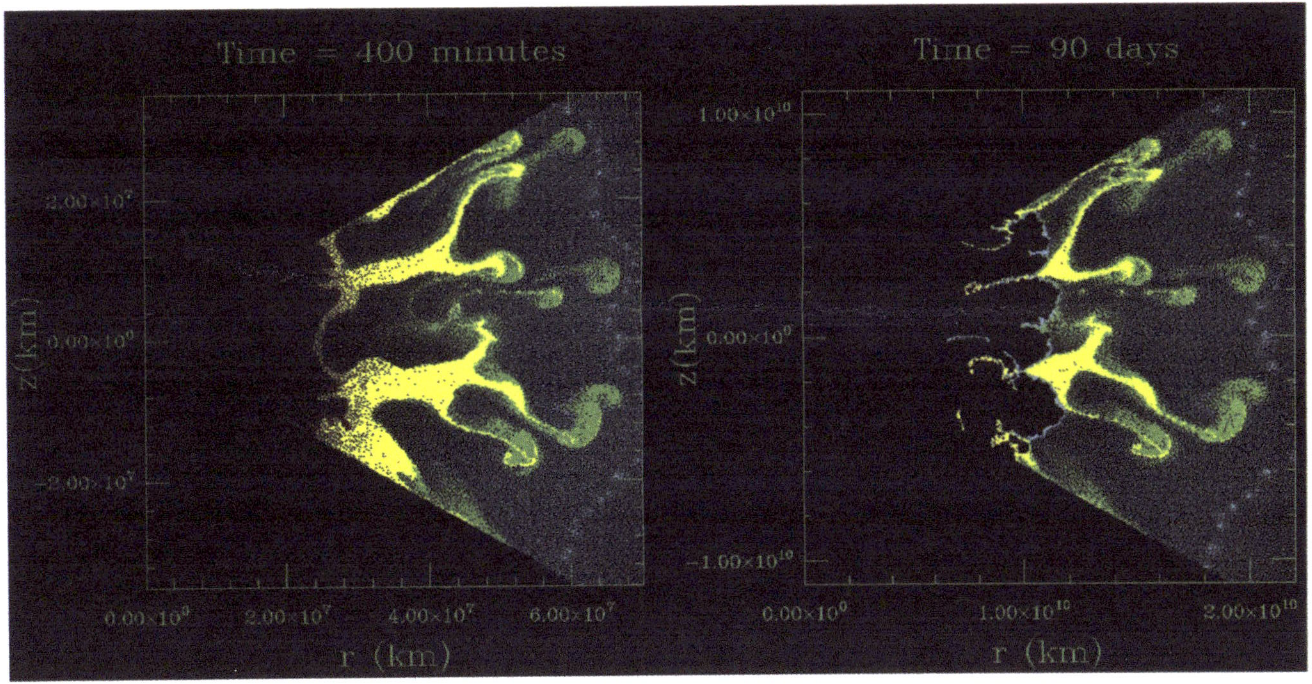

Time = 400 minutes
2.00×10^7
0.00×10^0
-2.00×10^7
z(km)
0.00×10^0
2.00×10^7
4.00×10^7
6.00×10^7
r (km)
Time = 90 days
1.00×10^10
0.00×10^0
-1.00×10^10
z(km)
0.00×10^0
1.00×10^10
2.00×10^10
r (km)

FIGURE 2

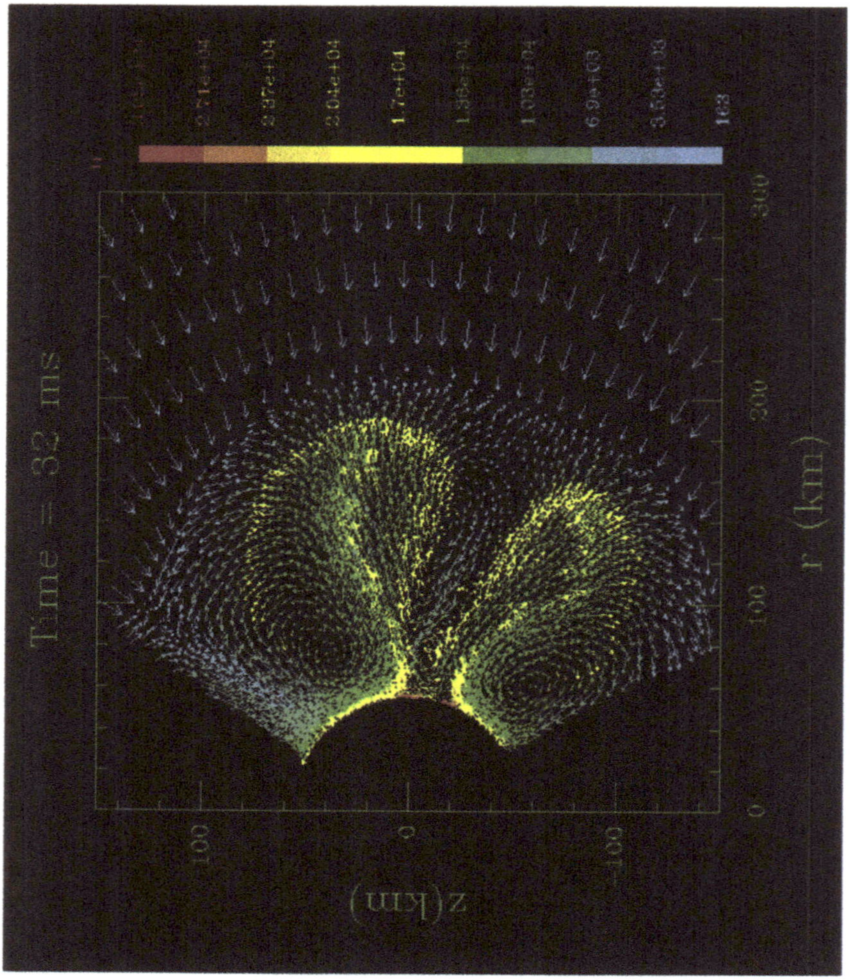

Growth, Localization and Competition between Faults by Spatio-Temporal Dynamics of Earthquakes

Didier Sornette, Patience A. Cowie, Anne Sornette and Christian Vanneste

Laboratoire de Physique de la Matière Condensée, CNRS URA 190
Université de Nice-Sophia Antipolis, Parc Valrose, 06108 NICE Cedex 2, FRANCE

Abstract : A simple dynamical model is proposed which captures both the spontaneous formation of complex fault structures by repeated earthquakes in systems presenting featureless small-scale initial disorder and an interesting spatio-temporal dynamics of earthquakes. Fault localization is related to an optimization problem in random media, which in 2D corresponds to a generalized random directed polymer with long range interactions. We observe a competition in the activity between several faults at long time scales which results from the mechanism of ground-state instability in random systems. The numerically derived Gutenberg-Richter power law is rationalized in this framework in terms the properties of fluctuations close to a global marginal instability.

The modelling of earthquake dynamics and of the mechanics of plate tectonic has recently received a strong stimulus from analogies with non-linear spatio-temporal complex dynamics [1]. The challenge is to understand the rich phenomenology of scaling laws observed in earthquakes and the associated self-similar structures of tectonic deformations and fault patterns [2], using concepts and tools borrowed from the statistical physics of critical phenomena, non-linear dynamics and random systems. The hope is then that this improved understanding of the fundamental physical principles will be instrumental in solving the long-standing earthquake prediction problem, which has important human and economical consequences.

Beginning with the block-spring stick-slip Burridge-Knopoff model [3], a variety of continuous or discrete "block-slider" models have been proposed as simulations for earthquake ruptures on a single fault surface (see for instance [1b-c,4]). These studies focus on the spatio-temporal organization of earthquakes within the single fault. Complementary to this modelization, it has been argued [2b,5] that crustal deformation is in fact self-organized on long (geologic) time scales as well as on seismic timescales as a consequence of the buildup of long-range elastic correlations. Along this idea, numerical crack propagation models have been introduced [6], which however are unable to account for stable geological fault structure over very long time scales, but rather obtain a diffuse seismic activity. Indeed, the existence of localized fault patterns in which the seismic activity is clustered is an ubiquitus observation [2,7] and seems to be a fundamental feature of the mechanics of the brittle crust. In return, one would like to understand the mechanisms by which this localization occurs and the possible role of earthquakes, as elementary deformation steps, in this process. It is important to realize that on time scales less than a few thousand years or more, the fault patterns which control the seismic activity are essentially quenched [2]. Faults evolve and grow on much larger time scales of the order of 10^{5-6} years [2,8]. Therefore, a reasonable model, which attempts to encompass both the short time earthquake dynamics and the long time growth of fault patterns, should address this separation of time scales.

The present work precisely focus on this point. We thus study a simple mechanical model, which can be viewed as an extension of both the models introduced in refs.[6] and of the thermal fuse model [9]. We thus bring together short and long time scale deformations using a statistical rupture model which includes elastic deformation, a rupture criterion and post-rupture healing. Our model seems to capture both the spontaneous formation of complex fault structures by repeated earthquakes in systems presenting featureless small-scale initial disorder and an interesting spatio-temporal dynamics of earthquakes. We show that, in 2D, fault localization corresponds approximately to the problem of choosing the best path spanning a random system with long range interactions, i.e. to a generalized random directed polymer problem. We observe a competition in the activity between several faults at long time scales which can be understood in terms of the drastic sensitibity of the best path to slight modifications of the system [10]. This model thus suggests a picture for the organization of the crust which is different from the spatio-temporal chaos view, embedded in the stick-slip block-spring approach [1b-c,4] : the organization of fault patterns and of seismicity clustering would result from the deterministic dynamics on the hierarchical or ultrametric family-tree structure of local optimal problems, i.e. essentially from the general problem of the dynamics of random systems [11,12].

Except for a crucial ingredient discussed below, our model is basically identical to that of ref.[6a] and is a direct extension of ref.[9] obtained by introducing healing after each rupture. In the present version, it simulates anti-plane shear deformation of an elastic material using a 2-D square lattice, which consists of a mesh of elements or plaquettes oriented at 45° with respect to the edges. We think of each plaquette as representing a block of typical size W by W, where W($\approx$15 km) is the thickness of the brittle crust. The 2D-lattice thus represents a tectonic plate at scales larger than its thickness. The model can also be viewed as a tectonic plate under extension which allows the formation of vertical shear or so-called normal faults.

A constant velocity boundary condition is used in this model to simulate deformation in a tectonic plate with one boundary fixed and the other moving at a velocity V. A constant displacement increment is added at equal time steps to the elements along the top boundary and the stress distribution throughout the network is instantaneously recalculated using a gradient conjugate method. To test whether this discretization method introduces artificial periodicity into the model, we also considered a random time step whilst increasing the displacement appropriately to give a constant rate of increase. All the results below are insensitive to the discretization procedure and remain robust for sufficiently small V. Periodic boundary conditions are applied along the two vertical edges of the lattice. Each element is characterized by an elastic constant g. The strain field is a scalar potential and the problem is equivalent to Kirchoff law of electric currents in a 2D network of conducting elements of conductance g [9]. The elasticity equations determine the strain and stress on each element of the network.

As the applied deformation increases, a rupture occurs in an element i when its stress σ_i reach a critical value σ_c. Then, the elastic stress in the element is instantaneously relaxed to a value $\sigma_c(1-\gamma)$, where γ is a parameter of the model. The stress distribution throughout the lattice is then adjusted instantaneously after each rupture, and at the positions of increased stress, further rupture can occur leading to a chain reaction, a model earthquake. Note that we do not describe the details of the dynamical process of rupture : this corresponds to treating the velocity of sound as infinitely large or better to considering time scales larger than the few seconds to minutes that an earthquake lasts, such that the stress has had time to equilibriate everywhere. In this second interpretation, a chain of rupture events is a cluster of earthquakes, localized in time and space. Note that such approximations are unavoidable in order to simplify the complexity of the full problem which involves three extremely different time scales [1) the dynamical rupture lasting second to minutes, ii) the interseismic period of correlated events, corresponding to time scales from minutes to hundred of years and iii) the tectonic deformation time scale of thousands to millions years]. In the present model, we keep the second and third time scales but embody the physics of the dynamical rupture in the parameter γ, controlling the stress relaxation induced by an earthquake [13]. The stress relaxation is introduced by an irreversible deformation $w_s=2\gamma w_e$ in the broken plaquette, where $w_e=\sigma_c/g$ is the elastic deformation of the plaquette just prior to the rupture. This irreversible deformation is intended to represent the slip associated with an earthquake. Imposing the slip w_s corresponds to adding a permanent force dipole $f=gw_s$ which opposes the stress. It turns out that the results presented below are insensitive to the choice of γ (we have worked with γ in the range [1-2]). Note that in the equivalent electrical formulation, the slip $w_s=2\gamma w_e$ corresponds to applying a voltage source in series with the element, of amplitude equal to 2γ times that of the initial voltage drop on the element and opposed to it. After rupture, the element which has slipped suffers no change in its material properties and it can support stress again in the future. As we will see, this is the key difference with previous models [6] which assigned a new random stress threshold after each rupture event.

To complete the description of the model, we have to define the nature of the heterogeneities. A first choice, similar to that made in the stick-slip block-spring models [1b-c,3,4], is to impose random values of initial slips w_s from elements to elements, while keeping all elastic constants g and stress thresholds σ_c identical. This corresponds to disorder on initial conditions and a deterministic dynamics on an homogeneous system. Preliminary computations indicate periodic cycles of seismic activity with periods increasing linearly with the size of the system. A separate study is called for in this special case and will be reported elsewhere. In the sequel, we have studied in details the two other case where the randomness is quenched and can be either in the values g_i (resp. σ_i) of the elastic constants (resp. rupture stress thresholds). The results are found equivalent for the two choices as expected, since in both cases the macroscopic shear stress and strain fields arising from the boundary condition becomes inhomogeneous on small scales due to the inhomogeneity of the elastic constants or to that of the accumulated slips which is induced by the stress thresholds randomness. The value g_i (resp. σ_i) for each element i is drawn from a probability distribution $P_g(g)$ (resp. $P_\sigma(\sigma)$) chosen in these computations to be uniform in the interval $[1-\Delta g/2, 1+\Delta g/2]$ (resp. $[1-\Delta\sigma/2, 1+\Delta\sigma/2]$) with the values of Δg and $\Delta\sigma$ between 0.1 and 1.99. Let us stress that if the system possesses quenched disorder, its properties are not allowed to evolve. This corresponds to a *deterministic dynamics* in a quenched random system. Previous similar models [6] assigned a new random stress threshold after each rupture event, thus corresponding a kind of Langevin description or annealed noise, thus embodying a completely different physics. The physics embedded in our assumption, that an element which has slipped suffers no change in its material properties, is that the slip is too small (a few meters for a fault of a few hundred kilometers) to induce a significant change of the mechanical properties of the gouge, of the asperities and barriers which control the stick-slip behavior of the element.

At time $t=0$, the constant velocity boundary condition is switched on to some small value V. Fig.1 shows the evolution of rupture activity at three different stages of the scaled deformation Vt/w_eL in a system of size L=180 by L=180, where $w_e=\sigma_c/g$. Are represented those elements which have slipped at least once, the light grey to black scale corresponding to increasing slips. At small deformations, we observe a diffuse damage where rupture reveals the weakest elements, distributed within the system in an uncorrelated way. As the deformation increases, clusters of elements which have slipped begin to appear, grow and coalesce. Elements on which the rupture activity is recurrent are found to organize in 1D lines, i.e. model faults, while the rupture activity progressively disappears in the other elements. Contrary to a common belief that localization appears as a result of the change (i.e. weakening) of mechanical properties of the ruptured elements which tends to favor additional rupture, the localization observed in our model appears as a result of a slowly convergent optimization process, through which the system tends to organize in a state of marginal stability reminiscent of self-organized criticality in sandpile models [14]. Indeed, fig.2 shows that the total elastic energy stored in the system rapidly grows and then tends to saturate at an average value : the system then seems to operate persistently in the vicinity of this maximum.

To undertand this idea, let us consider the two extreme cases where we divide the square system in 1) L parallel rows associated in series and 2) in L independent parallel culumns and treat as single entities the rows or columns respectively. We assume that each row (resp. column) is characterized by its own rupture threshold σ_c taken from a given distribution and that all elastic constants are equal. In case 1), as the deformation at the border increases, the failure threshold $\sigma_c{}^1$ for the weakest row is eventually reached for $Vt/L=\sigma_c{}^1 / g$ and this row slips by a given amount $w_s=2\gamma\sigma_c{}^1 / g$. As a consequence, the stress is partly relaxed in all rows. This system exhibits relaxation oscillations by periodic slipping of the weakest row with a period $T_1=2\gamma\sigma_c{}^1 / Vg$. This case illustrates that as soon as the system has found its weakest row, the deformation becomes localized in it and the system repeatedly visits its marginal stable point given by the stress threshold of the weakest row. This simple limit allows to rationalize the existence in fig.1 of localized lines on which all the slips occur. However, the complete square lattice is more complex than simple rows associated in series and one must consider the influence of the association in parallel. The corresponding physics is illustrated by the simple case 2). Since all L columns are independent, they must all accomodate the imposed deformation at the boundary. Each of them thus undergoes relaxation oscillations of increasing elastic deformations giving place to a sudden sliding which relaxes a part of the imposed displacement. Since the sliding thresholds are randomly distributed, each column possesses its own characteristic period of stick-slip activity, proportional to its stress threshold. It is interesting to note that the time evolution of the total stress at imposed deformation is thus quasiperiodic due to the incommensurabilities of the stress thresholds.

The complete square lattice system is a composite of these two extremes. Cases 1) and 2) teach us that each column must repeatedly slip as a result of the tectonic velocity and that the weakest element is each column would be sorted out in absence of lateral coupling. The lateral coupling between columns introduces an additional constraint that the set of persistently sliding elements must form a set connecting continuously the two vertical borders. Otherwise, there would exist elements in between the discontinuity which would be submitted to stresses above their thresholds due to the stress redistribution at the tip of the cracks. The localization of the slips thus corresponds to finding the best path connecting the two vertical borders. If we were to neglect the long range elastic forces, the problem would reduce to finding the path which minimizes the sum $\Sigma_n \sigma_c{}^n$ over the stress thresholds of the elements belonging to the path, i.e. would become equivalent to the random directed polymer problem [15]. In this case, the best path is self-affine with a roughness exponent $\zeta=2/3$ describing the average width $h \sim L^\zeta$ of the transverse fluctuations of the path of length L. However, long-range elastic interactions exist between the slipping elements for which both the stress and the space dependent increment of the elastic energy decay as r^{-s} as long distances with s=2. A general criterion [16] recently refined for the directed polymer problem [17] indicates that long range interactions are irrelevant (in the sense that ζ is unchanged) if ζ^{-1} is less than s, which is the case here. These results allow to rationalize the self-affine structure of the localized fault in the large disorder limit. A similar analogy between crack interfaces and random directed polymers have been proposed for the fuse model, which however breaks down for large disorder [18]. Note however that the progressive organization of a stationary self-affine localized fault in the present model is a completely different mechanism than the transient irreversible crack growth of the fuse model. Since nonlinear coupling between three or more oscillators with incommensurate periods gives a well-known route to chaotic behavior [19], it is an intriguing open problem to understand the spatio-temporal chaos obtained in our system as resulting from the coupling between the incommensurate oscillations of the elements of the system.

In fact, it turns out that the "seismic" activity may be clustered after the transient, not on a single fault, but on two or three faults, which are in general separated by distances a fraction of the system size. This is illustrated in fig.3 which shows, in the lower panel, the distribution of rupture events as a function of deformation after the initial transient has finished. In this system realization, three main "faults" have developed, illustrated in the inset, each having a self-affine roughness. These faults are separated by regions where few or no ruptures occur except during the initial transient phase. One fault is typically active for very long periods of time (much larger than the longest time scale $\sim V^{-1}$ introduced by the tectonic velocity V) until the locus of ruptures spontaneously switches to a different fault(s) and the previously active fault may become completely silent for a long time. In some cases the switch to another structure is brief, as for example at t = 11200 when the activity shifted from fault 1 to fault 3. At t = 15500, however, the activity on fault 1 stopped entirely while faults 2 and 3 were the new loci of the ruptures. This feature of alternate fault activity over long period of times has been described in several geological contexts and its origin is still a mystery.

The existence of such a long time scale much larger than all the relevant time scale of the problem may also appear a mystery in our model. In fact, these observations find a natural explaination within the analogy developed above with random directed polymers. Indeed, it seems to be a general property that when a random medium is allowed to undergo a slow drift, the best path can undergo long period of stability and be suddently drastically altemed [10]. This results from the ultrametric family-tree structure of local optimal paths [20]. In the present model, the slow drift can be envisonned to be due to the fluctuating nature of the release by earthquakes at short time scales of the elastic energy stored in the system (see top panel of figure 3). The earthquake dynamics at short times thus provides a mechanism for the variations of the stress field in the system, around an average characterizing the global stationary marginal state. It is interesting to note in fig.3 that correlations between shifts in activity and the total energy exist. Two features are noticed: the first is that each fault is in general characterized by a different energy level, fault 1 having the lowest. The second feature is that brief excursions in activity are often associated with spikes in the energy level. Closer observation indicates that these increases in energy are the result of a quiescence throughout the lattice. In the particular finite simulation shown in Figure 3, fault 1 was never re-activated later in the evolution. This gives the appearance that there is a long term increasing trend to the energy level although this is not in general true of other simulations.

We now turn briefly to the details of clusters of ruptures occuring on an active fault. We observe that cascades of events (earthquakes) occur simultaneously, typically involving neighboring elements, and these cascades are separated by short periods of time when few or no ruptures occur. Fig.5 shows the accumulation of displacement on fault 1 as a function of time. Three profiles are shown, taken across the fault at x=20, 40 and 60. Each profile shows the long term steady increase in displacement due to the applied boundary condition. In detail the profiles contain sharp increases in displacements that are irregular and do not conform to either a time-predictable or slip predictable pattern [21], but seem typical to earthquake sequences [2]. Each step increase in the displacement corresponds to an earthquake on the fault. The differential distribution of the size N(E) of the energy dissipated in these cascades follows a Gutenberg-Richter power law $N(E) \sim E^{1+\beta}$, with $\beta \approx 0.3$, as shown in fig.4. In the limit where the fault structure is completely controlled by the quenched randomness of the element stress thresholds, this distribution can be understood as resulting from the fluctuations of the stress released by the earthquakes, close to a global marginal instability [22]. A mean field solution developed in [22] gives $\beta=0.5$. We have no precise theory for the value 0.3 found in our model.

In summary, a fully deterministic statistical model of shear deformation in a thin disordered plate driven at a constant velocity along one boundary has been used to simulate both the long time-scale deformation and organization of the earth's crust and the intermediate seismic activity. The main conclusion is that the complexity of fault growth and earthquake clustering result from the spatio-temporal chaos in quenched heterogeneous systems. Three time scales appear in the model: the shortest time scale concerns a single earthquake or delayed multiple seismic rupture; the intermediate time scales concerns variations in the seismic moment release rate along a fault; the largest time scale concerns the periods of activity followed by quiescence of individual faults as the locus of active deformation shifts across the lattice.

We are presently working on extensions of the model using a more sophisticated criterion for rupture [9] in terms of a damage variable associated to each element, which accumulates gradually with time even at constant stress but higher stress levels enhances damage development. Preliminary results indicate that both models, qualitatively at least, exhibit the same rupture characteristics. Ruptures become localized on the same set of lattice elements over long periods of time. Our results thus appear robust with respect to the rupture criterion. However, the inclusion of a delayed damage process allows to vary this time scale relative to the time scale of the boundary condition, and to obtain a range of rupture phenomena, such as delayed multiple events, slow earthquakes, and variations in seismic moment release rate. Furthermore, the delay mechanism included in the damage process modifies the b-value of the Gutenberg-Richter power-law distribution of the energy dissipated in the earthquakes, which becomes a continuous function of the delay parameter [9]. These results and extension of the model to treat general tensorial deformations will be reported in the future.

Acknowledgments : We are grateful to L.Knopoff for stimulating discussions.

REFERENCES

[1] a) Sornette, A. and D. Sornette, Europhys. Lett, 9, 197-202, 1989.
b) Bak, P. and C. Tang, J. Geophys. Res., 94, 15635-15637, 1989.
c) Carlson, J. M. and J. S. Langer, Phys. Rev. A, 40, 6470-6484, 1989.

[2] a) Scholz C., The mechanics of earthquakes and faulting (Cambridge University Press, Cambridge, 1990)
b) Sornette D., Self-organized criticality in plate tectonics, in Proceeding of the NATO ASI, Spontaneous formation of space-time structures and criticality, eds. by Riste T. and Sherrington D., Geilo, Norway 2-12 April 1991 (Kluwer Academic Press, 1991), p.57-106

[3] Burridge, R. and L. Knopoff, Bull. Seismol. Soc. Am., 57, 341-371, 1967.

[4] a)Carlson J.M., Langer J.S., Shaw B. and Tang C., Phys.Rev.A 44, 884, 1991.
b)Nakanishi, H., Phys. Rev. A, 43, 6613-6621, 1991.
c)Matsuzaka M. and Takayasu H., J.Geophys.Res., 96, 19925-19931, 1991.
d)Brown S.R., Scholz C. and Rundle J.B., Geophys.Res.Lett., 18, 215-218, 1991.
e)Olami Z., Feder H.J.S. and Christensen K., Phys.Rev.Lett. 68, 1244, 1992.
f) Schmittbul, Vilotte J.-P. and Roux S., Europhys.Lett. 3 (1993)
g)Vasconcelos G.L., de Sousa Vieira M. and Nagel S.R., in Proc. Conference on Fractals and Disordered systems, ed. Bunde A., Hamburg, 1992.
h) Crisanti A., Jensen M.H., Vulpiani A. and Paladin G., Phys.Rev.A, 46, R7363, 1992.
i) Lomnitz-Adler J. and Lemus-Diaz P., Geophys.J.Int., 99, 183-194, 1989.

[5] Sornette, D., P. Davy, and A. Sornette, J. Geophys. Res., 95, 17353-17361, 1990.
Sornette, D. and J. Virieux, Nature 357, 401-403, 1992.

[6] a) Chen, K., P. Bak and S. P. Obukhov, Phys. Rev. A, 43, 625-630, 1992.
b) Xu, H-J., B. Bergersen and K. Chen, A new crack propagation model of earthquakes. In press, Phys. Rev. Letts.

[7] Kagan Y.Y. and Knopoff L., Geophys. J. Roy. Astron. Soc. 62, 303, 1980.
Kagan Y.Y., Geophys.J.Int., 110, 305-320, 1992.

[8] a)Sornette A., Davy P. and Sornette D., Phys.Rev.Lett. 67, 2266, 1990.
b)Davy P., Sornette A. and Sornette D., Nature 348, 56, 1990.
c)Chekunov A.V., Sollogub V.B., Gintov O.B., Tripol'ski A.A. and Isai V.M., Geophys.J, 9, 795-836, 1991.
d)Cowie P.A. and Scholz C., J.Geophys.Res., 97, 11085-11095, 1992.

[9] Sornette, D. and C. Vanneste, Phys. Rev. Lett., 68, 612-615, 1992.
Sornette, D., C. Vanneste and L. Knopoff, Phys. Rev. A, 45, 8351-8357, 1992.
Vanneste, C. and D. Sornette, J. Phys. I. France, 2, 1621-1644, 1992.

[10] Zhang Y.-C., Phys.Rev.Lett., 59, 2125, 1987.
Bray A.J. and Moore M.A., Phys.Rev.Lett., 58, 57, 1987.

[11] L.Knopoff has recently proposed a version of the Burridge-Knopoff model with quenched random slip thresholds.

[12] Souletie J., Vannimenus J. and Stora R., eds., Chance and Matter, Les Houches 1986 Session XLVI (North Holland, Amsterdam, 1987)

[13] Knopoff L., idée de prendre un overshot pour modéliser la dynamique temps court

[14] Bak P., Tang C. and Wiesenfeld K., Phys.Rev.Lett. 59, 381, 1987 ; Phys.Rev.A 38, 364, 1988.
Sornette D., Phys.Rev.Lett., 69, 1287, 1992; J.Phys. I France, 2, 2065-2073, 1992.

[15] Kardar M., Parisi G. and Zhang Y.-C., Phys.Rev.Lett. 56, 889, 1986.

[16] Burkhardt T.W., Phys.Rev.Lett. 48, 216, 1982; Cordery R., Phys.Rev.Lett. 48, 215, 1982.

[17] Igloi F., Europhys.Lett., 19, 305-310, 1992.

[18] Hansen A., Hinrichsen E. and Roux S., Phys.Rev.Lett. 66, 2476, 1991.

[19] Bergé P., Pomeau Y. and Vidal C., Order within Chaos, Wiley, New York, 1984.

[20] Kardar M. and Zhang Y.-C., Phys.Rev.Lett. 58, 2087, 1987.

[21] Shimazaki K. and Nakata T., Geophys.Res.Lett., 7, 279-82, 1980.

[22] Sornette D., J.Phys. I France, 2, 2089-2096, 1992.

FIGURE CAPTIONS

Figure 1. Evolution of rupture activity at three different stages of the scaled deformation Vt/w_eL in a system of size L=180 by L=180, where $w_e=\sigma_c/g$. Are represented those elements which have slipped at least once, the light grey to black scale corresponding to increasing slips.

Figure 2. Variation of the total strain energy stored in the lattice in units of $\Sigma[\sigma_c{}^2/g]$, where the sum is carried over the L^2 elements of the system, as a function of deformation; a moving average filter has been applied to suppress high frequency fluctuations.

Figure 3. Inset shows the border of the lattice and the structure of the active faults labeled 1, 2, and 3; the x-y origin is at bottom-left and the size of the lattice is $0 \leq x \leq 80$, $0 \leq y \leq 80$. Bottom panel: y coordinate (or row number) of ruptures as a function of time. The trace of each fault traverses approximately 15 to 20 rows of the lattice. Jumps in activity between the faults are correlated with changes in the mean energy level in the system (represented in the top panel), as indicated by the dashed lines.

Figure 4. Differential distribution of the size N(E) of the energy dissipated in clustered rupture events (earthquakes) : it follows a Gutenberg-Richter power law $N(E) \sim E^{1+\beta}$, with $\beta \approx 0.3$.

Figure 5. a) Trace on the x-axis of the rupture events;
b) Cumulative displacement (d_c) as a function of time (t) on fault 1 shown at three different points along the fault: x (or column number) =45, 90 and 135. The profiles are offset from each other to improve the clarity of the figure.

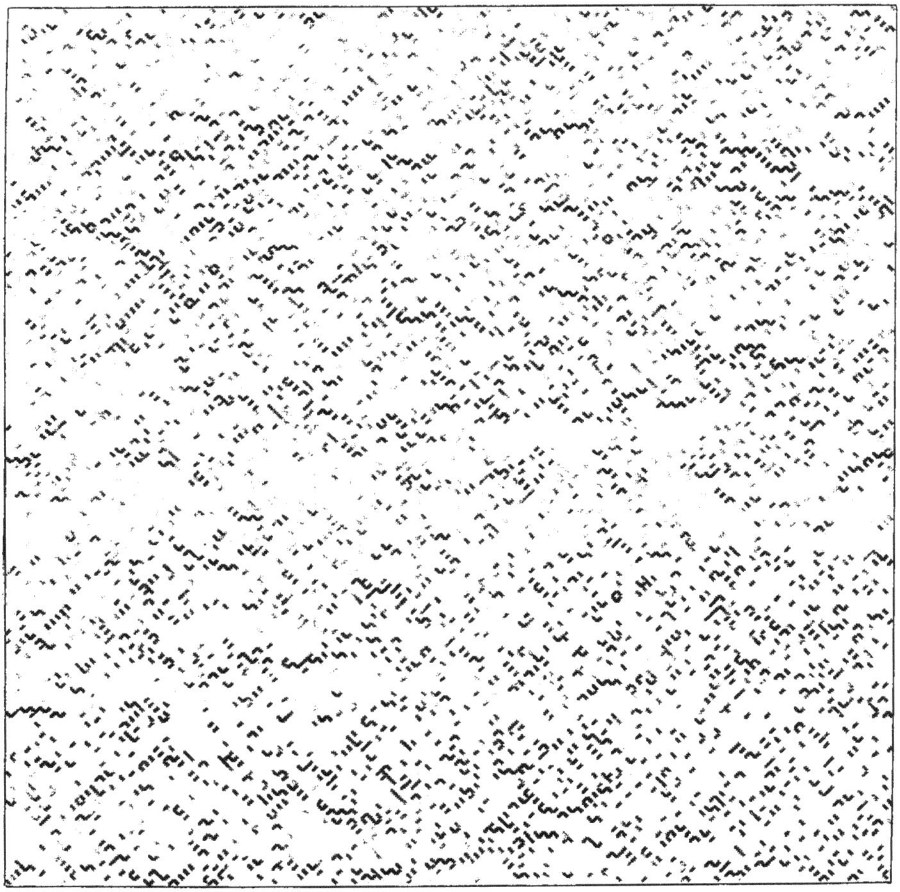

Fig. 1(a)

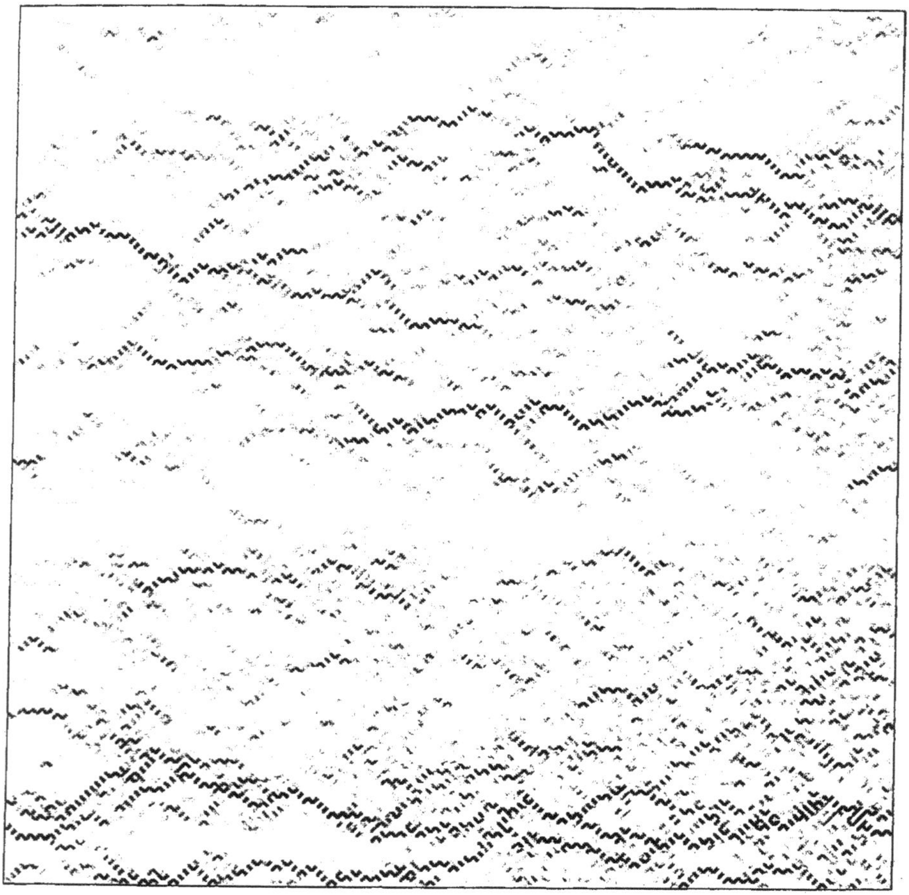

Fig. 1(b)

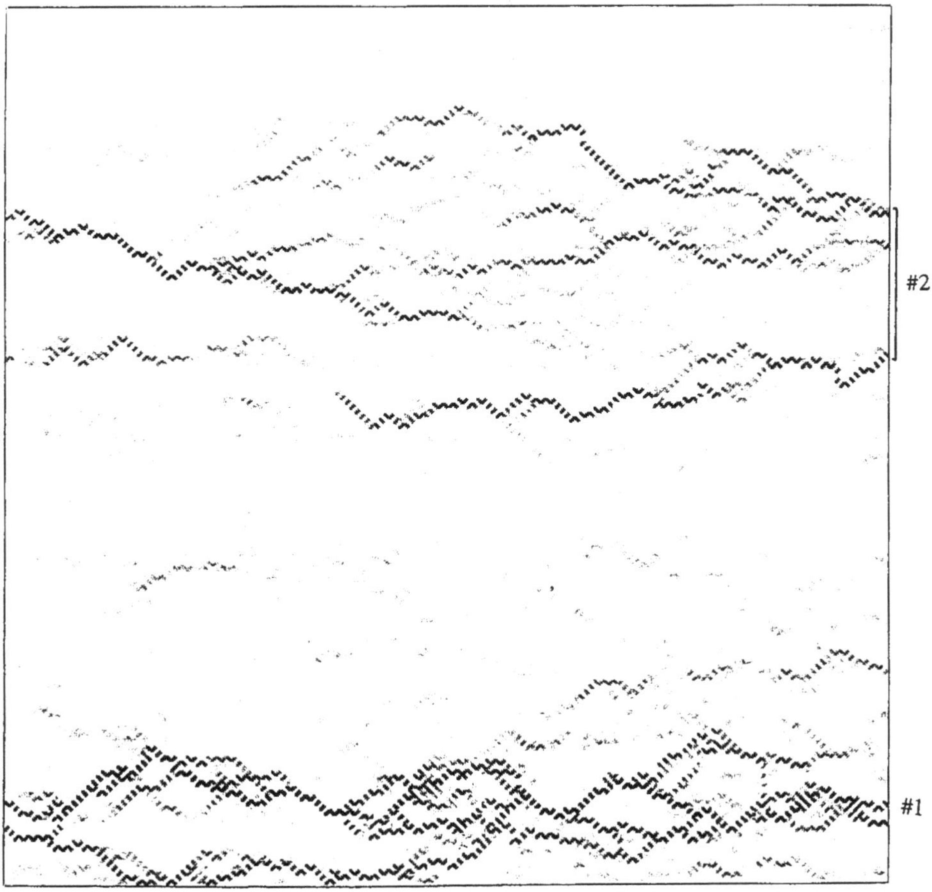

Fig. 1(c)

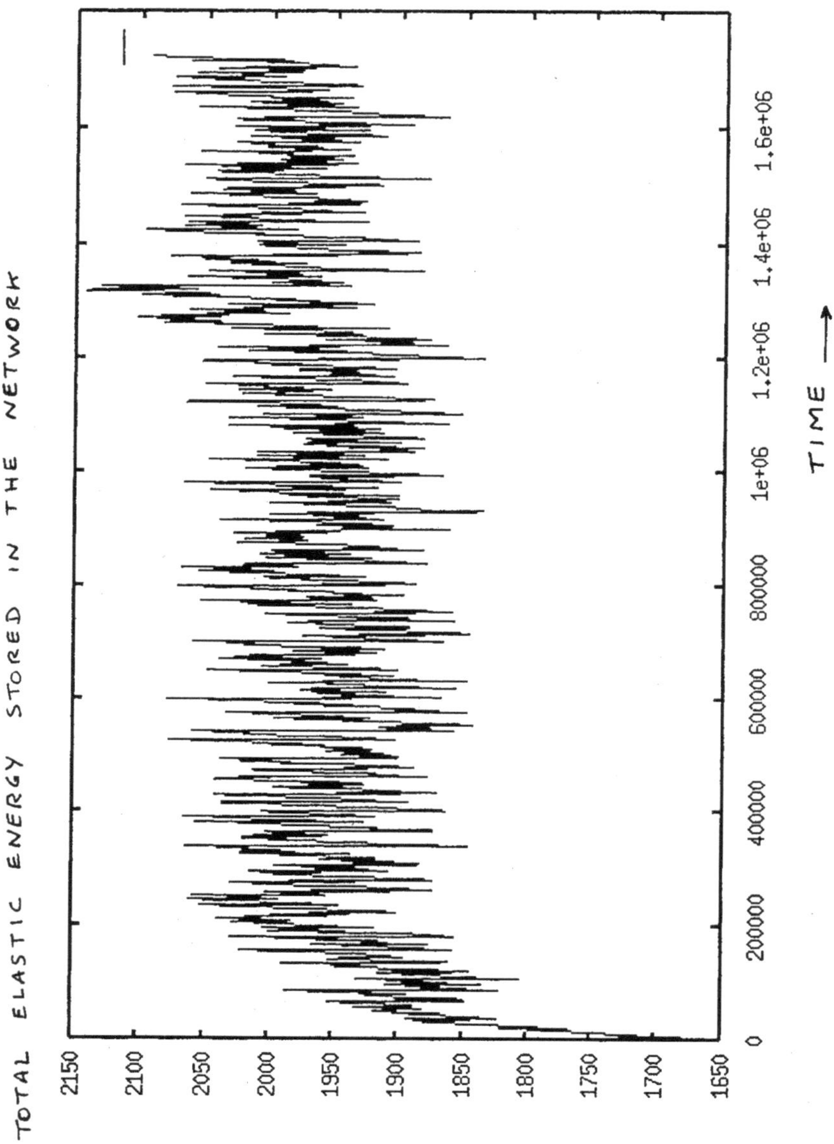
TOTAL ELASTIC ENERGY STORED IN THE NETWORK
2150
2100
2050
2000
1950
1900
1850
1800
1750
1700
1650
0
200000
400000
600000
800000
1e+06
1.2e+06
1.4e+06
1.6e+06
TIME

Fig. 2

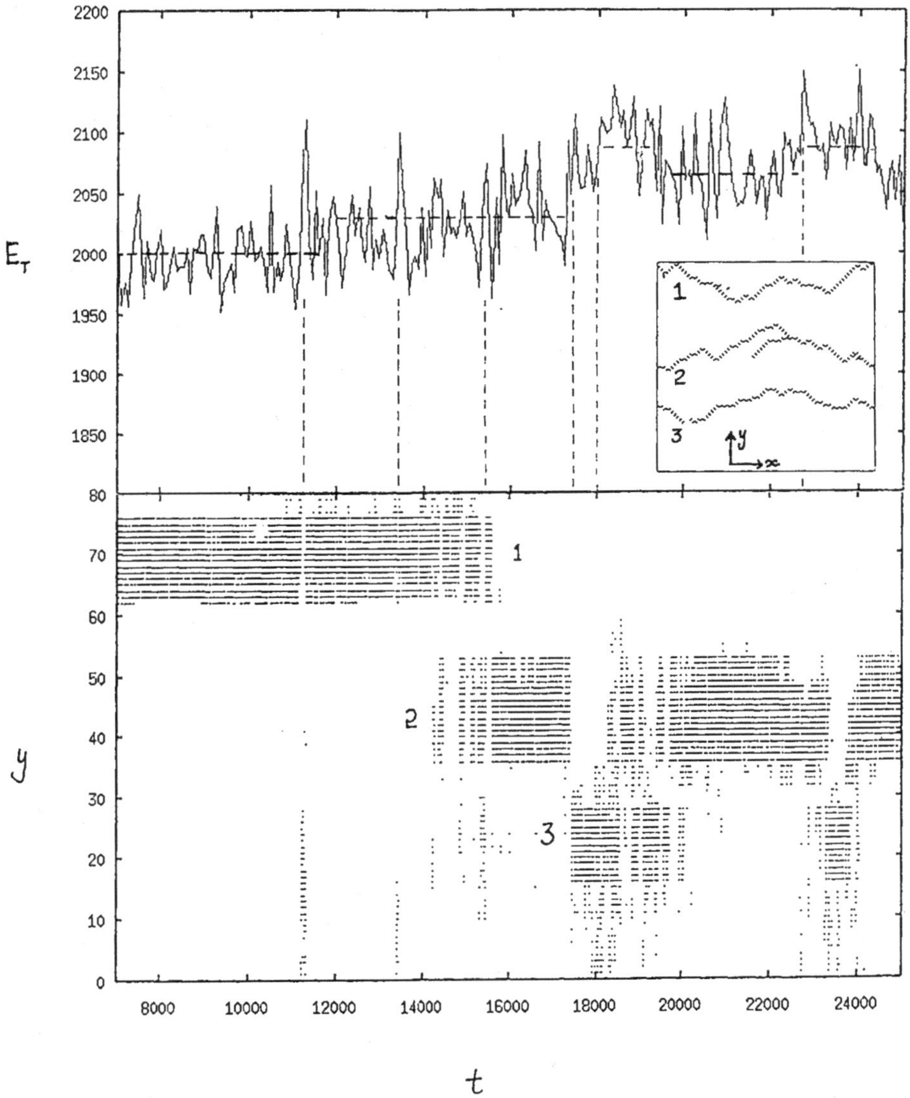

2200
2150
2100
2050
2000
1950
1900
1850
E_T
1
2
3
y
x
80
70
60
50
40
30
20
10
0
y
1
2
3
8000
10000
12000
14000
16000
18000
20000
22000
24000
t

Fig. 3

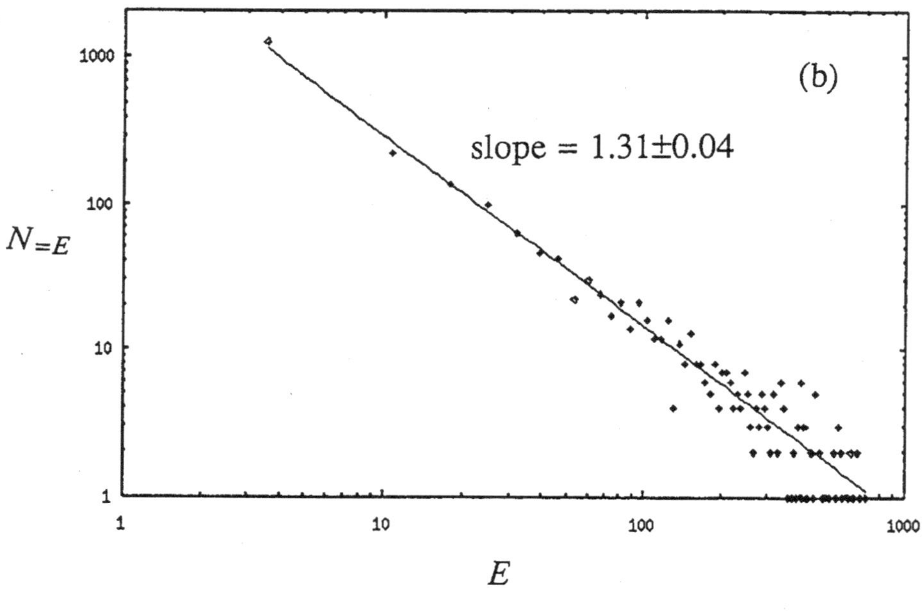

Fig. 4

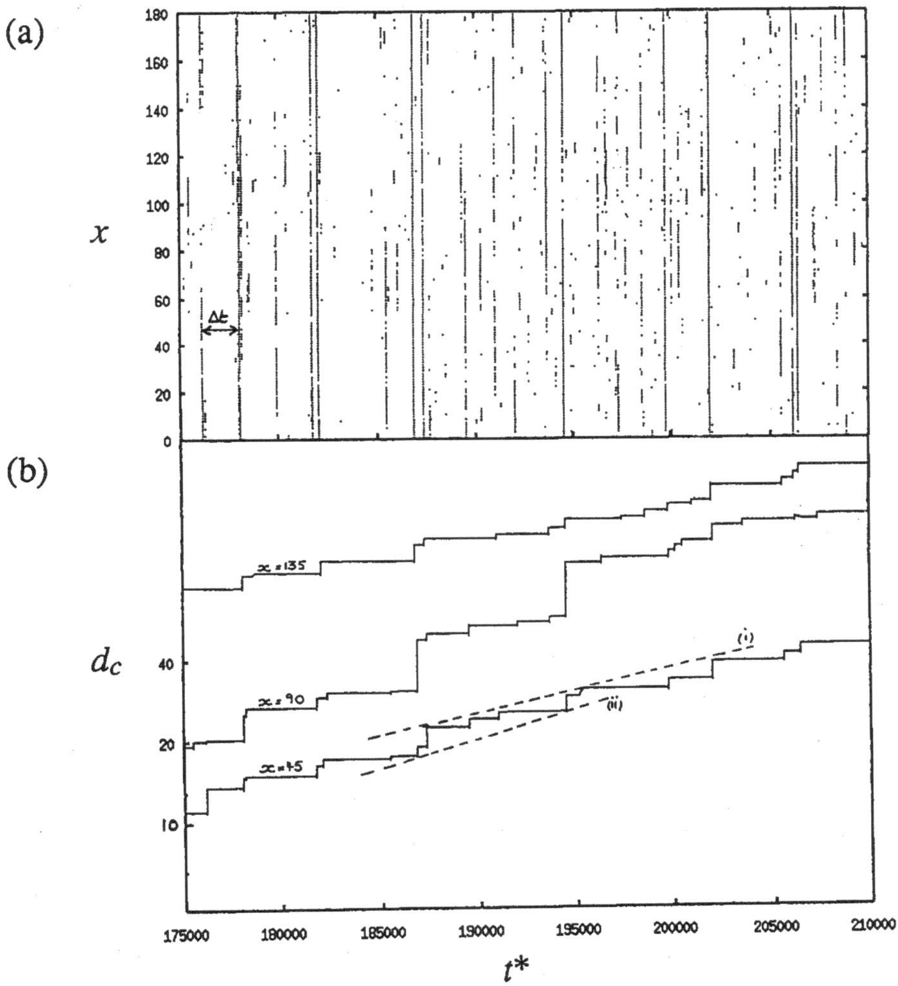
(a)
x
Δt
(b)
d_c
x=135
x=90
x=45
(i)
(ii)
t*
175000
180000
185000
190000
195000
200000
205000
210000

Fig. 5

A Video on CA Models in Astrophysics

A. Lejeune and J. Perdang

University of Liège
B-4000 Liège, Belgium

As a complementary demonstration of the astrophysical relevance of cellular automaton models, we have assembled a videotape containing a variety of animated simulations of phenomena in the fields of astrophysics as well as fluid dynamics and geophysics. With the exception of the film by Rothman and Kadanoff — a movie of a more geophysical flavour than the remaining contributions — these animations refer to physical processes analysed in detail in the papers of this book.

The existence of a huge number of degrees of freedom in a cellular automaton raises the problem of efficiency communicating the results in a synthetic, and at the same time sufficiently complete form. Since on the one hand the most attractive feature of cellular automata is their ability of modelling highly complex space and time patterns, and since on the other hand we typically resort to this class of models only when huge numbers of degrees of freedom are effectively involved, any compression of information supplied by these models consists in a genuine loss of information. As transpires indeed from the papers of this book, all models dealt with by the cellular automaton framework do show a high degree of complexity.

It is therefore imperative that information can be communicated with a minimum of compression. The best way of achieving this goal consists in actually displaying the time evolution of the totality of the degrees of freedom.

For 2 spatial dimensions this can be accomplished in principle by adopting a stereoscopic representation of a stack of frames at successive times (cf. Part I of this book), each frame visualising all individual discrete cellstates in a colour code; the limitation of this procedure is obvious. A more satisfying way consists in showing an animated sequence of the frames exhibiting the automaton state over an arbitrarily large number of successive timesteps. The majority of the animations of the video refer to 2D models and exhibit indeed directly the individual cellstates. In the context of fluid dynamic models the actual cellstates of the automaton are computational fictions playing an ancillary part only; a representation of the relevant macroscopic variables (density, temperature, velocity, ...) — accompanied by some compression of information due to smoothing — is performed in an essentially similar way.

In the case of full 3-dimensional CA models, the animations exhibit 2D spatial sections only.

These CA animated simulations show characteristic space and time patterns reminiscent of the complexity of the real natural phenomena they are devised to mimic.

Contents of the tape

The video contains 8 movies of roughly 4 minutes each, dealing with the following topics:

Fluid dynamics:

(1) Lattice Boltzmann simulation of thermal convection
(Briscolini M., Massaioli F., Santangelo P., Succi S.)
showing the evolution of the temperature field

(2) Lattice Boltzmann simulation of fluid flows
(Succi S., Higuera F., Cancellière A.)
simulating stream- and streaklines of flows past a cylinder

(3) Multifractal dynamics
(Brenier P., Schertzer D., Davis A., Sarma G., Lovejoy S., Lavallée D.)
simulating the dynamics of a turbulent cloud field

Stellar stability

(4) Smooth-particle hydrodynamics and supernova explosions
(Benz W., Herant M.)
exhibiting the non-radial expansion process accompanying the supernova explosion process

(5) Variable stellar variability
(Perdang J., Lejeune A.)
illustrating the occurrence of non-radial effects in stellar pulsations

Galaxies

(6) Galactic structure from induced star formation
(Lejeune A., Perdang J.)
showing the collective evolution of different types of galactic components

Waves

(7) The wave automaton
(Sornette D., Sebbah P., Vanneste C.)
simulating wave propagation in heterogeneous media

Geophysics

(8) Many bubble dynamics
(Rothman D.H., Kadanoff P.)
exhibiting diverse geophysical simulations (volcanoes, sedimentation, convection)

Availability

The tape is available in VHS (PAL, SECAM, NTSC) from the Multimedia Services of the University of Liège, at the price of $35 (1200 BEF) (until June 1994), including shipping. A U-Matic version can also be provided (price on request). For ordering, please specify the format and contact

Dr. A. Lejeune,
University of Liège, Institute of Physics B5,
B-4000 Liège 1, Belgium
(Fax: +32-41-563607, E-mail: ALEJEUNE@vm1.ulg.ac.be)

Payment: Bank Draft/Eurocheque/Postal Cheque/International Money Order.

We would like to thank the video contributors, as well as B. Monfort of the Multimedia Services Center for her assistance in shaping this project.